# α-Gal and Anti-Gal

## α1,3-Galactosyltransferase, α-Gal Epitopes, and the Natural Anti-Gal Antibody

**Subcellular Biochemistry**

**Volume 32**

# SUBCELLULAR BIOCHEMISTRY



---

*Recent Volumes in This Series*

Volume 23 **Physicochemical Methods in the Study of Biomembranes**
Edited by Herwig J. Hilderson and Gregory B. Ralston

Volume 24 **Proteins: Structure, Function, and Engineering**
Edited by B. B. Biswas and Siddhartha Roy

Volume 25 **Ascorbic Acid: Biochemistry and Biomedical Cell Biology**
Edited by J. Robin Harris

Volume 26 ***myo*-Inositol Phosphates, Phosphoinositides, and Signal Transduction**
Edited by B. B. Biswas and Susweta Biswas

Volume 27 **Biology of the Lysosome**
Edited by John B. Lloyd and Robert W. Mason

Volume 28 **Cholesterol: Its Functions and Metabolism in Biology and Medicine**
Edited by Robert Bittman

Volume 29 **Plant–Microbe Interactions**
Edited by B. B. Biswas and H. K. Das

Volume 30 **Fat-Soluble Vitamins**
Edited by Peter J. Quinn and Valerian E. Kagan

Volume 31 **Intermediate Filaments**
Edited by Harald Herrmann and J. Robin Harris

Volume 32 **α-Gal and Anti-Gal: α1,3-Galactosyltransferase, α-Gal Epitopes, and the Natural Anti-Gal Antibody**
Edited by Uri Galili and José Luis Avila

---

# α-Gal and Anti-Gal

## α1,3-Galactosyltransferase, α-Gal Epitopes, and the Natural Anti-Gal Antibody

## Subcellular Biochemistry Volume 32

**Edited by**

**Uri Galili**

*MCP-Hahnemann School of Medicine*
*Allegheny University of the Health Sciences*
*Philadelphia, Pennsylvania*

**and**

**José Luis Avila**

*Institute of Biomedicine*
*Caracas, Venezuela*

**Kluwer Academic / Plenum Publishers**
**New York, Boston, Dordrecht, London, Moscow**

The Library of Congress cataloged the first voume of this title as follows:

Sub-cellular biochemistry.
London, New York, Plenum Press.
v. illus. 23 cm. quarterly.
Began with Sept. 1971 issue. Cf. New serial titles.
1. Cytochemistry—Periodicals. 2. Cell organelles—Periodicals.
QH611.S84 574.8'76 73-643479

ISSN 0306-0225

ISBN 0-306-46103-X

This series is a continuation of the journal *Sub-Cellular Biochemistry*, Volumes 1 to 4 of which were published quarterly from 1972 to 1975


233 Spring Street, New York, N.Y. 10013

10 9 8 7 6 5 4 3 2 1

A C.I.P. record for this book is available from the Library of Congress



Printed in the United States of America

# Contributors

**José Luis Avila** Instituto de Biomedicina, Caracas 1010A, Venezuela

**Michael E. Breimer** Institute for Surgical Sciences, Department of Surgery, University of Göteborg, Sahlgrenska universitetssjukhuset, SE 413 45 Göteborg, Sweden

**Choa-Guang Chen** Immunology Research Centre, St Vincent's Hospital Melbourne, Fitzroy, Australia, 3065

**David K. C. Cooper** Transplantation Biology Research Center, Massachusetts General Hospital/Harvard Medical School, Boston, Massachusetts 02129

**Peter J. Cowan** Immunology Research Centre, St Vincent's Hospital Melbourne, Fitzroy, Australia, 3065

**Anthony J. F. d'Apice** Immunology Research Centre, St Vincent's Hospital Melbourne, Fitzroy, Australia, 3065

**Uri Galili** Department of Microbiology and Immunology, MCP Hahnemann School of Medicine, Philadelphia, Pennsylvania 19129

**Irwin J. Goldstein** Department of Biological Chemistry, University of Michigan, Ann Arbor, Michigan 48109-0606

**Timothy R. Henion** Department of Biomedical Sciences, Eunice Kennedy Shriver Center, Waltham, Massachusetts 02254

**Jan Holgersson** Division of Clinical Immunology, Karolinska Institute, Huddinge University Hospital, SE 141 86 Huddinge, Sweden

**David H. Joziasse** Department of Medical Chemistry, Vrije Universiteit, NL-1081 BT Amsterdam, The Netherlands

**Takaaki Kobayashi** Department of Surgery II, Nagoya University School of Medicine, Nagoya, Japan,

**Denise C. LaTemple** Department of Medicine, Division of Hematology and Oncology, University of Pennsylvania, Philadelphia, Pennsylvania

**Ian F. C. McKenzie** Molecular Immunogenetics Laboratory, Austin Research Institute, Austin and Repatriation Medical Centre, Heidelberg, Victoria, 3084, Australia

**Martin J. Pearse** Immunology Research Centre, St Vincent's Hospital Melbourne, Fitzroy, Australia, 3065

**Charalabos Pothoulakis** Division of Gastroenterology, Beth Israel Deaconess Medical Center, Harvard Medical School, Boston, Massachusetts 02215

**Marko Z. Radic** Department of Microbiology and Immunology, MCP Hahnemann School of Medicine, Philadelphia, Pennsylvania 19129

**Russell P. Rother** Molecular Development, Alexion Pharmaceuticals, New Haven, Connecticut 06511

**Lennart Rydberg** Institute of Laboratory Medicine, Department of Clinical Chemistry and Transfusion Medicine, Sahlgrenska universitetssjukhuset, SE 413 45 Göteborg, Sweden

**Bo E. Samuelsson** Institute of Laboratory Medicine, Department of Clinical Chemistry and Transfusion Medicine, Sahlgrenska universitetssjukhuset, SE 413 45 Göteborg, Sweden

**Mauro S. Sandrin** Molecular Immunogenetics Laboratory, Austin Research Institute, Austin and Repatriation Medical Centre, Heidelberg, Victoria, 3084, Australia

**Joel H. Shaper** Cell Structure and Function Laboratory, Oncology Center, and Department of Pharmacology and Molecular Sciences, The Johns Hopkins University School of Medicine, Baltimore, Maryland 21287-8937

**Nancy L. Shaper** Cell Structure and Function Laboratory, Oncology Center, The Johns Hopkins University School of Medicine, Baltimore, Maryland 21287-8937

**Trixie A. Shinkel** Immunology Research Centre, St Vincent's Hospital Melbourne, Fitzroy, Australia, 3065

**Aron D. Thall** BioTransplant, Inc., Charlestown Navy Yard, Charlestown, Massachusetts 02129

**Le Wang** Department of Microbiology and Immunology, MCP Hahnemann School of Medicine, Philadelphia, Pennsylvania 19129

**Harry G. Winter** Department of Biological Chemistry, University of Michigan, Ann Arbor, Michigan 48109-0606

# Preface

It has been 15 years since the first report on the isolation of anti-Gal from human serum and the demonstration that this antibody is the most prevalent antibody in humans (Galili *et al.*, *J.Exp. Med.* **160**: 1519, 1984). Subsequent interdisciplinary studies in immunology, carbohydrate biochemistry, molecular biology, and evolution demonstrated the highly restricted specificity of anti-Gal for the carbohydrate epitope Gal α1-3Galβ1-4GlcNAc-R, (termed here the α-gal epitope), the unprecedented evolutionary pattern of distribution of α-gal and anti-Gal in mammals, and explained the evolution of this antigen and antibody by analysis of the α1,3galactosyltransferase gene, the gene that encodes the enzyme that synthesizes the α-gal epitope. These studies have suggested that a major selection process that occurred in the course of evolution of ancestral Old World primates resulted in the inactivation of the α1,3galactosyltransferase gene and the subsequent appearance of anti-Gal in these primates. Other studies in immunoparasitology have demonstrated the possible physiologic significance of anti-Gal in protection against certain parasitic infections.

Major scientific attention was focused on α-gal and anti-Gal with the realization in the early nineties that the interaction between this antigen and antibody is the major obstacle to xenotransplantation. The success of immunosuppressive drugs, in the last two decades of the 20th century, in preventing allograft rejection, has raised hopes for cure in many patients in need of organ transplant. Because of limited supply of allografts, only 20% of patients receive the needed organ. The use of other species as an alternative source for organs, in particular pigs, is currently not possible because the interaction of the recipient's anti-Gal with α-gal epitopes on the xenograft causes rejection of the graft. Understanding of the various aspects of the α-gal/anti-Gal system and development of methods to overcome this interaction will help make xenotransplantation a reality in the 21st century.

This book is the first attempt to compile the interdisciplinary information that has been obtained in the last 15 years on α-gal and anti-Gal. The book is divided roughly into two parts. The first part contains information on much of the basic research that has been conducted in this system, including its evolution; the characteristics of α-gal, anti-Gal, and α1,3galactosyltransferase; methods for identifying the antigen, antibody, and enzyme; the structure of the α1,3galactosyltransferase gene; and the structure-function analysis of this enzyme. The second part includes chapters that discuss the various paraclinical aspects of this system including its significance in viral and parasitic infections, recognition of α-gal epitopes by bacterial toxins, and the role of α-gal and anti-Gal in xenotransplantation and in autoimmunity. The therapeutic potential of anti-Gal is discussed as well. The ubiquitous presence of anti-Gal in humans and the possible manipulation of this antibody by α-gal epitopes coupled to various molecules creates a unique opportunity for the exploitation of this antibody in multiple clinical areas, most of which have not been explored as yet. Examples of the exploitation of anti-Gal to increase the immunogenicity of tumor vaccines are discussed.

The contributors of the chapters are among the researchers who performed many of the pioneering studies in this system. Thus, we hope that the information included in this volume of Subcellular Biochemistry will be helpful to established researchers in this area and to young scientists who intend to pursue research in the various pathophysiologic and therapeutic aspects of α-gal and anti-Gal.

Finally, we would like to thank Ms. Ruthann Vendy for assisting in preparation of this book.

Uri Galili
José Luis Avila

*Philadelphia, Pennsylvania*
*Caracas, Venezuela*

# Contents

*Chapter 1*

**Evolution of α1,3Galactosyltransferase and of the α-Gal Epitope**

Uri Galili



*Chapter 2*

**The α1,3-Galactosyltransferase Gene**

David H. Joziasse, Joel H. Shaper, and Nancy L. Shaper

*Chapter 4*
## The Natural Anti-Gal Antibody
Uri Galili, Le Wang, Denise C. LaTemple, and Marko Z. Radic



*Chapter 5*
## α-Gal Epitopes in Animal Tissue Glycoproteins and Glycolipids
Lennart Rydberg, Jan Holgersson, Bo E. Samuelsson, and Michael E. Breimer

*Chapter 6*
**The *Griffonia simplicifolia* I - $B_4$ Isolectin: A Probe for α-D-Galactosyl End Groups**
Irwin J. Goldstein and Harry G. Winter



*Chapter 7*
**α-Gal Epitopes on Viral Glycoproteins**
Russell P. Rother and Uri Galili

*Chapter 8*
**α-Galactosyl-Bearing Epitopes as Potent Immunogens in Chagas' Disease and Leishmaniasis**

José Luis Avila

*Chapter 9*
**Enterotoxin A of *Clostridium difficile* and α-Gal Epitopes**
Charalabos Pothoulakis



*Chapter 10*
**Anti-Gal, α-Gal Epitopes, and Xenotransplantation**
Takaaki Kobayashi and David K. C. Cooper

*Chapter 14*

**Graves' Disease as a Model for Anti-Gal Involvement in Autoimmune Diseases**

Uri Galili

*Chapter 1*

# Evolution of α1,3Galactosyltransferase and of the α-Gal Epitope

Uri Galili

## 1. INTRODUCTION

Most of the genes studied in humans have homologues in other mammals; however, the α1,3galactosyltransferase (α1,3GT) gene represents a unique exception. Whereas it is highly active in nonprimate mammals, prosimians and New World monkeys (i.e., monkeys of South America), producing one of the most abundant cell surface carbohydrate epitopes, the epitope Galα1-3Galβ1-4GlcNAc-R (termed the α-gal or α-galactosyl epitope), this gene is completely inactive in Old World primates (i.e., humans, apes and Old World monkeys). As discussed below, the inactivation of α1,3GT in ancestral primates was likely to be associated with a catastrophic evolutionary event that led primates of the Old World (i.e., primates of Asia and Africa) into almost complete extinction. Because of this event, humans, apes and Old World monkeys produce very large amounts of a natural antibody against the α-gal epitope. This antibody, termed anti-Gal, constitutes 1% of circulating immunoglobulins and it prevents transplantation of organs or tissues from nonprimate mammals (e.g., pigs) into humans because it readily binds to the α-gal epitopes on such xenografts. It is impossible to determine the actual evolu-

**Uri Galili** Department of Microbiology and Immunology, MCP Hahnemann School of Medicine, Philadelphia, Pennsylvania 19129.

*Subcellular Biochemistry, Volume 32: α-Gal and Anti-Gal*, edited by Galili and Avila.
Kluwer Academic / Plenum Publishers, New York, 1999.

tionary events which have led to inactivation of the α1,3GT gene and suppression of α-gal epitope expression in ancestral Old World primates. However, information on the expression of this epitope in various species, the structure of the α1,3GT pseudogene in primates and the fossil record of primates, enable us to speculate on both the evolutionary period in which α1,3GT gene inactivation occurred, as well as possible causes for this event. The striking reciprocal evolution of α1,3GT and anti-Gal may provide a unique paradigm for the understanding of evolutionary changes driven by host-pathogen relationships. Description of α1,3GT, α-gal epitopes, and of anti-Gal is cursory in this chapter and are referred to here in the context of discussing evolutionary aspects of this system. Other chapters in this book provide detailed information on these issues.

## 2. DISTRIBUTION OF α-GAL EPITOPES AND OF α1,3GT IN MAMMALS

### 2.1. Distribution of α-Gal Epitopes in Mammals

Carbohydrate chains linked to asparagine (N-linked) on cell surface glycoproteins are synthesized in the endoplasmic reticulum where they are linked to the protein. Subsequently, glycoprotein molecules are transported to the Golgi apparatus. The carbohydrate chains that are to become of the complex type are trimmed by mannosidases in the Cis compartment of the Golgi, so that the two terminal mannosyl units serve as the base for elongation of the carbohydrate chain. This elongation process resembles that of an assembly line in a car plant (Figure 1). The carbohydrate units are sequentially transferred from sugar-nucleotide molecules (e.g., uridine diphosphate galactose [UDP-Gal] or cytidine monophosphate-sialic acid [CMP-SA]) to the growing carbohydrate chain by enzymes specific for each step (Schachter and Roseman, 1980; Kornfeld and Kornfeld, 1985). These enzymes are termed glycosyltransferases. The terminal sugars that cap a large proportion of the complex carbohydrate chains in mammals are sialic acid (SA) and galactose linked to N-acetyllactosaminyl (Galβ1-4GlcNAc-R). Similar epitopes are found on glycolipid chains of the neolacto series (i.e., glycolipids containing N-acetyllactosaminyl groups) (Macher and Sweeley, 1978). The epitope SAα2-3(6)Galβ1-4GlcNAc-R is present on carbohydrate chains of glycoproteins and glycolipids of all mammals including humans. However, the epitope Galα1-3Galβ1-4GlcNAc-R (i.e., the α-gal epitope) is absent in humans, apes and Old World monkeys because these species lack the glycosyltransferase that links the terminal galactosyl in an α1,3 linkage to the N-acetyllactosaminyl of the carbohydrate chain. This enzyme is the α1,3galactosyltransferase (α1,3GT) and it catalyzes the following reaction within the trans-Golgi network:

$$\underset{\text{N - acetyllactosamine}}{\text{Gal}\beta1\text{ - 4GlcNAc - R + UDP - Gal}} \xrightarrow{\alpha1,3\text{GT}} \underset{\alpha\text{ - gal epitope}}{\text{Gal}\alpha1\text{ - 3Gal}\beta1\text{ - 4GlcNAc - R + UDP}}$$

This enzyme was originally described in rabbit bone marrow by Basu and Basu (1973) and subsequently reported by Betteridge and Watkins (1983) in rabbit intestinal mucosa, in Ehrlich ascites (Blake and Goldstein, 1981; Elices et al., 1986), and in bovine thymus (Van den Eijnden et al., 1983; Blanken and van den Eijnden, 1985). From immunohistochemistry studies one could infer that this enzyme is present in many species because of the expression of its product, the α-gal epitope, on various cell surface glycoconjugates or on secreted glycoproteins. By using a lectin *Bandeiraea (Griffonia) simplicifolia* $IB_4$ ($BSIB_4$), which is specific for α-gal epitopes, Ekhardt and Goldstein (1983) reported the expression of α-gal epitopes on mouse Ehrlich ascites cells, while Cummings and Kornfeld (1984) described this epitope on mouse lymphoma cells, and Cummings and Mattox (1988) demonstrated its presence on mouse teratocarcinoma cells. α-Gal epitopes have also been reported on glycolipids from rabbit red cells (Eto et al., 1968; Stellner et al., 1973; Egge et al., 1985) and cow red cells (Umeura et al., 1978; Chien et al., 1979). Spiro and Bohyroo (1984) analyzed expression of α-gal epitopes on thyroglobulin molecules from various species and reported the expression of variable numbers of such epitopes per molecule on thyroglobulin from nonprimate mammals including cow, pig, dog, guinea pig, and rat but not on human thyroglobulin (Spiro and Bohyroo, 1984). Collectively, these observations suggested that α1,3GT is active in various nonprimate mammals, but not in humans. Our findings on the ubiquitous production of anti-Gal in humans (Galili et al., 1984) and on the high specificity of this antibody for α-gal epitopes (Galili et al., 1985; 1987) implied that α-gal epitopes are not likely to be expressed on human cells. Expression of such epitopes in humans would result in their interaction with anti-Gal and thus, the propagation of an autoimmune reaction (Galili, 1989).

Based on this information, it was of interest to determine whether the distribution of α-gal epitopes, α1,3GT, and of anti-Gal display any distinct characteristics in mammals and in other vertebrates. Our initial studies included measurement of α-gal epitope expression on red cells from various species by assessing the binding of anti-Gal isolated from normal human serum to these cells. In parallel, we studied the binding of the α-gal epitope specific lectin $BSIB_4$ (Wood et al., 1979), the characteristics of which are described in the chapter of Goldstein and Winter in this book. The presence of anti-Gal in sera of various species was determined by measuring indirect agglutination of rabbit red cells using corresponding anti-IgG antibodies as secondary antibodies. Rabbit red cells express an abundance of α-gal epitopes (Eto et al., 1968; Stellner et al., 1973; Egge et al., 1984) and thus are a convenient means for assessing anti-Gal activity. α-Gal epitopes are abundant on red cells from various nonprimate mammals including rat, rabbit, cow, pig, and dog

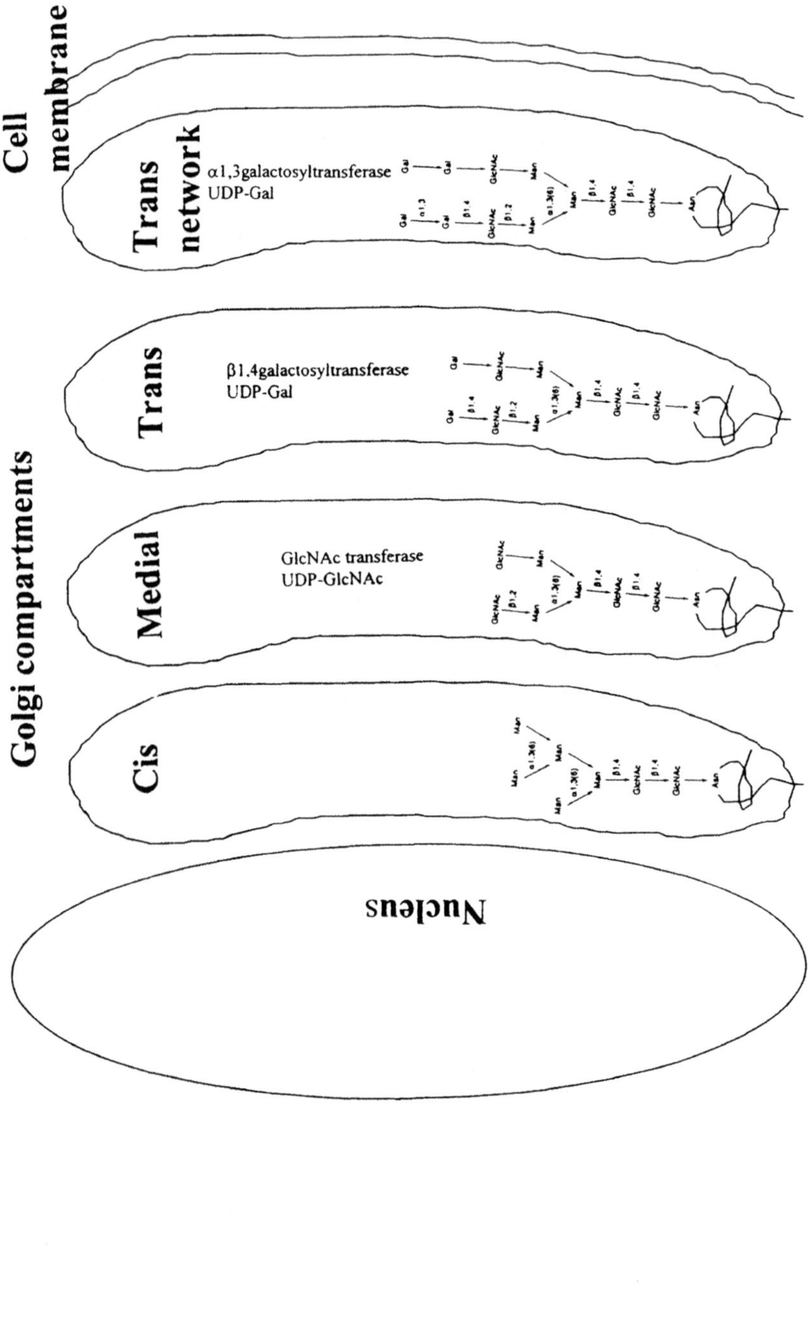
Cell membrane
Golgi compartments
Trans network
α1,3galactosyltransferase
UDP-Gal
Trans
β1,4galactosyltransferase
UDP-Gal
Medial
GlcNAc transferase
UDP-GlcNAc
Cis
Nucleus

**Figure 1.** Synthesis of α-gal epitopes on carbohydrate chains of glycoproteins in the Golgi apparatus of mammalian cells. Within the endoplasmic reticulum N-linked carbohydrate chains of glycoproteins are linked to the amino acid asparagine (Asn) regions of proteins that contain the sequence Asn-X, Ser, or Thr (asparagine—any amino acid—serine [threonine]). The glycoproteins are then transported into the Cis Golgi compartment where many of the carbohydrate chains are trimmed to expose the two Man (mannose) linked α1,3 or α1,6 to a penultimate mannose. Upon transfer to the Medial compartment of the Golgi, GlcNAc-(N-acetyllactosamine) is linked to Man by GlcNAc transferase. In the Trans-Golgi, Gal (galactose) in linked to GlcNAc by β1,4galactosyltransferase. In the Trans-Golgi network, α1,3GT caps the carbohydrate chain with Gal linked α1,3 to form the α-gal epitope. In this compartment, various glycosyltransferases can cap the carbohydrate chain with other sugars such as SA (sialic acid).

(Galili et al., 1987c). Similar analyses with red cells from primates demonstrated a striking distribution of α-gal epitopes. Studies on red cells of 13 prosimians including loris, galago, potto, and lemur (most of which were received from Duke University Primate Center) demonstrated abundance of α-gal epitope expression, whereas no anti-Gal was found in their sera (Galili, 1997). Similarly, analysis of blood samples from 23 New World monkeys of seven different species (received from a zoo in San Paulo, Brazil) demonstrated invariable expression of α-gal epitopes on the red cells and the absence of anti-Gal in their sera (Galili et al., 1987c). In contrast, studies on blood samples of Old World monkeys and apes (received from the San Francisco zoo) indicated that their red cells lack α-gal epitopes; however, anti-Gal is present in sera of these primates in titers comparable to those in humans (Galili et al., 1987c). We further analyzed α-gal epitopes on glycolipids isolated from various species by immunostaining on thin layer chromatography with human anti-Gal isolated from sera of blood type AB individuals (Galili et al., 1987c). In parallel, we immunostained the glycolipids with a mouse monoclonal anti-Gal antibody, designated Gal-13 (Galili et al., 1987a). We found α-gal epitopes on glycolipids of rabbit, cow, and New World monkey red cells, but not on glycolipids of Old World monkey or human red cells. Moreover, the glycolipid molecule ceramide penta hexoside (Galα1-3Galβ1-4GlcNAcβ1-3Galβ1-4Glc-Cer) was found in rabbit, cow, and New World monkey red cells (Galili et al., 1987a; 1987c).

These initial observations prompted the analysis of α-gal epitope expression on nucleated cells of various species by measuring binding of anti-Gal and of $BSIB_4$ lectin. No α-gal epitopes were detected on fibroblasts obtained from nonmammalian vertebrates including fish, amphibians, reptiles, or birds. In contrast, fibroblasts, epithelial cells, endothelial cells, smooth muscle cells, and lymphoid cells from marsupials, such as opossum and kangaroo, or from placental mammals including mouse, rabbit, pig, cow, horse, sheep, dog, mink, cat, bat, and dolphin, all expressed large amounts of α-gal epitopes (Galili et al., 1988b). Quantitation by Scatchard plot analysis of labeled $BSIB_4$ lectin binding indicated the expression of $1\times10^6$–$35\times10^6$ α-gal epitopes per cell (Galili et al., 1988b). This number of α-gal epitopes in various mammals is in accord with other studies that have estimated expression of $18\times10^6$ such epitopes per cell on mouse Ehrlich ascetis cells (Ekhardt and Goldstein, 1983) and with studies indicating that the α-gal epitope is the most abundant carbohydrate epitope on mouse 3T3 fibroblasts (Santer et al., 1989). Analysis of freshly obtained mammalian tissues also demonstrated ubiquitous distribution of α-gal epitopes in nonprimate mammals since these epitopes were found on thyroid cell membranes from all nonprimate mammals studied (Thall et al., 1991). Similarly, Hendriks et al (1990) demonstrated the presence of α-gal epitopes on glycolipids extracted from kidneys of various nonprimate mammals while Bäcker et al. (1998) reported α-gal epitope expression on glycolipids of various organs obtained from one pig. A study by Oriol et al.

(1996) examined the expression of α-gal epitopes in 23 species of nonprimate mammals by immunostaining tissue sections with $BSIB_4$ lectin and confirmed the ubiquitous presence of this carbohydrate epitope in mammals. Distribution of α-gal epitopes on primate nucleated cells is similar to that observed with red cells. Fibroblasts and lymphoid cells of lemurs and New World monkeys express millions of α-gal epitopes per cell, whereas similar cells from Old World monkeys, apes and humans all lack this epitope (Galili et al., 1988b). It is not clear as yet, whether all tissues in nonprimate mammals, prosimians and New World monkeys express α-gal epitopes. Negative staining of histological sections by $BSIB_4$ lectin may not necessarily indicate the absence of α-gal epitopes because if these epitopes are far apart from each other, the lectin molecule may not be able to bridge between α-gal epitopes, so that at least two of the four binding sites of the lectin are involved in the interaction. Binding via of only one binding site of the lectin is too weak and the lectin will be removed in the washing steps. Recently, we have developed a highly sensitive ELISA inhibition assay with a monoclonal anti-Gal antibody designated M86 (Galili et al., 1998). This assay can detect as few as $1 \times 10^5$ α-gal epitopes per cell and can be used for identification of α-gal epitopes in tissue homogenates (Galili et al., 1998; Stone et al., 1998). This assay or other high sensitivity detection assays are required for determination of small amounts of α-gal epitopes expressed on various tissues. Detection of these epitopes is important since the primate immune system reacts effectively against such epitopes by producing large amounts of high affinity anti-Gal; therefore, xenotransplantation of organs with even low number of α-gal epitopes would elicit a strong anti-Gal response in humans and in Old World monkeys (Galili et al., 1995, 1997).

## 2.2. α1,3Galactosyltransferase Activity in Mammals

We have found that the distribution pattern of α-gal epitopes in mammals, described above, parallels the activity of α1,3GT in various mammals. Comparative studies on this enzyme activity in lymphoid cells have demonstrated active enzyme in mice, cows and New World monkeys, but not in human cells (Galili et al., 1988b). Furthermore, our studies on α1,3GT activity in microsomal fractions isolated from fresh thyroid tissues demonstrated such catalytic activity in thyroids obtained from various nonprimate mammals including cow, pig, rat, rabbit, and guinea pig and in thyroid from squirrel monkeys (a New World monkey), but not from rhesus monkey (an Old World monkey) or in humans (Thall et al., 1991). These findings support the previous observations on the activity of this enzyme in cells or tissues obtained from mice, rabbits, and cows (Basu and Basu, 1973; Blake and Goldstein, 1981; Betteridge and Watkins, 1983; Blanken and van den Eijnden, 1985). This enzyme is active in the Trans-Golgi network, as described in Figure 1. Smith et al. (1990) have shown that α1,3GT competes with sialyltransferases in that compartment for capping carbohydrate chains of the complex type with terminal carbohydrate units. The variations in the amount of α-gal epitopes expressed on different cells are likely to be associated with such competition between various terminal glycosyltransferases for capping carbohydrate chains. Detailed discussions on the α1,3GT, its gene and other related glycosyltransferases are included in the chapters of Joziasse et al., and of Henion and Galili in this book.

## 3. EVOLUTION OF α1,3GALACTOSYLTRANSFERASE

The distribution patterns of α-gal epitopes and α1,3GT in vertebrates suggest that α1,3GT appeared early in mammalian evolution, before the divergence between marsupials and placentals, which is estimated to have occurred approximately 125 million years ago. Since no α-gal epitopes have been detected in other vertebrates including fish, amphibians, reptiles and birds, it is likely that α1,3GT appeared as a new glycosylation enzyme in the mammalian class. It is of interest to note that a carbohydrate structure Galα1-3Galβ1-4(Fucα1-3)GlcNAc-R, was reported to be the major oligosaccharide in cobra venom factor but not in venom of other snakes or other reptiles (Gowda et al., 1992; 1994). It is not clear, at present, whether the gene of the galactosyltransferase in cobra venom glands is related to that of the mammalian α1,3GT.

Data obtained on lemurs and New World monkeys suggests that ancestral primates, like other mammals, also had active α1,3GT and synthesized α-gal epitopes. Lemurs are prosimians that diverged from ancestral primates approximately 65 million years ago and evolved under geographic isolation in Madagascar. These primates express α-gal epitopes on both fibroblasts and red cells, and accordingly, lack anti-Gal in their sera (Galili et al., 1987c; 1988b). New World monkeys diverged from Old World monkeys 30–40 million years ago under geographic isolation in South America (Pilbeam, 1984). New World monkeys also have active α1,3GT and produce α-gal epitopes. In contrast, Old World monkeys, apes, and humans (termed here as Old World primates), all of which have evolved in Africa and Asia, lack this glycosyltransferase activity and accordingly lack α-gal epitopes. Unlike the rest of mammals, this group of primates produces large amounts of antibody against α-gal epitopes (i.e., the natural anti-Gal antibody) (Galili et al., 1987c). The ubiquitous expression of α-gal epitopes in nonprimate mammals, prosimians, and New World monkeys but its absence in Old World monkeys, apes, and humans (Figure 2) implies that the α1,3GT gene was inactivated in ancestral Old World primates after they diverged from New World monkeys.

Understanding the evolutionary fate of the α1,3GT gene in Old World primates became feasible after its cloning in cow (Joziasse et al., 1989) and mouse (Larsen et al., 1989). The α1,3GT gene was found to be present as a pseudogene in Old World primates (Joziasse et al., 1989; Laren et al., 1990; Galili and Swanson,

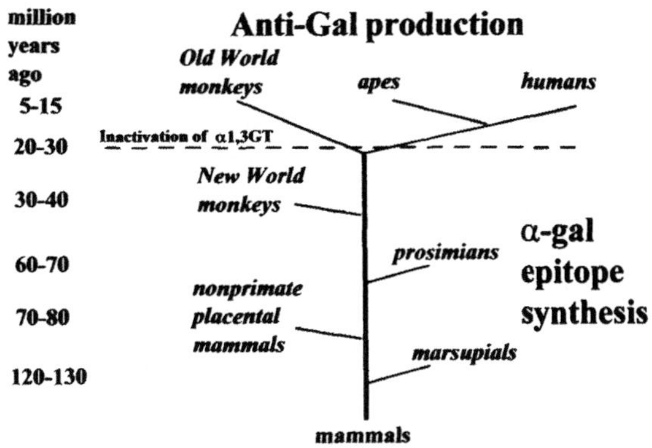


**Figure 2.** Evolution of α1,3GT and α-gal epitopes in mammals. The α-gal epitope has been produced in mammals by α1,3GT for more than 125 million years, since before the divergence between placentals and marsupials. Nonmammalian vertebrates all lack α1,3GT and do not produce α-gal epitopes. Production of this epitope was suppressed in ancestral Old World primates after they diverged from New World monkeys, and probably after apes and monkeys diverged from each other. Suppression of the α-gal epitope was followed by the production of the natural anti-Gal antibody, which is not made in nonprimate mammals, prosimians, or New World monkeys.

1991) implying that ancestral Old World primates indeed had an active α1,3GT gene. Collectively, these findings suggest ancestral Old World primates were subjected to a major evolutionary pressure for suppression of α1,3GT activity and elimination of α-gal epitopes.

### 3.1. Possible Cause for α1,3GT Suppression

Theoretically, there are two types of pathogens which might have induced the evolutionary suppression of α1,3GT in Old World primates.

#### a. Pathogens that Express α-Gal Epitopes

One could hypothesize that Old World primates might have been exposed to a pathogen that was endemic to the Old World (i.e., because of geographic barriers it did not spread to the South American continent). This pathogen was detrimental to primates and expressed α-gal epitopes. Such a pathogen could have caused the elimination of most primate populations and thus would have exerted a selective pressure for the evolution of primates in which α-gal epitope expression was suppressed. Consequently these primates lost immune tolerance to α-gal epitopes and gained the ability to produce anti-Gal. This antibody could have contributed to immune protection against such a putative pathogen. Presently, there are numerous pathogens that express α-gal epitopes and that can be affected by anti-Gal. For example, enveloped viruses that develop in cells containing α1,3GT express α-gal epitopes on the carbohydrate chains of the envelope glycoproteins. This is because the carbohydrate chains on the virus are synthesized by the glycosylation machinery of the host cell. We have shown that Eastern Equine Encephalitis virus propagated in mouse cells express α-gal epitopes, whereas the same virus propagated in Old World monkey cells lack this epitope (Repik et al., 1994). Similarly, influenza virus propagated in chicken cells lack α-gal epitopes, whereas influenza virus propagated in cow or dog cells expresses the epitope (Galili et al., 1996). Binding of anti-Gal IgM to viruses expressing α-gal epitopes results in complement activation and lysis of the virus (Rother et al., 1995; Takeuchi et al., 1996; Welsh et al., 1998). Furthermore, interaction of anti-Gal IgG molecules with the virus can prevent their adhesion to target cells (Repik et al., 1994). Thus, it is possible that anti-Gal was generated in ancestral Old World primates to protect against viruses with α-gal epitopes. Additional discussion on viruses expressing α-gal epitopes is included in the chapter of Rother and Galili in this book.

Bacteria also express various carbohydrate epitopes which have structures similar to α-gal epitopes and which bind anti-Gal. We have found that anti-Gal readily binds to various bacteria of the normal gastrointestinal flora (Galili et al., 1988a; see additional information in the chapter on anti-Gal by Galili). Furthermore, Hamadeh et al. (1992) showed that binding of anti-Gal to bacteria can activate complement and induce lysis of bacteria. There are also some protozoa that express α-gal epitopes such as *Trypanosoma cruzi* and *Leishmania mexicana* (Almeida et al., 1991, 1994; Avila et al., 1989; Couto et al., 1990). The interaction of anti-Gal with these protozoa is discussed in detail in the chapter of Avila in this book. Similar to the argument raised above with regard to enveloped viruses, a bacteria or protozoa with a carbohydrate make-up similar to these microbial agents could have selected primates of the Old World to evolve without α-gal epitopes and to produce anti-Gal.

Knock-out mice for α1,3GT have been instrumental in demonstrating that the loss of α-gal epitopes results in the gaining of the ability to produce anti-Gal. The gene for α1,3GT has been disrupted by homologous recombination in these mice (Thall et al., 1995; Tearle et al., 1996; see the chapter of Thall in this book). Whereas wild type mice can not produce anti-Gal, knock-out mice produce low amounts of this antibody (Thall et al., 1996). The titers of anti-Gal in these mice can be greatly increased by immunization with cells expressing α-gal epitopes, such as rabbit red cells (LaTemple and Galili, 1998). Moreover, the exposure of these mice to Leishmania, which expresses α-gal epitopes, also induces production of high amounts of anti-Gal (see the chapter of Pearse and D'Apice in this book).

### b. Pathogens Using α-Gal Epitopes as Receptors

The chapter of Pothoulakis in this book describes the specific interaction of enterotoxin A of *Clostridium difficile* with α-gal epitopes on cell membranes. The binding of this toxin to α-gal epitopes on cells induces cell death (Krivan et al., 1986). In animals, the binding of enterotoxin A to its receptor (i.e., the α-gal epitope on the intestinal mucosa cells) results in extensive diarrhea which may lead to the death of the animal (Pothoulakis et al., 1996). A pathogen that produced a toxin with similar receptor specificity and which was detrimental to Old World primates could have exerted a strong selective pressure for the evolution of primates which lack α-gal epitopes so they are not sensitive to such a toxin. As indicated above, the elimination of α-gal epitopes in such primates would have abolished immune tolerance to this epitope and the animals would have started producing anti-Gal because of antigenic stimulation by bacteria of the normal flora.

Although it is impossible to identify, at present, the pathogen(s) that induced the inactivation of α1,3GT in Old World primates, it is probable that such a pathogen was geographically confined to Asia and Africa. Thus, New World monkeys, which evolved in South America, were not subjected to its evolutionary pressure. Epstein Barr virus is an analogous example of a pathogen that evolved in the Old World and thus did not affect immune evolution of New World monkeys because of geographic separation. This virus evolved from the Herpes virus family after the separation of the African and South American continents (Frank et al., 1976). Epstein Barr virus (EBV) infects B lymphocytes and transforms them into immortalized lymphoid clones. The immune systems in humans and Old World monkeys have evolved to generate a very strong T cell response against B cells immortalized by EBV and thus prevent the expansion of such immortalized B cells which may cause multiclonal lymphomas. Thus, humans infected with EBV usually develop a mild disease, called infectious mononucleosis, in which there is massive proliferation of T cells that lyse the EBV transformed B cells (DeThe, 1980). In contrast, New World monkeys have not been exposed to EBV in the course of their evolution; therefore, they have not acquired the ability to generate an effective cellular immune response. Thus, New World monkeys infected with EBV develop multiclonal B cell lymphomas which cause the death of the infected animal (Frank et al., 1976). In humans, similar multiclonal lymphomas are observed only in individuals that are severely immunocompromised (Hanto et al., 1982).

There are also analogous examples of geographic selection for immunological resistance to pathogens in human populations. European populations have evolved with certain natural immune resistance to smallpox viruses. However, the introduction of virus to the New World, subsequent to the conquest of Mexico by Spain, resulted in the decimation of Indian populations by this disease in a relatively short period of time. This is because these populations have had no previous exposure to the smallpox virus, and thus, were not selected for the development of natural immunity against it. All these examples imply that host-pathogen relationship throughout evolution greatly affect the efficacy of the immune system in various species to cope with microbial infections. The evolutionary appearance of anti-Gal in ancestral Old World primates is likely to be an extreme example of microbial effects on immunologic evolution.

## 4. POSSIBLE MOLECULAR MECHANISMS FOR EVOLUTIONARY SUPPRESSION OF α1,3GALACTOSYLTRANSFERASE

The absence of α1,3GT activity in Old World primates raised the questions of how the α1,3GT gene was inactivated in ancestral Old World primates and whether this gene is still present in the human genome. Although the molecular mechanism(s) which led to the suppression of α1,3GT activity in ancestral Old World primates are not fully understood as yet, substantial information on these changes has been obtained from the cloning of the α1,3GT gene from murine (Larsen et al., 1989), bovine (Joziasse et al., 1989), and New World monkey origin (Henion et al., 1994). By using the bovine α1,3GT cDNA as a probe, Joziasse et al. (1989) demonstrated production of α1,3GT mRNA in cells of nonprimate mammal and New World monkey origin, but not in cells of Old World monkey or human origin. However, probing the human genomic DNA by Southern blot analysis demonstrated the presence of the α1,3GT gene as a pseudogene in humans (Joziasse et al., 1989). Subsequent studies showed that this gene was present on human chromosome 9 and a second spliced α1,3GT pseudogene on chromosome 12 (Joziasse et al., 1991; Shaper et al., 1992). Comparative sequences of the human pseudogene with the bovine or mouse α1,3GT genes demonstrated 78% sequence identity (Larsen et al., 1991; Joziasse et al., 1991; Shaper et al., 1992). We have further found the α1,3GT pseudogene in apes and Old World monkeys (Galili and Swanson, 1991). This presence of the α1,3GT gene as a pseudogene and the absence of α1,3GT mRNA in humans and Old World primates (Joziasse et al., 1989; Larsen et al., 1990) implies that the α1,3GT gene was not eliminated from ancestral Old World primates, but was conserved as a pseudogene that is not transcribed.

In general, transcription of evolutionary conserved genes is prevented by regulatory mutations which destroy functional promoters and/or enhancers, resulting in the inability of RNA polymerases to transcribe the gene in question (Wilson, 1985). The identity of these regulatory mutations in the α1,3GT pseudogene is not known at present, but it may be discerned by studying the promoter of α1,3GT in species in which the enzyme is active, and comparing these sequences and function of the promoter with the homologous region in humans.

Apes have acquired additional structural mutations (i.e., deletions in the α1,3GT pseudogene) which ensure prevention of α1,3GT activity in case the gene is aberrantly transcribed. Larsen et al. (1990) isolated a human genomic clone that contained the largest exon of the α1,3GT gene, termed Exon IX (see the Chapter by Joziasse et al., for detailed description of the α1,3GT gene). Exon IX within the active gene was found to encode most of the catalytic domain of the enzyme (Joziasse et al., 1992; Henion et al., 1994). Exon IX of the human α1,3GT pseudogene was found to include two deletions which cause a shift of the reading frame and generation of premature stop codons (Larsen et al., 1990). Chimpanzees, like humans, have the same two deletions, whereas other apes (i.e., gorilla and orangutan) have only one of the two deletions, but no such deletions were found in Old World monkeys (Galili and Swanson, 1991). It is possible that these structural mutations (i.e., mutations within the open reading frame of the gene) act as a back-up mechanism for preventing production of α1,3GT. It could be hypothesized that aberrant expression of the α1,3GT gene in Old World primates is detrimental because interaction of the circulating anti-Gal with the *de novo* synthesized α-gal epitopes can cause an autoimmune disease (Galili, 1989). Such aberrant expression may theoretically be the result of viral infection or spontaneous revertant mutations. The deletions in the human and ape α1,3GT pseudogene ensure that in case of its abberant expression, the protein produced will be truncated and will lack catalytic activity (Henion et al., 1994). The effects of truncation of the primate α1,3GT gene on the catalytic activity of the enzyme are described in detail in the chapter of Henion and Galili.

## 5. ESTIMATING THE EVOLUTIONARY PERIOD OF α1,3GALACTOSYLTRANSFERASE SUPPRESSION

The ability of New World monkeys to synthesize α-gal epitopes by α1,3GT and the absence of this enzyme in humans clearly indicates that the inactivation of the α1,3GT gene occurred in an evolutionary period during which New World monkeys and Old World primates were geographically separated. As indicated above, paleontologists estimate that Old World primates and New World monkeys diverged from a common ancestor 30 - 40 million years ago (Pilbeam, 1984) (see Figure 2). Thus, it is likely that α1,3GT inactivation occurred less than 30 million years ago. Further information on the period during which α1,3GT was inactivated could be obtained by sequencing the α1,3GT gene in primates, as described below.

Studies on molecular aspects of evolution have indicated that, in general, the evolutionary history of organisms is inscribed in the DNA of contemporary species (Wilson et al., 1977). This principle has been demonstrated in the phylogenetic relationships among primate species, as indicated by similarities in DNA sequences of homologous genes (Koop et al., 1986; Rouvolo et al., 1991). Therefore, we assumed that the α1,3GT gene and pseudogene sequences in primates might provide clues to the time of evolutionary suppression of this gene in Old World primates. We further assumed that the smaller the difference between Old World primate α1,3GT pseudogene sequences and the corresponding sequences in New World monkeys where the gene is active, the later this gene must have been inactivated in the course of evolution of ancestral Old World primates. This assumption was based on a basic principle in molecular evolution which indicates that sequences in genes evolve in a "clock-wise" manner (i.e. the rate of mutation in a given gene is usually constant in time). In addition, the rate of mutation is faster in silent DNA (i.e., pseudogenes or introns) than in coding sequences because it is not impeded by natural selection (Wilson et al., 1977). The rate of mutations in a functional gene is slowed by functional constraints (i.e., many mutations affecting the function of the gene will not survive). In contrast, the mutation rate in pseudogenes is not affected by any function and thus they can freely occur. Studies on the η globulin gene, which is a pseudogene in all mammals, have shown that the maximal rate of mutations of the evolutionary clock in silent DNA is approximately 0.5% mutations per million years. This rate is somewhat slower in primates (Koop et al., 1986).

In order to apply these considerations in the study of α1,3GT evolution, we cloned a 370 bp region of exon IX and compared the sequences of the various pseudogenes in Old World monkeys and apes, with those obtained from the active gene in New World monkeys (Galili & Swanson, 1991). The cloning was performed by PCR with primers that corresponded to sequences with high similarity in the mouse cDNA (Larsen et al., 1989), cow cDNA (Joziasse et al., 1989) and the human pseudogene (Larsen et al., 1990). The aligned DNA sequences are described in Figure 3. The Old World primates studied were the three ape species: chimpanzee, gorilla and orangutan; and three Old World monkeys: rhesus monkey, African green monkey and Patas monkey. In addition, three New World monkeys were studied, including: spider monkey, squirrel monkey and howler monkey. These sequences were compared to the human sequence (Larsen et al., 1989). The cow sequence (Joziasse et al., 1989), served as an external reference.

As shown in Figure 3, the sequences of the human and chimpanzee α1,3GT pseudogenes are identical, with the exception of one base. The small differences in α1,3GT sequences among Old World monkeys (1.8–3.6%), or apes (0.3–3.2%) indicate that these differences indeed reflect phylogenetic relationships between the various species. The 3.2% difference in sequences human/orangutan is similar to that observed in the η-globin pseudogene (Koop et al., 1986) and suggests that α1,3GT has been inactive before the divergence of the orangutan lineage from the human lineage, which is estimated to have occurred 12–18 million years ago (Pilbeam, 1984). The 7.6–8.1% human/Old World monkey differences are also equal to that of the η-globin pseudogene, suggesting that for a substantial part of the period since the divergence of apes from Old World monkeys (estimated to be 28 million years ago [Pilbeam, 1984]), the α1,3GT gene, has been conserved in Old World primates in an inactive form (i.e., as a pseudogene).

Detailed comparison of the sequences in the various primates suggests that the inactivation of the α1,3GT gene occurred after the divergence of ape and Old World monkey lineages from each other, rather than immediately after the divergence between New World monkeys and Old World primates (Galili and Swanson, 1991; Galili and Andrews, 1995). This suggestion is based on several findings. **1.** The deletion at base 904, which prevents the production of active α1,3GT in case of aberrant expression of the gene (Henion et al., 1994), is common to apes and humans, but is not found in Old World monkeys (Figure 3). The absence of this dele-

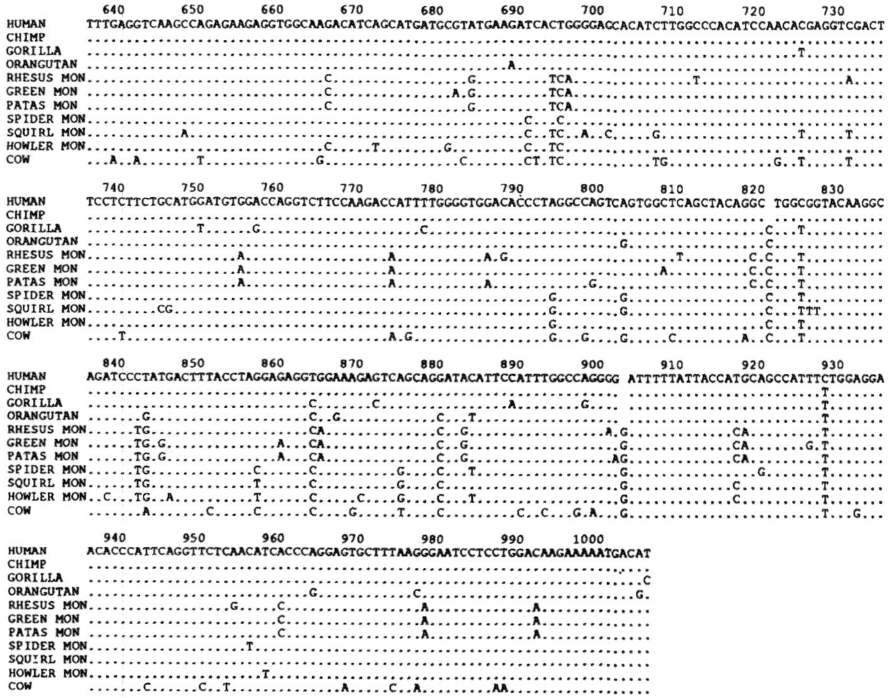


**Figure 3.** Aligned DNA sequence of a 370 base pair region in the 3′ portion of the α-galactosyltransferase (α1,3GT) gene from human (Larsen et al., 1990), apes (chimpanzee, gorilla, and orangutan), Old World monkeys (rhesus monkey, African green monkey and Patas monkey), New World monkeys (spider monkey, squirrel monkey and howler monkey) and domestic cow (Joziasse et al., 1989). The base numbers in this figure are according to the open reading frame of the mouse α1,3GT cDNA described by Larsen et al. (1989). The numbered base is under the second digit. Dots represent sequences identical to that of the human α1,3GT gene. Minus signs represent base deletions. From Galili and Swanson (1991), with permission.

tion, also observed in additional Old World monkeys as baboon and langur (Galili and Swanson, 1991), suggests that this deletion had occurred in apes after they diverged from monkeys and could have occurred in response to the evolutionary pressure for preventing production of catalytically active α1,3GT. **2.** Old World monkeys share specific sequences that are not found in either apes or in New World monkeys. Such sequences include C in position 667 (C-667), G-865, A-775, C-820, A-866, G-884, A-919, C-961, A-979 and A-993 (Figure 3). These sequences, which are unique to Old World monkeys (found also in baboon and langur: data not shown), suggest that ancestral Old World monkeys carried the active α1,3GT gene for a period after divergence from apes and that the functional constraints on this gene resulted in the conservation of many unique sequences shared by the different Old World monkey species at present. **3.** Various nucleotides, such as C-696, are shared by New World monkeys and Old World monkeys, but not found in apes. Other nucleotides, such as C-844 and C-881, are shared by New World monkeys, Old World monkeys, orangutan and gorilla; whereas T-929 is shared by all primates except for humans. This indicates that many of the substitution mutations reflect a sequential, rather than a random pattern, further supporting the concept that the α1,3GT gene has been subjected to functional constraints for long periods in the course of Old World primate evolution. These findings suggest, but do not prove, that α1,3GT gene inactivation occurred less than 28 million years ago, after Old World monkey and ape lineages diverged from each other (Figure 2). As discussed below, the fossil record, which demonstrates a decline in diversity of ape species during the Miocene, is also in accord with the suggestion that inactivation of the α1,3GT gene in ancestral Old World primates occurred after divergence between apes and monkeys.

## 6. CORRELATION WITH THE FOSSIL RECORD

The evolutionary transition of ancestral Old World primates from a state of abundant synthesis of α-gal epitopes to the state of absence of this epitope and production of large amounts of antibody against it, is an unprecedented process in evolution. It is probable that this transition was not a gradual process characteristic to neutral evolution. In neutral evolution, which underlies many evolutionary traits, random drift of neutral or nearly neutral mutations gradually changes the characteristics of populations (Kimura, 1983). It is unlikely that ancestral primates gradually reduced α-gal epitope synthesis and increased anti-Gal production, because individuals who had this antigen could not produce the antibody. If, for some reason, anti-Gal production was initiated before complete elimination of α-gal epitopes, individuals producing both antigen and antibody would have become extinct because of autoimmune diseases. Thus, it is more likely that the transition of ancestral primates from α-gal epitope expression to anti-Gal production was the result of a massive Darwinian selective process, rather than a random drift. Since, at present, there is no known example of polymorphism in α-gal epitope expression in mammals (e.g., 100% of pigs tested express α-gal epitopes [Alvarado et al., 1995]), it is probable that the proportion of ancestral primates which successfully suppressed α-gal epitope expression was extremely small. This suggests that the selection process for transition of Old World primates from α-gal epitope expression to anti-Gal production might have been a catastrophic evolutionary event that caused an almost complete extinction of these primates. Therefore, it was of interest to determine whether the fossil record of primates suggests the occurrence of such an event.

It is evident that the immunologic and biochemical studies on anti-Gal and α-gal epitopes can not be directly associated with the fossil record. Nevertheless, there is an interesting parallel between the information obtained from the fossil record and the evolutionary events discussed above. Fossil evidence from the early Miocene (17–22 million years ago) shows that there were many more ape species at that time than at present. The relative number of individuals ape fossils discovered from that period is high and these ancestral apes were restricted to Africa where they formed a conspicuous part of the fauna (Andrews 1981, 1992). After this period, apes begin to disappear from the fossil record, and by 7–8 million years ago, they completely vanish from Africa, but are found in significant numbers in Asia and Southern Europe. After this time apes are extremely rare, a situation which still exists at present time. Thus, the ape fossil record suggests that a massive extinction process occurred within the ape population causing complete absence of ape fossils from both Asia and Africa 5–7 million years ago (Andrews, 1981; 1992). It may be possible that these evolutionary changes in ape populations reflect the event of α1,3GT suppression and the appearance of anti-Gal which might have occurred approximately 10–20 million years ago (Galili and Andrews, 1995).

The fossil record of Old World monkeys is harder to associate with this event since up to 7–8 million years ago no more than two species of monkeys could have been identified (Andrews, 1992). At that time monkeys start to become abundant and their fossils are commonly found in all Old World late Miocene sites. It is difficult to establish whether suppression of α1,3GT also played a role in maintaining the low diversity of Old World monkeys up to 7 million years ago, and if the great radiation of these monkeys took place after α-gal epitopes were eliminated.

## 7. CHANGES IN CARBOHYDRATE EXPRESSION IN HUMAN POPULATIONS

It is impossible to assess the length of time necessary for the transition of Old World primate populations from α-gal epitope expression to elimination of this epitope and production of anti-Gal. There is, however, one example of changes in carbohydrate expression in human populations suggesting that such events may occur during a surprising short period of time. Most human populations are polymorphic and include variable proportions of blood group O, A, B and AB phenotypes (Mourant et al., 1976). However, the Indians of the Amazon basin all have blood group O (Salzano, 1957). These Indians, who migrated to South America some 10,000–20,000 years ago, are believed to share common ancestors with North American Indians, many of whom have blood type A (i.e., GalNAcα1-3[Fucα1-2]Gal-β). Furthermore, blood type A antigens has been identified in human bones from 1500 year old Amazonian burial sites (Pereira et al., 1984). If one excluded the possibility that the migrating populations were exclusively of blood group O, the absence of blood group A among Amazonian Indians may be attributed to the ineffective immune response of individuals with blood group A against environmental bacteria expressing a similar antigen. Therefore, within a span of less than 20,000 years, this Indian population might have been selected for individuals with only blood group O, who could produce both anti-A and anti-B antibodies.

This example of shift in carbohydrate antigen expression within a population is not analogous to that of α-gal epitope suppression, since genes for blood group antigens (i.e., the corresponding A and B transferases) are allelic polymorphic within human populations. In contrast, α1,3GT is either expressed in 100% of some primate populations (i.e., lemurs and New World monkeys), or is completely absent in others (i.e., Old World primates). Nevertheless, the apparent elimination of individuals with blood group A transferase in the Amazonian Indian populations suggests that changes in carbohydrate expression as a result of host-pathogen relationship can be a relatively rapid process on an evolutionary time-scale.

## 8. CONCLUDING REMARKS

The observations on α-gal epitope and α1,3GT expression suggest that α1,3GT first appeared in mammals prior to the divergence of marsupials and placentals and that it was suppressed in ancestral Old World primates after their divergence from New World monkeys. Data obtained from comparative studies on sequences of the α1,3GT pseudogene in Old World primates suggest that this event occurred in the Old World after monkeys and apes diverged from each other. This evolutionary event, which possibly brought Old World primates to the verge of extinction might have occurred 10–20 million years ago and resulted in elimination of α-gal epitopes and the appearance of anti-Gal in all Old World monkeys, apes and humans. This evolutionary event has formed a physiologic barrier that currently prevents the transplantation of organs from nonprimate mammals, such as pigs into humans.

It is presently established that the rapid immune rejection of pig xenografts in monkeys or humans (Auchincloss, 1988), is the result of interaction between the natural anti-Gal antibody and α-gal epitopes on the xenograft (Good et al., 1992; Galili, 1993; Sandrin et al., 1993). This barrier is discussed in several chapters of this book including those of Kobayashi and Cooper; Sandrin and McKenzie; and Pearce and D'Apice. However, as discussed in chapter of LaTemple and Galili, the immunologic potential of anti-Gal may be exploited for beneficial clinical use as in improving immunogenicity cancer and viral vaccines.

## REFERENCES

Almeida, I.C., Ferguson, M.A., Schenkman, S., and Travassos, L.R., 1994, Lytic anti-α-galactosyl antibodies from patients with chronic Chagas' disease recognize novel O-linked oligosaccharides on mucin-like glycosyl-phosphatidylinositol-anchored glycoproteins of *Trypanozoma cruzi. Biochem. J.* **304**:793–802.

Almeida, I.C., Milani, S.R., Gorin, A.J., and Travassos, L.R., 1991, Complement mediated lysis of *Trypanosoma cruzi* trypomastigotes by human anti α-galactosyl antibodies. *J. Immunol.* **146**:2394–2401.

Alvarado, C.G., Cotterell, A.H., McCurry, K.R., Collins, B.H., Magee, J.C., Berthold, J., Logan, J.S., and Platt, J.L., 1995, Variation in the level of xenoantigen expression in porcine organs, *Transplantation* **59**:1589–1596.

Andrews, P., 1981, Species diversity and diet in monkeys and apes during the Miocene. In: *Aspects of human evolution*, Ed., Stringer, C.B., pp. 25–61.

Andrews, P., 1992, Evolution and environment in the hominoidea. *Nature* **360**:641–646.

Auchincloss, H., 1988, Xenogeneic transplantation. *Transplantation* **45**:1–20.

Avila, J.L., Rojas, M., and Galili, U., 1989, Immunogenic Galα1-3Gal carbohydrate epitopes are present on pathogenic American Trypanosoma and Leishmania. *J. Immunol.* **142**:2828–2834.

Bäcker, A.E., Holgersson, J., Samuelsson, B.E., and Karlsson, H., 1998, Rapid and sensitive GC/MS characterization of glycolipid released Galα1,3Gal-terminated oligosaccharides from small organ specimens of a single pig. *Glycobiology* **8**:33–545.

Basu, M., and Basu, S., 1973, Enzymatic synthesis of blood group related pentaglycosyl ceramide by an α-galactosyltransferase. *J. Biol. Chem.* **248**:1700–1706.

Betteridge, A., and Watkins, W.M., 1983, Two α-3-D galactosyltransferases in rabbit stomach mucosa with different acceptor substrate specificities. *Eur. J. Biochem.* **132**:29–35.

Blake, D.D., and Goldstein, I.J.,1981, An α-D-galactosyltransferase in Ehrlich ascites tumor cells: Biosynthesis and characterization of a trisaccharide (α-D-galacto(1-3)-*N*-acetyllactosamine). *J. Biol. Chem.* **256**:5387–5393.

Blanken, W.M., and van den Eijnden, D.H., 1985, Biosynthesis of terminal Galα1-3Galβ1-4GlcNAc-R oligosaccharide sequence on glycoconjugates: Purification and acceptor specificity of a UDP-Gal: *N*-acetyllactosamine α1,3galactosyltransferase. *J. Biol. Chem.* **260**:12972–12934.

Chien, J.L., Li, S.C., and Li, Y.T., 1979, Isolation and characterization of a heptaglycosylceramide from bovine erythrocyte membrane. *J. Lipid Res.* **20**:669–.

Couto, A.S., Conclaves, M.F., Colli, W., and de Lederkremer, R.M., 1990, The N-linked carbohydrate chain of the 85-kilodalton glycoprotein from *Trypanosoma cruzi* trypmastigotes contains sialyl, fucosyl and galactosyl (α1-3) galactose units. *Molec. Biochem. Parasitol.* **39**:101–107.

Cummings, R.D., and Kornfeld, S., 1984, The distribution of repeating (Galβ1→4GlcNAcβ1→3) sequences in asparagine-linked oligosaccharides of the mouse lymphoma cell lines BW 147 and PHA 2.1. *J. Biol. Chem.* **259**:6253–6260, 1984.

Cummings, R.D., and Mattox, S.A., 1988, Retinoic acid-induced differentiation of the mouse teratocacinoma cell line F9 is accompanied by an increase in the activity of UDP-galactose: β-D-galctosyl-α1,3-galactosyltrainsferase, *J. Biol. Chem.* **263**:511–519.

DeThe, G., 1980, Role of Epstein–Barr virus in human diseases: Infectious mononucleosis, Burkitt's lymphoma and nasopharyngeal carcinoma. In: *Viral Oncology*, Ed., Klein, G. Raven Press, New York, pp. 169–197.

Eckhardt, A.E., and Goldstein, I.J., 1983, Isolation and characterization of α-galactosyl containing glycopeptides from Ehrlich ascites tumor cells. *Biochemistry* **22**:5290–5303.

Egge, H., Kordowicz, M., Peter-Katalinic, J., and Hanfland, P., 1985, Immunochemistry of I/i-active oligo- and polyglycosylceramides from rabbit erythrocyte membranes. *J. Biol. Chem.* **260**:4927–4935.

Elices, M.J., Blake, D.A, and Goldstein, I.J., 1986, Purification and characterization of a UDP-Gal: beta-D-Gal (1,4)-D-GlcNAc α1,3galactoysltransferase from Ehrlich ascites tumor cells, *J. Biol. Chem.* **261**:6064–6072.

Eto, T., Iichikawa, Y., Nishimura, K., Ando, S., and Yamakawa, T., 1968, Chemistry of lipids of the posthemolytic residue or stroma of erythrocytes. XVI. Occurance of ceramide pentasaccharide in the membrane of erythrocytes and reticulocytes in rabbit. *J. Biochem. (Tokyo)* **64**:205–213.

Frank, A., Andeman, W.A., and Miller, G., 1976, Epstein–Barr virus and nonhuman primates. Natural and experimental infection. *Adv. Cancer Res.* **23**:171–201.

Galili, U., 1989, Abnormal expression of α-galactosyl epitopes in man: A trigger for autoimmune processes? *Lancet* **ii**:358–361.

Galili, U., 1997, The α-galactosyl epitope (Galα1-3Galβ1-4GlcNAc-R) and the natural anti-Gal antibody. In: *Molecular Biology and Evolution of Blood Groups and MHC Antigens in Primates*, Eds., Socha, W.W., and Blancher, A. Springer Verlag, pp. 236–253.

Galili, U., and Andrews, P., 1995, Suppression of α-galactosyl epitopes synthesis and production of the natural anti-Gal antibody: A major evolutionary event in ancestral Old World primates. *J. Human Evolution* **29**:433–442.

Galili, U., Basbaum, C.B., Shohet, S.B., Buehler, J., and Macher, B.A., 1987a, Identification of erythrocyte Galα1-3Gal glycosphingolipids with a mouse monoclonal antibody. *J. Biol. Chem.* **262**:4683–4687.

Galili, U., Buehler, J., Shohet, S.B., and Macher, B.A., 1987b, The human natural anti-Gal IgG. III. The subtlety of immune tolerance in man as demonstrated by crossreactivity between natural anti-Gal and anti-B antibodies. *J. Exp. Med.* **165**:693–704.

Galili, U., Clark, M.R., Shohet, S.B., Buehler, J., and Macher, B.A., 1987c, Evolutionary relationship between the anti-Gal antibody and the Galα1-3Gal epitope in primates. *Proc. Natl. Acad. Sci. USA* **84**:1369–1373.

Galili, U., LaTemple, D.C., and Radic, M.Z., 1998, A sensitive assay for measuring α-gal epitope expression on cells by a monoclonal anti-Gal antibody. *Transplantation* **65**:1129–1132.

Galili, U., LaTemple, D.C., Walgenbach, A.W., and Stone, K.R., 1997, Porcine and bovine cartilage transplants in cynomolgus monkey: II. Changes in anti-Gal response during chronic rejection. *Transplantation* **63**:646–651.

Galili, U., Macher, B.A., Buehler, J., and Shohet, S.B., 1985, Human natural anti-α-galactosyl IgG. II. The specific recognition of α(1-3)-linked galactose residues. *J. Exp. Med.* **162**:573–582.

Galili, U., Mandrell, R.E., Hamadeh, R.M., Shohet, S.B., and Griffis, J.M., 1988a, Interaction between human natural anti-α-galactosyl immunoglobulin G and bacteria of the human flora. *Infect. Immun.* **56**:1730–1737.

Galili, U., Rachmilewitz, E.A., Peleg, A., and Flechner, I., 1984, A unique natural human IgG antibody with anti-α-galactosyl specificity. *J. Exp. Med.* **160**:1519–1531.

Galili, U., Repik, P.M., Anaraki, F., Mozdzanowska, K., Washko, G., and Gerhard, W., 1996, Enchancement of antigen presentation of influenza virus hemagglutinin by the natural anti-Gal antibody. *Vaccine*: 321–328.

Galili, U., Shohet, S.B., Kobrin, E., Stults, C.L.M., and Macher, B.A., 1988b, Man, apes, and Old World monkeys differ from other mammals in the expression of α-galactosyl epitopes on nucleated cells. *J. Biol. Chem.* **263**:17755–17762.

Galili, U., and Swanson, K., 1991, Gene sequences suggest inactivation of α1,3 galactosyltransferase in catarrhines after the divergence of apes from monkeys. *Proc. Natl. Acad. Sci. USA* **88**:7401–7404.

Galili, U., Tibell, A., Samuelsson, B., Rydberg, L., and Groth, C.G., 1995, Increased anti-Gal activity in diabetic patients transplanted with fetal porcine islet cell clusters. *Transplantation* **59**:1549–1556.

Good, A.H., Cooper, D.C.K., Malcolm, A.J., Ippolito, R.M., Koren, E., Neethling, F.A., Ye, Y., Zuhdi, N., and Lamontage, L.R., 1992, Identification of carbohydrate structures which bind human anti-porcine antibodies: implication for discordant xenografting in man. *Transplant. Proc.* **24**:559–562.

Gowda, D.C., Petrella, E.C., Raj, T.T., Bredehorst, R., and Vogel, C.W.,1994, Immunoreactivity and function of oligosaccharides in cobra venom factor. *J. Immunol.* **152**:2977–2986.

Gowda, D.C., Schultz, M., Bredehorst, R., and Vogel, C.W.,1992, Structure of the major oligosaccharide of cobra venom factor. *Mol. Immunol.* **29**:335–344.

Hamadeh, R.M., Jarvis, G.A., Galili, U., Mandrell, R.E., Zhou, P., and Griffis, J.M., 1992, Human natural anti-Gal IgG regulates alternative complement pathway activation on bacterial surfaces. *J. Clin. Invest.* **89**:1223.

Hanto, D.W., Frizzera, G., Gajl-Plczalska, K.J., Sakamoto, K., Purtilo, D.T., Balfour, H.H., Simmons, R.L., and Najarian, J.S., 1982, Epstein–Barr virus induced B-cell lymphoma after renal transplantation. *N. Engl. J. Med.* **306**:913–918.

Hendricks, S.P., He, P., Stults, C.L., and Macher, B.A., 1990, Regulation of the expression of Galα1-3Galβ1-4GlcNAc glycosphingolipids in kidney. *J. Biol. Chem.* **265**:17621–17626.

Henion, T.R., Macher, B.A., Anaraki, F., and Galili, U., 1994, Defining the minimal size of catalytically active primate α1,3galactosyltransferase: Structure function studies on the recombinant truncated enzyme. *Glycobiology* **4**:193–201.

Joziasse, D.H., Shaper, J.H., Jabs, E.W., and Shaper, N.L., 1991, Characterization of an α1-3galactosyltransferase homologue on human chromosome 12 that is organized as a processed pseudogene. *J. Biol. Chem.* **266**:6991–6998.

Joziasse, D.H., Shaper, N.L., Kim, D., Van den Eijnden, D.H., and Shaper, J.H., 1992, Murine α1,3-galactosyltransferase. A single gene locus specifies four isoforms of the enzyme by alternative splicing. *J. Biol. Chem.* **267**:5534–5541.

Joziasse, D.H., Shaper, J.H., Van den Eijnden, D.H., Van Tunen, A.H., and Shaper, N.L., 1989, Bovine α1-3galactosyltransferase: Isolation and characterization of a cDNA clone. Identification of homologous sequences in human genomic DNA. *J. Biol. Chem.* **264**:14290–14297.

Kimura, M., 1983, *The neutral theory of molecular evolution.* Cambridge University Press, Cambridge.

Krivan, H.C., Clark, G.F., Smith, D.F., and Wilkins, D.T., 1986, Cell surface binding site for *Clostridium difficile* enterotoxin: Evidence of a glycoconjugate containing the sequence Galα1→3Galβ1→4GlcNAc. *Infect. Immun.* **53**:573–581.

Koop, B.F., Goodman, M., Xu, P., Chan, K., and Slighton, J.L., 1986, Primate η-globin DNA sequences and man's place among great apes. *Nature* **319**:234–238.

Kornfeld, R., and Kornfeld, S., 1985, Assembly of asparagine-linked oligosaccharides. *Ann. Rev. Biochem.***54**:631–634.

Larsen, R.D., Rajan, V.P., Ruff, M., Kukowska-Latallo, J., Cummings, R.D., and Lowe, J.B., 1989, Isolation of a cDNA encoding murine UDP galactose: βD-galactosyl-1,4-N-acetyl-D-glucosaminide α1,3-galactosyltransferase: Expression cloning by gene transfer. *Proc. Natl. Acad. Sci. USA* **86**:8227–8231.

Larsen, R.D., Rivera-Marrero, C.A., Ernst, L.K., Cummings, R.D., and Lowe, J.B., 1990, Frameshift and nonsense mutations in a human genomic sequence homologous to a murine UDP-Galβ-D-Gal(1,4)-D-GlcNAcα(1,3) galactosyltransferase cDNA. *J. Biol. Chem.* **265**:7055–7062.

LaTemple, D.C., and Galili, U., 1998, Adult and neonatal anti-Gal response in knock-out mice for α-galactosyltransferase. *Xenotransplantation*, in press.

Macher, B.A., and Sweeley, C.C., 1978, Glycosphingolipids: Structure, biological source and properties. *Methods Enzymol.* **50**:236–251.

Mourant, A.E., Kopéc, A.C., and Domaniewska-Sobczak, O., 1976, *The Distribution of Blood Groups and Other Polymorphisms*, Vol. 1, 2nd edition. Oxford Press, p. 1050.

Oriol, R., Candelier, J.J., Taniguchi, S., Peters, L., and Cooper, D.K., 1996, Major oligosaccharide epitopes found in tissues of 23 animal species, potential donors for organ xenotransplantation. *Transplant. Proc.* **28**:794.

Pereira, M., Rohr, P.J.A., Langyel, I., and Barretto, O.C.O.P., 1984, Os grupos sanguineos ABO em esqueletos pré-históricos de aborigines de Ilha de Santa Catarina, Brasil. *Cienca e Cultura* **36**:1597–1599 (in Portuguese, abstract in English).

Pilbeam, D., 1984, The decent of hominoids and hominids. *Sci. Am.* **March**: 84–95.

Pothoulakis, C., Galili, U., Castagliuolo, I., Kelly, S., Nikulasson, P.K., Brasitus, T.A., and Lamont, J.T., 1996, Human anti-Gal IgG binds to the same receptors and mimics the effects of *C. difficile* Toxin A in rat colon. *Gastroenterology* **98**:641–649.

Repik, P.M., Strizki, J.M., and Galili, U., 1994, Differential host dependent expression of α-galactosyl epitopes on viral glycoproteins: A study of eastern equine encephalitis virus as a model. *J. General Virol.* **75**:1177–1181.

Rother, R.P., Fodor, W.L., Springhorn, J.P., Birks, C.W., Setter, E., Sandrin, M.S., Squinto, S.P., and Rollins, S.A., 1995, A novel mechanism of retrovirus inactivation in human serum mediated by anti-α-galactosyl natural antibody. *J. Exp. Med.* **182**:1345–1355.

Rouvolo, M., Disotell, T.R., Allard, M.W., Brown, W.M., and Honeycutt, R.L., 1991, Resolution of the African hominoid trichotomy by the use of a mitochondrial gene sequence *Proc. Natl. Acad. Sci. USA* **88**:1570–1574.

Salzano, F.M., 1957, The blood groups of South American Indians. *Am. J. Phys. Anthropol.* 555–579.

Sandrin, M., Vaughan, H.A., Dabkowski, P.L., and McKenzie, I.F.C., 1993, Anti-pig IgM antibodies in human serum react predominantly with Galα1-3Gal epitopes. *Proc. Natl. Acad. Sci. USA* **90**:11391–11395.

Santer, U.V., DeSantis, R., Hard, K.J., van Kuik, J.A., Vliegenthart, J.F.G., Won, B., and Glick, M.C., 1989, N-linked oligosaccharide changes with oncogenic transformation require sialylation of multiantennae. *Eur. J. Biochem.* **181**:249–260.

Schachter, H., and Roseman, S., 1980, Mammalian glycosyltransferases. Their role in the synthesis and function of complex carbohydrates and glycolipids. In: *The Biochemistry of Glycoproteins and Proteoglycans*, Ed., Lennarz, W.J., New York, Plenum, pp. 85–160.

Shaper, N.L., Lin, S.P., Joziasse, D.H., Kim, D.Y., and Yang-Feng, T.L., 1992, Assignment of two human α1,3galactosyltransferase gene sequences (GGTA1 and GGTA1P) to chromosome 9q33–q34 and 12q14–q15. *Genomics* **12**:613–615.

Smith, D.F., Larsen, R.D., Mattox, S., Lowe, J.B., and Cummings, R.D., 1990, Transfer and expression of a murine UDP-Gal: β-D-Gal-α1,3-galactosyltransferase gene in transfected Chinese hamster ovary cells. Competition reactions between the α1,3-galactosyltransferase and the endogenous α2,3-sialytransferase, *J. Biol. Chem.* **265**:6225–6234.

Spiro, R.G., and Bhoyroo, V.D., 1984, Occurance of α-D-galactosyl residues in the thyroglobulin from several species. Localization in the saccharide chains of the complex carbohydrate units. *J. Biol. Chem.* **259**:9858–9866.

Stellner, K., Saito, H., and Hakomori, S., 1973, Determination of aminosugar linkage in glycolipids by methylation. Aminosugar linkage of ceramide pentasaccharides of rabbit erythrocytes and of Forssman antigen. *Arch. Biochem. Biophys.* **133**:464–472.

Stone, K.R., Ayala, G., Goldstein, J., Hurst, R., Walgenbach, A., and Galili, U., 1998, Porcine cartilage transplants in cynomolgus monkey: III. Transplantation of α-galactosidase treated porcine cartilage. *Transplantation* **65**:1577–1583.

Takeuchi, Y., Porter, C.D., Strahan, K.M., Preece, A.F., Gustafsson, K., Cosset, F.L., Weiss, R.A., Collins, M.K., 1996, Sensitization of cells and retroviruses to human serum by (α1-3) galactosyltransferase, *Nature* **379**:85–88.

Teale, R.G., Tange, M.J., Zannettino, Z.L., Katerelos, M., Shinkel, T.A., Van Denderen, B.J., Lonie, A.J., Lyons, I., Nottle, M.B., Cox, T., Becker, C., Peura, A.M., Wigley, P.L., Crawford, R.J., Robins, A.J., Pearse, M.J., and d-Apice, A.J., 1996, The α1,3-galactosyltransferase knockout mouse. Implications for xenotransplantation, *Transplantation* **61**:13–19.

Thall, A., Etienne-Decerf, J., Winand, R., and Galili, U., 1991, The α-galactosyl epitope on mammalian thyroid cells. *Acta Endocrin.* **124**:692–699.

Thall, A.D., Maly, P., and Lowe, J.B., 1995, Oocyte Galα1-3Gal epitopes implicated in sperm adhesion to the zona pellucida glycoprotein ZP3 are not required for fertilization in the mouse. *J. Biol. Chem.* **270**:21437–21442.

Thall, A., Murphy, H., and Lowe, J.B., 1996, α1,3galactosyltransferase deficient mice produce natural anti-Gal antibodies. *Transplant. Proc.* **28**:561–562.

Umeura, K., Yuzawa, M., and Taketomi, T., 1978, Characterization of glycolipids in bovine erythrocyte membrane. *J. Biochem.* (Tokyo) **83**:463-.

Van den Eijnden, D.H., Blanken, W.M., Winterwerp, H., and Schiphorst, W.E., 1983, Identification and characterization of an UDP-Gal: N-acetyllactosaminide α1,3-galactosyltransferase in calf thymus, *Eur. J. Biochem.* **134**:523–530.

Welsh, R.M., O'Donnell, C.L., Reed, D.J., and Rother, R.P.,1988, Evaluation of the Galα1-3Gal epitope as a host modification factor eliciting natural humoral immunity to enveloped viruses. *J. Virol.***72**:4650–4656.

Wilson, A.C., 1985, The molecular basis of evolution. *Sci. Amer.* **Oct**:164–173.

Wilson, A.C., Carlson, S.S., and White, T.J., 1977, Molecular evolution. *Annu. Rev. Biochem* **46**:573–639.

Wood, C., Kabat, E.A., Murphy, L.A., and Goldstein, I.J., 1979, Immunochemical studies of the combining sites of two isolectins A4 and B4 isolated from *Bandeiraea simplicifolia*. *Arch. Biochem. Biophys.* **198**:1–8.

*Chapter 2*

# The α1,3-Galactosyltransferase Gene

David H. Joziasse, Joel H. Shaper, and Nancy L. Shaper

## 1. INTRODUCTION

Cell surface glycoconjugates (glycoproteins and glycolipids) are well positioned topologically to be involved in cell-cell interactions and transmembrane signaling. Their glycans (sugar chains) protrude from the surface, and display considerable structural complexity. The glycan structures assembled in a given cell type are dictated in part by the subset of glycosyltransferases that are translationally expressed. As the relative levels of the various glycosyltransferases determine which carbohydrate structures are carried by eukaryotic glycoconjugates, they indirectly affect intercellular interactions. In particular, the terminal sugars of N- and O-linked glycans are known to play a role in receptor-mediated recognition and binding.

Sugar chains that are O-linked to a protein via serine or threonine are assembled in the Golgi apparatus by the stepwise addition of individual monosaccharide units. Sugar chains that are N-linked to asparagine are initiated in the endoplasmic

**David H. Joziasse** Department of Medical Chemistry, Vrije Universiteit, NL-1081 BT Amsterdam, The Netherlands. **Joel H. Shaper** Cell Structure and Function Laboratory, Oncology Center, and Department of Pharmacology and Molecular Sciences, The Johns Hopkins University School of Medicine, Baltimore, Maryland 21287-8937. **Nancy L. Shaper** Cell Structure and Function Laboratory, Oncology Center, The Johns Hopkins University School of Medicine, Baltimore, Maryland 21287-8937

*Subcellular Biochemistry, Volume 32: α-Gal and Anti-Gal*, edited by Galili and Avila.
Kluwer Academic / Plenum Publishers, New York, 1999.

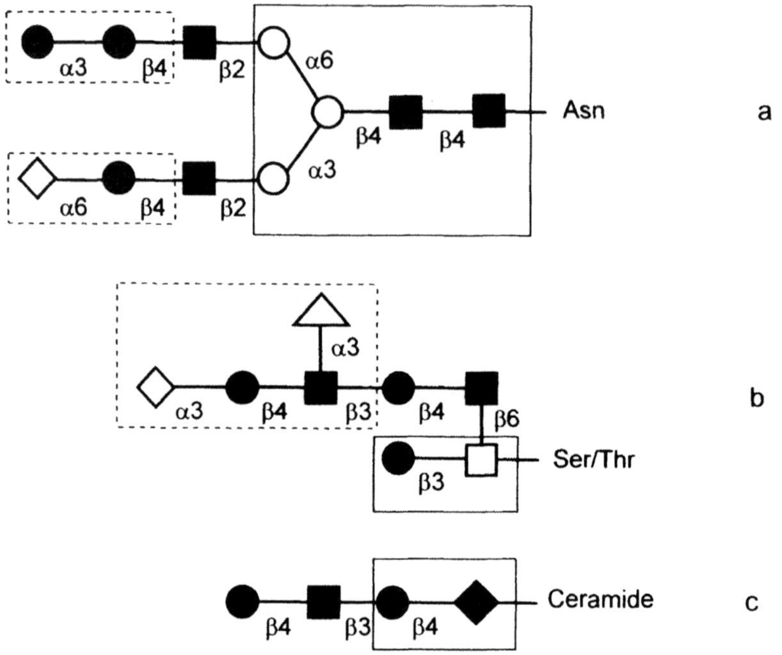


**Figure 1.** Typical structures of glycoprotein and glycolipid glycans. (a) A diantennary N-glycan. In mammals, the core structure ($Man_3GlcNAc_2$, *box* with *solid* line) is usually extended by the attachment of one or more branches, initiated by a GlcNAc residue, to the α1,3- and α1,6-linked mannose residues (Schachter, 1991). The non-substituted core structure is often found in insect cell glycoproteins. (b) A branched O-glycan. Various core structures have been described for O-glycans (Brockhausen and Schachter, 1997); these sugar chains are initiated by the transfer of a GalNAc residue in α-linkage to a Ser or Thr. The frequently occurring "core-1" (Galβ1,3GalNAc-Ser/Thr) is *boxed*. (c) Many glycosphingolipids contain as the core the disaccharide Galβ1,4Glc *(boxed)*, which is linked to the lipid ceramide. Terminal structures (*boxes* with *dashed line*) shown are (a): the "α-gal epitope" (Galα1,3Gal) and the NeuAcα2,6Gal structure, and (b): the sialyl-Lewis$^x$ determinant. Alternatively, N- and O-glycans can be terminated by blood group determinants (Galα1,3[Fucα1,2]Galβ1-R, GalNAcα1,3[Fucα1,2]Galβ1-R, or Fucα1,2Galβ1-R), polysialic acid, or sulphate groups. The same terminal structures can be found on both glycoprotein N- and O-glycans and glycolipids. Symbol key: ■, N-acetylglucosamine;O, mannose; ●, galactose; □, N-acetylgalactosamine; ◆, glucose; ◇, sialic acid; Δ, fucose.

reticulum by the *en bloc* transfer of a preformed oligosaccharide chain to the nascent polypeptide. The precursor is then trimmed by processing glycosidases, and subsequently modified by the sequential addition of various monosaccharides by specific glycosyltransferases. Only a small number (12–15) of structurally different monosaccharides are used in the assembly of mammalian glycoconjugates. The structural diversity of their glycans arises from the large number of possible linkages between the individual sugar residues. The glycans of glycoproteins and glycolipids are assembled on a limited number of different core structure (Figure 1). Variation is found in the outer-chain, terminal structures, and in the degree of branching.

In mammals, only a limited number of terminal glycan structures are known to date. In general, addition of an α-linked sugar residue such as sialic acid or galactose to a terminal β-linked Gal blocks further elongation of the growing sugar chain (Fig. 1). Sulfation of the β-linked Gal usually has the same effect. The mammalian glycosyltransferases responsible for chain termination are the α2,3- and α2,6-sialyltransferases, and α1,3-galactosyltransferase (α1,3GT). However, in specific stages of development, and in specific tissues, sialic acid terminated glycans may be extended by the addition of sialic acid residues forming polysialic acids. This reaction is catalyzed by polysialyltransferase, an α2,8-sialyltransferase. The same terminal structures can occur on N- and O-linked glycans, and on lipid-bound sugar chains. In several instances, they are elaborated by the same glycosyltransferases. For example, a single α1,3-galactosyltransferase is sufficient to produce the Galα1,3Gal structure (here termed "the α-gal epitope") that may terminate both glycoprotein glycans and the glycolipid Galα1,3Galβ1,4GlcNAcβ1,3Galβ1,4Glc-ceramide (Blanken and Van den Eijnden, 1985).

In this review we shall focus on the protein structure and the gene organization of α1,3GT, the various other members of the α1,3GT gene family, and the evolution of this family. The first member of the gene family to be purified was bovine thymus α1,3GT. Since the properties of this enzyme form the base-line for all subsequent studies, the first section of this review will deal with the distinguishing characteristics of bovine α1,3GT and the molecular cloning of the corresponding cDNA. In sections 2. and 3., the organization of the α1,3GT gene and its expression will be discussed.

### 1.1. Bovine α1,3-Galactosyltransferase: Enzymatic Properties

The enzyme UDP-Gal:Galβ1,4GlcNAc α1,3-galactosyltransferase (α1,3GT, EC 2.4.1.87) catalyzes the reaction:

$$\text{UDP-Gal} + \text{Gal}\beta1,4\text{GlcNAc-R} \xrightarrow{Mn^{2+}} \text{Gal}\alpha1,3\text{Gal}\beta1,4\text{GlcNAc-R} + \text{UDP},$$

in which R may be glycoprotein or glycolipid (Blanken and Van den Eijnden, 1985). Lactose (Galβ1,4Glc) is recognized as an acceptor, but the $K_m$ for this substrate is only 9.0 mM, whereas that for N-acetyllactosamine (Galβ1,4GlcNAc) is 1.4 mM. Both protein- and lipid-linked lactosaminoglycans are excellent substrates for α1,3GT. However, the enzyme cannot transfer to fucosylated substrates such as the blood group H structure (Fucα1,2Galβ1,3/4GlcNAc), and in this re-

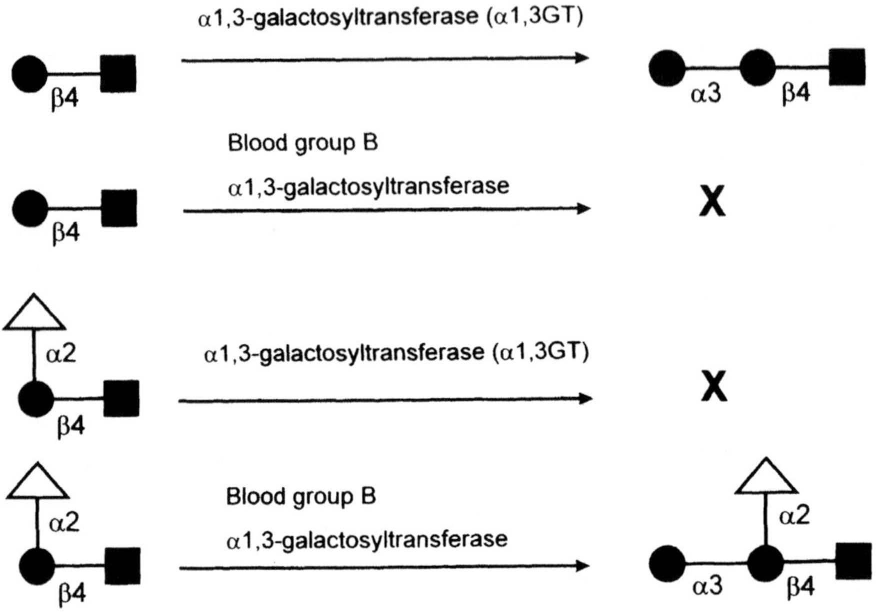


**Figure 2.** Reactions catalyzed by α1,3GT and the blood group B α1,3-galactosyltransferase. Fucosylated acceptors such as Fucα1,2Gal or Galβ1,4[Fucα1,3]GlcNAc cannot be α1,3-galactosylated by α1,3GT. In contrast, a non-fucosylated acceptor such as Galβ1,4GlcNAc is not a substrate for the blood group B transferase. The X indicates a reaction that does not occur. Symbol key: see Figure 1.

spect is different from the blood group B α1,3-galactosyltransferase. Also, the $Lewis^x$ structure (Galβ1,4[Fucα1,3]GlcNAc) is not an acceptor substrate for α1,3GT. Figure 2 shows the reaction products of the B transferase and α1,3GT.

A good glycolipid acceptor substrate for α1,3GT is lacto-N-*neo*tetraosylceramide (Galβ1,4GlcNAcβ1,3Galβ1,4Glc-ceramide), but lactosylceramide (Galβ1,4Glc-ceramide) cannot be used. This distinguishes the enzyme from an α1,3-galactosyltransferase detected in rat spleen and bone marrow, which is involved in the synthesis of trihexosylceramide (Stoffyn *et al.*, 1973).

The α1,3GT enzyme shows a certain degree of fine-specificity with regard to its acceptor substrate. It has been shown that the Manα1,6Man arm of diantennary complex-type oligosaccharides is preferred over the Manα1,3Man arm (see Figure 1a), and that there is also a distinct branch preference for more highly branched acceptors (Elices and Goldstein, 1989; Van den Eijnden *et al.*, 1985). It is not yet known if the protein portion of a glycoprotein acceptor can interact with α1,3GT and affect galactosyl transfer.

Based on lectin analysis and immunostaining, it was found that the product of the reaction catalyzed by α1,3GT, the "α-gal epitope" (Galα1,3Gal), is present on most mammalian cells. Only cells and tissues derived from catarrhines (humans, apes and Old World monkeys) lack this α-gal epitope (Galili *et al.*, 1988). Interestingly, New World monkeys, which separated from the catarrhines only ~ 40 million years ago, express both the epitope and the corresponding α1,3GT activity.

## 1.2. Molecular Cloning of Bovine α1,3GT

In general, glycosyltransferases show low expression levels and, being membrane-bound, require lengthy purification procedures. As a result, progress in the analysis of glycosyltransferases was hampered by the lack of DNA sequence information. The first two mammalian glycosyltransferase cDNAs (β1,4-galactosyltransferase (Shaper *et al.*, 1986), and α2,6-sialyltransferase (Weinstein *et al.*, 1987)) were cloned in 1986–1987. The third cDNA clone isolated encoded bovine α1,3GT.

α1,3GT was purified from bovine thymus by use of affinity chromatography on UDP-Sepharose (Blanken and Van den Eijnden, 1985). Antibodies were raised against the purified enzyme, and partial amino acid sequence was obtained. The α1,3GT cDNA was isolated from a bovine thymus cDNA expression library in λgt11 by immunoscreening using the antibody and α1,3GT-specific oligonucleotide probes (Joziasse *et al.*, 1989). A 1836-bp cDNA was cloned that encoded a 368 amino acid protein, and contained a relatively long 5′-untranslated region (468 nt). The coding sequence predicted a type II transmembrane protein, consisting of a 6 amino acid amino-terminal cytoplasmic tail, a transmembrane domain (signal anchor sequence) of 19 residues, a 50–60 amino acid linker peptide termed the "stem region," and a large, carboxy-terminal catalytic domain of approximately 280 amino acids (see Figure 3). More recent studies by Henion *et al.* (Henion *et al.*, 1994) have confirmed that the α1,3GT catalytic domain is likely to be contained in the terminal ~286 amino acid residues.

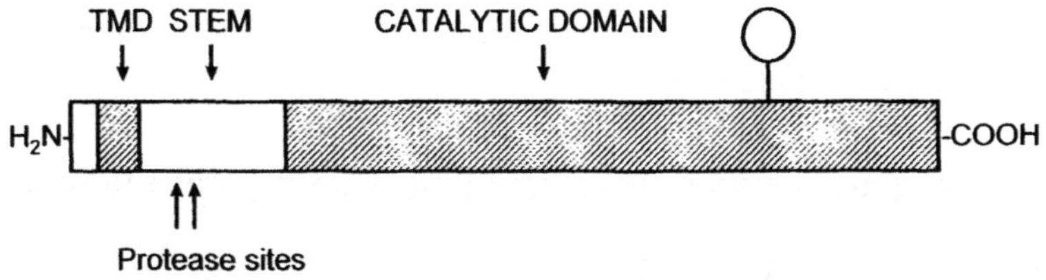


**Figure 3.** Schematic of the protein domain structure of α1,3GT. The cytoplasmic tail, the transmembrane domain (TMD), the stem region and the catalytic domain are indicated. A single N-glycan is attached to the catalytic domain at Asn293 (indicated as O, not drawn to scale). The stem region contains several protease-sensitive sites; proteolysis in this area will result in the secretion of a truncated, soluble, enzymatically active protein.

## 2. MAMMALIAN α1,3GT ORTHOLOGUES

### 2.1. Murine α1,3GT

A mouse cDNA was obtained from F9 cells by Larsen *et al.*, via an expression cloning strategy (Larsen *et al.*, 1989). Briefly, transfected COS cells expressing α1,3GT were enriched by panning on lectin coated dishes. The lectin used was *Griffonia simplicifolia* I-B4 (GSI-B4), which has an affinity for α1,3-linked terminal galactose residues. A second mouse cDNA, clone 4A, was isolated by homology screening of a cDNA library using the bovine cDNA as a probe (Joziasse *et al.*, 1992). The two murine cDNAs predict polypeptides that differ in the length of the stem region (41 and 29 amino acid residues, respectively), which occurs as a result of alternative splicing (see 3.3). Comparison of the coding sequence contained in cDNA 4A with the bovine α1,3GT cDNA revealed a 78% identity at the nucleotide level. The identity is 73% at the amino acid level when the full length murine protein, derived from the genomic sequence, is compared with bovine α1,3GT (Figure 5). A large number of differences are present in the stem region, whereas a high degree of similarity (79% identical amino acids in 286 residues) is found in the carboxy-terminal portion of the enzyme which contains the putative catalytic domain.

### 2.2. Other Mammalian α1,3GTs

Homologous sequences from marmoset (a New World monkey) (Henion *et al.*, 1994), and pig (Strahan *et al.*, 1995; Dabkowski *et al.*, 1993), were cloned by homology screening using the murine α1,3GT cDNA as a probe. Again, there is a high degree of similarity in the carboxy-terminal 250 amino acid residues when this region is compared to α1,3GTs from other mammalian species. Partial sequences were obtained from various species of primates by use of PCR in which oligonucleotide primers were employed that corresponded to highly conserved regions of the α1,3GT sequence (Galili and Swanson, 1991). The human genome was found to contain two homologues of the α1,3GT gene (Joziasse *et al.*, 1991; Larsen *et al.*, 1990). However, both genes are pseudogenes, and apparently neither gives rise to a transcript (see below).

## 3. MURINE α1,3GT: GENE ORGANIZATION AND EXPRESSION

### 3.1. Structure of the α1,3GT Gene

The gene structure of both the murine α1,3GT (Joziasse *et al.*, 1992) and the ABO blood group transferases (Yamamoto *et al.*, 1995) (see section 4.1) has been analyzed. Murine α1,3GT genomic sequences were first isolated by Joziasse *et al.*, (Joziasse *et al.*, 1992), by screening a genomic library using the bovine α1,3GT

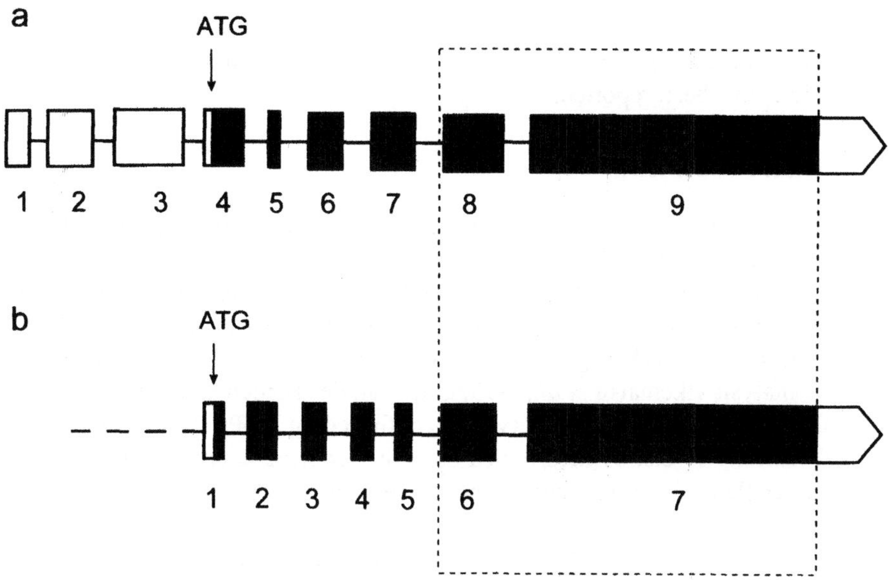


**Figure 4.** The organization of the α1,3GT and the blood group transferase genes.

cDNA as a probe. Murine α1,3GT is encoded by a multi-exon gene which spans at least 80 kb (Figure 4). Exons 1 to 3 encode 5′-untranslated sequence, and the protein coding sequence is distributed over exons 4 to 9. The start codon is located in exon 4, whereas exons 5–7 encode the "stem region," a hydrophilic peptide devoid of secondary structure that links the catalytic domain to the transmembrane domain. The largest exon, exon 9, contains almost the entire catalytic domain, and in addition encodes ~1.9 kb of 3′ untranslated sequence.

The schematic shows the exon structure of (a) the α1,3GT gene, and (b) the blood group ABO transferase genes. Exons are presented as *boxes*, introns as *lines*. The size of the introns is not to scale. The α1,3GT gene is distributed over 9 exons; exons 1–3 and part of exon 4 contain 5′-untranslated sequence (*open boxes*). Protein coding sequence is indicated by *closed boxes*. Of the ABO genes only the protein coding exons have been isolated (exons 1–7). It is not known how many exons encoding 5′-untranslated sequence precede exon 1, as the upstream sequence has not been fully determined. The region of greatest similarity between the two genes, which corresponds to the putative catalytic domain, has been *boxed*.

### 3.2. Expression of Murine α1,3GT

As described in section 1.1, α1,3GT shows a species-specific pattern of expression. Expression seems to be restricted mainly to mammals, although it is possible that related enzymes are active in certain species of fish, snakes and amphibians. In addition, several examples of tissue-specific and cell type-specific expression have been reported.

### 3.2.1. Tissue Specificity

In rats and mice, α1,3GT is expressed in most tissues and organs as determined by Northern blot and RT-PCR analysis, and enzyme activity assays. However, in mouse liver, α-gal epitopes (Galα1,3Gal) are undetectable by *G. simplicifolia* I-B4 lectin staining, as shown by Peters and Goldstein (1979). Also, transcription of the α1,3GT gene in liver is below the level of detection by Northern blot analysis (Ikematsu *et al.*, 1993; C.G. Van der Linden and D.H. Joziasse, unpublished observation), an observation confirmed by RT-PCR. As anticipated, α1,3GT activity levels, as determined by direct enzyme assay, are minimal in mouse liver (D.H. Joziasse, unpublished observation). Similar results were obtained for rat liver.

Interestingly, although transgenic mice, ectopically expressing α1,3GT under a strong general promoter, express high levels of α1,3GT mRNA in their liver, binding sites for GSI-B4 lectin were not detected. Occasionally, the livers of later generations of these mice contained a number of hepatocytes reacting with the lectin. This feature seemed to be associated with early death of the animals (Ikematsu *et al.*, 1993). This suggests that α1,3GT expression in hepatocytes may interfere with an as yet unknown physiological process, and that decreased hepatic function may cause abnormalities in the transgenic animal.

### 3.2.2. Cell-Type Specificity

Mouse liver does not stain with GSI-B4 lectin (Peters and Goldstein, 1979), and as hepatocytes constitute the main cell type of this organ (~70%) these cells are predicted to lack α-gal epitopes. Indeed, Northern blot analysis of RNA isolated from purified hepatocytes failed to detect an α1,3GT transcript (C.G. Van der Linden and D.H. Joziasse, unpublished observation). On the other hand, liver also contains cell types such as tissue macrophages and endothelial cells, which may carry the α-gal epitope. This might explain the residual low level of α-gal expression in liver that sometimes is reported (U. Galili, personal communication), and would also be consistent with our observation that murine liver contains traces of α1,3GT enzyme activity. Lack of α1,3GT expression in hepatocytes would predict that α-galactosylated glycoproteins are absent in murine serum, as the hepatocyte is the biosynthetic source of circulating glycoproteins.

It has also been found that α1,3GT is not expressed in murine male germ cells (Johnston *et al.*, 1995; Joziasse *et al.*, 1992). Initially, α1,3GT expression in murine male and female germ cells was investigated in view of the various alternative models concerning the role of cell surface carbohydrates in sperm-egg binding during fertilization. It has been suggested that terminal α1,3-linked galactose on O-linked glycans of zona pellucida glycoprotein ZP3 is recognized and bound by a sperm surface lectin, SP56, and in this way mediates initial events during fertilization (Bleil and Wassarman, 1988). Based on RT-PCR analysis it was shown that both pre- and postmeiotic male germ cells are devoid of α1,3GT transcripts, whereas both oocytes and eggs express normal levels of α1,3GT mRNA (Johnston *et al.*, 1995). This expression pattern may be explained if an α-galactose specific lectin on the head of the sperm binds an α1,3-galactosylated glycan on the zona pellucida. Suppression of α1,3-galactosylation of sperm glycoproteins would be predicted to prevent sperm aggregation. However, this hypothesis seems to be incompatible with studies by Thall *et al.* (1995) who showed that α1,3GT knockout mice, which do not carry α1,3-galactosylated glycans on their zona pellucida, are nevertheless fertile. Recent studies have suggested that in addition to a terminal α1,3-linked galactose, α1,3-linked fucose is part of the sperm binding epitope (Johnston *et al.*, 1998), and in fact is essential to form a high-affinity ligand.

Even though a hamster genomic DNA fragment homologous to α1,3GT was identified by Southern blot analysis, a transcript cannot be detected in Chinese Hamster Ovary (CHO) tissue culture cells, and enzyme activity levels are below the detection limit (Smith *et al.*, 1990). However, under specific conditions, or perhaps in a specific subclone of these cells, expression does occur as judged from the presence of non-reducing terminal Galα1,3Gal structures on recombinant human soluble CD4 produced in CHO cells, as shown by Ashford *et al.* (1993). This would imply that CHO cells contain an α1,3GT gene which is functional, but for some reason is usually not expressed.

### 3.2.3. Differentiation and Activation

Multiple studies, mostly on whole mice and murine tissue culture cells, have shown that the expression of α1,3GT can be induced under specific conditions such as cellular differentiation and activation. The injection of thioglycollate into the peritoneal cavity of a mouse or rat leads to an influx of blood monocytes. These exudate macrophages are "elicited," and show a ~5–10-fold increase in steady-state α1,3GT mRNA levels concomitant with an upregulation of enzyme activity, and increased α1,3-galactosylation of cell surface glycoconjugates when compared with resident peritoneal macrophages (Sato and Hughes, 1994; Joziasse *et al.*, 1992; Cummings and Mattox, 1988; Sheares and Mercurio, 1987; Mercurio and Robbins, 1985). It has been suggested that increased α-galactosylation creates binding sites for the cell surface lectin Mac-2 (galectin-3), which is followed by an increase in Mac-2 binding to the cell surface. The physiological relevance of increased Mac-2 binding is presently unknown, although it is possible that it may allow autocrine signaling.

Ordered, sequential changes in the expression of specific glycan structures at the cell surface may provide a mechanism for controlling cell-cell contacts and fate during development. Such changes during embryogenesis have been documented. For example during the differentiation of the inner cell mass into parietal and visceral endoderm, expression of α-galactosylated structures increases (Pennington *et al.*, 1985). Similarly, retinoic acid treatment of murine F9 embryonal carcinoma cells, which induces their differentiation *via* primitive endoderm into parietal endoderm, results in an induction of α1,3GT expression. This induction seems to be linked directly to increased transcription, and results in an increased α-galactosylation of the cell surface, masking Lewis$^x$ determinants by the formation of an α1,3-Gal-Lewis$^x$ epitope (Cho *et al.*, 1996; Cummings and Mattox, 1988).

### 3.2.4. Transcriptional Regulation

While the α1,3GT gene appears to be constitutively expressed in most mouse tissues, there are specific examples of upregulation of α1,3GT transcription during cell differentiation and activation (see 3.2.3). In initial studies it has been demonstrated that retinoic acid treatment of mouse F9 embryonal carcinoma cells results in increased transcription of the α1,3GT gene (Cho *et al.*, 1996). Recently, additional 5′ flanking sequence of the murine α1,3GT gene was isolated from a P1 genomic library (C.G. Van der Linden and D.H. Joziasse, unpublished results). By 5′-deletion mapping it was shown that sequences between -113 and +31 (transcription initiation site +1) are sufficient for basal transcription in mouse tissue culture cells. This DNA fragment also contains elements that confer retinoic acid responsiveness on the gene. However, the mechanism of retinoic acid stimulation has not been established in detail. The analysis of the transcriptional regulation of the α1,3GT gene is an important and essential area for future studies. It will be interesting to compare the results with those obtained in an analysis of the regulatory sequences of other glycosyltransferases (*e.g.* β1,4-galactosyltransferase (Rajput *et al.*, 1996; Harduin-Lepers *et al.*, 1993), various α2,3- and α2,6-sialyltransferases (Kurosawa *et al.*, 1996; Yoshida *et al.*, 1996; Svensson *et al.*, 1992; Wang *et al.*, 1990; Svensson *et al.*, 1990), and β1,2-GlcNAc-transferase-II (Chen *et al.*, 1998)).

### 3.3. Alternative Splicing

The stem region of α1,3GT (Figure 3) is encoded in at least three exons (5, 6, and 7). It has been reported previously that the murine α1,3GT mRNA is alternatively spliced in the region that encodes the stem region (Joziasse *et al.*, 1992). Protein isoforms produced from alternatively spliced mRNAs differ in the length of the stem region due to the absence of either exon 5 or 6, or both. Similar splicing patterns have been detected in the porcine α1,3GT (Vanhove *et al.*, 1997; Vanhove *et al.*, 1996). Although the exact function of the stem region of a glycosyltransferase is unknown, it may play a role in determining the half-life of the enzyme within the cell. Recently, it has been reported that α1,3GT is continuously secreted from murine, canine and bovine tissue culture cells into the growth medium, as a result of proteolysis occurring at specific sites in the stem region of the enzyme (Cho and Cummings, 1997; Cho *et al.*, 1997), Figure 3. The region that is sensitive to protease activity is encoded by exon 6, and it is conceivable that elimination of this exon by alternative splicing may render α1,3GT less sensitive to proteolysis. Differences (between different cell types) with regard to the extent of alternative splicing may thus target α1,3GT to the Golgi apparatus or to the secretory pathway. Previously, we could not obtain evidence for a regulation of the stem length of bovine α1,3GT by alternative splicing, as the bovine transcript contains all three exons 5–7 (Joziasse *et al.*, 1992). Interestingly, α1,3GT appears to be secreted more efficiently from bovine than from canine and murine tissue culture cells (Cho *et al.*, 1997). Secretion controlled by proteolysis may thus form an additional, posttranslational mechanism to regulate intracellular α1,3GT levels.

## 4. OTHER MEMBERS OF THE α1,3GT GENE FAMILY

Computer assisted sequence comparison has detected extensive similarities between the mammalian α1,3-galactosyltransferase orthologues (Strahan *et al.*, 1995; Henion *et al.*, 1994; Dabkowski *et al.*, 1993; Joziasse *et al.*, 1992; Galili and Swanson, 1991; Larsen *et al.*, 1989; Joziasse *et al.*, 1989), the ABO blood group transferases (Yamamoto *et al.*, 1990a; Yamamoto *et al.*, 1990b), Forssman α1,3-GalNAc-transferase (Haslam and Baenziger, 1996), and various related pseudogenes (Yamamoto *et al.*, 1991; Joziasse *et al.*, 1991). Together these genes form the α1,3GT gene family (*cf.* Breton *et al.*, 1998b). The high degree of similarity in the putative catalytic domains facilitates sequence alignment, even though there is little similarity between the transmembrane domains and the stem regions of the various transferases (Figure 5). The extent of amino acid sequence conservation between the mammalian α1,3GT orthologues is in the order of 73 to 88%. The degree of sequence conservation between the Forssman transferase and α1,3GT and the ABO transferases is 35% and 42%, respectively. The degree of similarity with the Forssman transferase increases if only the catalytic domains are compared (to 48% and 50% identical amino acid residues, respectively).

### 4.1. The ABO Blood Group Transferases

Enzymes with a substrate specificity related to that of α1,3GT, have been found in human and other mammalian tissues. For example, the blood group B transferase also produces a Galα1,3Gal linkage, but efficient acceptor substrates for this enzyme have to contain a fucose residue α1,2-linked to the terminal Gal, as in the blood group H structure (Figure 2). The highly similar blood group A transferase transfers GalNAc in an α1,3-linkage to the Fucα1,2Galβ1–3/4GlcNAc structure to form the blood group A determinant.

The various human blood group transferase cDNAs have been cloned using PCR-assisted cDNA library screening based on partial amino acid sequence obtained for the purified A transferase (Yamamoto *et al.*, 1990; Yamamoto *et al.*, 1990). In a series of elegant studies, Yamamoto and Hakomori have unraveled the molecular basis for the various A/B blood group subtypes. Mostly, these subtypes are based on one or more point mutations in the genes encoding the A or B blood group transferases, as in the cis-AB subtype (Yamamoto *et al.*, 1994) and in the $A^2$ subtype (Yamamoto *et al.*, 1992). The AB transferases are ~ 33% similar to α1,3GT; when only the catalytic domain is compared the similarity increases to 46%. On the other hand, the N-terminal 100 amino acids (encompassing the N-terminal cytoplasmic tail, the transmembrane domain and the stem region) share no similarity. The A and B transferases are 99% identical. Nucleotide differences at 7 positions result in a difference of 4 amino acids between the two enzymes. Yamamoto and Hakomori (1990) showed by extensive mutation analysis that only two of these four amino acid differences were required to confer either the A or the B phenotype.

The gene structure of the human ABO genes has been reported (Yamamoto *et al.*, 1995) and, as anticipated based on extensive sequence similarity, it is reminiscent of the α1,3GT gene. The coding sequence of the ABO genes is distributed over seven different exons, and again the largest exon (exon 7) encodes almost all (229 amino acids) of the carboxy-terminal domain (Figure 4). Sequence similarity with the α1,3GTs is concentrated in ABO exons 6 and 7, which are equivalent to α1,3GT exons 8 and 9. Therefore, these exons are likely encode the catalytic domain.

Even though there is a consensus that the three-dimensional structure of different glycosyltransferases may show a similar folding (Breton *et al.*, 1998a; Breton *et al.*, 1998b; Breton *et al.*, 1996), the organization of the corresponding genes can be quite different. Various GlcNAc-transferases and fucosyltransferases have their coding sequence contained in a single exon, whereas other glycosyltransferase such as α2,6-sialyltransferase and β1,4-galactosyltransferase have their coding sequence contained in multiple exons, (reviewed in (Joziasse, 1992)). In this respect the α1,3GT gene and the blood group transferase genes are similar. Comparison of the α1,3GT genomic structure with that of the ABO transferases showed that both have their coding sequence distributed over multiple exons (α1,3GT, six; ABO, seven), whereas the majority of the coding sequence is present on a single large exon. It is possible that after a gene duplication event, exon-shuffling has occurred, allowing the fusion of the region that is involved in UDP-Gal(NAc) binding and the transfer of Gal(NAc) to form a Gal(NAc)α1,3-Gal

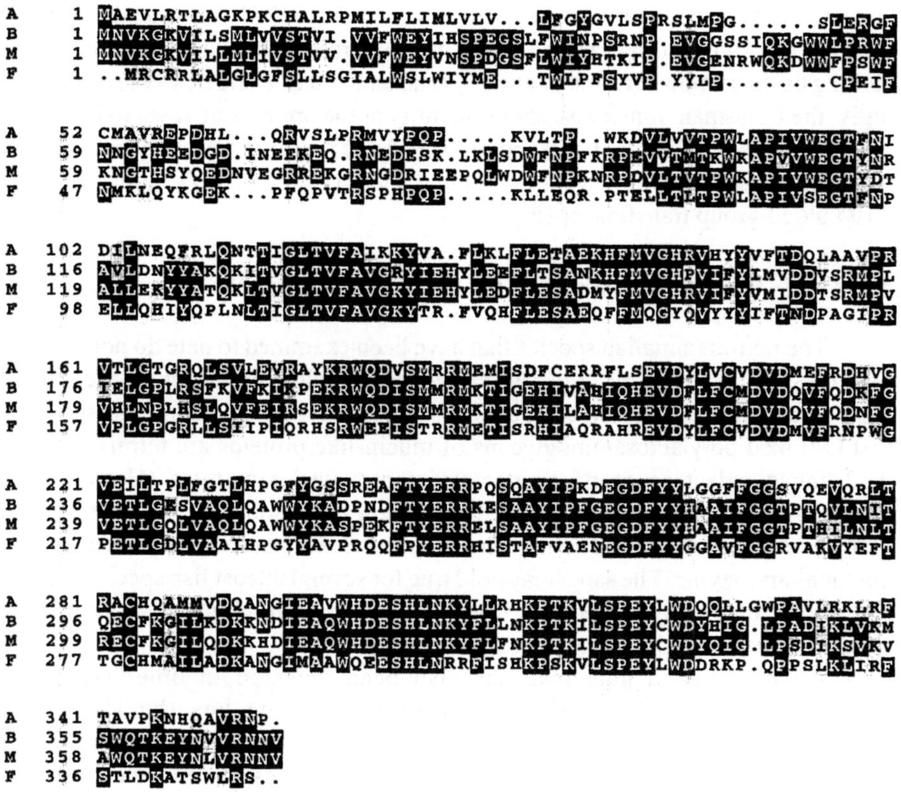

**Figure 5.** Sequence comparison of members of the α1,3GT gene family. Amino acid sequence alignment of the A blood group α1,3-GalNAc-transferase (A), bovine (B) and murine (M) α1,3GT, and the Forssman α1,3-GalNAc-transferase (F). The GenBank accession numbers of these sequences are J05175, J04989, M85153, and U66140, respectively.

linkage, with a different N-terminal sequence. This hypothesis is consistent with the observation that most of the intron-exon boundaries in the ABO genes, as in the α1,3GT gene, are of the "phase-2" type. Previously we have discussed the possibility of repeated use of the same 275 to 300-amino acid building block in different glycosyltransferases (Joziasse, 1992; Joziasse *et al.*, 1992), and this discussion may be extended to include the Forssman transferase.

## 4.2. The Forssman α1,3-GalNAc-Transferase

The Forssman transferase, UDP-GalNAc:GalNAcβ1,3Gal-R α1,3-GalNAc-transferase (EC 2.4.1.88), acts on glycosphingolipids. Like α1,3GT, this enzyme forms an α1,3-linkage, but transfers GalNAc to GalNAc instead of Gal to Gal. The cDNA encoding canine Forssman transferase was isolated by expression cloning using an anti-Forssman antibody (Haslam and Baenziger, 1996). Interestingly, the Forssman transferase shows identity at the amino acid level to both the AB-transferases (42% identical) and to α1,3GT (35% identical), which may point to the existence of a common ancestor that is shared by the α1,3GT family and the ABO blood group transferases (Figure 5).

### 4.3. Other Potential Members of the α1,3GT Gene Family

The non-mammalian species that have been examined to date do not appear to express α1,3GT (Galili *et al.*, 1988) with the possible exception of the cobra *Naja naja kaouthia*. Carbohydrate analysis of cobra venom glycoproteins showed that O-linked polylactosaminoglycans of mucin-like proteins are terminated by α1,3-linked galactose or α1,3-galactosylated Lewis$^x$ structures (Gowda and Davidson, 1994; Gowda *et al.*, 1992; Gowda and Davidson, 1992). Therefore, it seems likely that an α1,3GT is expressed in the venom gland that is related to the mammalian enzyme. The same may hold true for several teleost fish species. Their egg glycoproteins (hyosophorins) were found to contain N-glycans terminated with the structure Galα1,3Gal (Taguchi *et al.*, 1995).

An α1,3-linked galactose has also been detected in other types of glycoconjugates. For example, trihexosylceramide has the structure Galα1,3Galβ1,4Glc-ceramide. However, α1,3GT cannot use lactosylceramide (Galβ1,4Glc-ceramide) as a substrate. This distinguishes the enzyme from an α1,3-galactosyltransferase detected in rat spleen and bone marrow, which is involved in the synthesis of trihexosylceramide (Stoffyn *et al.*, 1973). Similarly, the globoside α1,4-galactosyltransferase also appears to be restricted in its action to glycolipids. It forms the Galα1,4Gal linkage in glycosphingolipids of the globoside series (Galα1,4Galβ1,4Glc-ceramide). The cDNA sequence encoding this α1,4-galactosyltransferase has not been isolated.

A globoside containing the structure Galα1,3Galα1,4-R, and more extended globosides with internal repeats of α1,3-linked Gal residues, are present in rat small intestine (Chen *et al.*, 1998). Similar glycolipids were found in rat pheochromocytoma cells (PC12h) (Ariga *et al.*, 1989). The α1,3-galactosyltransferase involved in the synthesis of such glycosphingolipids has been characterized in detail (Pal *et al.*, 1992). It may be related, but is not necessarily identical to α1,3GT. In *Leishmania* and Trypanosomes, indirect evidence suggests that Galα1,3Gal structures may be present both on glycolipids and glycoproteins (Avila *et al.*, 1989; Towbin *et al.*, 1987). The role of Galα1,3Gal structures in parasite physiology is still unknown. Also, in *Leishmania*, lipophosphoglycans may contain internal α1,3-linked Gal in the glycan core which is linked to a phosphatidyl-inositol anchor (Turco *et al.*, 1989). Furthermore, in glycophospholipids of both Trypanosomes and *Leishmania* a Galα1,3Man structure has been found (Avila and Rojas, 1990).

In addition, other animal species may contain glycosyltransferases related to α1,3GT. An analysis of several amphibian egg jelly glycans by Strecker et al. (Strecker *et al.*,1992; Strecker *et al.*,1995) has suggested that an α1,3-GalNAc-transferase that can act on non-fucosylated acceptor structures, an α1,4-galactosyltransferase that acts on the H-substance (Fucα1,2Galβ1-R), and a galactosyltransferase that transfers Gal in an α1,3-linkage to the GalNAc residue of the blood group A structure, may be present in the oviduct.

### 4.4. Relationship with Other Glycosyltransferase Gene Families

Limited similarity was observed between α1,3GT and the members of the mammalian β1,4-galactosyltransferase family. Also, a human pseudogene having similarity to both α1,3GT and the β1,4-galactosyltransferases was identified by Yamamoto *et al.*, using PCR (Yamamoto *et al.*,1991). This gene may form a possible link between the α1,3-galactosyltransferase family and that of the β1,4-galactosyltransferases. Whereas the α1,3GTs are involved in the synthesis of the α-gal epitope, the β1,4-galactosyltransferases catalyze the transfer of Gal in a β1,4-linkage to the non-reducing terminal GlcNAc residues of the carbohydrate chains of glycoconjugates. The presence of two short, conserved peptide motifs (DVD and DKKN) in both the eukaryotic β1,4-galactosyltransferases and the α1,3GTs has been reported previously (Joziasse *et al.*,1989). Use of hydrophobic cluster analysis allowed Breton *et al.* (1998b) to detect additional local homologies between these two groups of enzymes, one of which was located at the N-terminus of the putative catalytic domain. It was suggested that conserved regions in the catalytic domain, characterized by the presence of conserved aspartic acid residues, might be involved in UDP-Gal binding. This analysis suggests that α1,3GT and β1,4-galactosyltransferase once shared an ancestor, and that these proteins may be folded in a similar way. On the other hand, no similarity could be detected between mammalian α1,3GT and the bacterial α1,3-galactosyltransferases (Breton *et al.*, 1998b).

## 5. EVOLUTION OF THE α1,3GT GENE FAMILY

### 5.1. Human α1,3GT and Evolution of the Gene Family

The human genome contains two α1,3GT homologues. Both genes contain multiple frame-shift mutations (deletions and insertions), and internal stop codons in the protein coding sequence (Joziasse *et al.*,1991a; Larsen *et al.*,1990), and are considered to be pseudogenes. One gene, HGT-10, contains multiple introns, and may be the once-functional orthologue of the α1,3GT gene present in other mammals. This gene was mapped to human chromosome 9q34 by Southern analysis and fluorescent *in situ* hybridization of somatic cell hybrids (Shaper *et al.*, 1992). The intron-exon boundaries of the human gene that have been established correspond exactly with those in the mouse gene. The second pseudogene, HGT-2, appears to be the product of a reverse transcription event, which was followed by insertion into the genome (Joziasse *et al.*, 1991a). This gene has been mapped to chromosome 12q13–14. It contains the entire coding sequence on a 1.2-kb genomic fragment devoid of intron sequence, and is flanked by partial repeats. These features are characteristic of a processed pseudogene.

The mouse α1,3GT orthologue, termed GGTA1, has been localized to the centromeric region of mouse chromosome 2 (Joziasse *et al.*, 1991b), in the vicinity of the markers *Hc* and *Abl*. This region shows linkage homology to human chromosome 9q22-ter. The porcine α1,3GT has been mapped to pig chromosome 1 to the region q2.10-q2.11 that also contains the pig blood group locus EAA, the equivalent of the human ABO (Strahan *et al.*, 1995; Ellegren *et al.*, 1994). This region is syntenic with human chromosome 9q34. Interestingly, it has been established that the ABO gene locus also maps to human chromosome 9q34 (Watkins, 1987). In view of the position of both the mouse and pig α1,3GT gene within a region with linkage homology to human chromosome 9q33–34, and the shared localization of human α1,3GT and the ABO locus on chromosome 9q, it appears that the human homologue of the α1,3GT gene is linked to the ABO locus, and that these genes originated by tandem duplication from a single ancestral gene (Figure 6). This hypothesis is consistent with the observed sequence similarity between the ABO blood group transferases and α1,3GT, and their similar genomic organization. After duplication of the ancestral gene approximately 400 million years ago (Saitou and Yamamoto, 1997), one gene copy might have evolved into the ABO genes, and the other one into the human α1,3GT gene sequence on chromosome 9 represented by HGT-10. At some point in time, after the divergence of New World monkeys from the catarrhines, the human α1,3GT gene may have given rise to an mRNA that, after reverse transcription, was inserted into chromosome 12.

The evolution of the α1,3GT gene family may be an example of the "semiconservative model" of gene duplication (*cf.* Ruddle, 1997). This model states that when genes undergo duplication, one product retains the functional attributes of the precursor gene, whereas its counterpart is free to diverge and take on other functional attributes. To understand what kind of evolutionary pressure has driven divergence, it would be of interest to answer the question as to the substrate preferences of the enzyme encoded by the ancestral gene that upon gene duplication gave rise to the mammalian α1,3GT gene as we know it.

The lack of α1,3GT expression in humans, apes and Old World monkeys does not seem to have a deleterious effect. Transgenic mice in which the functional α1,3GT gene has been inactivated by homologous recombination, also appear to

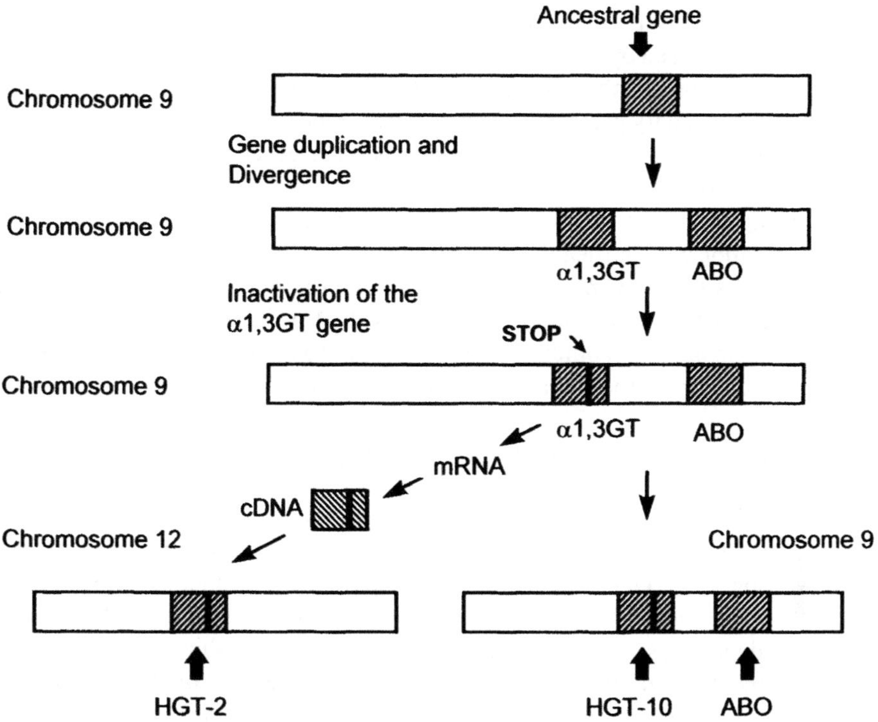


**Figure 6.** A model of the evolution of the α1,3GT gene family and the inactivation of the α1,3GT gene in catarrhines. Duplication of the ancestral gene may have occurred ~ 400 million years ago. Subsequent divergence of the two gene copies has given rise to the ABO blood group genes and α1,3GT.

develop normally, although they do develop eye cataracts (Tearle *et al.*, 1996). If α1,3GT is not required for normal development, why have most mammalian species maintained the functional gene? And conversely, what has been the selective pressure to inactivate the α1,3GT gene in the catarrhines? Host-pathogen interactions may provide the answer. Non-reducing terminal saccharide structures are functionally important in the binding of pathogenic microorganisms to mammalian cells. A change in these glycan structures that are expressed at the mammalian cell surface may abolish specific pathogen binding. Consequently, it is reasonable to speculate that continuous exposure to specific pathogens dictated a switch in cell surface glycosylation at multiple times during mammalian evolution, with concomitant suppression or alternatively, *de novo* expression of individual (terminal) glycosyltransferases. Also, exposure to microorganisms has probably resulted in the continuous presence within the host population of a certain degree of polymorphism in terminal glycan structures. This situation would be advantageous in that at least a subset of the species would assemble cell surface glycans that are not reconized by a particular pathogen. The necessity of flexibility in cellular glycosylation and the maintenance of polymorphism may have exerted considerable evolutionary pressure to both inactivate specific terminal glycosyltransferase genes, and introduce new glycosyltransferase genes. This situation may be similar to that observed with the ABO polymorphism. Even though the A and O blood groups seems to be preferred in certain geographical regions, there is no clear correlation with the health of an individual. Nevertheless, both the functional A and B genes and the O pseudogene, are well conserved. Individuals with the O phenotype may have an advantage over those with A or B phenotypes by being devoid of certain pathogen binding sites, or by being able to produce a broad spectrum of natural anticarbohydrate antibodies (Saitou and Yamamoto, 1997).

### 5.2. Events that Led to α1,3GT Gene Inactivation

It has been shown by Galili and co-workers that during the course of evolution, the α1,3GT gene has been inactivated in several mammalian species (Galili and Swanson, 1991; Galili *et al.*, 1988). Whereas α1,3GT is functionally expressed in most species including the New World monkeys, it is absent in the catarrhines (Old World monkeys, apes and man) (Galili *et al.*, 1988). As shown by PCR analysis, the homologues of the α1,3GT gene present in the genome of the latter species are pseudogenes. Their coding sequence contains multiple frame shift mutations; as a result, these genes are not anticipated to encode an enzymatically active protein. In addition, mutation(s) have apparently occurred in the upstream regulatory sequence, because α1,3GT transcripts can no longer be detected.

Based on comparative sequence analysis, Galili *et al.* have concluded that separate gene inactivating events have occurred in the Old World monkeys on the one hand, and in apes and humans on the other (Galili and Swanson, 1991). A single-base deletion resulting in a frame shift in the sequence encoding the catalytic domain of α1,3GT was noted to be shared by apes and humans, but absent in Old World monkeys. To explain gene inactivation in the latter species, it was suggested that an additional event had independently inactivated the Old World monkey gene. Alternatively, it was proposed that a series of minor changes occurred in the sequence, leading to a gradual inactivation of the encoded enzyme. It was concluded that gene activation in apes was a relatively recent event, and that the separate inactivation events in apes and Old World monkeys had occurred after the two groups diverged from each other.

However, a simpler, alternative model can be constructed which requires only a single inactivating event (compare Figure 6). This alternative model assumes that the α1,3GT gene was inactivated in the lineage of the catarrhines after their divergence from the New World monkeys by the introduction of a stop codon at amino acid position Tyr-263 (bovine sequence). Comparative DNA sequence analysis has shown that this specific nonsense mutation (TAG) interrupts the coding sequence in all Old World monkey and ape sequences analyzed thus far (Gustafsson *et al.*, 1994; Galili and Swanson, 1991). The same mutation is present in the human α1,3GT pseudogene described by Larsen et al. (Larsen *et al.*, 1990). The reported partial sequence is likely to be part of human pseudogene HGT-10, located on chromosome 9 (Joziasse *et al.*, 1991a). Its structure and chromosomal localization supports the hypothesis that HGT-10 constitutes the original, once functional α1,3GT gene in hominids. This nonsense mutation leads to a premature termination of translation, and by itself is sufficient for gene inactivation. The mutation is absent in all species that express an active α1,3GT homologue (New World monkeys, cow, pig and mouse), where the codon reads TAC or TAT (Tyr). Importantly, the same mutation is present in the human processed pseudogene HGT-2, which may suggest that transcription of the gene continued for a certain period of time even though translation of the mRNA would have yielded a truncated, enzymatically inactive protein. This situation is similar to what is observed today for the blood group type O. A frameshift is introduced into the protein encoded by the ABO locus as the result of a point mutation (a single base deletion in the 6th coding exon of the ABO gene) in the coding sequence. We have estimated the age of the processed pseudogene HGT-2 to be ≈25 million years, which means that the insertion of HGT-2 into the genome may have taken place 15 million years after the divergence between the catarrhines and New World monkeys (Joziasse *et al.*, 1991a). At this time the source gene continued to be transcribed, but may already have contained the stop codon at amino acid position 263. A more precise timing of the event would require the analysis of the Old World monkey genome for the presence of a HGT-2 homologue. An additional mutation in the regulatory sequences of the source gene that suppressed transcription must have occurred at a later point in time.

From the sequence differences between the human pseudogene HGT-10 and the New World monkey sequences (4.6–7.0%) it can be calculated that gene inactivation may have taken place between 22 and 32 million years ago (Galili and Swanson, 1991). This would be consistent with a scenario in which gene activation through the insertion of a stop codon occurred prior to the divergence of apes from Old World monkeys (20–28 million years ago), and be consistent with the presence of this stop codon in all α1,3GT pseudogenes. The fact that HGT-2 has diverged further than HGT-10 may be due to the lower mutation rate of (pseudo)genes that continue to be transcribed as compared with (pseudo)genes such as HGT-2 that have lost their regulatory sequences. The biological importance of α1,3GT gene inactivation is discussed elsewhere in this volume.

ACKNOWLEDGMENTS. This work was supported in part by The Human Frontiers Science Program Grant RG-414/94M (D.H.J.), EC Framework Program "Biotechnology" grants BIO4-CT95-0004 and BIO4-CT97-2242 (D.H.J.), and the National Institutes of Health Grant CA45799 (J.H.S.).

## REFERENCES

Ariga, T., Suzuki, M., Yu, R.K., Kuroda, Y., Shimamda, I., Inagaki, F., and Miyatake, T., 1989, Accumulation of unique globo-series glycolipids in PC12h pheochromocytoma cells, *J. Biol. Chem.* **264**:1516–1521.

Ashford, D.A., Alafi, C.D., Gamble, V.M., Mackay, D.J.G., Rademacher, T.W., Williams, P.J., Dwek, R.A., Barclay, A.N., Davis, S.J., Somoza, C., Ward, H.A., and Williams, A.F., 1993, Site-specific glycosylation of recombinant rat and human soluble CD4 variants expressed in Chinese Hamster Ovary cells, *J. Biol. Chem.* **268**:3260–3267.

Avila, J.L. and Rojas, M., 1990, A galactosylα1,3mannose epitope on phospholipids of *Leishmania mexicana* and *L. brasiliensis* is recognized by trypanosomatid-infected human sera, *J. Clin. Microbiol.* **28**:1530–1537.

Avila, J.L., Rojas, M., and Galili, U., 1989, Immunogenic Galα1,3Gal carbohydrate epitopes are present on pathogenic American Trypanosoma and Leishmania, *J. Immunol.* **142**:2828–2834.

Blanken, W.M. and Van den Eijnden, D.H., 1985, Biosynthesis of terminal Galα1,3Galβ1,4GlcNAc structures. Purification of a UDP-Gal:Galβ1,4GlcNAc α1,3-galactosyltransferase from bovine thymus, *J. Biol. Chem.* **260**:12927–12934.

Bleil, J.D. and Wassarman, P.M., 1988, Galactose at the nonreducing terminus of O-linked oligosaccharides of mouse egg zona pellucida glycoprotein ZP3 is essential for the glycoprotein's sperm receptor activity, *Proc. Natl. Acad. Sci.* **85**:6778–6782.

Breton, C., Oriol, R., and Imberty, A., 1996, Sequence alignment and fold recognition of fucosyltransferases, *Glycobiology* **6**:vii-xii.

Breton, C., Oriol, R., and Imberty, A., 1998a, Conserved structural features in eukaryotic and prokaryotic fucosyltransferases, *Glycobiology* **8**:87–94.

Breton, C., Bettler, E., Joziasse, D.H., Geremia, R.A., and Imberty, A., 1998b, Sequence-function relationships in prokaryotic and eukaryotic galactosyltransferases, *J. Biochem. Tokyo* in press.

Brockhausen, I. and Schachter, H., 1997, Glycosyltransferases involved in N- and O-glycan biosynthesis, in: *Glycosciences. Status and Perpectives*, (H.-J. Gabius and S. Gabius, eds.), Chapman & Hall, Weinheim, pp. 79–113

Chen, S., Zhou, S., Tan, J., and Schachter, H., 1998, Transcriptional regulation of the human UDP-GlcNAc:α-D-mannoside β1,2-N-acetylglucosaminyltransferase II gene (MGAT2) which controls complex N-glycan synthesis, *Glycoconjugate J.* **15**:301–308.

Cho, S.K. and Cummings, R.D., 1997, A soluble form of α1,3-galactosyltransferase functions within cells to galactosylate glycoproteins, *J. Biol. Chem.* **272**:13622–13628.

Cho, S.K., Yeh, J., Cho, M., and Cummings, R.D., 1996, Transcriptional regulation of α1,3-galactosyltransferase in embryonal carcinoma cells by retinoic acid. Masking of Lex antigens by α-galactosylation, *J. Biol. Chem.* **271**:3238–3246.

Cho, S.K., Yeh, J., and Cummings, R.D., 1997, Secretion of α1,3-galactosyltransferase by cultured cells and presence of enzyme in animal sera, *Glycoconjugate J.* **14**:809–819.

Cummings, R.D. and Mattox, S.A., 1988, Retinoic-acid-induced differentiation of the mouse teratocarcinoma cell line F9 is accompanied by an increase in the activity of UDP-galactose:β-D-galactosyl α1,3-galactosyltransferase, *J. Biol. Chem.* **263**:511–519.

Dabkowski, P.L., Vaughan, H.A., McKenzie, I.F.C., and Sandrin, M.S., 1993, Characterisation of a cDNA clone encoding the pig α1,3-galactosyltransferase: Implications for xenotransplantation, *Transplantation Proc.* **25**:2921.

Elices, M.J. and Goldstein, I.J., 1989, Biosynthesis of bi-, tri-, and tetraantennary oligosaccharides containing α-D-galactosyl residues at their nonreducing termini. Branch specificity of the Ehrlich tumor cell α1,3-galactosyltransferase, *J. Biol. Chem.* **264**:1375–1380.

Ellegren, H., Chowdhary, B.P., Fredholm, M., Hoyheim, B., Johansson, M., Nielsen, P.B., Thomsen, P.D., and Andersson, L., 1994, A physically anchored linkage map of pig chromosome 1 uncovers sex- and position-specific recombination rates, *Genomics* **24**:342–350.

Galili, U. and Swanson, K., 1991, Gene sequences suggest inactivation of α1,3-galactosyltransferase in catarrhines after the divergence of apes from monkeys, *Proc. Natl. Acad. Sci.* **88**:7401–7404.

Galili, U., Shohet, S.B., Kobrin, F., Stults, C.L.M., and Macher, B.A., 1988, Man, apes and Old World monkeys differ from other mammals in the expression of α-galactosyl epitopes on nucleated cells, *J. Biol. Chem.* **263**:17755–17762.

Gowda, D.C. and Davidson, E.A., 1992, Structural features of carbohydrate moieties in snake venom glycoproteins, *Biochem. Biophys. Res. Commun.* **182**:294–301.

Gowda, D.C. and Davidson, E.A., 1994, Isolation and characterization of novel mucin-like glycoproteins from cobra venom, *J. Biol. Chem.* **269**:20031–20039.

Gowda, D.C., Schultz, M., Bredehorst, R., and Vogel, C.-W., 1992, Structure of the major oligosaccharide of cobra venom factor, *Mol. Immunol.* **29**:335–342.

Gustafsson, K., Strahan, K., and Preece, A., 1994, α1,3-Galactosyltransferase: A target for *in vivo* genetic manipulation in xenotransplantation, *Immunol. Rev.* **141**:59–70.

Harduin-Lepers, A., Shaper, J.H., and Shaper, N.L., 1993, Characterization of two cis-regulatory regions in the murine β1, 4-galactosyltransferase gene. Evidence for a negative regulatory element that controls initiation at the proximal site, *J. Biol. Chem.* **268**:14348–14359.

Haslam, D.B. and Baenziger, J.U., 1996, Expression cloning of Forssman glycolipid synthetase: A novel member of the histo-blood group ABO gene family, *Proc. Natl. Acad. Sci.* **93**:10697–10702.

Henion, T.R., Macher, B.A., Anaraki, F., and Galili, U., 1994, Defining the minimal size of catalytically active primate α1,3-galactosyltransferase: structure-function studies on the recombinant truncated enzyme, *Glycobiology* **4**:193–201.

Ikematsu, S., Kaname, T., Ozawa, M., Yonezawa, S., Sato, E., Uehara, F., Obama, H., Yamamura, K., and Muramatsu, T., 1993, Transgenic mouse lines with ectopic expression of α1,3-galactosyltransferase: production and characteristics, *Glycobiology* **3**:575–580.

Johnston, D.S., Shaper, J.H., Shaper, N.L., Joziasse, D.H., and Wright, W.W., 1995, The gene encoding murine α1,3-galactosyltransferase is expressed in female germ cells but not in male germ cells, *Dev. Biol.* **171**:224–232.

Johnston, D.S., Wright, W.W., Shaper, J.H., Hokke, C.H., Van den Eijnden, D.H., and Joziasse, D.H., 1998, Murine sperm-zona binding: A fucosyl residue is required for a high affinity sperm-binding ligand. A second site on sperm binds a nonfucosylated, β-galactosyl-capped oligosaccharide, *J. Biol. Chem.* **273**:1888–1895.

Joziasse, D.H., 1992, Mammalian glycosyltransferases: genomic organization and protein structure, *Glycobiology* **2**:271–277.

Joziasse, D.H., Shaper, J.H., Van den Eijnden, D.H., Van Tunen, A.J., and Shaper, N.L., 1989, Bovine α1,3-galactosyltransferase: isolation and characterization of a cDNA clone. Identification of homologous sequences in human genomic DNA, *J. Biol. Chem.* **264**:14290–14297.

Joziasse, D.H., Shaper, J.H., Jabs, E.W., and Shaper, N.L., 1991a, Characterization of an α1,3-galactosyltransferase homologue on human chromosome 12 that is organized as a processed pseudogene, *J. Biol. Chem.* **266**:6991–6998.

Joziasse, D.H., Shaper, N.L., Shaper, J.H., and Kozak, C.A., 1991b, Gene for murine α1,3-galactosyltransferase is located in the centromeric region of chromosome 2, *Somatic Cell Mol. Genet.* **17**:201–205.

Joziasse, D.H., Shaper, N.L., Kim, D., Van den Eijnden, D.H., and Shaper, J.H., 1992, Murine α1,3-galactosyltransferase. A single gene locus specifies four isoforms of the enzyme by alternative splicing, *J. Biol. Chem.* **267**:5534–5541.

Kurosawa, N., Inoue, M..L., Yoshida, Y., and Tsuji, S., 1996, Molecular cloning and genomic analysis of mouse Galβ1,3GalNAc-specific GalNAc α2,6-sialyltransferase, *J. Biol. Chem.* **271**:15109–15116.

Larsen, R.D., Rajan, V.P., Ruff, M.M., Kukowska-Latallo, J., Cummings, R.D., and Lowe, J.B., 1989, Isolation of a cDNA encoding a murine UDP-Gal:Galβ1,4GlcNAc-R α1,3-galactosyltransferase: Expression cloning by gene transfer, *Proc. Natl. Acad. Sci.* **86**:8227–8231.

Larsen, R.D., Rivera-Marrero, C.A., Ernst, L.K., Cummings, R.D., and Lowe, J.B., 1990, Frameshift and nonsense mutations in a human genomic sequence homologous to a murine UDP-Gal:β-D-Gal1,4GlcNAc α1,3-galactosyltransferase cDNA, *J. Biol. Chem.* **265**:7055–7061.

Mercurio, A.M. and Robbins, P.W., 1985, Activation of mouse peritoneal macrophages alters the structure and surface expression of protein-bound lactosaminoglycans, *J. Immunol.* **135**:1305–1312.

Pal, S., Saito, M., Ariga, T., and Yu, R.K., 1992, UDP-Gal:globotriaosylceramide α-galactosyltransferase activity in rat pheochromocytoma (PC12h) cells, *J. Lipid Res.* **33**:411–417.

Pennington, J.E., Rastan, S., Roelcke, D., and Feizi, T., 1985, Saccharide structures of the mouse embryo during the first eight days of development. Inferences from immunocytochemical studies using monoclonal antibodies in conjunction with glycosidases, *J. Embryol. Exp. Morphol.* **90**:335–361.

Peters, B.P. and Goldstein, I.J., 1979, The use of fluorescein-conjugated *Bandeiraea simplicifolia* $B_4$-isolectin as a histochemical reagent for the detection of α-D-galactopyranosyl groups, *Exp. Cell Res.* **120**:321–334.

Rajput, B., Shaper, N.L., and Shaper, J.H., 1996, Transcriptional regulation of murine β1,4-galactosyltransferase in somatic cells. Analysis of a gene that serves both a housekeeping and a mammary gland-specific function, *J. Biol. Chem.* **271**:5131–5142.

Ruddle, F.H., 1997, Vertebrate genome evolution - the decade ahead, *Genomics* **46**:171–173.

Saitou, N. and Yamamoto, F., 1997, Evolution of primate ABO blood group genes and their homologous genes, *Mol. Biol. Evol.* **14**:399–411.

Sato, S. and Hughes, R.C., 1994, Regulation of secretion and surface expression of Mac-2, a galactoside-binding protein of macrophages, *J. Biol. Chem.* **269**:4424–4430.

Schachter, H., 1991, The 'yellow brick road' to branched complex N-glycans, *Glycobiology* **1**:453–461.

Shaper, N.L., Shaper, J.H., Meuth, J.L., Fox, J.L., Chang, H., Kirsch, I.R., and Hollis, G.F., 1986, Bovine galactosyltransferase: Identification of a clone by direct immunological screening of a cDNA expression library, *Proc. Natl. Acad. Sci.* **83**:1573–1577.

Shaper, N.L., Lin, S.-P., Joziasse, D.H., Kim, D., and Yang-Feng, T.L., 1992, Assignment of two human α1,3-galactosyltransferase sequences (GGTA1 and GGTA1P) to chromosomes 9q33-q34 and 12q14-q15, *Genomics* **12**:613–615.

Sheares, B.T. and Mercurio, A.M., 1987, Modulation of two distinct galactosyltransferase activities in populations of mouse peritoneal macrophages, *J. Immunol.* **139**:3748–3752.

Smith, D.F., Larsen, R.D., Mattox, S., Lowe, J.B., and Cummings, R.D., 1990, Transfer and expression of a murine UDP-Gal:β-D-Gal α1,3-galactosyltransferase gene in transfected Chinese Hamster Ovary cells. Competition reactions between the α1,3-galactosyltransferase and the endogenous α2,3-sialyltransferase, *J. Biol. Chem.* **265**:6225–6234.

Stoffyn, P., Stoffyn, A., and Hauser, G., 1973, Structure of trihexosylceramide biosynthesized *in vitro*, *J. Biol. Chem.* **248**:1920–1923.

Strahan, K.M., Gu, F., Preece, A.F., Gustavsson, I., Andersson, L., and Gustafsson, K., 1995, cDNA sequence and chromosome localization of pig α1,3-galactosyltransferase, *Immunogenetics* **41**:101–105.

Strecker, G., Wieruszeski, J., Michalski, J.-C., Alonso, C., Leroy, Y., Boilly, B., and Montreuil, J., 1992, Primary structure of neutral and acidic oligosaccharide-alditols derived from the jelly coat of the Mexican axolotl. Occurrence of oligosaccharides with Fucα1,3Fucα1,4–3-deoxy-D-glycero-D-galacto-nonulosonic acid and Galα1,4[Fucα1,2]Galβ1,4GlcNAc sequences, *Eur. J. Biochem.* **207**:995–1002.

Strecker, G., Wieruszeski, J., Plancke, Y., and Boilly, B., 1995, Primary structure of 12 neutral oligosaccharide-alditols released from the jelly coats of the anuran *Xenopus laevis* by reductive β-elimination, *Glycobiology* **5**:137–146.

Svensson, E.C., Conley, P.B., and Paulson, J.C., 1992, Regulated expression of α2,6-sialyltransferase by the liver-enriched transcription factors HNF-1, DBP, and LAP, *J. Biol. Chem.* **267**:3466–3472.

Svensson, E.C., Soreghan, B., and Paulson, J.C., 1990, Organization of the β-galactoside α2,6-sialyltransferase gene. Evidence for the transcriptional regulation of terminal glycosylation, *J. Biol. Chem.* **265**:20863–20868.

Taguchi, T., Kitajima, K., Muto, Y., Inoue, S., Khoo, K., Morris, H., Dell, A., Wallace, R.A., Selman, K., and Inoue, Y., 1995, A precise structural analysis of a fertilization-associated carbohydrate-rich glycopeptide isolated from the fertilized eggs of euryhaline killi fish (*Fundulus heteroclitus*). Novel penta-antennary N-glycan chains with a bisecting N-acetylglucosaminyl residue, *Glycobiology* **5**:611–624.

Tearle, R.G., Tange, M.J., Zannetinno, Z.L., Katerelos, M., Shinkel, T.A., Van Denderen, B.J., Lonie, A.J., Lyons, I., Nottle, M.B., Cox, T., Becker, C., Peura, A.M., Wigley, P.L., Crawford, R.J., Robins, A.J., Pearse, M.J., and d'Apice, A.J., 1996, The α1,3-galactosyltransferase knockout mouse. Implications for xenotransplantation, *Transplantation* **61**:13–19.

Thall, A., Maly, P., and Lowe, J.B., 1995, Oocyte Galα1,3Gal epitopes implicated in sperm adhesion to the zona pellucida glycoprotein ZP3 are not required for fertilization in the mouse, *J. Biol. Chem.* **270**:21437–21440.

Towbin, H., Rosenfelder, G., Wieslander, J., Avila, J.L., Rojas, M., Szarfman, A., Esser, K., Nowack, H., and Timpl, R., 1987, Circulating antibodies to mouse laminin in Chagas disease, American cutaneous Leishmaniasis, and normal individuals recognize terminal Galα1,3Gal epitopes, *J. Exp. Med.* **166**:419–432.

Turco, S.J., Orlandini, P.A.,Jr., Homans, S.W., Ferguson, M.A.J., Dwek, R.A., and Rademacher, T.W., 1989, Structure of the phophosaccharide-inositol core of the *Leishmania donovani* lipophosphoglycan, *J. Biol. Chem.* **264**:6711–6715.

Van den Eijnden, D.H., Joziasse, D.H., Koenderman, A.H.L., Blanken, W.M., Schiphorst, W.E.C.M., and Koppen, P.K., 1985, in: *Glycoconjugates, Proceedings of the VIIIth International Symposium, Houston*, (E.A. Davidson, J.C. Williams, N.M. Di Ferrante, eds.), Praeger Publishers, New York, p. 285.

Vanhove, B., Goret, F., Mirenda, V., Soulillou, J.P., and Pourcel, C., 1996, Variability of α1,3-galactosyltransferase splicing isoforms in pig tissues, *Transplantation Proc.* **28**:622–623.

Vanhove, B., Goret, F., Soulillou, J.P., and Pourcel, C., 1997, Porcine α1,3-galactosyltransferase: tissue-specific and regulated expression of splicing isoforms, *Biochim. Biophys. Acta* **1356**:1–11.

Wang, X., O'Hanlon, T.P., Young, R.F., and Lau, J.T.Y., 1990, Rat β-galactoside α2,6-sialyltransferase genomic organization: alternate promoters direct the synthesis of liver and kidney transcripts, *Glycobiology* **1**:25–31.

Watkins, W.M., 1987, Biochemical genetics of blood group antigens: retrospect and prospect, *Biochem. Soc. Trans.* **15**:620–624.

Weinstein, J., Lee, E.U., McEntee, K., Lai, P.-H., and Paulson, J.C., 1987, Primary structure of β-galactoside α2,6-sialyltransferase. Conversion of membrane-bound enzyme to soluble forms by cleavage of the NH2-terminal signal anchor, *J. Biol. Chem.* **262**:17735–17743.

Yamamoto, F. and Hakomori, S., 1990, Sugar-nucleotide donor specificity of histo-blood group A and AB transferases is based on amino acid substitutions, *J. Biol. Chem.* **265**:19257–19262.

Yamamoto, F., Marken, J., Tsuji, T., White, T., Clausen, H., and Hakomori, S., 1990a, Cloning and characterization of DNA complementary to human UDP-GalNAc:Fucα1,2Gal α1,3-GalNAc-transferase (histo-blood group A transferase) mRNA, *J. Biol. Chem.* **265**:1146–1151.

Yamamoto, F., Clausen, H., White, T., Marken, J., and Hakomori, S., 1990b, Molecular genetic basis of the histo-blood group ABO system, *Nature* **345**:229–233.

Yamamoto, F., McNeill, P.D., and Hakomori, S., 1991, Identification in human genomic DNA of the sequence homologous but not identical to either the histo-blood group ABH genes or α1,3-galactosyltransferase pseudogene, *Biochem. Biophys. Res. Commun.* **175**:986–994.

Yamamoto, F., McNeill, P.D., and Hakomori, S., 1992, Human histo-blood group $A^2$ transferase coded by $A^2$ allele, one of the A subtypes, is characterized by a single base deletion in the coding sequence, which results in an additional domain at the carboxyl terminal, *Biochem. Biophys. Res. Commun.* **187**:366–374.

Yamamoto, F., McNeill, P.D., Kominato, Y., Yamamoto, M., Hakomori, S., Ishimoto, S., Nishida, S., Shima, M., and Fujimura, Y., 1994, Molecular genetic analysis of the ABO blood group system: 2. cis-AB alleles, *Vox Sang.* **64**:120–123.

Yamamoto, F., McNeill, P.D., and Hakomori, S., 1995, Genomic organization of human histo-blood group ABO genes, *Glycobiology* **5**:51–58.

Yoshida, Y., Kurosawa, N., Kanematsu, T., Kojima, N., and Tsuji, S., 1996, Genomic structure and promoter activity of the mouse polysialic acid synthase gene (mST8Sia II). Brain-specific expression from a TATA-less GC-rich sequence, *J. Biol. Chem.* **271**:30167–30173.

*Chapter 3*

# Structure Function Studies of a New World Monkey $\alpha$1,3Galactosyltransferase

## Analysis of Alternative Splicing and Expression in Baculovirus System

T. R. Henion and U. Galili

### 1. ORGANIZATION OF GLYCOSYLTRANSFERASES

The $\alpha$1,3galactosyltransferase ($\alpha$1,3GT) is one of a growing number of cloned glycosyltransferase enzymes which exhibit a remarkable degree of similarity in their overall structure. The $\alpha$1,3GT, which was simultaneously cloned from the cow (Joziasse *et al.*, 1989) and the mouse (Larsen *et al.*, 1989), was one of the earliest glycosyltransferases for which complete sequence information was available. Since then, over 40 distinct glycosyltransferase genes have been cloned in

---

**T. R. Henion** Department of Biomedical Sciences, Eunice Kennedy Shriver Center, Waltham, Massachusetts 02254. **U. Galili** Department of Microbiology and Immunology, MCP-Hahnemann School of Medicine, Philadelphia, Pennsylvania 19129.

*Subcellular Biochemistry, Volume 32: α-Gal and Anti-Gal*, edited by Galili and Avila.
Kluwer Academic / Plenum Publishers, New York, 1999.

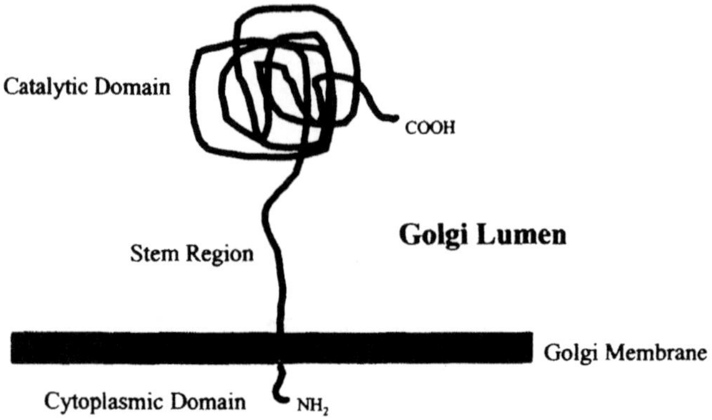


**Figure 1.** The four domains shared by all glycosyltransferases: 1) The N-terminal cytoplasmic domain, 2) the transmembrane domain, which also contains the signal sequences for membrane insertion and Golgi localization, 3) the stem region, and 4) the globular catalytic domain.

humans alone. Without exception, each is organized as a type II transmembrane protein, with an N-terminal end within the cytosol and the C-terminal end protruding into the lumen of the endoplasmic reticulum or the Golgi apparatus.

In addition to their identical topology, there also exists a remarkable similarity in the organization of the functional domains within the enzymes themselves (Figure 1). All glycosyltransferases cloned to date contain four domains: a short N-terminal cytoplasmic tail of typically 6 to 10 amino acids, a single transmembrane domain of 12–17 amino acids, and a flexible extracellular stem region of variable length that joins directly to a C-terminal catalytic domain (Paulson and Colley, 1989; Joziasse, 1992a). The theme of a common domain organization in the various glycosyltransferases reflects the structural requirements needed to achieve the positioning of the catalytic domains of the enzymes in a proper orientation within the lumen of the endoplasmic reticulum or the Golgi, where they carry out oligosaccharide biosynthesis on elongating carbohydrate chains of glycoproteins and glycolipids.

Of these domains, the most intensely studied from a structure-function perspective has been the transmembrane domain, since it is mostly within this region that the signal-anchor sequence is located, which is responsible for insertion of the nascent polypeptide sequence into the membrane of the endoplasmic reticulum in the correct orientation, and its proper localization to the Golgi apparatus. The exact motifs and structural requirements that are responsible for this targeting have not been determined, but it is clear from the many studies published on the subject that residues within the transmembrane domain and, in some cases, the adjacent stem region are critical for this localization (reviewed in Colley, 1998).

Several lines of evidence have suggested that the stem region performs mainly a structural role for positioning of the catalytic domain within the Golgi lumen. It has been known for some time that enzymatically active forms of many glycosyltransferases exist in a soluble, non-membrane bound state (Chester *et al.*, 1969; Whitehead *et al.*, 1974; Paulson *et al.*, 1977; Powell *et al.*, 1977). The cloning of several of these enzymes was facilitated by using N-terminal amino acid sequence data from soluble, active forms of the glycosyltransferases generated during the preparation of the enzymes from tissue sources. A comparison of N-terminal peptide sequences generated from the free enzyme with the deduced amino acid sequence of the cDNA clones convincingly showed that the soluble forms are produced by cleavage of the membrane bound protein within the stem region, with release of the active catalytic domain (Yamamoto *et al.*, 1990; Weinstein *et al.*, 1987; Gillespie *et al.*, 1992, White *et al.*, 1995) as had been suggested earlier (Roth *et al.*, 1985). Thus, a considerable portion of the N-terminus of glycosyltransferases was dispensable for the maintenance of catalytic function.

A substantially higher rate of amino acid sequence divergence is found within the stem region relative to the C-terminal catalytic domain when members of glycosyltransferase families are aligned, or when the sequences of a specific enzyme, expressed in different species, are aligned. This is presumed to be an indication of a lack of functional constraints for the maintenance of enzyme activity. Furthermore, it has been shown that protein isoforms for bovine α1,3GT exist as a result of alternative splicing of several exons within the presumed stem region (Joziasse, 1992b). It was later confirmed with porcine α1,3GT that alternative splicing of these same exons, encoding a combined 33 amino acids, had no obvious effects on enzyme activity (Vanhove *et al.*, 1997). Protein structural determinations for many glycosyltransferases have indicated that the stem domain is largely devoid of secondary structure and has a relatively high proline content, which may provide a degree of rigidity for maintaining the enzyme's catalytic domain in an extended confirmation within the Golgi lumen (Sarkar *et al.*, 1991, Joziasse 1992a).

We have been interested in studying the characteristics of the α1,3GT in primates (i.e. New World monkeys) and determining whether they have been evolutionarily conserved in primates and rodents (i.e. mice). In the course of these studies we identified the exact boundary of the stem region with the catalytic domain. In addition, we aimed to determine the role of amino acids at the C-terminal end towards maintaining the functional activity of the α1,3GT. To study these issues, we cloned the α1,3GT gene from the marmoset, a New World monkey. By expressing soluble forms of recombinant truncated marmoset α1,3GT, we were able for the first time to precisely define the size of the stem region and to show the absolute requirement for amino acids at the C-terminus for the maintenance of enzyme activity (Henion *et al.*, 1994).

### 1.1. Cloning of Marmoset α1,3GT

Several cDNAs of marmoset α1,3GT were cloned from a cDNA library prepared from oligo-dT primed mRNA extracted from the marmoset lymphoid cell line B95.8, and inserted into the vector pcDNA1 (Invitrogen, San Diego, CA). This plasmid contains an SV40 origin of replication, which enables vector replication to occur following transient transfection in cells expressing the SV40 large T antigen, such as COS cells. These cells originate from African green monkey Vero cells and do not endogenously express α1,3GT activity. Positive clones were initially identified by screening with a probe from the murine α1,3GT cDNA (Larsen *et al.*, 1989), obtained from Dr. John Lowe, which encompassed several hundred nucleotides near the C-terminal end of the catalytic domain. To ensure that the isolated clones contained α1,3GT sequences comprising most, if not all, of the coding region, a PCR-based screen was employed in which primers spanning the predicted murine α1,3GT coding region were used to discriminate near full-length clones from those with only short regions of overlap with the murine probe. Given that most glycosyltransferases have exceptionally long 3′ untranslated regions, and that reverse transcription of poly(A) mRNA beyond 2,000–3,000 base pairs is generally inefficient, we expected to find many truncated cDNA clones.

Of fifteen possible candidates, three distinct cDNA clones gave PCR products of the predicted sizes and were isolated for further characterization. The longest insert found contained a near full-length sequence when compared with that of the murine α1,3GT, but was short of the predicted start codon by 18 base pairs. It did, however, encode the entire transmembrane, stem and catalytic domains.

The remainder of the 5′ end of the marmoset gene was obtained by amplifying plasmid preps of the whole B95.8 cDNA library with an amplimer set of an upstream vector anchored primer from within the T7 promoter of the pcDNA1 plasmid, and several different downstream primers within the marmoset α1,3GT gene. PCR products amplified by these reactions were subcloned, identified with probes from within the α1,3GT insert, and sequenced. By this method, the N-terminal end of the α1,3GT coding sequence was determined, as well as a portion of the 5′ untranslated region (5′ UTR) of the α1,3GT gene.

As expected, the overall organization of the marmoset enzyme was similar to that of the mouse α1,3GT with 82% identity between the nucleotide coding sequence (Figure 2). Analysis of the marmoset cDNA revealed a predicted cytoplasmic tail of 6 amino acids, a transmembrane domain of 16 amino acids and a lumenal stem/catalytic domain of 354 amino acids. Homology to a human pseudo gene for α1,3GT (Larsen *et al.*, 1990) was even more substantial with only 5.6% differences by comparison. As speculated earlier, when short fragments of other New World monkey α1,3GT coding regions were compared with this human pseudogene, this high degree of conservation supports the theory of a relatively late suppression of α1,3GT expression during the evolution of Old World mon-

```
                                                            GAGAAAATA

  1   M  N  V  K  G  K  V  I  L  S  M  L  V  V  S  T  V  I  V  V     20
    ATGAATGTCAAAGGAAAAGTAATTCTGTCGATGCTGGTTGTCTCAACTGTGATTGTTGTG
    ...........G........G..C....T.......A..........C...G....C...

 21   F  W  E  Y  I  N  S  P  E  G  S  F  L  W  I  Y  H  S  K  N     40
    TTTTGGGAATATATCAACAGCCCAGAAGGCTCTTTCTTGTGGATATATCACTCAAAGAAC
    ............G.............C........................A....A.TT

 41   P  E  V  D  D  S  S  A  Q  K  D  W  W  F  P  G  W  F  N  N     60
    CCAGAAGTTGATGACAGCAGTGCTCAGAAGGACTGGTGGTTTCCTGGCTGGTTTAACAAT
    .....G....G...G.A...ATGG.................C..AA..........A...

 61   G  I  H  N  Y  Q  Q  E  E  E  D  T  D  K  E  K  G  R  E  E     80
    GGGATCCACAATTATCAACAAGAGGAAGAAGACACAGACAAAGAAAAAGGAAGAGAGGAG
    ....C.....G.......G....CA.C.T...AGG.CGG.G......G..T...A.T.GA

 81   E  Q  K  K  E  D  D  T  T  E  L  R  L  W  D  W  F  N  P  K    100
    GAACAAAAAAAGGAAGATGACACAACAGAGCTTCGGCTATGGGACTGGTTTAATCCAAAG
    .AT.GC.TTG.---------------.....C..A...............C.........

101   K  R  P  E  V  M  T  V  T  Q  W  K  A  P  V  V  W  E  G  T    120
    AAACGCCCAGAGGTTATGACAGTGACCCAATGGAAGGCGCCGGTTGTGTGGGAAGGCACT
    ..C.....G..T...T............CG............A.................

121   Y  N  K  A  I  L  E  N  Y  Y  A  K  Q  K  I  T  V  G  L  T    140
    TACAACAAAGCCATCCTAGAAAATTATTATGCCAAACAGAAAATTACCGTGGGGTTGACG
    ..TG...C...TC.G..G.....G..C..C....C.......C.C..T......C....A

141   V  F  A  I  G  R  Y  I  E  H  Y  L  E  E  F  V  T  S  A  N    160
    GTTTTTGCTATTGGAAGATATATTGAGCATTACTTGGAGGAGTTCGTAACATCTGCTAAT
    ..G......G.G....AG..C..............A..A..C..TC.GGAG......G.C

161   R  Y  F  M  V  G  H  K  V  I  F  Y  V  M  V  D  D  V  S  K    180
    AGGTACTTCATGGTCGGCCACAAAGTCATATTTTATGTCATGGTGGATGATGTCTCCAAG
    .T............T.....TCGG...........C......A.A..C..CAC....CG.

181   A  P  F  I  E  L  G  P  L  R  S  F  K  V  F  E  V  K  P  E    200
    GCGCCGTTTATAGAGCTGGGTCCTCTGCGTTCCTTCAAAGTGTTTGAGGTCAAGCCAGAG
    AT...TG.CG.GC.C...AAC.....A.A......AC....C......A...G.T.T...

201   K  R  W  Q  D  I  S  M  M  R  M  K  T  I  G  E  H  I  L  A    220
    AAGAGGTGGCAAGACATCAGCATGATGCGTATGAAGACCATCGGGGAGCACATCTTGGCC
    ...........G..T..............C...........T............C.....

221   H  I  Q  H  E  V  D  F  L  F  C  M  D  V  D  Q  V  F  Q  D    240
    CACATCCAACACGAGGTTGACTTCCTCTTCTGCATGGATGTGGACCAGGTCTTCCAAGAC
    ........G........C....................C.....T..A.....T......
```

**Figure 2.** Sequence of marmoset α1,3GT (top) and comparison to murine α1,3GT nucleotide sequence (bottom). Mouse nucleotides that are the same as the marmoset sequence are indicated by dots. Dashes represent nucleotides present in the marmoset, but absent in the mouse. Deduced amino acid sequence of marmoset α1,3GT appears in single letter code above the marmoset nucleotide sequence. Amino acids comprising the predicted transmembrane domain are underlined. Reprinted from Henion *et al.*, 1994 with permission. (*Continues on next page.*)

```
241  H  F  G  V  E  T  L  G  Q  S  S  A  Q  L  Q  A  W  W  Y  K   260
     CATTTTGGGGTAGAGACCCTGGGCCAGTCGGTGGCTCAGCTACAGGCCTGGTGGTACAAG
     A.C..C.....G..A..T.........CT...A..A.....C..................

261  A  D  P  D  D  F  T  Y  E  R  R  K  E  S  A  A  Y  I  P  F   280
     GCAGATCCTGATGACTTTACCTATGAGAGGCGGAAAGAGTCGGCAGCATATATTCCATTT
     ..CAG...C..GA....C...............G..CT......C..G..C........C

281  G  Q  G  D  F  Y  Y  H  A  A  I  F  G  G  T  P  I  Q  V  L   300
     GGCCAGGGGGATTTTTATTACCATGCAGCCATTTTTGGAGGAACACCGATTCAGGTTCTC
     ..AG.............C.....C..G.................G..T.C...CA.....

301  N  I  T  Q  E  C  F  K  G  I  L  L  D  K  K  N  D  I  E  A   320
     AACATCACCCAGGAGTGCTTTAAGGGAATCCTCCTGGACAAGAAAAATGACATAGAAGCC
     ...C.....AG...............G.......A..........C..............

321  E  W  H  D  E  S  H  L  N  K  Y  F  L  L  N  K  P  S  K  I   340
     GAGTGGCATGATGAAAGCCACCTAAACAAGTATTTCCTTCTCAACAAACCCTCTAAAATC
     C.............G........C.....A..C......T...........A........

341  L  S  P  E  Y  C  W  D  Y  H  I  G  L  P  S  D  I  K  T  V   360
     TTATCTCCAGAATACTGCTGGGATTATCATATAGGCCTGCCTTCAGATATTAAAACTGTC
     C..........G..T........C....G..........................G....

361  K  L  S  W  Q  T  K  E  Y  N  L  V  R  K  N  V  *
     AAGCTATCATGGCAAACAAAAGAGTATAATTTGGTTAGAAAGAATGTCTGA
     ...G..G.T.....G..........................T.........
```

**Figure 2.** (*Continued*)

keys and apes (Galili and Swanson, 1991; see the chapter of Galili on evolution of α1,3galactosyltransferase and of the α-gal epitope).

## 1.2. Structure-Function Analysis

It is well established that in the course of evolution there is an inverse relationship between the rate at which nucleic acids within a gene mutate and the functional constraints placed on the protein product for the maintenance of activity (Wilson, 1985). Because of this, the stem domain of the marmoset α1,3GT, when compared with the enzyme in different species, would be predicted to accumulate more amino acid differences, while the catalytic domain would have fewer.

An indication of the functional importance of C-terminal amino acids within the lumenal portion of the enzyme can be observed by examining the relative degree of homology which exists in the various regions between the marmoset and the mouse sequences (Figure 3). A histogram of amino acid differences evaluated

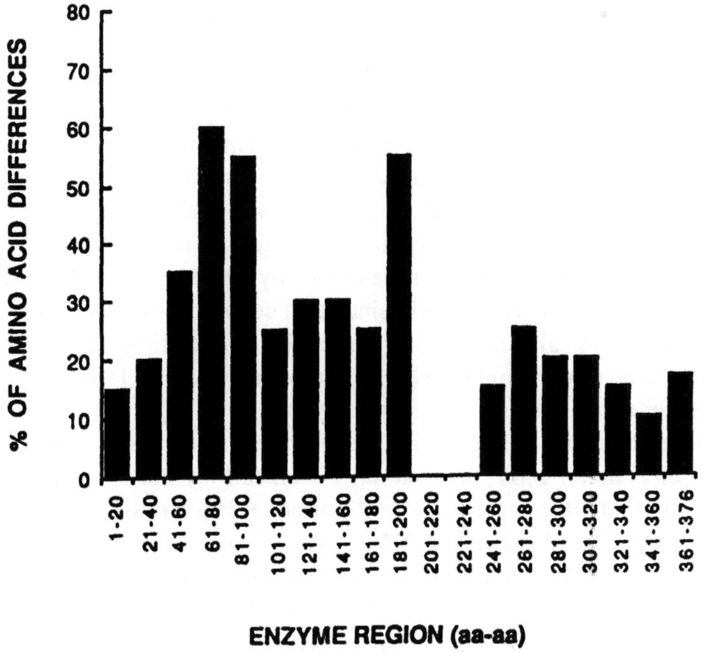


**Figure 3.** A comparison of the overall similarity between marmoset and mouse α1,3GT over the length of their coding regions, presented as a percent of amino acid differences in blocks of 20 amino acids. The last block of amino acids at the C-terminus has only 16 residues. Reprinted from Henion *et al.*, 1994 with permission.

in blocks of 20 amino acids from the N-terminal end to the C-terminus shows that a proportionally high number of amino acid substitutions have occurred between the two species immediately downstream of the transmembrane domain, especially between amino acids 41 and 100, in the region expected to contain most of the stem domain. Because of the lack of functional constraints on this portion of the molecule for the maintenance of enzyme activity, it would be predicted that this region would contain a higher proportion of amino acid substitutions relative to parts of the protein that are involved in catalysis. Accordingly, the complete absence of substitutions between amino acids 201 and 240, and the relatively low rate of substitutions throughout the remainder of the C-terminal end of the enzyme suggests that these regions are important for the enzymatic activity of the protein and thus mutate at a slower rate than the stem region.

To demonstrate that the cloned marmoset cDNA encodes an active enzyme, a recombinant expression system was utilized in which the lumenal portion of α1,3GT was synthesized as a fusion protein downstream of the IgG binding domain of *Staphylococcus aureus* protein A. The lumenal portion of the marmoset α1,3GT, without the cytoplasmic and transmembrane domains, was amplified by PCR and inserted into the vector pPROTA (Sanchez-Lopez *et al.*, 1988). The protein A-enzyme chimera in this vector is expressed under the control of a strong SV40 promoter and contains an N-terminal transin signal peptide for cellular export following translation. Constructs were transiently transfected into COS cells and the secreted chimeric enzyme was collected from the culture medium by coincubation with IgG-Sepharose, which binds with high affinity to the protein A portion of the molecule. The use of such protein A chimeras for glycosyltransferase studies was first described by Lowe and colleagues (Larsen *et al.*, 1989) and has been used widely since as a convenient method of expressing and purifying cloned glycosyltransferases, and characterizing their enzymatic activity (Bierhuizen and Fukuda, 1992; Kitagawa and Paulson, 1994, Holmes *et al.*, 1995; Bakker *et al.*, 1997; Sasaki *et al.*, 1997).

The endogenous α1,3GT enzyme, purified from tissue sources such as Ehrlich ascites tumor cells (Blake *et al.*, 1981), calf thymus (Van den Eijnden *et al.*, 1983), and rabbit stomach mucosa (Betteridge, 1983) utilizes the soluble type II acceptor β-D-Gal-(1→4)-D-GlcNAc (N-acetyllactosamine) efficiently to synthesize the α-Gal terminated trisaccharide structure. Lactose (β-D-Gal-(1→4)-D-Glc) also serves as acceptor for the endogenous enzyme, but is utilized less efficiently (Van den Eijnden *et al.*, 1983). Studies on both recombinant and purified bovine α1,3GT from calf thymus recorded a five fold decrease in kinetic efficiency when lactose was used as an acceptor relative to N-acetyllactosamine (Joziasse *et al.*, 1990).

These specificities clearly differentiate α1,3GT from other Golgi-bound galactosyltransferases, such as the blood group B-transferase, which utilizes α-L-Fuc(1→2)-β-D-Gal(1→4)D-Glc (2′fucosyllactose), or 2′fucosyllactosamine, but not N-acetyllactosamine. It is also different from both β1,3 and β1,4-galactosyltransferase which require terminal N-acetylglucosamine residues and initiate synthesis of type I and type II carbohydrate chains respectively. Based on this, we used N-acetyllactosamine as the primary acceptor when measuring activity of the chimeric marmoset enzyme.

To assess the activity of the secreted fusion protein immobilized on IgG Sepharose, the beads were washed and incubated with N-acetyllactosamine (0.5 mM) as sugar acceptor, and with 200nM [$^3$H] UDP-Gal as sugar donor in MES buffer containing 25mM $MgCl_2$. After incubation for 2 hours at 37°C the synthesized [$^3$H]Galα1–3Galβ1-4GlcNAc was isolated from the radiolabeled sugar donor by anion exchange chromatography with Dowex-1 columns. The resin strongly binds the negatively charged UDP-Gal, but not the neutrally charged, radiolabeled trisaccharide product, which is counted in the column effluent.

We first measured the catalytic activity of the marmoset α1,3GT cDNA that was amplified from amino acids 47 and 376 by PCR. This fragment was inserted into the pPROTA vector and transfected into COS cells. The secreted fusion protein product was harvested and subjected to the catalytic activity assay. This sequence is lacking only a small portion of the N-terminal lumenal portion of the enzyme and would be expected to harbor full catalytic function. As a qualitative assessment of recombinant enzyme activity, assays were conducted using limited quantities of [$^3$H] UDP-Gal. After transient expression in COS cells, the secreted chimeric α1,3GT attached to the IgG-Sepharose beads was able to synthesize [$^3$H]-labeled trisaccharide at a rate corresponding to 1.2 pmoles of the α-galactosylated product per hour (Figure 4).

The acceptor specificity of this recombinant enzyme was assayed by replacing N-acetyllactosamine with either lactose or galactose. Like the mouse α1,3GT, N-acetyllactosamine was the best acceptor, while lactose was only utilized 50% as efficiently. Galactose, which is the terminal acceptor sugar of both lactose and N-acetyllactosamine, was not effective as a monosaccharide acceptor. These results are in good agreement with those obtained using the purified α1,3GT from calf thymus tissue and mouse tumor cells (Blanken and Van den Eijnden, 1985; Elices *et al.,* 1986).

Following the confirmation of enzymatic activity of the fusion protein, we next examined the size of the catalytic domain of the marmoset α1,3GT by truncating amino acids from the 5′ or the 3′ end of the enzyme. These truncated con-

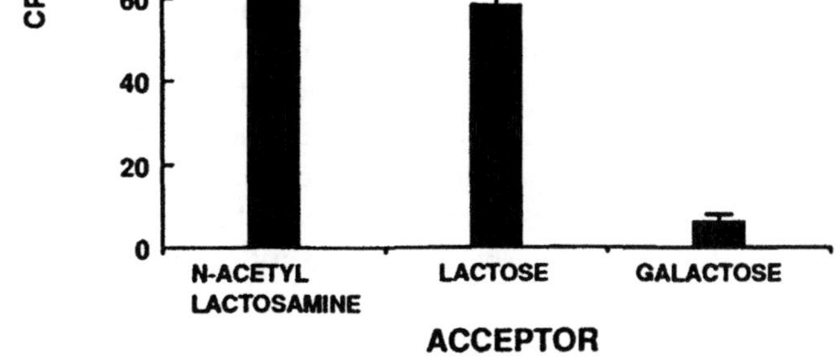


**Figure 4.** Acceptor specificity of the chimeric marmoset α1,3GT enzyme expressed as a protein A fusion protein in COS cells. The enzyme immobilized on IgG-Sepharose beads was coincubated with [$^3$H] UDP-Gal and the designated acceptors. The reaction products were counted by scintillation following anion exchange chromatography. Data represents the mean ± SD from five different transfection experiments. Reprinted from Henion *et al.*, 1994 with permission.

structs were generated by PCR amplification of the marmoset cDNA clone 14 m using primer sets that spanned variable lengths of the coding region of the marmoset enzyme. Fragments generated were subcloned into pPROTA, sequenced over their entire length to confirm the absence of amplification errors, and transfected into COS cells as previously described to assess enzymatic activity.

Relative to the initial construct 47–376, a construct containing the entire lumenal domain, 23–376, had only 70% the enzymatic activity (Figure 5A). With the exception of this small decrease, the truncation of as many as 90 amino acids from the N-terminal end of the marmoset α1,3GT did not have a significant effect on the level of enzymatic activity. Subsequent sequential truncation of single amino acids resulted in a progressive loss of catalytic function, until the construct 94–376, which had no catalytic activity. Smaller constructs 97–376 and 102–376 were also inactive.

To determine if the removal of stem region amino acids had an effect on the rate of enzyme activity, we compared the synthesis of α-gal-trisaccharide by the construct that displayed the highest level of enzymatic activity, i.e. amino acids 47–376, with a construct lacking 40 additional N-terminal amino acids within the stem region, i.e., amino acids 87–376. As shown in Figure 5A, both constructs had a similar level of activity with N-acetyllactosamine when a prolonged incubation period was used. During the reaction, aliquots of the samples were removed and the activity determined. Product formation at ten minute intervals was the same throughout the time course, indicating that the truncation of additional stem region amino acids did not affect the rate of enzyme activity with N-acetyllactosamine (data not shown).

Whereas several lines of evidence had previously suggested that amino acids of the N-terminal stem region of glycosyltransferases were dispensable for catalytic function, the importance of amino acids at the C-terminal end was unknown. We addressed this question using the marmoset α1,3GT and the same truncation strategy used to analyze the stem domain. In contrast to the stem region truncations, the removal of amino acids at the C-terminal end caused a precipitous decline in catalytic function (Figure 5B). Truncation of two C-terminal amino acids had no effect on the catalytic activity. However, the removal of one additional amino acid completely destroyed enzyme function. Further C-terminal truncations were likewise inactive. These results indicate that amino acids very close to the C-terminal tail of the marmoset α1,3GT are critical to the proper functioning of the protein.

We confirmed that the observed lack of catalytic activity of the truncated enzymes was not a result of defective secretion due to defects in synthesis, transport, or stability of the fusion proteins. This was performed by metabolic labeling of the expressed enzyme with [$^{35}$S]-methionine and analysis of the products on SDS-PAGE. No decrease in either synthesis or secretion of the fusion proteins was observed in the catalytically inactive forms (data not shown, Henion *et al.*, 1994).

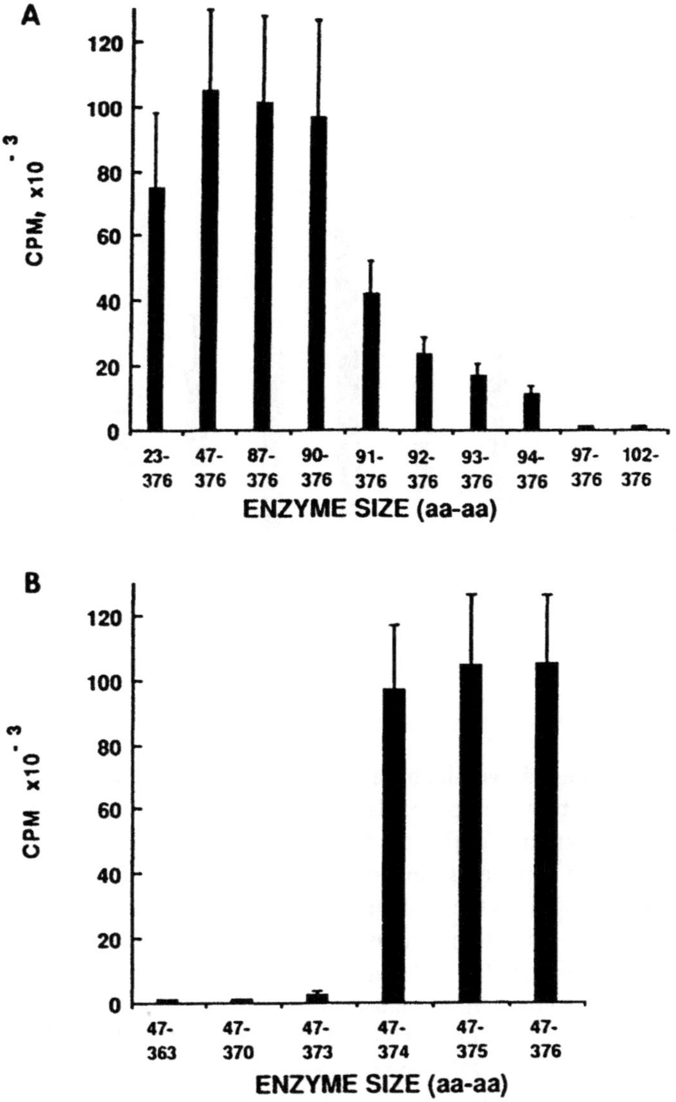


**Figure 5.** Effect of N-terminal (A) and C-terminal (B) truncations on catalytic activity of the marmoset α1,3GT. Numbers correspond to the amino acids in Figure 2 and represent the length of the α1,3GT portion of each construct. (C) Conservation of catalytic domain size for α1,3GT in different species. The mouse α1,3GT was subjected to the same N-terminal truncation as the marmoset cDNA. The amino acid numbers correspond to those of the marmoset sequence. Values in all cases are the mean ± SD of 3–5 experiments. Reprinted from Henion *et al.*, 1994 with permission.

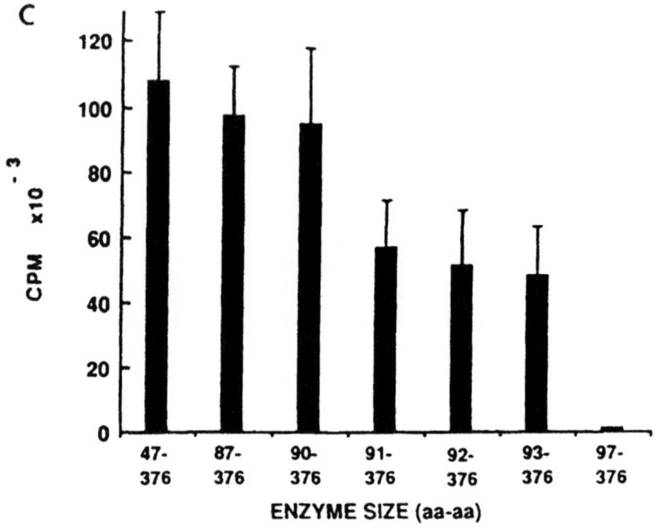


**Figure 5.** (*Continued*)

As mentioned above, the homology between the mouse and the marmoset enzyme is very substantial, particularly in the regions corresponding to the marmoset catalytic domains. Within the stem region a relatively higher degree of variability was found, including an insertion of 15 nucleotides in the marmoset α1,3GT which lengthens the protein by 5 amino acids. Such deletions and insertions are common in the stem regions of the same glycosyltransferase from different species (Bendiak *et al.*, 1993; Ihara *et al.*, 1993; Sarkar *et al.*, 1998), or when members of glycosyltransferase families having a high degree of homology and similar specificities are compared with each other (Weston *et al.*, 1992; Almeida *et al.*, 1998; Hennet *et al.*, 1998, Nomura *et al.*, 1998).

This stem region heterogeneity suggested that the length of the mouse catalytic domain could be similar, if not identical to that of the marmoset. To confirm this assumption, the analogous deletions were made in the catalytic domain of the mouse α1,3GT cDNA, and the constructs were cloned into the pPROTA vector and expressed in COS cells as above (Figure 5C). Immediately downstream of the 5 amino acid deletion in the mouse, in the position corresponding to marmoset amino acid 91, the loss of single amino acids began to progressively destroy enzyme activity. This is the same location where catalytic activity declined after marmoset $\alpha$1,3GT truncation. The same rapid decrease in $\alpha$1,3GT activity following marmoset C-terminal truncation was found in the mouse as well. Thus, the number of amino acids which encode the minimal catalytic domains are exactly conserved between the mouse and the marmoset enzyme. The stem regions, with the exception of the 5 amino acid insertion found in the marmoset, are also of the same size.

### 1.3. Structure-Function Studies with Other Glycosyltransferases

The results of these experiments demonstrated for the first time the exact proportions of a glycosyltransferase catalytic unit. Since the publication of our studies on the marmoset and mouse $\alpha$1,3GT (Henion *et al.*, 1994) several other studies have addressed this topic as well. A similar approach was recently applied to the analysis of the catalytic regions of the $\alpha$1,3-fucosyltransferase (FUT) family (Xu *et al.,* 1996). This closely related group of enzymes consists, to date, of five members, termed FUT III-VII, each with subtle differences in acceptor specificity.

Two of the most closely related members of this family, FUT III and FUT V, share greater than 90% amino acid identity, and have only a single amino acid difference within the C-terminal halves of their coding regions. There are, however, 26 amino acid differences within the N-terminal portion of the lumenal domain, which led to the initial speculation that substitutions within the stem region were influencing the acceptor specificity of the respective glycosyltransferases (Weston *et al.,* 1992). When structure-function analysis was conducted on recombinant FUT enzymes, a minimal catalytic domain of 299 amino acids for each was necessary to maintain enzyme function. Domain swapping studies between the two FUT sequences identified a cluster of eight amino acid substitutions just downstream of the stem region at the N-terminal end of the minimal catalytic unit as critical in deciding the specificity of the recombinant FUT genes (Xu *et al.,* 1996). An additional similarity that exists between $\alpha$1,3GT and the FUT genes is the critical function performed by amino acids that lie close to the C-terminus of the enzyme, as the loss of even two amino acids from this region completely destroyed FUT V activity (Xu *et al.,* 1996).

The determinants of specificity for the blood group ABO transferases, which share sequence homology with $\alpha$1,3GT, also seem to lie well within the catalytic domain. In this case, two C-terminal amino acid substitutions account for the differences in the sugar nucleotide specificity between the blood group A and B transferase (Yamamoto and Hakomori, 1990b). It thus appears that in each of these enzymes, as with the marmoset $\alpha$1,3GT, the amino acids which determine specificity lie within the catalytic domains themselves, and are not influenced by the high degree of amino acid variability inherent to the stem region.

While the primary sequence of the stem region does not appear to grossly alter $\alpha$1,3GT acceptor specificity directly, it should be pointed out that as a functional tether for the catalytic domain, the stem region may influence more subtle aspects of glycoconjugate preference. A recent study has shown that soluble murine α1,3GT, expressed in recombinant form in human 293 cells, could in most respects glycosylate glycoprotein acceptors as efficiently as the membrane-bound form (Cho and Cummings, 1997). The truncated construct, from which the cytoplasmic, transmembrane, and a large portion of the stem domain was deleted, was fully capable of glycosylating soluble glycoprotein acceptors, but appeared less efficient at utilizing membrane-associated glycoproteins than the full length enzyme. It will be interesting to see whether the absence of these domains from other glycosyltransferases also influences recognition of membrane-associated glycoconjugates. The soluble α1,3GT form also appeared to be secreted slowly from transfected cells, raising the possibility that the stem region may have an effect on Golgi retention as well (Cho and Cummings, 1997).

Interestingly, the loss of enzyme function as a result of truncation may not be caused by the same mechanism when comparing different glycosyltransferases. Recent studies on the rabbit β-1,2-N-acetylglucosaminyltransferase I (GnTI) demonstrated that N-terminal truncation of 120 amino acids led to a loss of activity that was correlated with the absence of secreted enzyme in their baculovirus expression system (Sarkar *et al.*, 1998). The exact defect in the GnTI was not determined, but was likely the result of enzyme instability and degradation, although defective transcription/translation, or secretion are possibilities as well. No such defects were observed with the marmoset α1,3GT, where truncated forms of the chimeric enzyme, which were inactive, were still secreted at comparatively high levels.

Glycosyltransferases in general are fairly consistent in their overall size. With rare exception, they fall within a range of approximately 350–500 amino acids. The limited heterogeneity that occurs within this family of enzymes, however, cannot be ascribed solely to differences in the length of their stem regions. The structure-function results described in this report and the references cited within, as well as those of truncated β1,4-galactosyltransferase (lactose synthase) expressed in *E.coli* (Boeggeman *et al.*, 1993), suggest that the minimal size of glycosyltransferase catalytic domains varies from less than 274 amino acids (bovine β1,4-galactosyltransferase) to greater than 327 amino acids (rabbit GnT I). These differences are likely to reflect the variability necessary for accommodating different acceptor carbohydrate structures within the enzyme active site for the process of sugar nucleotide transfer to the elongating oligosaccharide chain.

## 2. ANALYSIS OF MARMOSET α1,3GT 5′ UNTRANSLATED REGIONS

In addition to conserved domain organizations, many glycosyltransferases also share the common feature of having 5′ untranslated regions (UTRs) which are spread over multiple exons. In many cases these exons are alternatively spliced

(Saito *et al.*, 1995, Furukawa *et al.*, 1996) or initiate from multiple promoters (Harduin-Lepers *et al.*, 1993; Wen *et al.*, 1992; Lo and Lau, 1996) which give rise to 5′ UTRs of different lengths and contribute to the heterogeneous size of glycosyltransferase transcripts. Alternative splicing also occurs in the 5′ UTR of members of the fucosyltransferase and N-acetylglucosaminyltransferase family for which the coding regions themselves are retained on a single exon (Cameron *et al.*, 1995; Koyama *et al.*, 1996).

The full-length murine α1,3GT transcript has been determined to be distributed over nine exons (Joziasse *et al.*, 1992b) (for clarity, the exon numbering of the mouse genomic organization will be applied to the marmoset as well and expanded where necessary; see the chapter of Joziasse on the "α1,3galactosyltransferase gene"). The murine 5′ UTR sequence was found to be spread over four exons and totaled greater than 440 bps. Similarly, the 5′ UTR of the bovine α1,3GT was reported to be quite long, totaling 468 bps. A comparison of untranslated regions between these two species revealed that some of the exon sequences identified in the murine upstream were conserved in the cow, although there are several unique areas of the bovine mRNA as well (Figure 6A). Although Joziasse *et al.* were able to

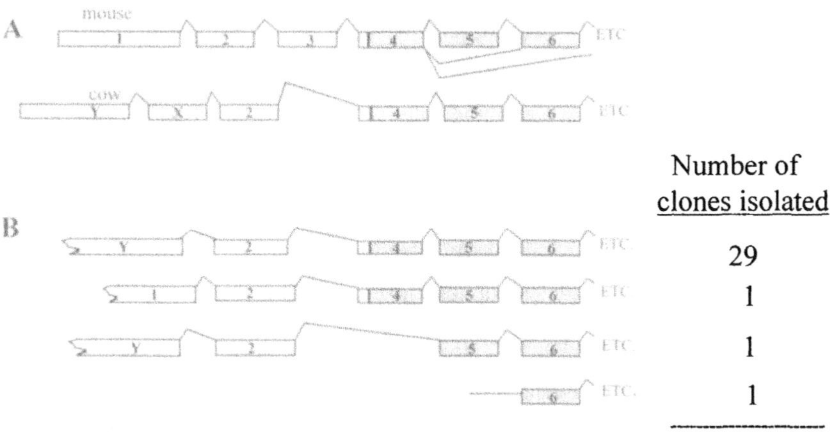


**Figure 6.** A comparison of the 5′ UTR from cDNA clones of different species. (A) Schematic organization of the murine (Larsen *et al.*, 1989; Joziasse *et al.*, 1992b) and bovine (Joziasse *et al.*, 1989) α1,3GT upstream regions. The dotted sequence marked as X in the bovine upstream is distinct from sequences found in the other species. (B) Comparison of the organization of the 5′ UTR and N-terminal end of α1,3GT from different marmoset PCR sequences obtained by the cDNA library PCR and 5′ RACE (Rapid Amplification of cDNA Ends) analysis. Alignment of these sequences with exons reported from the known genomic organization in the mouse (Joziasse *et al.*, 1992b) and with the bovine cDNA (Joziasse *et al.*, 1989) allowed the assignment of the exon junctions within the marmoset upstream. All coding portions of exons are shaded. 5′ exons, for which sequences are probably incomplete, are depicted with jagged borders. The bottom sequence terminated within the intron upstream of exon 6.

identify alternatively spliced exons within the coding region of the murine α1,3GT, none of the 5′ noncoding exons appeared to display alternative splicing (Joziasse *et al.*, 1992b). Our analysis of the marmoset upstream sequences demonstrates the occurrence of alternative splicing in this region.

## 2.1. Structure of Marmoset 5′ UTR and Comparison with Other Species

We obtained some of the marmoset α1,3GT 5′ UTR sequence initially by the vector anchored PCR strategy we employed to complete the 5′ end of the coding sequence. A complementary approach using 5′ RACE PCR (Rapid Amplification of cDNA Ends) was also utilized to attempt to map the transcriptional start site. When compared with the known mouse genomic organization, it was apparent that sequences absent in the marmoset corresponded to whole exons, indicating that the splice junctions were conserved between the two species, and that alternative splicing events might occur in the marmoset (Figure 6B). Most notably, an equivalent of murine exon 3 was not found in any of the marmoset PCR sequences obtained, suggesting that this exon is either spliced out in a tissue specific manner, or not utilized at all. There were also sequences present in the marmoset and bovine untranslated region which were not shared with the mouse, suggesting that additional exons may be present in other species which are not found in the mouse cDNA (see Figure 6).

Additionally, heterogeneity was found at the 5′ end of the different marmoset sequences generated by PCR when compared with each other. In almost all cases in which the available sequence extended 5′ of exon 2, the murine equivalent of exon 1 was not present immediately upstream. In its place, a sequence homologous to a region of the bovine 5′ UTR, which we termed exon Y, was present (see Figure 6B). In at least one marmoset clone, however, the organization of the 5′ end followed that of the mouse, with a marmoset homologue of exon 1 located immediately upstream of exon 2. This analysis of the 5′ end of the PCR clones suggested that the marmoset α1,3GT initiated with several different sequences, implying the presence of multiple transcription initiation sites, or the presence of alternatively spliced exons downstream of an unidentified common start site. To attempt to differentiate these possibilities, additional upstream information would be necessary.

We were unable to identify by 5′ RACE longer PCR products which could give us additional information about the relative positioning of the two different 5′ sequences we had identified upstream of exon 2. This was undoubtedly made more difficult by a high GC content that was encountered within exon Y (Figure 7), and which hindered the reverse transcription step during the synthesis of cDNA for the PCR reactions. This elevated GC content is not unusual for exons in the 5′ UTR of glycosyltransferases and has been suggested previously to be a possible mechanism for translational control of transcripts (Aasheim *et al.*, 1993; Harduin-Lepers *et al.*, 1993).

```
                                                                    -301
                          GG  AGCTAGGCGC  GGTCCTGCCC  AGCCCCGATC

                                                 Y | 1              -251
GCCGCGGCCG  GCCCTCGGCG  CCCGGGTTCT  GCGGATCAGG  CGTGTCTCAG

                                                                    -201
GAGGCAGGAG  ATGTAGGGTG  GAGGCCTGGC  TTGGCCGCTT  CCAGTAAGGA

                                                                    -151
GCCACCCTCC  CAGACAGAGC  CTCAGTGTCC  CCGTATCGTG  AAACAAGGAT

                                     1 | 2                          -101
ACCGCTGCAG  ACCTCCCTTC  CAGTCTCTGA  GAGTTGAAAC  CAGCACCTCC

                                                                     -51
CCTTCCTCTG  AGTCCTGCCT  TCTCCTGCAG  AAGCAAGCTC  AAAAGAACTT

                                                 2 | 4                -1
TGTTTTCCTT  TTACTCTGGG  GTGAAAGCAG  CAGACGATGA  GGAGAAAATA
```

**Figure 7.** 5′ UTR sequence of marmoset α1,3GT, obtained by RACE analysis and sequencing of genomic inverse PCR products. The positions of the exons, as determined by comparison with the known mouse genomic organization (Joziasse *et al.*, 1992b), are marked with arrows. A homologue of mouse exon 3 was not found in any of the marmoset PCR sequences analyzed.

## 2.2. Analysis of Alternative Splicing

To determine whether exons Y and 1 could occur together in the same transcript, and to assess their relative orientation, we amplified B95.8 mRNA with different combinations of exon-specific primers from within each of these different sequences. PCR products were obtained when a primer within exon Y was used as the upstream primer with an exon 1 specific downstream primer, but not vice versa (data not shown). This demonstrated that exon Y was 5′ of exon 1 in the genomic organization of the marmoset and that these two sequences could occur together in the same transcript. Although this could be demonstrated by direct PCR, the fact that no RACE PCR products were obtained which contained both exons Y and exon 1 illustrates that the combination of both of these exons together in a single mRNA must be a relatively rare event.

From these results it was apparent that exon 1 was alternatively spliced out from the majority of marmoset α1,3GT transcripts. To investigate this further and screen for other sequence heterogeneities, we conducted a PCR analysis with primers spanning the entire known 5′ UTR. PCR reactions using an upstream primer from within exon Y and a downstream primer from within the coding re-

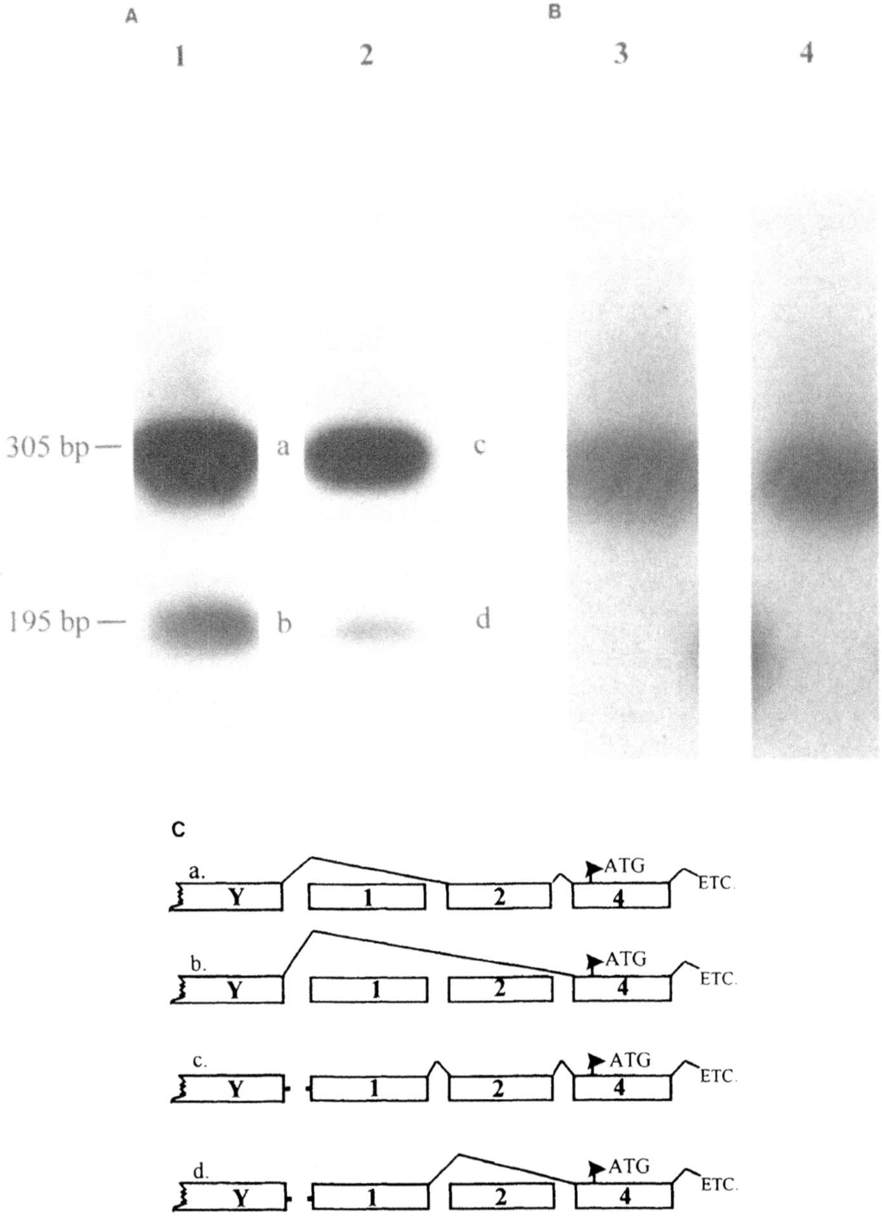


**Figure 8.** Alternative splicing within the 5′ UTR of the marmoset α1,3GT. (A) Southern analysis of the products amplified using primers spanning exons Y and downstream exon 6 (lane 1) or exons 1 and 6 (lane 2). Blots were hybridized with an oligonucleotide probe from within exon 4. (B) Southern blot of same PCR products as in (A), hybridized with an oligonucleotide within exon 2. (C) The four major species detected by RT-PCR of B95.8 mRNA in (A), labeled a-d. The relationship between exons Y and 1 for transcript c and d has not been definitively determined, and is therefore depicted with dots.

gion resulted in the amplification of two bands (Figure 8). The first of these corresponded in size to a PCR product lacking exon 1 only. The second band corresponded to a PCR product lacking exons 1 and 2. Hybridization with oligonucleotide primers from within exon 2 confirmed that this was indeed the case. The absence of longer PCR products in these reactions suggests that a mouse exon 3 homologue is not present in the marmoset 5′ UTR, in agreement with the analysis of the PCR clones.

The complete absence of detectable PCR products containing exon 1 in these reactions was further indication that in this marmoset lymphoid cell line, transcripts which include both exons Y and 1 are extremely rare, and seem only to be detectable by direct amplification with primers from within both exons. A second set of reactions was made using the same downstream primer, but now with an upstream primer from within exon 1 itself. In this case, PCR products were obtained which corresponded in size to either a transcript containing the full complement of exons 1 through 6, or a smaller product which lacked exon 2 (Figure 8). This indicates that exon 2 can be alternatively spliced independently of exon 1.

Previously, the only exons in α1,3GT which were reported to be alternatively spliced were 5 and 6 of the murine (Joziasse *et al.*, 1992) and porcine (Vanhove *et al.*, 1997) gene. These are coding exons from within the stem region of the enzyme. Since the marmoset α1,3GT genomic organization appears to be conserved with that of the mouse, we examined whether these same exons were alternatively spliced in the marmoset as well. The cDNA from B95.8 cells was amplified by PCR using upstream primers within either exon Y or exon 1, and a downstream primer in exon 8. It is unlikely that this downstream exon would itself be spliced out, since it encodes 46 amino acids in the middle of the catalytic domain of the enzyme. In each case, two PCR products were amplified which corresponded in size to the inclusion or exclusion of only exon 2. This was confirmed by hybridization with oligonucleotide probes from within exons 4, 6 and 7, all of which hybridized to both of the PCR products amplified, demonstrating that the shortened transcripts were not the product of alternative splicing of any of these exons.

## 2.3. Significance of Alternative Splicing

Our data imply that there are species differences in exon usage for the α1,3GT gene. In the mouse and pig, four isoforms of the α1,3GT are generated by the splicing of exon 5 and 6 (Joziasse *et al.*, 1992b; Vanhove *et al.*, 1997). However, in several bovine tissues examined, and in the marmoset lymphoid cell line analyzed in this study, the coding region appears to be unspliced. It is possible that this could be the result of tissue differences in the generation of splicing isoforms, a phenomenon observed with porcine α1,3GT (Vanhove *et al.*, 1997). However, the tissue specific splicing of the stem region appears to be species dependent, since in the mouse, unlike the pig, transcripts corresponding to all four isoforms occur in all tissues examined with the exception of male spermatids, where the α1,3GT mRNA does not appear to be expressed at all (Joziasse *et al.*, 1992b).

The significance of the alternative usage of exons 5 and 6 is also unclear at present. These exons encode up to amino acid 61, while our structure-function analysis indicates that the truncation of up to 85 N-terminal amino acids has no obvious effect on the enzymatic activity of the mouse α1,3GT towards N-acetyllactosamine (Figure 5C). Thus, the alternative splicing would not markedly perturb catalytic activity. Additionally, the differential presence of the exons did not noticeably alter the juxtanuclear localization of porcine α1,3GT following stable transfection in HeLa cells (Vanhove *et al.*, 1997). Recent studies though, on the fine localization of α1,3GT and α(1,2)fucosyltransferase within the Golgi, have highlighted the potential importance of amino acids outside of the transmembrane domain for fine control of temporal positioning of glycosyltransferases within the Golgi stacks, leading to alterations in cell surface glycosylation profile (Osman *et al.*, 1996). Based on this, a role for stem region splicing in altering more subtle aspects of targeting of the α1,3GT gene within the Golgi cannot be ruled out.

But in contrast to the situation found for the coding region, the 5′ UTR of the mouse α1,3GT did not display alternative splicing while that of the marmoset in lymphoid cells has at least two exons which are differentially utilized. The reasons for differences in exon utilization in the 5′ UTR between different species are unclear. One possibility is that the lack of functional constraints on this noncoding region, which results in a high rate of sequence divergence relative to the coding regions of an enzyme, has lead to alterations in the splice donor and acceptor sites for the exons. Such mutations could lead to sub-optimal splicing signals that suppress the inclusion of an exon that might otherwise be used constituitively.

An indication that this possibly could occur for the marmoset α1,3GT comes from exon flanking sequences that we generated using inverse PCR on genomic DNA isolated from the B95.8 cell line. A short stretch of sequence obtained from the intron preceding exon 1 revealed the presence of an atypical 3′ splice acceptor site upstream of exon 1. This unusual splice site deviated from the consensus splice site sequence by the presence of several purine residues adjacent to what is typically a polypyrimidine tract (data not shown). It is possible that this unusual sequence could be a factor in the relatively rare inclusion of exon 1 in the mature transcripts of the B95.8 α1,3GT.

The biological significance of alternative splicing within the 5′ noncoding regions in general is unknown. Most speculations have centered on potential effects on translation efficiency, since long, GC-rich sequences can form stable hairpin secondary structures which have been implicated in ribosome stalling and decreased translation (Kozak, 1991). This has been addressed previously with the human α2,6sialyltransferase, where a transcript with a truncated leader sequence was found to be translated somewhat more efficiently than mRNAs with longer 5′

UTRs (Aasheim *et al.*, 1993). The lack of other supporting studies makes it difficult to generalize this as a regulatory mechanism for other glycosyltransferases. In the case of the marmoset α1,3GT, although exons 1 and 2 are alternatively spliced and contain a combined 252 base pairs, they are unlike exon Y in that they are not particularly GC-rich. A study of the translation efficiencies of leader sequences lacking one or both of these exons may provide insight as to the significance of these transcript isoforms that do not otherwise alter the coding sequence of the α1,3GT.

We have provided evidence that exon 1 is alternatively spliced from transcripts that initiate upstream of exon Y. This includes direct amplification by PCR of transcripts containing both exons Y and 1, as well as the presence of an apparently weak splice site in the intron preceding exon 1. Joziasse has recently characterized the 5′ flanking region preceding exon 1 in the mouse as containing promoter activity sufficient for basal transcription of α1,3GT transcripts (unpublished observations). It may be that in the marmoset B95.8 cell line, this exon is directly preceded by a transcription initiation site, but is also alternatively spliced in longer transcripts that initiate upstream of exon Y.

## 3. EXPRESSION OF α1,3GT IN BACULOVIRUS

The expression of marmoset α1,3GT-protein A chimeras in COS cells has proven a convenient method for demonstrating the enzyme activity of the native enzyme and the truncation constructs. This expression system is not practical, however, for generating the type of enzyme quantities necessary for use in large scale glycosylation studies. For this purpose we have tried several different recombinant expression systems, including *E. coli* (Galili and Anaraki, 1995) and baculovirus (Henion *et al.*, 1997). While both systems produced large amounts of recombinant enzyme capable of glycosylating oligosaccharide and glycoprotein acceptors, the baculovirus system offered several advantages (see below) and was chosen as the method of choice for recombinant expression. The need for large amounts of recombinant α1,3GT is described in the chapter of LaTemple and Galili on "Enhancement of autologous tumor vaccine immunogenicity by anti-Gal."

### 3.1. Synthesis of Baculovirus α1,3GT

The marmoset α1,3GT coding sequence was prepared for the baculovirus system by amplifying the longest cDNA clone obtained from our library screen with PCR primers which spanned from amino acids 23 to the stop codon at position 376. As the truncation studies have demonstrated, the loss of the cytoplasmic and transmembrane domains, and much of the lumenal domain does not alter enzymatic activity. The truncated coding region was cloned in frame into the expression vector pBlueBacHisC for cotransfection with wild type *Autographa californica* multiple nuclear polyhedrosis virus (AcMNPV) into *Spodoptera frugiperda* (Sf9) insect cells. Homologous recombination between the transfer vector and the wild type virus generates a recombinant virus containing the marmoset α1,3GT gene under the control of the strong polyhedrin gene promoter.

Since the coding sequence of the α1,3GT insert has no signal sequence, the synthesized polypeptide accumulates in the cytoplasm instead of being cotranslationally inserted into the lumen of the endoplasmic reticulum. To facilitate purification, the enzyme is expressed with an N-terminal metal binding domain consisting of 6 histidine residues. After plaque purification of recombinant virus and several amplification rounds to generate a high titer virus stock, Sf9 cells infected for 46–63 hours with the virus are lysed, and the recombinant enzyme collected by chromatography on nickel-Sepharose affinity columns.

Under these conditions we were able to generate up to 5μg of the recombinant enzyme per $10^6$ infected Sf9 cells, which corresponded to a yield of 10–20 mg per liter of infected Sf9 spinner flask culture. Both of these figures are typical of proteins expressed in the baculovirus system (Fraser *et al.*, 1992; Ogonah *et al.*, 1996). This amount of enzyme is also higher than the 2 mg per liter obtained with the enzyme expressed in *E. coli.* and exceeded the 1.5–3.5 mg/L yield obtained with the bovine enzyme expressed in baculovirus (Joziasse *et al.*, 1990). In the latter case, however, the enzyme was expressed in the insect cells in its full-length, Golgi-bound form, and had to be purified on UDP-hexanolamine-Sepharose and

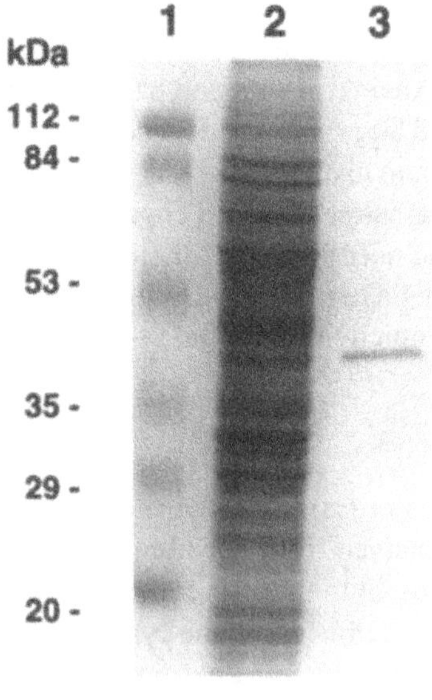


**Figure 9.** SDS PAGE analysis of marmoset α1,3GT expressed in baculovirus, stained with Coomassie Blue. The size of the MW markers in lane 1 is shown on the far left. Lane 2 is the whole cell lysate from 5 × $10^4$ Sf9 cells infected with the recombinant virus. Lane 3 is the recombinant enzyme from the equivalent number of infected cells following affinity chromatography on nickel-Sepharose and elution with 200mM imidazole. Reprinted from Henion *et al.*, 1997, with permission from Elsevier Science.

solubilized with Triton X-100, which may have contributed to the somewhat decreased recovery.

Analysis of the expressed protein, isolated from the cell lysates by affinity chromatography on nickel-Sepharose and eluted with 0.2M imidazole, indicated that the enzyme was obtained in a relatively pure form, with a molecular weight of approximately 41 kd (Figure 9). No additional bands were visible after staining the gels with Coomassie Blue. In contrast, a relatively impure product was obtained in the *E. coli* expression system, where a number of contaminating proteins accompanied the purified enzyme, even after extensive washing (Galili and Anaraki 1995).

### 3.2. Activity of Recombinant α1,3GT Expressed in Baculovirus System

We tested the activity of the expressed α1,3GT with oligosaccharide and glycoprotein acceptors. Similar to both the native enzyme purified from mammalian tissue sources and the chimeric protein A fusion protein, baculovirus-expressed α1,3GT utilized N-acetyllactosamine most efficiently, and to a lesser degree, lactose (Figure 10). Both galactose and N-acetylglucosamine monosaccharides were not able to serve as acceptors.

The specific activity was 1.15 units/mg protein, where 1 unit is defined as the amount of enzyme transferring 1 μmole of galactose to N-acetyllactosamine per minute at 37°. This is somewhat less than the activity obtained for the full-length baculovirus-expressed bovine α1,3GT, which was 3 units/mg, using lactose instead of N-acetyllactosamine as an acceptor (Joziasse *et al.*, 1990). The $K_m$ value of 83 μM derived for the marmoset enzyme for UDP-Gal was significantly different from that of the native bovine enzyme from tissue sources as well as the bovine enzyme expressed in baculovirus (approximately 2 mM). This increase in efficiency could well be due to factors such as species differences and glycosylation, since the lack of a signal sequence in the marmoset α1,3GT prevents insertion into the endoplasmic reticulum and Golgi where these reactions occur.

### 3.3. Glycosylation of Glycoprotein Acceptors for Improving Immunogenicity of Viral Vaccines

A high capacity for protein expression, such as that of α1,3GT in the baculovirus system, is valuable for the large-scale production of recombinant enzymes to be used for synthesizing carbohydrate chains on glycoproteins that can be exploited therapeutically. We have an interest in α-galactosylation of glycoprotein vaccines as a means of enhancing their immunogenicity through interactions *in vivo* with natural anti-Gal antibodies (Galili *et al.*, 1996; LaTemple *et al.*, 1996). The ability of antigen presenting cells to take up antigens for processing

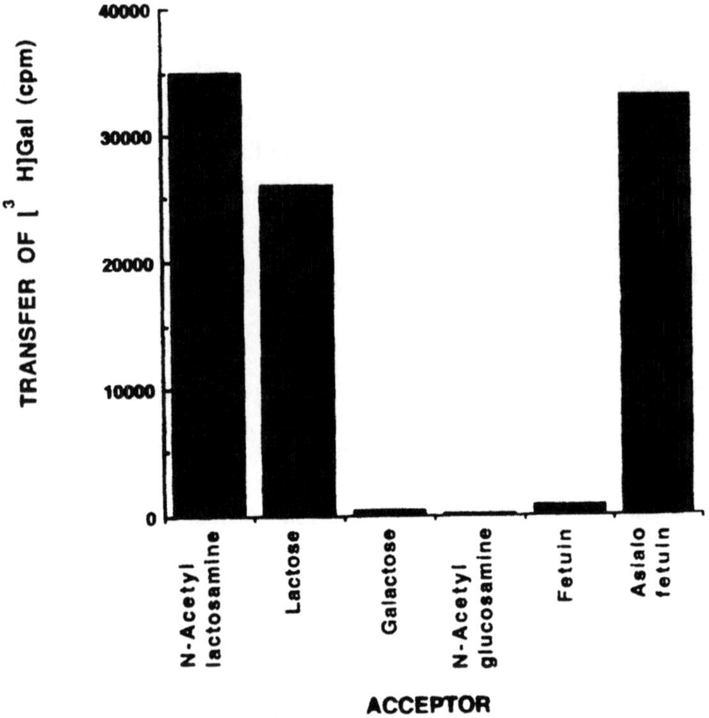


**Figure 10.** Acceptor specificity of the recombinant marmoset α1,3GT expressed in the baculovirus system. Activity was measured by coincubation of the recombinant enzyme with [$^3$H] UDP-Gal and the designated acceptors. The free carbohydrate was assayed at a concentration of 0.5mM and the fetuin and asialofetuin at 2μg per reaction. Reprinted from Henion *et al.*, 1997, with permission from Elsevier Science.

and presentation to antigen-specific T-helper cells is an important factor in generating an effective immune response to many vaccines, such as influenza. Since many types of antigen presenting cells express cell surface immunoglobulin Fc receptors, complexing α-galactosylated vaccines with anti-Gal targets the antigens to the antigen presenting cells more efficiently, and increases their processing and presentation to T-helper cells.

To demonstrate the principle of *in vitro* synthesis of α-gal epitopes by recombinant α1,3GT for immune enhancement, we chose to use whole influenza virus and isolated hemagglutinin as our model system. Influenza viruses in particular are excellent candidates for the synthesis of α-gal epitopes by recombinant α1,3GT, since the glycoproteins of the virion are unmodified by terminal sialic acid. The multiple N-acetyllactosamine acceptor sites on the major

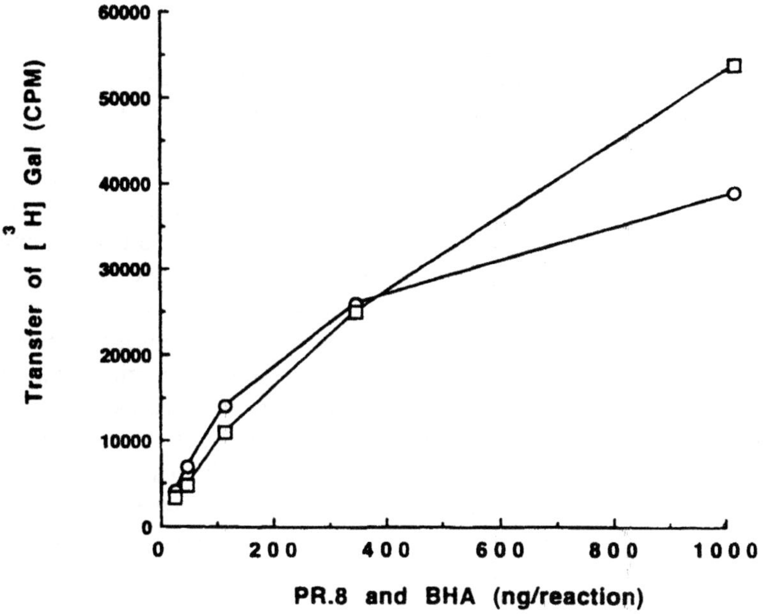


**Figure 11.** Buildup of α-gal epitopes on influenza virus glycoproteins by the recombinant α1,3GT expressed in baculovirus. Synthesis was determined for both intact PR8 strain influenza virus (circles) and for the isolated bromelin fragment of hemagglutinin (squares). Each reaction used 80nM [$^3$H] UDP-Gal, and acceptors diluted 1:3, starting from 1μg per reaction. Reprinted from Henion *et al.*, 1997, with permission from Elsevier Science.

glycoprotein of the viral membrane, the hemagglutinin molecule, are endogenously desialylated by the activity of the envelope-associated neuraminidase enzyme (Nakamura *et al.*, 1980; Keil *et al.*, 1985). For this reason, pretreatment with exogenous neuraminidase, which is normally a prerequisite to expose underlying N-acetyllactosamine, is unnecessary.

When either intact PR8 influenza virus, or the isolated bromelin-cleaved hemagglutinin molecule (BHA) are used as acceptor, the recombinant enzyme efficiently synthesizes α-gal epitopes in a dose dependent fashion (Figure 11). This level of enzyme activity corresponds to the synthesis of at least 3,000 α-gal epitopes per influenza virion, even under these nonsaturating substrate conditions (Henion *et al.*, 1997). This is the first demonstration of the use of a recombinant glycosyltransferase for modifying the carbohydrate profile of viral vaccine antigens. This approach may prove effective in other systems as well, where the interaction of natural anti-carbohydrate antibodies with *de novo* synthesized carbohydrate epitopes may facilitate the targeting of vaccine components to antigen presenting cells.

## 4. CONCLUSIONS

Recombinant DNA technology has made the high capacity production of glycosyltransferase enzymes a readily available technology. We have summarized in this chapter our work over the past several years on structure-function aspects of marmoset α1,3GT and the use of a recombinant form of marmoset α1,3GT, expressed in baculovirus, for altering the glycosylation profile of vaccine antigens. In the case of the marmoset α1,3GT and probably all other forms of membrane bound glycosyltransferases, the ability to remove cytoplasmic, transmembrane and stem domains makes it possible to express catalytic units which retain full enzymatic function. We have shown with the recombinant marmoset enzyme that the absence of these domains does not markedly alter catalytic function. This enables glycosyltransferases to be produced in recombinant systems under the control of strong promoters that direct high level expression with targeting signals that facilitate the isolation and final purification of the enzyme activities. The ability to synthesize α-gal epitopes on vaccines may prove to be a powerful technique for targeting antigens to presenting cells that process these antigens and activate immune effector cells.

## REFERENCES

Almeida, R., Amado, M., David, L., Levery, S.B., Holmes, E.H., Merkx, G., Geurts van Kessel, A., Rygard, E., Hasssan, H., Bennet, E., and Clausen, H., 1997, A family of β4-galactosyltransferases. Cloning and expression of two novel UDP-galactose:β-N-acetylglucosamine β1,4-galactosyltransferases, β4Gal-T2 and β4Gal-T3, J. Biol. Chem., **273**: 31979–31991.

Bakker, H., Friedmann, I., Oka, S., Kawasaki, T., Nifant'ev, N., Schachner, M., and Matei, N., 1997, Expression cloning of a cDNA encoding a sulfotransferase involved in the biosynthesis of the HNK-1 carbohydrate epitope **272**: 29942–29946.

Bendiak, B., Ward, L.D., and Simpson, R.J., 1993, Proteins of the Golgi apparatus. Purification to homogeneity, N-terminal sequence, and unusually large Stokes radius of the membrane bound form of UDP-galactose: N-acetylglucosamine β1-4galactosyltransferase from rat liver, Eur. J. Biochem., **216**: 405–417.

Betteridge, A., and Watkins, W.M., 1983, Two α-3-D-galactosyltransferases in rabbit stomach mucosa with different acceptor substrate specificities, Eur. J. Biochem. **132**: 29–35.

Bierhuizen, M.F.A., and Fukuda, M., 1992, Expression cloning of a cDNA encoding UDP-GlcNAc:Galβ1–3-GalNAc-R (GlcNAc to GalNAc) β1–6GlcNAc transferase by gene transfer into CHO cells expressing polyoma large tumor antigen, PNAS **89**: 9326–9330.

Blake, D.A., and Goldstein, I.J., 1983, An α-D-galactosyltransferase activity in Ehrlich ascites tumor cells, J. Biol. Chem. **256**: 5387–5393.

Blanken, W.M., and Van den Eijnden, 1985 Biosynthesis of terminal Galα1→3Galβ→4GlcNAc-R oligosaccharide sequences on glycoconjugates. Purification and acceptor specificity of a UDP-Gal: N-acetyllactosaminide α1→3-galactosyltransferase from calf thymus, J. Biol. Chem. **260**: 12927–12934.

Boeggeman,E.E., Balaji, P.V., Sethi, N., Masiby, A., and Qasaba, P.K., 1993, Expression of deletion constructs of bovine β-1,4-galactosyltransferase in *Escherichia coli*: importance of Cys134 for its activity, Protein Engineering **6**: 779–785.

Cameron, H.S., Szczepaniak, D., and Weston, B.W., 1995, Expression of human chromosome 19p α(1,3)-fucosyltransferase genes in normal tissues. Alternative splicing, poladenylation, and isoforms, J. Biol. Chem. **270:** 20112–20122.

Chester, M.A., and Watkins, W.M., 1969, Alpha-L-fucosyltransferases in human submaxillary gland and stomach tissues associated with the H, Lea and Leb blood-group characters and the ABH secretor status, Biochem. Biophys. Res. Commun. **34**: 835–842.

Cho, S.K., and Cummings, R.D., 1997, A soluble form of α1,3-galactosyltransferase functions within cells to galactosylate glycoproteins, J. Biol. Chem. **272**: 13622–13628.

Colley, K.J., 1997, Golgi localization of glycosyltransferases: more questions than answers, Glycobiology **7**: 1–13.

D'Agostaro, G., Bendiak, B., and Tropak, M., 1989, Cloning and expression of the membrane-bound form of bovine β1,4-galactosyltransferase, EJB **183**: 211–217.

Elices, M.J., Blake, D.A., and Goldstein, I.J., 1986, Purification and characterization of a UDP-Gal:β-D-Gal(1,4)-D-GlcNAc α(1,3)-galactosyltransferase from Ehrlich ascites tumor cells, J. Biol. Chem. **261**: 6064–6072.

Fraser, M.J., 1992, The baculovirus-infected insect cell as a eukaryotic gene expression system, Curr. Topics in Microbiol. Immunol. **158**: 131–172.

Furukawa, K., Soejima, H., Niikawa, N., and Furukawa, K., 1996, Genomic organization and chromosomal assignment of the human β1,4-N-acetylgalactosaminyltransferase gene. Identification of multiple transcription initiation sites, J. Biol. Chem. **271**: 20836–20844.

Galili, U., and Swanson K., 1991, Gene sequences suggest the inactivation of α-1,3-galactosyltransferase in cattarrhines after the divergence of apes from monkeys, PNAS **88**: 7401–7404.

Galili, U., and Anaraki, F., 1995, α-galactosyl (Galα1-3Galβ1-4GlcNAc-R) epitopes on human cells: synthesis of the epitope on human red cells by recombinant primate α1,3galactosyltransferase expressed in *E.coli*, Glycobiology **5**: 775–782.

Galili, U., Repik, P.M., Anaraki, F., Mozdzanowska, K., Washko, G., and Gerhard, W., 1996, Enhancement of antigen presentation of influenza virus hemagglutinin by the natural anti-gal antibody, Vaccine **14**: 321–328.

Gillespie, W., Kelm, S., and Paulson, J.C., 1992, Cloning and expression of the Galβ1,3GalNAcα2,3-sialyltransferase, J. Biol. Chem. **267**: 21004–21010.

Harduin-Lepers, A., Shaper, J.H., and Shaper, N.L., 1993, Characterization of two *cis*-regulatory regions in the murine β1,3-galactosyltransferase gene. Evidence for a negative regulatory element that controls initiation at the proximal site, J. Biol. Chem. **268**: 14348–14359.

Henion, T.R., Macher, B.A., Anaraki, F., Galili, U., 1994, Defining the minimal size of catalytically active primate α1,3 galactosyltransferase: structure-function studies on the recombinant truncated enzyme, Glycobiology **4**: 193–201.

Henion, T.R., Gerhard, W., Anaraki, F., and Galili, U., 1997, Synthesis of α-Gal epitopes on influenza virus vaccines, by recombinant α1,3galactosyltransferase, enables the formation of immune complexes with the natural anti-gal antibody, Vaccine **15**: 1174–1182.

Hennet, T., Dinter, A., Kuhner, P., Mattu, T.S., Rudd P.M., and Berger, E.G., 1998, Genomic cloning and expression of three murine UDP-galactose: β-N-acetylglucosamine β1,3-galactosyltransferase genes, J. Biol. Chem. **273**: 58–65.

Holmes E.H., Xu, Z., Sherwood, A.L., and Macher, B.A., 1995, Structure-function analysis of human α1→3fucosyltransferase, J. Biol. Chem. **270**: 8145–8151.

Ihara, Y., Nishikawa, A., Tohma, T., Soejima, H., Niikawa, N., and Taniguchi, N., 1993, cDNA cloning, expression, and chromosomal assignment of human N-acetylglucosaminyltransferase III (GnT-III), J. Biochem. **113**: 692–698.

Joziasse, D.H., Shaper, J.H., Van den Eijnden, D.H., Van Tunen, A.J., and Shaper, N.L., 1989, Bovine α1→3-galactosyltransferase: Isolation and characterization of a cDNA clone. Identification of homologous sequences in human genomic DNA, J. Biol. Chem. **264**: 14290–14297.

Joziasse, D.H., Shaper, N.L., Salyer, L.S., Van den Eijnden, D.H., vander Spoel, A.C., and Shaper, J.H., 1990, α1→3-galactosyltransferase: the use of recombinant enzyme for the synthesis of α-galactosylated glycoconjugates, Eur. J. Biochem. **191**: 75–83.

Joziasse, D.H., 1992a, Mammalian glycosyltransferases: genomic organization and protein structure, Glycobiology **2**: 271–277.

Joziasse, D.H., Shaper, N.L., Kim, D., Van den Eijnden, D.H., and Shaper, J.L., 1992b, Murine α1,4-galactosyltransferase. A single gene locus specifies four isoforms of the enzyme by alternative splicing. J. Biol. Chem., **267**: 5534–5541.

Keil, W., Geyer, R., Dabrowski, J., Nieman, H., Stirm, S., and Klenk, H., (1985). Carbohydrates of influenza virus. Structural elucidation of the individual glycans of the FPV hemagglutinin by two-dimensional $^1$H n.m.r. and methylation analysis, EMBO J. **4**: 2711–2720.

Kitagawa, H., and Paulson, J.C., 1994. Cloning and expression of a novel α2,3-sialyltransferase that sialylates glycoprotein and glycolipid carbohydrate groups. J. Biol. Chem. **269**: 1394–1401.

Koyama, N., Miyoshi, E., Ihara, Y., Kang, R., Nishikawa, A., and Taniguchi, N., 1996, Human N-acetylglucosaminyltransferase III gene is transcribed from multiple promoters, Eur. J. Biochem. **238:** 853–861.

Kozak, M., 1991, Structural features of eukaryotic mRNAs that modulate translation. J. Cell. Biol. **107**: 19867–19870.

Larsen, R.D., Rajan, V.P., Ruff, M.M., Kukowska-Lattallo, J., Cummings, R.D., and Lowe, J.B., 1989, Isolation of a cDNA encoding a murine UDPgalactose:β-D-galactosyl-1,4-N-acetyl-D-glucosaminide α-1,3-galactosyltransferase: Expression cloning by gene transfer. PNAS **86**: 8227–8231.

Larsen, R.D., Rivera-Marrero, C.A., Ernst, L.K., Cummings, R.D., and Lowe, J.B., 1990, Frameshift and nonsense mutations in a human genomic sequence to a murine UDP-Gal:β-D-Gal(1,4)-D-GlcNAc α(1,3)-galactosyltransferase cDNA, J. Biol. Chem. **265**: 7055–7061.

LaTemple, D.C., Henion, T.R., Anaraki, F., and Galili, U., 1996, Synthesis of α-galactosyl epitopes by recombinant α1,3galactosyltransferase for opsonization of human tumor cells vaccines by anti-gal, Cancer Res. **56**: 3069–3074.

Lo, N., and Lau, J.T., 1996, Transcription of the β-galactoside α2,6-sialyltransferase gene in B lymphocytes is directed by a separate and distinct promoter. Glycobiology **6**: 271–279.

Nakamura, K., Brown, A.S., and Compans, R.W., 1980, Glycosylation sites of influenza viral glycoproteins. Tryptic glycopeptides from A/WSN (H0N1) hemagglutinin glycoprotein, Virol. **107**: 208–221.

Nomura, T., Takizawa, M., Aoki, J., Arai, H., Inoue, K., Wakisaka, E., Yoshizuka, N., Imokawa, G., Dohmae, N., Takio, K., Hattori, M., and Matsuo, N., 1998, Purification, cDNA cloning and expression of UDP-Gal: glucosylceramide β-1,4-galactosyltransferase from rat brain, J. Biol. Chem. **273**: 13570–13577.

Ooganah, O.W., Freedman, R.B., Jenkins, N., Patel, K., and Rooney, B.C., 1996, Isolation and characterization of an insect cell line able to perform complex N-linked glycosylation on recombinant glycoproteins, Biotech. **14**: 197–202.

Osman, N., McKenzie, I.F.C., Mouhtouris, E., and Sandrin, M., 1996, Switching amino-terminal cytoplasmic domains of α(1,2)fucosyltransferase and α(1,3)galactosyltransferase alters the expression of H substances and galα(1,3)gal. J. Biol. Chem. **271**: 33105–33109.

Paulson, J.C., Beranek, W.E., and Hill, R.L., 1977, Purification of the sialyltransferase from bovine colostrum by affinity chromatography on CDP-agarose, J. Biol. Chem. **252**: 2356–2362.

Paulson, J.C., and Colley, K.J., 1989, Glycosyltransferases: structure, localization, and control of cell type-specific glycosylation, J. Biol. Chem. **264**: 17615–17618.

Powell, J.T., Jarlfors,U., and Brew, K., 1977, Enzymatic characterization of fat globule membranes from the bovine colostrum and bovine milk, JCB **72**: 617–627.

Roth, J., Taatjees, D.J., Lucocq, J.M., Weinstein, J., and Paulson, J.C., 1985, Demonstration of an extensive *trans*-tubular network continuous with the Golgi apparatus stack that may function in glycosylation, Cell **43**: 287–295.

Saito, H., Gu, J., Nishikawa, Y., Fujii, J., Kohgo, Y., and Taniguchi, N., 1995, Organization of the human N-acetylglucosaminyltransferase V gene, Eur. J. Biochem. **233**: 18–26.

Sanchez-Lopez, R., Nicholson, R., Gesnel, M., Matrisian, L.M., and Breathnach, R., 1988, Structure-function relationships in the collagenase family member transin, J. Biol. Chem. **263:** 11892–11899.

Sarkar, M., Pagny, S., Unligil, U., Joziasse, D.H., Muha, J., Glossl, J., and Schachter, H., 1998, Removal of 106 amino acids from the N-terminus of UDP-GlcNAc: α-3-D-mannoside β-1,2-N-acetylglucosaminyltransferase I does not inactivate the enzyme, Glycoconjugate J. **15**: 193–197.

Sasaki, K., Kurata-Miura, K., Ujita, M., Angata, K., Nakagawa, S., Sekine, S., Nishi, T., and Fukuda, M., 1997, Expression cloning of cDNA encoding a human β-1,3-N-acetylglucosaminyltransferase that is essential for poly-N-acetyllactosamine synthesis, PNAS **94**: 14294–14299.

Van den Eijnden, D.H., Blanken, W.M., Winterwerp, H., and Schiphorst, E.C.M., 1983, Identification and characterization of an UDP-Gal: N-acetyllactosaminide α-1,3-D-galactosyltransferase in calf thymus, Eur. J. Biochem. **134**: 523–530.

Vanhove, B., Goret, F., Soulillou, J., and Pourcel, C., 1997, Porcine $\alpha_{1,3}$-galactosyltransferase: tissue-specific and regulated expression of splicing isoforms, Biochimica et Biophysica Acta **1356**: 1–11.

Weinstein, J., Lee, E.U., McEntee, K., Lai, P., and Paulson, J.C., 1987, Primary structure of β-galactoside α2,6-sialyltransferase; conversion of membrane-bound enzyme to soluble forms by cleavage of the $NH_2$-terminal signal anchor, J. Biol. Chem. **262**: 17735–17743.

Wen, D.X., Svensson, E.C., and Paulson, J.C., 1992, Tissue-specific alternative splicing of the β-galactoside α2,6-sialyltransferase gene, J. Biol. Chem. **267**: 2512–2518.

Weston, B., Nair, R.P., Larsen, R.D., and Lowe, J.B., 1992, Isolation of a novel α1,3GT(1,3)fucosyltransferase gene and molecular comparison to the human Lewis blood group α(1,3/1,4)fucosyltransferase genes. Syntenic, homologous, nonallelic genes encoding enzymes with distinct acceptor substrate specificities, J. Biol. Chem. **267**: 4152–4160.

White, T., Bennett, E.P., Takio, K., Sorenson, T., Bonding, N., and Clausen, H., 1995, Purification and cDNA cloning of a human UDP-N-acetylgalactosamine:polypeptide N-acetylgalactosaminyltransferase, J. Biol. Chem. **270**: 24156–24165.

Wilson, AC, 1985, The molecular basis of evolution, Scientific American **253**: 164–173.

Xu, Z., Vo, L., and Macher, B.A., 1996, Structure-function analysis of human α1,3-fucosyltransferase. Amino acids involved in acceptor substrate specificity, J. Biol. Chem. **271**: 8818–8823.

Yamamoto, F., Marken, J., Tsuji, T., White, T., Clausen, H., and Hakomori, S., 1990, Cloning and characterization of DNA complementary to human UDP-GalNAc: Fucα1→2Galα1 → 3GalNAc transferase (histo-blood group A transferase) mRNA, J. Biol. Chem. **265**: 1146–1151.

Yamamoto, F., and Hakomori, S., 1990b, Sugar-nucleotide donor specificity of histo-blood group A and B transferases is based on amino acid substitutions, J. Biol. Chem. **265**: 19257–19262.

*Chapter 4*

# The Natural Anti-Gal Antibody

Uri Galili, Le Wang, Denise C. LaTemple, and
Marko Z. Radic

## 1. INTRODUCTION

Anti-Gal is the most abundant antibody found in humans. It constitutes 1% of circulating antibodies and interacts specifically with the α-gal epitope (Galα1-3Galβ1-4GlcNAc-R). This chapter describes the studies which have led to the identification of the unique specificity of this antibody, the mechanisms inducing production of this antibody and regulating its affinity, the genes encoding this antibody, and the effects of immune tolerance on anti-Gal specificity. Understanding the various aspects of anti-Gal activity in humans and in mice that lack α-gal epitopes (i.e. knock-out mice for α-galactosyltransferase) provides paradigms for a general understanding of anti-carbohydrate immune response.

## 2. IDENTIFICATION OF ANTI-GAL AND DEMONSTRATION OF ITS CARBOHYDRATE SPECIFICITY

The occurrence of a human IgG antibody with specificity to carbohydrate chains containing terminal galactosyl carbohydrates was first suggested in a study

**Uri Galili, Le Wang, and Marko Z. Radic** Department of Microbiology and Immunology, MCP Hahnemann School of Medicine, Philadelphia, Pennsylvania 19129. **Denise C. LaTemple** Department of Medicine, Division of Hematology and Oncology, University of Pennsylvania, Philadelphia, Pennsylvania 19129.

*Subcellular Biochemistry, Volume 32: α-Gal and Anti-Gal*, edited by Galili and Avila.
Kluwer Academic / Plenum Publishers, New York, 1999.

on *in vivo* binding of antibodies to pathologic red blood cells of patients with β-thalassemia (Galili et al., 1983). In that study, IgG molecules that bind to thalassemia red blood cells *in situ* were readily eluted by incubation with galactose, but not with glucose, fructose, mannose or arabinose. Subsequent studies demonstrated similar IgG molecules on human normal senescent red blood cells (Galili et al., 1986a) and on red blood cells of patients with sickle cell anemia (Galili et al., 1986b). Based on these observations, it was of interest to determine whether normal human sera contain such an antibody. For this purpose, normal human sera from individuals of blood type AB (i.e., individuals lacking anti-blood group A or B antibodies) were subjected to affinity chromatography on α-galactosyl columns (e.g., melibiose [Galα1-6Glc]-sepharose) and eluted by a low pH buffer or with free melibiose, followed by dialysis. This procedure resulted in the isolation of an IgG antibody moiety that corresponded to approximately 1% of circulating IgG, i.e., 30–100 μg IgG per ml serum (Galili et al., 1984).

The specificity of anti-Gal for carbohydrate antigens containing terminal α-galactosyl units was first demonstrated by its binding to rabbit red blood cells (Galili et al., 1984). These red cells were used for this purpose because previous studies of Eto et al. (1968) and of Stellner et al. (1973) indicated that the two major glycolipid molecules on these red cells are ceramide trihexoside (CTH) with the terminal Galα1-4Galβ1-4Glc-R structure and ceramide pentahexoside (CPH) with terminal Galα1-3Galβ1-4GlcNAc-R structure (see Table I). Anti-Gal isolated from blood group AB sera readily bound to rabbit red blood cells and this interaction, which was demonstrated by agglutination or rosette formation with K562 cells, could be inhibited by melibiose and α-methylgalactoside, but not by lactose (Galβ1-4Glc), glucose, fructose, mannose, or fucose (Galili et al., 1984). In most blood donors studied, anti-Gal was capable of binding to rabbit red blood cells at dilutions up to 1:500 to 1:2000. In cord blood, maternal anti-Gal IgG molecules are present in titers comparable to those in the maternal blood. Its level decreases to approximately 20% the maximal titer by the age of 6–12 months. Subsequently, autologous anti-Gal is continuously produced throughout life (Galili et al., 1984). Anti-Gal is also found in elderly individuals; however, after the age of 70–75, its activity is lower than in young individuals (Wang et al., 1995a).

An accurate identification of anti-Gal specificity was achieved by immunostaining glycolipids on thin layer chromatography (TLC) plates (Galili et al., 1985). Since the carbohydrate chains of glycolipids have molecular structures that are well defined by a variety of physicochemical methods, the analysis of antibody binding to glycolipids separated on TLC plates can provide accurate information on the specificity of the antibody in question (Magnani et al., 1981). Immunostaining of rabbit red blood cell glycolipids isolated on a TLC plate demonstrated binding of anti-Gal to CPH but not to CTH. These findings implied that anti-Gal displays a very narrow specificity for the carbohydrate structure Galα1-

**Table I**
**Glycolipids Tested for Anti-Gal Binding**

| Name | Source | Structure | Anti-Gal binding |
|---|---|---|---|
| CPH (ceramide penthahexoside) | Rabbit | Galα1→3Galβ1→4GlcNAcβ1→3Galβ1→4Glc1→3Cer | + |
| CHH (ceramide heptahexoside) | Cow | Galα1→3(Galβ1→4GlcNAcβ1→3)$_2$Galβ1→4Glc1→1Cer | + |
| Glucosyl ceramide | Human | Glc1→1Cer | – |
| Lactosyl ceramide | Human | Galβ1→4Glc1→1Cer | – |
| CTH (cermide trihexoside) | Rabbit | Galα1→4Galβ1→4Glc1→1Cer | – |
| Globoside | Human/pig | GalNAcβ1→3Galα1→4Galβ1→4Glc1→1Cer | – |
| Paragloboside | Human | Galβ1→4GlcNAcβ1→3Galβ1→4Glc1→1Cer | – |
| Forsman | Sheep | GalNAcα1→3GalNAcβ1→3Galα1→4Galβ1→4Glc1→1Cer | – |
| $P_1$ | Human | Galα1→4Galβ1→4GlcNAcβ1→3Galβ1→4Glc1→1Cer | – |
| $B_1$ | Human | Galα1→3(Fucα1→2)Galβ1→4GlcNAcβ1→3Galβ1→4Glc1→1Cer | – |
| $GM_3$ | Human | NeuAcα2→3Galβ1→4Glc1→1Cer | – |
| Sialylparagloboside | Human | NeuAcα2→3Galβ1→4GlcNAcβ1→3Galβ1→4Glc1→1Cer | – |

3Galβ1-4GlcNAc-R on glycolipids and it does not interact with closely related structures, such as Galα1-4Galβ1-4Glc-R. These findings were further confirmed by studies in which various glycolipids with structurally defined carbohydrate chains were immunostained (Table I), indicating that anti-Gal exclusively binds to glycolipids with the terminal Galα1-3Galβ1-4GlcNAc-R structure (i.e., α-gal epitopes) and not to other carbohydrate structures (Galili et al., 1987a). The subsequent observations on this anti-Gal interaction with α-gal epitopes on CPH from cow or squirrel monkey red blood cells (Galili et al., 1987b) greatly helped to elucidate the unique pattern of distribution of α-gal epitopes in mammals (see Galili's chapter on "Evolution of α-galactosyltransferase and of the α-gal epitope"). The findings on specific interactions of anti-Gal with α-gal epitopes were supported by observations of antibodies within human serum that bind to α-gal epitopes on glycolipids (Suzuki and Naiki, 1984) and glycoproteins (Towbin et al., 1987; Gabrielli et al., 1991a). Independent studies on the isolation of anti-Gal by affinity chromatography on melibiose-sepharose (Davin et al., 1987) further confirmed that this antibody is present in the serum at high concentrations corresponding to approximately 1% of circulating IgG. The interaction of anti-Gal with various glycolipids is discussed in greater detail in the chapter of Rydberg et al. in this book.

Recent studies analyzing the interaction of anti-Gal with a very large series of glycolipids isolated from various mammalian tissues confirmed the high specificity of anti-Gal for α-gal epitopes, but also demonstrated binding of anti-Gal to a relatively rare glycolipid termed $x_2$ having the structure GalNAcβ1-3Galβ1-4GlcNAcβ1-3Galβ1-4Glc-Cer (Teneberg et al., 1996). The binding of anti-Gal to $x_2$ was 50 fold less effective as the binding to α-gal epitopes on CPH. Nevertheless, this interaction may explain the observations on *in situ* anti-Gal binding to normal senescent red blood cells (Galili et al., 1986a), thalassemic red blood cells (Galili et al., 1983; 1984), or sickle red blood cells (Galili et al., 1986b). The $x_2$ carbohydrate chains which in young red cells are likely to be cryptic and inaccessible to anti-Gal, may become exposed on the cells because of gradual removal of cell surface glycoprotein molecules resulting from "wear and tear" processes in these red blood cell membranes during the course of normal or pathologic senescence (Galili, 1988).

## 3. ANTI-GAL INTERACTION WITH GLYCOPROTEINS

As discussed in other chapters in this book, α-gal epitopes are synthesized on mammalian glycolipids, as well as on mammalian glycoproteins, primarily on N-linked carbohydrate chains. The binding of purified anti-Gal to α-gal epitopes on such carbohydrate chains was readily demonstrated with two glycoproteins that were previously shown to have this epitope. These include mouse laminin, estimated to have 50–70 α-gal epitopes per molecule (Shibata et al., 1982; Arumugham et al., 1986) and bovine thyroglobulin, found to have 11 α-gal epitopes per molecule (Spiro and Bhoyroo, 1984). Anti-Gal readily bound to mouse laminin and bovine thyroglobulin in ELISA but it did not bind to these glycoproteins when treated with α-galactosidase, an enzyme which removes the terminal galactosyl unit from α-gal epitopes (Galili, 1993a). Specific binding of anti-Gal to α-gal epitopes on mouse laminin and bovine thyroglobulin has also been observed by other investigators (Towbin et al., 1987; Satake et al., 1994; Parker et al., 1994). As expected, anti-Gal did not bind to human thyroglobulin or human laminin, which as all other human glycoconjugates, completely lack α-gal epitopes (Spiro and Bhoyroo, 1984; Thall and Galili, 1990; Galili, 1993a). We have recently further demonstrated this specificity of anti-Gal by measuring its binding to fetuin. This glycoprotein from fetal bovine serum has three N-linked carbohydrate chains which are sialylated. Anti-Gal does not bind to fetuin or to desialylated fetuin. However, anti-Gal readily binds to desialylated fetuin on which α-gal epitopes were *de novo* synthesized by recombinant α1,3galactosyltransferase (α1,3GT) (LaTemple et al., 1996).

It is of interest to note that by analysis of anti-Gal interaction with peptide libraries, this antibody was reported to interact with a peptide sequence that is also found in mucine molecules on malignant cells such, such as mammary carcinoma (Sandrin et al., 1997; see the chapter of Sandrin and McKenzie in this book). These observations can be associated with previous reports of Castronovo et al. (1989) who observed binding of anti-Gal to mammary carcinoma cells in histologic sections. The significance of this interaction *in vivo* remains to be elucidated since there are no indications that anti-Gal facilitates *in vivo* destruction of tumor cells expressing such mucine molecules.

## 4. ANTI-GAL INTERACTION WITH SYNTHETIC CARBOHYDRATES

The narrow specificity of anti-Gal for α-gal epitopes was also demonstrated by interaction with synthetic oligosaccharides. These oligosaccharides were either linked to silica beads or to bovine serum albumin. The interaction of anti-Gal with synthetic carbohydrate chains linked to silica beads (Synsorb beads) was evaluated by immunoadsorption studies. Normal human sera were passed through Synsorb columns with various carbohydrate chains and anti-Gal activity was determined in the effluents and in low pH eluates from the washed columns. The antibody activity was determined by agglutination of rabbit red blood cells, using rabbit anti-human IgG as a secondary antibody (Galili, 1993). Adsorption of human serum on Synsorb columns with synthetic α-gal epitopes resulted in a decrease of more than 90% in agglutinating activity, whereas other synthetic carbohydrates did not alter the agglutination titer. Accordingly, only the eluate from the column with α-gal epitopes displayed hemagglutinating activity with rabbit red blood cells (Table II). The efficacy of α-gal epitopes in removal of anti-Gal from monkey blood was further demonstrated in recent studies aimed to prevent anti-Gal mediated rejection of xenografts (Xu et al., 1998; Kozlowski et al., 1998). A similar restricted specificity of anti-Gal to synthetic α-gal epitopes was also demonstrated by Weislander et al. (1990) and Gabrielli et al. (1991a; 1991b) who showed that anti-Gal interacts with synthetic Galα1-3Gal structures linked to albumin and to no other synthetic carbohydrate structure. Further analysis of anti-Gal with free synthetic oligosaccharides have indicated that the trisaccharide Galα1-3Galβ1-4GlcNAc interacts with the antibody ten fold as well as the disaccharide Galα1-3Gal, implying that the combining site of this antibody is of a size that recognizes at least a trisaccharide (Galili and Matta, 1996).

This very narrow specificity of anti-Gal differentiates this natural anti-carbohydrate antibody from natural antibodies to various proteins. The latter antibodies, which are found in very small concentrations in human sera usually display cross-reactivity and bind to various peptide antigens (Guilbert et al., 1982; Lutz et al., 1984). In contrast, anti-Gal, in spite of its high concentration in the serum, displays high specificity. It is probable that this specificity is associated with the lack of ionic bonds between anti-Gal and the α-gal epitope since there are no electrostatical charges on this epitope. Carbohydrate epitopes which lack sialic acid or sulfated groups, usually lack electrostatical charges; therefore, they interact with antibodies only via hydrogen bonds, van der Walls forces, and hydrophobic bonds. Such interactions, although of relatively low affinity, require precise

**Table II**
**The Specific Interaction of the Natural Anti-Gal with Synthetic Oligosaccharides on Synsorb Beads[a]**

| | Anti-Gal titer | |
|---|---|---|
| Synsorb oligosaccharides | Effluent | Eluate |
| Synsorb (no oligosaccharide) | 1:1280 | <1:10 |
| Galα1→3Galβ1→4GlcNAc-R (α-gal epitope) | 1:20 | 1:1280 |
| Galα1→4Galβ1→4GlcNAc-R (P1 antigen) | 1:1280 | <1:10 |
| Galβ1→4GlcNAc-R (N-acetyllactosamine) | 1:1280 | <1:10 |
| Galβ1→4GlcNAcα1→2Man-R (N-acetyllactosamine-mannosyl) | 1:1280 | <1:10 |
| GlcNAc→R (N-acetylglucosamine) | 1:1280 | <1:10 |
| Galβ1→3GalNAc-R (T antigen) | 1:1280 | <1:10 |
| Fucα1→2Galβ1→4GlcNAc-R (blood group H antigen) | 1:1280 | <1:10 |

[a]Human serum was chromatographed on different Synsorb columns. The columns were subsequently washed extensively with PBS and bound antibodies eluted with glycine-HCl buffer, pH 2.6. Anti-Gal titers were determined in the effluents and the eluates by rabbit red cell agglutination with rabbit anti-human Ig as a secondary antibody (from Galili, 1993, with permission).

"fitting" between the combining site of the antibody and the antigen to enable them to come close enough for water exclusion and bond formation (Kabat, 1976). In contrast, antibodies interacting with peptide epitopes also use ionic bonds between opposing charges on the antigen and antibody molecules. This most likely causes nonspecific interactions between arrays of opposing charges leading the cross-reactivity observed with natural anti-peptide antibodies.

## 5. VARIATIONS IN AFFINITY OF ANTI-GAL

Analysis of anti-Gal activity in healthy populations demonstrated significant differences from one individual to another in the ability of anti-Gal to bind to α-gal epitopes on solid phase antigens in ELISA (Wang et al., 1995a; Galili and Matta, 1996). However, the concentration of anti-Gal in most serum samples was found to be within the order of 1% of IgG. This raised the possibility that the differences in activity of serum anti-Gal may be partly associated with differences in the affinity of this antibody in various individuals. This is shown in Figure 1, where affinity-purified anti-Gal preparations from two individuals were tested for binding to α-gal epitopes linked to BSA solid phase antigen. Anti-Gal from donor #1 was 30 fold more effective in binding to α-gal epitopes than anti-Gal from donor #2, suggesting a significant difference in affinity of the two antibodies (Galili and Matta, 1996).

Direct analysis of anti-Gal affinity was performed in equilibrium dialysis assays with a radiolabelled free haptenic form of α-gal epitopes, i.e. the trisaccharide [$^3$H]Galα1-3Galβ1-4GlcNAc. This trisaccharide is formed by the following reaction catalyzed by recombinant α1,3GT:

$$\underset{\text{N-acetyllactosamine}}{\text{Gal}\beta\text{1-4GlcNAc}} + \text{UDP-}[^3\text{H}]\text{Gal} \xrightarrow{\text{r}\alpha\text{1,3GT}} \underset{\text{free }\alpha\text{-gal epitope}}{[^3\text{H}]\text{Gal}\alpha\text{1-3Gal}\beta\text{1-4GlcNAc}} + \text{UDP}$$

A dialysis bag containing anti-Gal and the labeled α-gal epitope hapten was immersed in an equal volume of phosphate buffered saline (PBS) within a tube. After 72 h incubation at 4°C, the concentration (i.e., radioactivity) of the hapten was measured inside and outside the dialysis bag. The affinity of anti-Gal was calculated as:

$$\text{Ka} = \frac{[\text{Hapten : Ab complex}]}{[\text{unbound hapten}]\,[\text{Ab}]} = \frac{[\text{inside cpm} - \text{outside cpm}]}{[\text{outside cpm}]\,[\text{anit-Gal}]}$$

Based on these calculations, the affinity anti-Gal from donor #1 was $6\text{x}10^6\text{M}^{-1}$ (Table III), whereas that of anti-Gal from donor #2 was 30-fold lower,

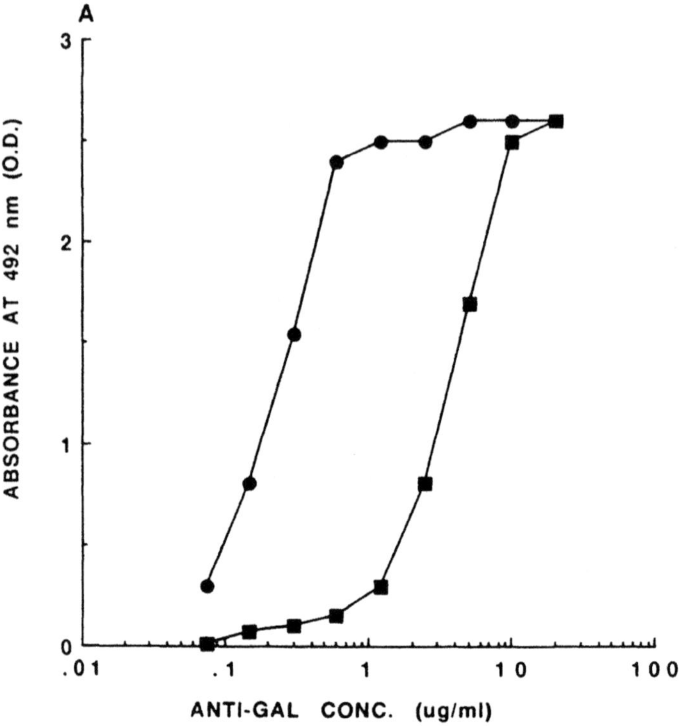


**Figure 1.** Differential binding of anti-Gal from two healthy individuals to α-gal epitopes in ELISA. Anti-Gal from donor #1 (●); anti-Gal from donor #2 (■). Note that anti-Gal of donor #1 is 30 fold as effective in binding to α-gal epitopes as that of donor #2. HRP-conjugated rabbit anti-human IgG was used as secondary antibody (from Galili and Matta, 1996 with permission).

$2x10^5M^{-1}$. These differences correlate with differences in the activity of the two antibody preparations by ELISA. Anti-Gal affinity in most human sera tested ranged from $2\times10^5$ to $6\times10^6M^{-1}$ (Wang et al., 1995). These affinities are much lower than those observed with antibodies to peptides and proteins because there are no ionic bonds between anti-Gal and α-gal epitopes. Nevertheless, as indicated below, this affinity can increase several fold following the immune response against α-gal epitopes on xenografts.

The technique of equilibrium dialysis also provided insight into the interaction pattern of anti-Gal with α-gal epitopes. The affinity of the antibody for the disaccharide [$^3$H]Galα1-3Gal was eight fold lower than that to the trisaccharide [$^3$H]Galα1-3Galβ1-4GlcNAc (Table III; Galili and Matta, 1996). This implies that the "pocket" within the combining site of anti-Gal is of a size that complements the three sugar units of the α-gal epitope. Moreover, the presence of the small N-acetyl group on the glucose at the reducing end affects the affinity, as indicated by the

**Table III**
**Affinities of Isolated Anti-Gal and Serum Anti-Gal to Radiolabelled Oligosaccharides as Measured in Equilibrium Dialysis**

| Source of anti-Gal | Oligosaccharide | Affinity |
|---|---|---|
| 1. Anti-Gal 1 | [$^3$H]Galα1-3Galβ1-4GlcNAc (free α-gal epitope) | $6\times10^6$ M$^{-1}$ |
| 2. Anti-Gal 1 | [$^3$H]Galα1-3Galβ | $9\times10^5$ M$^{-1}$ |
| 3. Anti-Gal 1 | [$^3$H]Galα1-3Galβ1-4Glc | $4\times10^6$ M$^{-1}$ |
| 4. Anti-Gal 2 | [$^3$H]Galα1-3Galβ1-4GlcNAc | $2\times10^5$ M$^{-1}$ |

(From Galili and Matta, 1996, with permission.)

30% lower affinity to [$^3$H]Galα1-3Galβ1-4Glc in comparison to that of [$^3$H]Galα1-3Galβ1-4GlcNAc (Table III). Furthermore, as shown below, anti-Gal is a polyclonal antibody and the combining site of the antibody differs in various clones. In some clones it may combine with the fucosylated form of the α-gal epitope, i.e., blood group B antigen, whereas other clones within the same individual, do not bind to blood group B antigen but interact exclusively with the α-gal epitope.

## 6. MOST OF HUMAN ANTI-BLOOD GROUP B ANTIBODIES HAVE ANTI-GAL SPECIFICITY

This section demonstrates the association between the blood type of various individuals and the specificity of anti-Gal they produce. The notion that anti-Gal from blood group A or O individuals can interact with the blood group B antigen is important for preventing false positive results on the interaction between anti-Gal and human cells. Blood group B antigen is very similar in structure to the α-gal epitope. The only difference between the two structures is the fucose linked to the penultimate galactose in blood group B antigen (Figure 2). It should be stressed that the linking of the terminal galactose in the two structures is catalyzed by two different glycosyltransferases, the α1,3galactosyltransferase (α1,3GT) for the α-gal epitope and the B transferase for blood group B antigen (see the chapters of Joziasse et al. and of Sandrin and McKenzie).

Immunostaining of glycolipids with purified anti-Gal from individuals with blood group B or AB demonstrated binding to α-gal epitopes, but not to blood group B epitope (Figure 3). This is to be expected since blood group B is a self antigen in these individuals. The binding of this anti-Gal to blood group B is prevented by the fucose linked to the penultimate galactose (Figure 2). This is supported by the finding that removal of the fucose from blood group B antigen resulted in the formation of α-gal epitope which readily bound this anti-Gal (Figure 3) (Galili et al., 1987a).

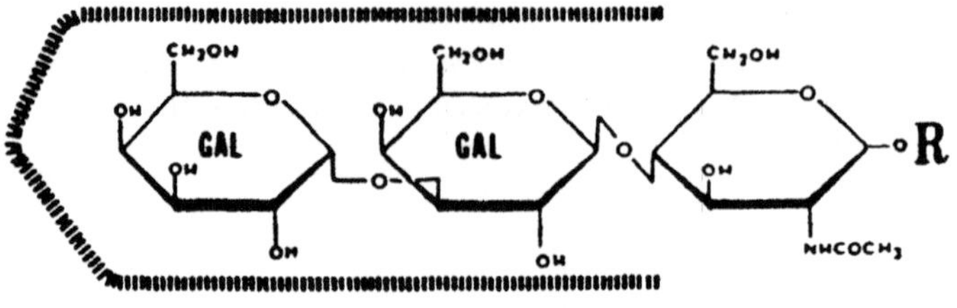


α-gal epitope
Galα1-3Galβ1-4GlcNAc-R

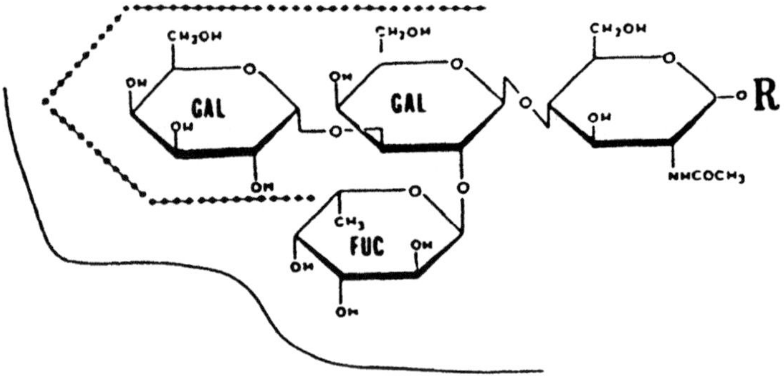


blood group B antigen
Galα1-3(Fucα1-2)Galβ1-4GlcNAc-R

**Figure 2.** Representation of the interaction of anti-Gal (hatched line), anti-Gal B (dotted line), and pure anti-B (solid line) with α-gal epitopes on rabbit or New World monkey red cell and with Galα1-3(Fucα1–2)Gal residue on human B-type red cell. Anti-Gal is present in all human sera. Anti-Gal B and pure anti-B are found only in individuals with blood types A and O. Anti-Gal interacts with α-gal epitopes, whereas anti-Gal B binds to α-gal epitope, with or without the fucosyl residue linked to the penultimate galactosyl (Galili et al., 1987a with permission).

Anti-Gal isolated from sera of individuals of blood group A or O was less restricted in its specificity and could bind to α-gal epitopes as well as to blood group B antigen (Figure 3). This implied that anti-Gal and anti-B antibodies have an overlapping specificity in blood group A and O individuals. To determine the extent of overlap, sera from blood group A or O individuals were adsorbed on an α-gal epitope column and the changes in anti-B activity were determined by hemagglutination assays with blood group B red cells. Approximately 90% of anti-blood group B antibodies were adsorbed on α-gal epitope columns and only 10% of the so called anti-B antibodies interacted exclusively with blood group B antigen and did not bind to α-gal epitopes (Table IV) (Galili et al. 1987a, Galili 1988b). In accord with these data, anti-blood group B antibodies isolated by ad-

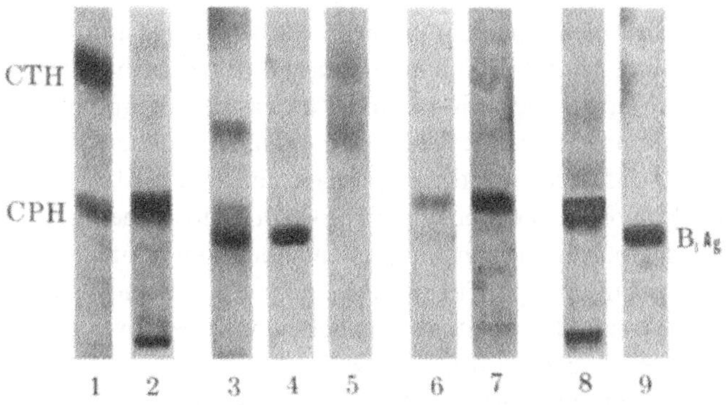


**Figure 3.** Thin-layer chromatogram of red cell glycosphingolipids immunostained with human anti-Gal and anti-B. Rabbit glycolipids stained nonspecifically with orcinol (Lane 1); rabbit glycolipids immunostained with anti-Gal from AB serum (Lane 2); human $B_1$ glycolipid from type B red cells stained with orcinol (Lane 3); as in Lane 3, immunostained with purified anti-B antibodies (Lane 4); as in Lane 3, immunostained with anti-Gal from AB serum (Lane 5); defucosylated human $B_1$ glycolipid stained with orcinol (Lane 6); as in Lane 6, immunostained with anti-Gal from AB serum (Lane 7); rabbit glycolipids as in lane 1, immunostained with anti-Gal from A serum (Lane 8); human $B_1$ glycosphingolipid as in lane 3, immunostained with anti-Gal from A serum (Lane 9) (from Galili et al., 1987a with permission).

sorption of blood type A or O sera on B type red blood cells followed by elution in low pH, readily bound to glycolipids with α-gal epitopes as well as to glycolipids with blood group B antigen (Galili et al. 1987a; Galili 1988b).

These observations suggested that anti-Gal, being a polyclonal antibody, can recognize multiple "facets" of the α-gal epitope (Figure 2). In blood group A and O individuals, approximately half of the anti-Gal antibody clones can interact with α-gal epitopes, regardless of the presence of a fucosyl linked to the penultimate galactose. Therefore, these clones can both interact with α-gal epitopes and blood group B antigen (Table IV). The finding that 90% of anti-blood group B antibody activity can be removed by adsorption on α-gal epitopes implies that most of

**Table IV**
**Specificities of Anti-Gal and Anti-B Antibodies to Carbohydrate Antigens**

| Antibody source | α-Gal epitope Galα1-3Galβ1-4GlcNAc-R | Blood group B antigen Galα1-3(Fucα1-2) Galβ1-4GlcNAc-R |
|---|---|---|
| Anti-Gal from blood group B or AB individuals | + | − |
| Anti-Gal from blood group A or O individuals | + | + |
| Human anti-B from blood group A individuals | + | + |
| Rabbit anti-human blood group B | − | + |

the so-called "anti-B antibodies" are, in fact, anti-Gal antibodies that also can bind to blood group B antigen. These antibodies were designated anti-Gal B antibodies (Galili et al. 1987a). In contrast, in blood group B or AB individuals, where blood B antigen is a self antigen, no anti-Gal clones are capable of binding to blood group B because of immune tolerance. These individuals produce anti-Gal antibodies only against the nonfucosylated α-gal epitopes, i.e., the Galα1-3Galβ1-4GlcNAc-R epitope (see Figure 2). The fucosyl unit linked α1,2 to the penultimate galactose forms the blood group B antigen and prevents the binding of anti-Gal from B or AB individuals to this epitope. The presence of anti-Gal B antibodies in the serum of blood group A and O individuals has been confirmed by other investigators (Parker et al., 1996; McMorrow et al., 1997).

These observations explain a 74 year old question posed by Landsteiner and Miller (1925) on the identity of the "B like" antigen on red cells of New World monkeys. In their seminal studies on blood group antigens in primates, these investigators observed that anti-B antibodies are, in fact, heterogenous antibodies recognizing more than one antigen. They found that New World monkey red cells are agglutinated by human anti-B antibodies, but these red cells do not bind anti-B antibodies from sera of rabbits immunized with human blood group B red cells. Therefore, Landsteiner and Miller suggested that New World monkey red cells express an antigenic structure that closely resembles blood group B antigen, but differs from it since it does not bind rabbit anti-B antibodies. They designated this antigen as the "B-like" antigen and indicated that it is present on New World monkey red cells, but is absent from human, ape, and Old World monkey red cells (Landsteiner and Miller, 1925; Soccha and Ruffie, 1983).

Our studies on anti-Gal B antibodies described above (Galili et al. 1987a) indicate that the B-like antigen is, in fact, the α-gal epitope, which is abundant on New World monkey red cells and on rabbit red cells (Galili et al. 1987b). Since most of anti-B antibodies are anti-Gal B, these antibodies can agglutinate human blood group B and New World monkey red cells (Table IV). However, anti-B antibodies produced in rabbit after immunization with human blood group B red cells can interact exclusively with the fucosylated form of the α-gal epitope (i.e., blood group B antigen). Rabbit anti-B antibodies do not bind to the α-gal epitope since the α-gal epitope is an autologous structure in rabbits and they are immunotolerant to it. Therefore, the rabbit anti-B antibodies can interact with blood group B red cells but not with New World monkey red cells, which lack blood group B antigen but express the α-gal epitope (Figure 3, Table IV). All these data imply that anti-Gal from blood group A or O individual can bind to blood group B antigens on various human cells. Based on such interaction, interpretation of α-gal epitopes on the human cells tested would be false. To prevent such results, the human anti-Gal used should be of blood group B or AB donor. Alternatively the isolated anti-Gal antibody preparation should be adsorbed on blood group B red cells in order to remove the anti-Gal B antibody activity.

## 7. ANTI-GAL CLASSES, SUBCLASSES, AND ISOTYPE SWITCH

Analysis of isotypes of anti-Gal isolated on affinity chromatography columns demonstrated that most of this natural antibody is of the IgG class (Figure 4) (Galili et al., 1995). These IgG molecules can bind to various cells that express α-gal epitopes, and opsonize them for phagocytosis by macrophages (LaTemple et al., 1996; Galili and LaTemple, 1997) or induce the destruction of the cells by antibody dependent cell cytotoxicity (ADCC) with macrophages and granulocytes (Galili, 1993b). Anti-Gal has, however, an important IgM component which is not readily isolated by affinity chromatography but which activates complement and induces cytolysis of cells expressing α-gal epitopes (Good et al., 1992; Sandrin et al., 1993). This interaction, followed by complement activation, induces in humans or Old World monkeys the hyperacute rejection of xenografts from species producing α-gal epitopes (see the chapters of Kobayashi and Cooper; Sandrin and McKenzie; and of Pearce and D'Apice).

Analysis of anti-Gal classes binding to α-gal epitopes in ELISA have shown that there are usually more IgG than IgM anti-Gal molecules, whereas the IgA class of this antibody is the least prevalent in the circulation (Hamadeh et al., 1995; Galili et al., 1995). Parker et al. (1994) reported that anti-Gal IgM constitutes approximately 10% of circulating IgM antibodies. Anti-Gal IgA molecules were found in large amounts in secretory fluids such as milk, colostrum, saliva, and bile (Hamadeh et al., 1995). With the exception of saliva, which contained exclusively

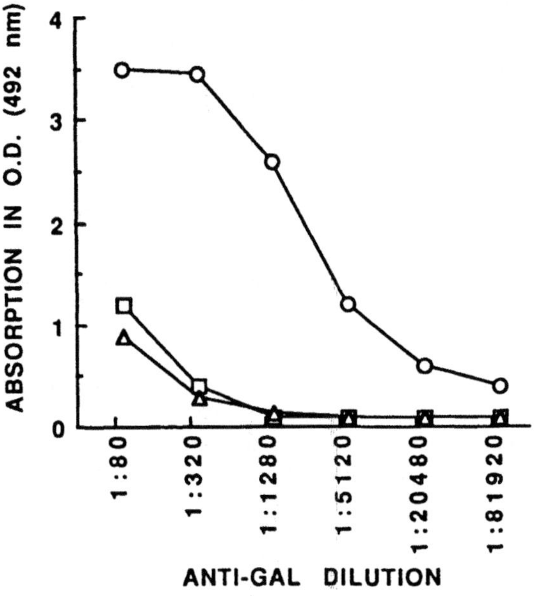


**Figure 4.** The different isotypes in anti-Gal isolated from pooled human sera, as tested by ELISA with α-gal epitopes linked to BSA as solid phase antigen. Isotype specific rabbit anti-human secondary antibodies were conjugated to horseradish peroxidase (HRP). (○) anti-Gal IgG; (□) anti-Gal IgM; (Δ) anti-Gal IgA.

anti-Gal IgA, other secretions also contained IgG and IgM isotypes of this antibody (Hamadeh et al., 1995). Studies on anti-Gal IgA have shown that in Henoch Schönlein purpurea there is a major increase in titer of anti-Gal IgA antibody within the serum (Davin et al., 1987). These IgA molecules may have an important protective role against bacteria in the gastrointestinal tract; however, their pathologic role in Henoch Schönlein purpurea is not clear.

Analysis of anti-Gal IgG in human sera demonstrated the presence of all four subclasses, with anti-Gal $IgG_1$ being the most prevalent, followed by $IgG_2$, $IgG_3$, and $IgG_4$, respectively (Ravindran et al., 1988; Gabrielli et al., 1991). It was further reported that in patients with malaria there is increased production of anti-Gal $IgG_3$, which become the most prevalent subclass in patients with this disease

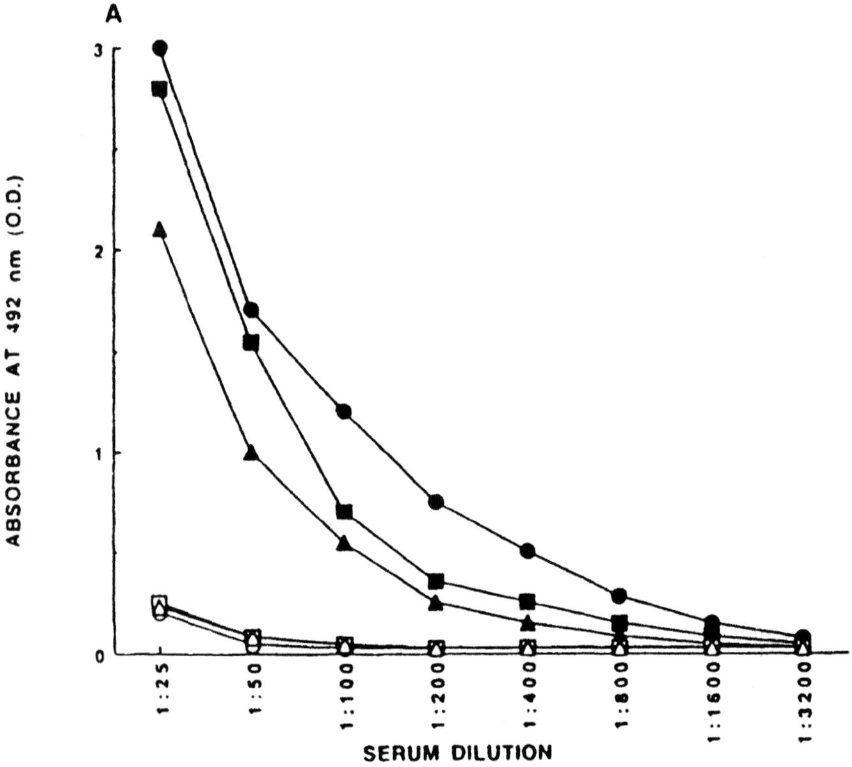


**Figure 5.** Anti-Gal activity in newborn baboons as assayed by ELISA with synthetic α-gal epitopes linked to BSA as solid-phase antigen. (A) Baboon without a transplant: (○) day 30; (□) day 44; (Δ) day 54. (B) Baboon with transplant of pig heart (○) pretransplantation, day 30; (□) day of rejection, day 38; (Δ) six days post-rejection, day 42. Open symbols—anti-Gal IgM; closed symbols—anti-Gal IgG (from Galili et al., 1997 with permission).

(Ravindran et al., 1988). Recent studies on anti-Gal IgG subclasses reported that $IgG_{2a}$ rather than $IgG_1$, is the most prevalent subclass in humans (Parker et al., 1994). It is not clear, as yet, whether the differences reported in various studies result from differences between reagents, different methods of measurement, or differences in the subclasses of this antibody in various individuals.

Anti-Gal undergoes isotype switch according to the general pattern of antibody response. This was demonstrated in studies on anti-Gal production in newborn baboons (i.e., 26–30 days old) that were heterotopically transplanted with baby pig hearts (Galili et al., 1997b). As shown in Figure 5, pre-transplantation serum contained anti-Gal IgG from the maternal source, and no anti-Gal IgM molecules. Porcine hearts transplanted into baby baboons were rejected 4 to 6 days post-transplantation (Minanov et al., 1997). At the time of rejection, no anti-Gal activity was detected in the serum, since the antibody was adsorbed on the graft. Six days post-rejection of the graft, distinct anti-Gal IgM and IgG antibodies were

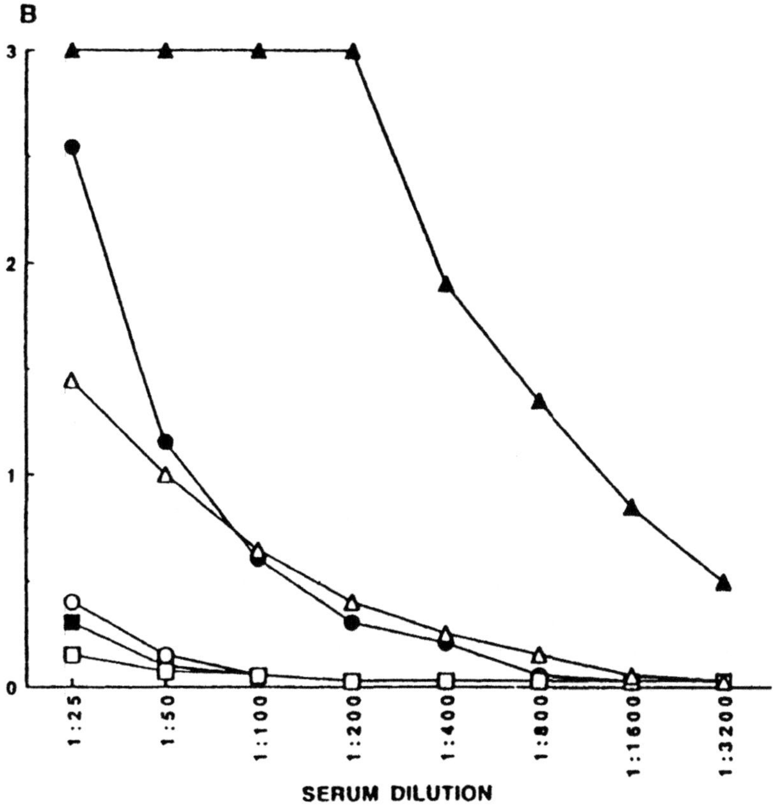


**Figure 5.** (*Continued*)

detected. After an additional week, anti-Gal IgG activity increased, whereas that of the IgM isotype decreased. These findings suggest that in course of xenograft rejection by maternal anti-Gal IgG molecules, the immune system of the neonatal baboon responds to α-gal epitopes on the xenografts, first producing anti-Gal IgM and subsequently anti-Gal IgG antibody molecules. Since isotype switching has been considered to be a T cell dependent process, it is possible that anti-Gal IgG antibody production is associated with T cell activity and that in the absence of T cells, no IgG or IgA isotypes of this antibody will be produced, and only the IgM isotype might be present.

## 8. PHYSIOLOGIC PRODUCTION OF ANTI-GAL

The continuous production of anti-Gal throughout life suggests that this antibody is produced in response to a chronic antigenic stimulation. It has been argued by Weiner (1951) that natural antibodies, such as anti-blood group antibodies are likely to be produced in response to bacterial antigens within the normal gastrointestinal flora. As a result of the understanding of the carbohydrate structure of blood group antigens (Kabat, 1976; Watkins, 1966), it became apparent that natural anti-blood group antibodies interact with a variety of carbohydrate structures shared between bacteria and some human blood group antigens (Springer, 1971). Furthermore, ingestion or inhalation of *E. coli* strain 086 was reported to increase anti-blood group B antibody activity in the individuals studied (Springer and Horton, 1969). All these observations raised the possibility that anti-Gal is also produced in response to antigenic stimulation by bacteria of the gastrointestinal flora. To analyze this possibility, the interaction of anti-Gal with various gastrointestinal bacteria was assayed by direct immunostaining and by ELISA. Anti-Gal readily bound to a variety of *E. coli*, *Klebsiella*, *Serratia*, and *Salmonella* isolates from normal stool (Galili et al., 1988). Furthermore, anti-Gal was found to interact with carbohydrate epitopes on bacterial lipopolysaccharides (LPS) molecules (Galili et al., 1988). Western blots of LPS from various bacteria were immunostained with anti-Gal isolated from normal human serum (Figure 6). The antibody readily bound to LPS molecules of *Klebsiella pneumonia*, *Serratia marcescens*, *Salmonella minnesota,* and *E. coli*. The ladder pattern of the bands is due to the incremental increase in size of the LPS molecules by the repeating carbohydrate units. These observations suggest that bacterial LPS serves as immunogen for the production of anti-Gal. All these findings are in accord with previous studies that have identified oligosaccharides with various α-galactosyl terminal structures on LPS of bacteria, such as *Salmonella* (Luderitz et al., 1965), *Klebsiella* (Björndal et al., 1971; Curval et al., 1973), *E. coli* (Jann and Jann, 1984; Jansson et al., 1981) and *Clostridia* (Gerwig et al., 1989). All these bacterial epitopes were reported to contain Galα1-terminal structures linked to various penultimate sugar units. Although

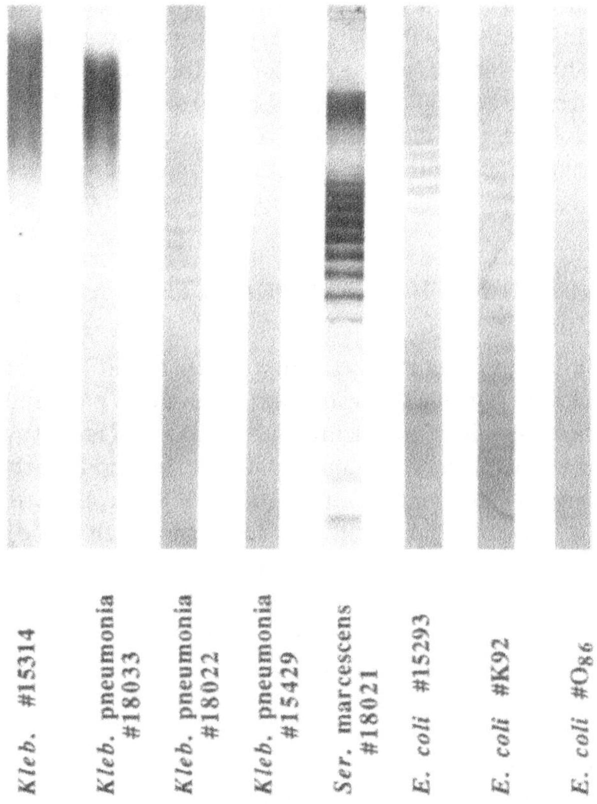


**Figure 6.** Binding of anti-Gal to LPS from various bacteria separated by gel electrophoresis. Anti-Gal binding in the western blot was detected by a secondary rabbit anti-human IgG antibody conjugated to alkaline-phosphatase. No immunostaining was observed with the secondary antibody alone (from Galili et al., 1988 with permission).

the exact bacterial structures that stimulate the human immune system to produce anti-Gal have not been identified as yet, it is likely that anti-Gal is produced as part of the immune response to a variety of bacterial antigens containing α-galactosyl structures.

In view of the high reactivity of anti-Gal in serum, milk, and saliva, it is possible that this antibody contributes to protection against some bacteria of the normal flora. Nevertheless, many bacteria in blood isolates were also found to interact with anti-Gal *in vitro* (Galili et al., 1988; Hamadeh et al., 1992). This suggests that some bacteria have developed mechanisms that protect them from anti-Gal. An example for such a mechanism has been demonstrated by Hamadeh et al. (1992) who showed that the binding of anti-Gal to LPS of a *Serratia marcescens* strain isolated from blood conferred increased protection from lysis by the alternative pathway of complement. In contrast, stool isolates of *Serratia marcescens* lacked this mechanism. The ability of anti-Gal to lyse enveloped viruses expressing α-gal epitopes (see the chapter of Rother and Galili) and protozoa such as *Trypanosoma cruzi* (Almeida et al., 1991) (see the chapter of Avila) suggests that anti-Gal may contribute to the natural protection against a variety of microbial agents.

## 9. ANTI-GAL PRODUCING B LYMPHOCYTES

The finding that anti-Gal constitutes about 1% of circulating immunoglobulins in humans suggested that a significant proportion of human B lymphocytes are capable of producing this antibody. To identify these lymphocytes, human peripheral blood B lymphocytes were immortalized by Epstein Barr virus (EBV), grown in microtiter wells, and assayed for production of anti-Gal by measuring agglutination of rabbit red cells. In all individuals tested approximately 1% of circulating B lymphocytes produced anti-Gal upon transformation, whereas only 0.2–0.3% of B lymphocytes produced anti-A or anti-B-antibodies (Galili et al., 1993). Thus, there are approximately four-fold more anti-Gal producing lymphocytes than lymphocytes producing anti-A or anti-B antibodies. Moreover, no anti-A or anti-B producing clones were found in individuals of the corresponding blood type (i.e., blood type A or B, correspondingly) because of immune tolerance to the self-blood group antigen.

Whereas anti-Gal producing B lymphocytes comprise 1% of blood B lymphocytes, the proportion of these lymphocytes in the spleen and lymph nodes is not known, as yet. It is probable, however, that these lymphoid organs contain many anti-Gal producing B cells. This is suggested from the rapid increase of anti-Gal activity in humans and monkeys transplanted with xenografts. B lymphocytes within the spleen and lymph nodes of the graft recipients become activated by α-gal epitopes of the xenograft, within a relatively short period of time (Galili et al., 1995; 1997a).

## 10. GENES ENCODING ANTI-GAL

The isolation of clones of EBV transformed lymphocytes that produce anti-Gal enabled us to determine the identity of Ig genes that encode this antibody. RNA was isolated from nine individual anti-Gal clones that were found to maintain anti-Gal secretions after their expansion. The RNA samples were subjected to reverse transcriptase (RT)-PCR, then the $V_H$ regions (i.e., the variable region of immunoglobulin heavy chain genes) were amplified using a mixture of 5′ immunoglobulin sequence primers and 3′ primers within the upstream domain of the constant region of Cγ or Cμ genes (Wang et al., 1995b).

Sequencing of the PCR products and comparison of the sequences with those of germ line Ig genes have shown that in eight of the nine clones studied the $V_H$ genes belonged the $V_H$ 3 family. These clones displayed sequence similarities among themselves which clustered them within closely related sequence regions in this family. The finding that anti-Gal is encoded by various Ig genes, was consistent with the polyclonality of this antibody on isoelectric focusing gels (Galili et al., 1984). Further comparison between the cloned anti-Gal $V_H$ genes and corresponding germ line genes indicated the presence of somatic mutations both in the frame work and the complementarity defining regions (CDR). These observations suggested that there is a pool of anti-Gal producing B lymphocytes. These lymphocytes contain somatic mutations and produce the polyclonal natural anti-Gal in the gastrointestinal tract in humans. As described below, the concentration and affinity of anti-Gal greatly increase after transplantation of xenografts into humans and monkeys. It is possible that within a short period post-transplantation, B lymphocyte clones containing somatic mutations that confer high affinity to this antibody proliferate within the lymph nodes and spleen and produce increased amounts of high affinity anti-Gal, as observed in recipients of xenografts (Galili et al., 1995; 1997a; 1997b).

The identification of several $V_H$ genes encoding for anti-Gal also explains the observation that anti-Gal is a polyclonal antibody that can recognize different facets of the α-gal epitope. This multi-facet recognition results in production of some clones in blood group A or O individuals that recognize α-gal epitopes exclusively, whereas other clones produced in these individuals can interact both with the α-gal epitope and with the blood group B antigen, as discussed above.

## 11. PRODUCTION OF ANTI-GAL IN XENOGRAFT RECIPIENTS

The physiologic level of anti-Gal is sufficient to cause hyperacute rejection of xenografts. However, this physiologic level, which reflects the immune response to gastrointestinal flora, is only a fraction of the full anti-Gal response observed in the course of chronic xenograft rejection. Elevated anti-Gal production in response to α-gal epitopes on xenografts was first observed in diabetic patients who were transplanted simultaneously with fetal porcine pancreas islets and with an allogeneic kidney (Groth et al., 1994). The pig islet cells were not immediately rejected and could be found in the recipients three weeks post-transplantation. Analysis of anti-Gal IgG activity by ELISA revealed a 20–100 fold increase in the antibody activity in the transplanted patients, 25–50 days post-transplantation (Galili et al., 1995; Satake et al., 1994). It should be stressed that this increase in anti-Gal activity occurred despite massive immunosuppressive regimens administered which were successful in preventing rejection of the kidney allografts in the patients. This implied that activation and expansion of anti-Gal producing lymphocytes by α-gal epitopes on the pig cells are powerful enough to overcome the inhibitory effects of various immunosuppressive drugs including glucocorticoids, cyclosporin A, deoxypergualin and anti-lymphocytic serum (Groth et al., 1994).

Since total IgG concentrations in the serum of the transplanted patients displayed no significant changes after transplantation, it is probable that the increase in anti-Gal activity may be associated with the appearance of high affinity anti-Gal (Galili et al., 1995). As indicated above, such clones are likely to emerge because of the immune response to α-gal epitopes on the pig islet cells. Analysis of the affinity of anti-Gal in sera of these patients, as performed by equilibrium dialysis with radiolabelled trisaccharide of the α-gal epitope (i.e., [$^3$H]Galα1-3Galβ1-4GlcNAc), demonstrated a 5–100 fold increase in affinity. Thus, the increase in the measured anti-Gal activity in these patients seems to be the result of a combination between increased concentration of anti-Gal and selective production of anti-Gal by B lymphocytes containing genes that encode for high affinity anti-Gal molecules.

We reproduced this phenomenon of production of high affinity anti-Gal in an experimental model of cynomolgus monkeys (Old World monkeys) transplanted with porcine cartilage (Stone et al., 1997; Galili et al., 1997). Monkeys were transplanted with approximately 0.5 gm of porcine meniscus cartilage into the suprapatellar pouch. Within two weeks post-transplantation, anti-Gal titers increased by 20–100 fold, followed by an additional two to four fold increase, one month post-transplantation. Anti-Gal activity remained very high as long as the xenograft was not removed, implying that α-gal epitopes on the xenograft continuously stimulate the production of high affinity anti-Gal. Monkeys transplanted with cartilage, in which α-gal epitopes were destroyed by treatment with α-galactosidase, did not display increased anti-Gal activity post-transplantation (Stone et al., 1998). This indicated that the observed anti-Gal response in the transplanted monkeys resulted from specific reaction against α-gal epitopes on the xenograft, rather than a nonspecific immune response to the xenograft due to polyclonal activation.

## 12. ANTI-GAL PRODUCTION IN KNOCK-OUT MICE FOR α1,3-GALACTOSYLTRANSFERASE

Anti-Gal is exclusively produced by humans, apes, and Old World monkeys, whereas other mammals are immunotolerant to the α-gal epitope since they synthesize this carbohydrate structure as a self antigen (Galili et al., 1987, 1988). We have previously hypothesized that anti-Gal appeared in ancestral apes and Old World monkeys subsequent to the suppression of α-gal epitope expression (Galili et al., 1988; Galili and Andrews, 1995). This raised the question of whether mammals that express α-gal epitopes have genes within their immunoglobulin gene repertoire, that enable the production of anti-Gal in the absence of α-gal epitopes, or whether additional evolution of this repertoire is required for the production of the antibody. This question is of particular interest with regard to the nature of the immunoglobulin gene repertoire. Anti-Gal production would have been detrimental in any mammal expressing α-gal epitopes because of autoimmunity. Therefore, it could be hypothesized that throughout evolution of nonprimate mammals, there might have been a continuous selective pressure for the prevention of anti-Gal production by eliminating immunoglobulin genes that encode this antibody.

The question whether anti-Gal can be produced in an animal synthesizing α-gal epitopes was addressed in knock-out mice for α1,3galactosyltransferase (α1,3GTKO mice) described in detail in the chapters of Thall and of Pearce and D'Apice in this book. These mice produced low amounts of anti-Gal (Thall et al., 1995; 1996; Shinkle et al., 1997). However, anti-Gal activity could be readily increased by immunizing the mice with rabbit red cell membranes since these cells express an abundance of α-gal epitopes (LaTemple and Galili, 1998). Wild type mice immunized with rabbit red cells failed to produce anti-Gal probably because of their immune tolerance to this carbohydrate structure.

Anti-Gal IgG produced by α1,3GTKO mice receiving three immunization with rabbit red cell membranes displayed titers and specificities similar to those of anti-Gal in humans. In addition, the serum of these mice contained anti-Gal IgM, but not anti-Gal IgA molecules. Subclass analysis of anti-Gal IgG in α1,3GTKO mice revealed the occurrence of $IgG_1$, $IgG_{2a}$, $IgG_{2b}$, and $IgG_3$ molecules with this specificity (LaTemple and Galili, 1998). These findings implied that the evolutionary pressure that has prevented anti-Gal production in mice and in their ancestors, did not eliminate from the Ig gene repertoire genes that are capable of encoding this antibody. Once the α-gal epitope was eliminated from these mice, anti-Gal production could ensue. This anti-Gal production by α1,3GTKO mice exemplifies the basic immunological law of Karl Landsteiner which states that "the immune system can produce antibodies against any antigen that is not a self-antigen."

The study of α1,3GTKO mice also enabled us to determine whether neonatal exposure of the immune system to α-gal epitopes results in induction of tolerance. Eight day old α1,3GTKO mice were immunized with large amounts of rabbit red cell membranes or with wild type mouse splenocytes, both expressing many α-gal epitopes. Subsequently, the mice were immunized twice with rabbit red cell membranes, starting at five weeks of age. These mice effectively produced anti-Gal IgG, suggesting that it may be difficult to induce immune tolerance to α-gal epitopes after birth (LaTemple and Galili, 1998). The mechanism preventing anti-Gal production in wild type mice is not clear as yet. This tolerance may be induced in early stages of mouse development *in utero*.

## 13. MONOCLONAL ANTI-GAL ANTIBODIES

### 13.1. Gal-13 Binds to α-Gal Epitopes on Glycolipids

Early attempts to produce monoclonal anti-Gal involved the immunization of Balb/c mice with rabbit red cells and adjuvant, followed by fusion with SP2/0 myeloma cells, and selection of clones which interact with rabbit CPH (i.e., Galα1-3Galβ1-4GlcNAcβ1-3Galβ1-4Glc-Cer) by ELISA. We found only one of the 700 clones tested to be positive (Galili et al., 1987b). This antibody, designated Gal-13, is of the IgG subclass and is highly specific for the α-gal epitope on glycolipids and for synthetic α-gal epitopes, such as synthetic epitopes on silica beads (Synsorb) or epitopes coupled to BSA. However, Gal-13 does not interact with α-gal epitopes on natural glycoproteins (e.g., mouse laminin and bovine thyroglobulin). In addition, Gal-13 does not bind to cells expressing α-gal epitopes on their surface glycoproteins.

Binding to α-gal epitopes on glycolipids, but not to the same epitope on glycoproteins implies that the mannosyl groups within the carbohydrate chain core structure of the glycoprotein prevent the binding of Gal-13. The mannosyl units are absent within glycolipids; therefore, binding is not prevented (Figure 7). This differential binding is likely to reflect the rigorous immune tolerance mechanism that prevents antibody production to self antigens. Gal-13 was produced by a wild type mouse (i.e., a mouse synthesizing α-gal epitopes) and is incapable of binding to self α-gal epitopes abundantly expressed on cell surface glycoproteins. However, the antibody can bind to α-gal epitopes on glycolipid chains, which are usually short, are much less protruding from the cell surface membrane. On the cell surface, these glycolipids usually can not bind Gal-13 because of the many surrounding glycoprotein chains. Since α-gal epitopes on glycolipids are inaccessible to antibody binding, these epitopes are not recognized by the immune system as self-antigens, and the Gal-13 could be produced against it. Thus, Gal-13 is an excellent reagent for detecting α-gal epitopes on glycolipids separated on thin layer chromatography plates (Galili et al., 1987b; Buehler et al., 1987; Teneberg et al., 1996).

### 13.2. M86 Interacts with α-Gal Epitopes on Glycolipids and Glycoproteins

We have recently produced a second monoclonal anti-Gal by fusion of spleen lymphocytes from α1,3GTKO mice with the mouse fusion partner x63AgA-653. We isolated a clone designated M86 that readily bound anti-Gal both on glycolipids and glycoproteins (Figure 7) (Galili et al., 1998). This antibody, which is of the IgM class, also binds to α-gal epitopes produced by hybridoma cells. Thus, the cells grow in clumps because of self-agglutination by M86. Addition of small amounts of complement in diluted normal rabbit serum in-

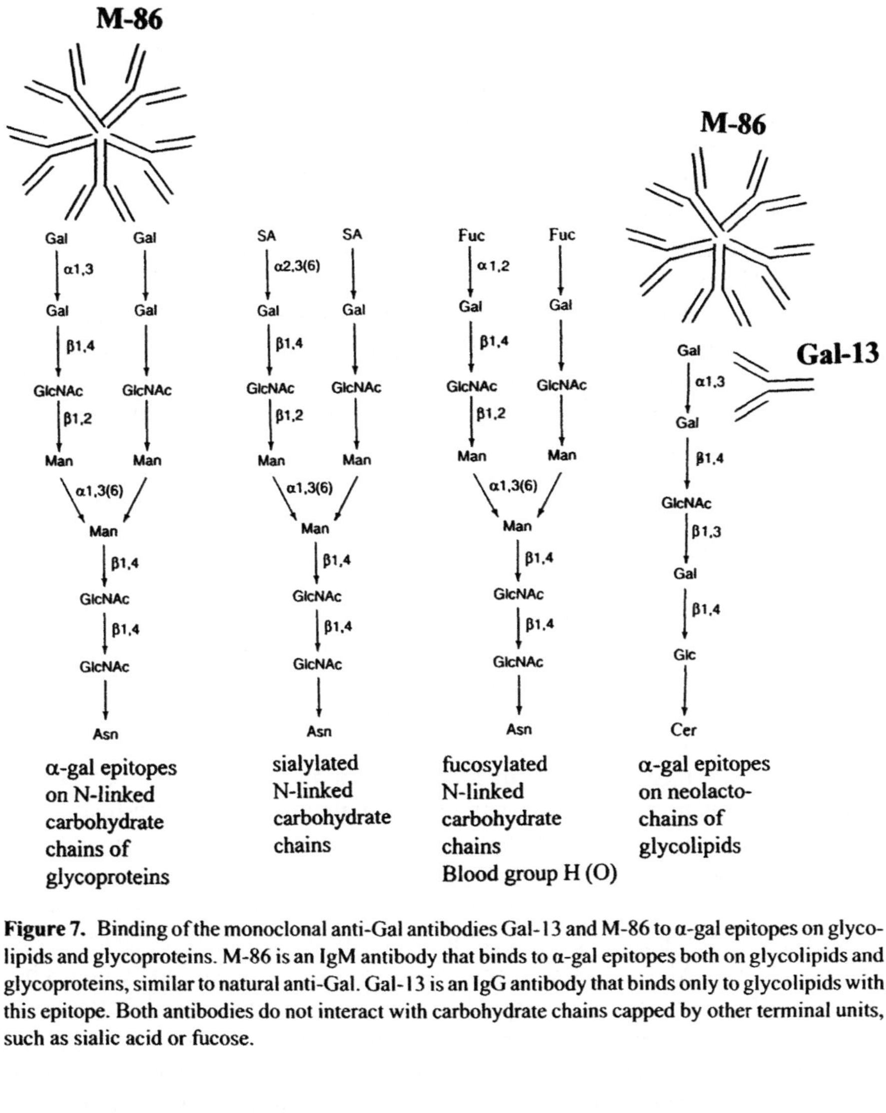


**Figure 7.** Binding of the monoclonal anti-Gal antibodies Gal-13 and M-86 to α-gal epitopes on glycolipids and glycoproteins. M-86 is an IgM antibody that binds to α-gal epitopes both on glycolipids and glycoproteins, similar to natural anti-Gal. Gal-13 is an IgG antibody that binds only to glycolipids with this epitope. Both antibodies do not interact with carbohydrate chains capped by other terminal units, such as sialic acid or fucose.

duces lysis of M86 cells because of complement activation by cell bound monoclonal antibody molecules. Therefore, M86 can not grow *in vivo* because of its complement-mediated autolysis (Galili et al., 1998).

The difference between Gal-13 and M86 is in the ability to interact with α-gal epitopes on glycoproteins. M86 is capable of such interaction because, unlike Gal-13, it is produced by a mouse that was not immunotolerant to α-gal epitopes i.e., α1,3GTKO mouse. These mice lack the ability to synthesize α-gal epitopes, and thus are not restricted in the production of antibodies to α-gal epitopes expressed on glycoproteins. However, since the fusion partner of this hybridoma is derived from wild type mouse origin, it contributes the α1,3GT enzyme that synthesizes α-gal epitopes on the hybridoma cells. Because of this synthesis of α-gal epitopes, the antibodies produced by M86 bind to the cells producing them and in presence of complement, this antibody induces autolysis.

Because M86 binds to α-gal epitopes on cell surface glycoproteins and glycolipids, it can be used to quantitate these epitopes on various cells, as well as on membranes from homogenates of various tissues (Galili et al., 1998; Stone et al., 1998). The assay is based on the overnight interaction between M86 and α-gal epitopes on cells or on membrane homogenates, and assessment of the remaining M86 activity in the supernatant by ELISA after removal of cells or membrane homogenates complexed with M86. There is an inverse correlation between the amount of remaining M86 in the supernatant and the extent of α-gal epitope expression on the cells or on tissue homogenates studied.

## 14. CONCLUSIONS

Anti-Gal is a unique natural antibody, in that it is the most abundant antibody in humans, apes and Old World monkeys and in that it is highly specific for the α-gal epitope. It is present in all classes and subclasses in human serum. Anti-Gal is a polyclonal antibody produced by a number of different Ig genes, mostly of the $V_H3$ family. These genes contain somatic mutations both in the framework and in the CDR regions. As many as 1% of circulating B lymphocytes can produce anti-Gal. Its production seems to result from antigenic stimulation by gastrointestinal bacteria. The specificity of anti-Gal slightly differs in individuals of various blood groups. In blood group A and O individuals, anti-Gal can also interact with blood group B antigen, whereas in individuals of blood group B or AB, anti-Gal interacts exclusively with α-gal epitopes. The affinity of anti-Gal is low and ranges between $2\times10^5$ and $5\times10^6$ $M^{-1}$. This low affinity is because of the lack of ionic bonds between anti-Gal and the α-gal epitope. Despite its low affinity, anti-Gal effectively prevents the transplantation of xenografts by binding to α-gal epitopes on xenograft cells and mediating the destruction of such cells, either by complement activation or by antibody dependent cell cytotoxicity. Furthermore, in individuals who are transplanted with xenografts, anti-Gal activity increases many fold because of the selective expansion of B cell clones producing high affinity anti-Gal. The prevention of anti-Gal binding to α-gal epitopes on xenografts is a prerequisite for progress in the area of xenotransplantation. This will require either elimination of α-gal epitopes from xenografts or induction of immunologic tolerance to it, thus preventing anti-Gal production. Anti-Gal may also have beneficial clinical effects. As discussed in the chapter of LaTemple and Galili, anti-Gal may be exploited for clinical use, such as the enhancement of tumor vaccine immunogenicity.

## REFERENCES

Almeida, I.C., Milani, S.R., Gorin, A.J., and Travassos, L.R., 1991, Complement mediated lysis of Trypanosoma cruzi trypomastigotes by human anti-α-galactosyl antibodies. *J. Immunol.* **146**:2394–2401.

Arumugham, R.G., Hsieh, T.C.Y., Tanzer, M.L., and Laine, R.A., 1986, Structures of the asparagine-linked sugar chains of laminin. *Biochem. Biophys. Acta* **883**:112–126.

Björndal, H., Lindberg, B., and Nimmich, W., 1971, Structural studies on *Klebsiella* O group 1 and 6 lipopolysaccharides. *Acta Chem. Scand.* **25**:750–762.

Buehler, J., Galili, U., and Macher, B.A., 1987, Use of enzyme-linked immunoadsorbent assay to monitor the purification of glycosphingolipid antigens by high-performance liquid chromatography. *Anal. Biochem.* 164:521–525.

Castronovo, V., Parent, B., Steeg, P.S., Colin, C., Foidart, J.M., Lambotte, R., and Mahieu, P., 1989, Human natural anti-Gal antibodies may play a role in the natural anti-tumor defense system. *J. Natl. Cancer Inst.* 81:212–216.

Curval, M., Linberg, B., Löngren, J., Ruden, U., and Nimmich, W., 1973, Structural studies on the *Klebsiella* O group 8 lipopolysaccharides. *Acta Chem. Scand.* 27:4019–4028.

Davin, J.C., Malaise, M., Foidart, J.M., and Mahieu, P., 1987, Anti-α-galactosyl antibodies and immune complexes in children with Henoch-Schönlein purpura of IgA nephropathy. *Kidney Int.* 31:1132–1139.

Eto, T., Iichikawa, Y., Nishimura, K., Ando, S., and Yamakawa, T., 1968, Chemistry of lipids of the posthemolytic residue or stroma of erythrocytes. XVI. Occurance of ceramide pentasaccharide in the membrane of erythrocytes and reticulocytes in rabbit. *J. Biochem. (Tokyo)* 64:205–213.

Gabrielli, A., Candel, M., Ricciatti, A.M., Caniglia, M.L., and Wieslander, J., 1991, Antibodies to mouse laminin in patients with systemic sclerosis (scleroderma) recognize galactose (α1,3)-galactose epitopes. *Clin. Exp. Immunol.* 86:367–373.

Gabrielli, A., Leoni, P., Danielli, G., Herrmann, K., Krieg, T., and Wieslander, J., 1991, Antibodies against galactosyl α(1-3)galactose in connective tissue disease. *Arthritis Rheum.* 34:375–376.

Galili, U., 1988a, The natural anti-Gal antibody, the B-like antigen, and human red cell aging. *Blood Cells* 14:205–220.

Galili, U., 1988b, The two antibody specificities within human anti-blood group B antibodies. *Transfusion Med. Rev.* 2:112–121.

Galili, U., 1993a, Evolution and pathophysiology of the human natural anti-Gal antibody. *Springer Seminars in Immunopathology* 15:155–171.

Galili, U., 1993b, Interaction of the natural anti-Gal antibody with α-galactosyl epitopes: A major obstacle for xenotransplantation in humans. *Immunology Today* 14:480–482.

Galili, U., Anaraki, F., Thall, A., Hill-Black, C., and Radic, M., 1993, One percent of circulating B lymphocytes are capable of producing the natural anti-Gal antibody. *Blood* 82:2485–2493.

Galili, U., Basbaum, C., Shohet, S.B., Buehler, J., and Macher, B.A., 1987b, Identification of erythrocyte Galα1→3Gal glycosphingolipids with a mouse monoclonal antibody Gal-13. *J. Biol. Chem.* 262:4683–4688.

Galili, U., Buehler, J., Shohet, S.B., and Macher, B.A., 1987a, The human natural anti-Gal IgG. III. The subtlety of immune tolerance in man as demonstrated by crossreactivity between natural anti-Gal and anti-B antibodies. *J. Exp. Med.* 165:693–704.

Galili, U., Clark, M.R., and Shohet, S.B., 1986b, Excessive binding of the natural anti-α-galactosyl IgG to sickle red cells may contribute to extravascular cell destruction. *J. Clin. Invest.* 77:27–33.

Galili, U., Clark, M.R., Shohet, S.B., Buehler, J., and Macher, B.A., 1987b, Evolutionary relationship between the anti-Gal antibody and the Galα1→3Gal epitope in primates. *Proc. Natl. Acad. Sci (USA)* 84:1369–1373.

Galili, U., Flechner, I., Kniszinski, A., Danon, D., and Rachmilewitz, E.A., 1986a, The natural anti-α-galactosyl IgG on human normal senescent red blood cells. *Br. J. Haematol.* 62:317–324.

Galili, U., Korkesh, A., Kahane, I., and Rachmilewitz, E.A., 1983. Demonstration of a natural anti-galactosyl IgG antibody on thalassemic red blood cells. *Blood* **61**:1258–1264.

Galili, U. and LaTemple, D.C., 1997, The natural anti-Gal antibody as a universal augmenter of autologous vaccine immunogenicity. *Immunology Today* 18:281–285.

Galili, U., LaTemple, D.C., and Radic, M.Z., 1998, A sensitive assay for measuring α-gal epitope expression on cells by a monoclonal anti-Gal antibody. *Transplantation* 65:1129–1132.

Galili, U., LaTemple, D.C., Walgenbach, A.W., and Stone, K.R., 1997a, Porcine and bovine cartilage transplants in cynomolgus monkey: II. Changes in anti-Gal response during chronic rejection. *Transplantation* 63:646–651.

Galili, U., Macher, B.A., Buehler, J., and Shohet, S.B., 1985, Human natural anti-α-galactosyl IgG. II. The specific recognition of α(1-3)-linked galactose residues. *J. Exp. Med.* 162:573–582.

Galili, U., Mandrell, R.E., Hamahdeh, R.M., Shohet, S.B., and Griffis, J.M., 1988, Interaction between human natural anti-α-galactosyl immunoglobulin G and bacteria of the human flora. *Infect. Immun.* 56:1730–1737.

Galili, U. and Matta, K.L., 1996, Inhibition of anti-Gal IgG binding to porcine endothelial cells by synthetic oligosaccharides. *Transplantation* 62:356–262.

Galili, U., Minanov, O., Michler, R.E., and Stone, K.R., 1997b, High affinity anti-Gal IgG in chronic rejection of xenografts. *Xenotransplantation* **4**:127–131.

Galili, U., Rachmilewitz, E.A., Peleg, A., and Flechner, I., 1984, A unique natural human IgG antibody with anti-α-galactosyl specificity. *J. Exp. Med.* 160:1519–1531.

Galili, U., Tibell, A., Samuelsson, B., Rydberg, L., and Groth, C.G., 1995, Increased anti-Gal activity in diabetic patients transplanted with fetal porcine islet cell clusters. *Transplantation* 59:1549–1556.

Gerwig, G.S., deWaard, P., Kamerling, J.P., Vliegenthart, J.F.G., Morgenstern, E., Lamed, R., and Bayer, E.A., 1989, Novel O-linked carbohydrate chains in the cellulase complex Cellulosome of *Clostridium thermocellum*. *J. Biol. Chem.* 264:1027–1035.

Good, A.H., Cooper, D.C.K., Malcolm, A.J., Ippolito, R.M., Koren, E., Neethling, F.A., Ye, Y., Zuhdi, N., and Lamontage, L.R., 1992, Identification of carbohydrate structures which bind human anti-porcine antibodies: implication for discordant xenografting in man. *Transplant. Proc.* 24:559–562.

Groth, C.G., Korsgren, O., Tibell, A., Tollerman, J., Möller, E., Bolinder, J., Ostman, J., Reinholt, F.P., Hellerstrom, C., and Andersson, A., 1994, Transplantation of fetal porcine pancreas to diabetic patients: biochemical and histological evidence for graft survival. *Lancet* 344:1402–1404.

Guilbert, B., Dighiero, G., and Avrameas, S., 1982, Naturally occurring antibodies against nine common antigens in human sera. I. Detection, isolation and characterization. *J. Immunol.* **128**:2779–2790.

Hamadeh, R.M., Galili, U., Zhou, P., and Griffis, J.M., 1995, Human secretions contain IgA, IgG and IgM anti-Gal (anti-α-galactosyl) antibodies. *Clin. Diagnos. Lab. Immunol.* 2:125–131.

Hamadeh, R.M., Jarvis, G.A., Galili, U., Mandrell, R.E., Zhou, P., and Griffiss, J.M., 1992, Human natural anti-Gal IgG regulates alternative complement pathway activation on bacterial surfaces. *J. Clin. Invest.* 89:1223–1235.

Jann, K., and Jann, B., 1984, Structure and biosynthesis of O-antigens. In: Riestcel, E.T. (ed). *Handbook of endotoxins, vol I.* Chemistry of endotoxins. Elsevier, Amsterdam, p. 138.

Jansson, P.E., Lindberg, A.A., Lindberg, B., and Wollin, R., 1981, Structural studies on the hexose region of the core lipopolysaccharides from Enterobacteriaceae. *Eur. J. Biochem.* 115:571–577.

Kabat, E.A., 1976, *Structural concepts in immunology and immunochemistry (2nd Ed).* Hold, Rinehart & Winston. New York, 174–189.

Kozlowski, T., Ierino, F., Lambrigts, D., Foley, A., Andrews, D., Awwad, M., Monroy, R., Cosimi, A.B., Cooper, D.K.C., and Sachs, D.H., 1998, Depletion of anti-Galα1-3Gal antibody in baboons by specific α-gal immunoaffinity columns. *Xenotransplantation* 5:122–131.

Landsteiner, K., and Philip-Miller, C., 1925, Serological studies on the blood of the primates. III. Distribution of serological factors related to human isoagglutinations in the blood of lower monkeys. *J. Exp. Med.* 42:863–875.

LaTemple, D.C., and Galili, U., 1998, Adult and neonatal anti-Gal response in knock-out mice for α-galactosyltransferase. *Xenotransplantation* 1998, in press.

LaTemple, D.C., Henion, T.R., Anaraki, F., and Galili, U., 1996, Synthesis of α-galactosyl epitopes by recombinant α1,3galactosyltransferase for opsonization of human tumor cell vaccines by anti-Gal. *Cancer Research* 56:3069–3074.

Lüderitz, O., Simmons, D.A.R., and Westphal, O., 1965. The immunochemistry of *Salmonella* chemotype VI O-antigen. The structure of oligosaccharides from *Salmonella* group U 043 lipopolysaccharides. *Biochem. J.* 97:820–832.

Lutz, H.U., Flepp, R., and Stringnaro-Wipf, G., 1984, Naturally occurring antibodies to exoplasmic and cryptic regions of band 3 protein, the major integral protein of human red blood cells. *J. Immunol.* **133**:2610–2618.

Magnani, J.L., Brockhaus, M., Smith, D.F., Ginsburg, V., Blaszczyk, M., Mitchell, K.F., Steplewski, Z., and Koprowski, H., 1981, A monosialoganglioside is a monoclonal antibody-defined antigen of colon carcinoma. *Science* **212**:55–56.

McMorrow, I.M., Comrack, C.A., Nazarey, P.P., Sachs, D.H., and DerSimonian, H., 1997, Relationship between ABO blood group and levels of αGal1,3galactose-reactive human immunoglobulin G. *Transplantation* **64**:546–549.

Minanov, O.P., Itescu, S., Neethling, F.A., Morgenthau, A.S., Kwiatkowski, P., Cooper, D.K.C., and Michler, R.E., 1997, Anti-Gal IgG antibodies in sera of newborn humans and baboons and its significance in pig xenotransplantation. *Transplantation* **63**:182–186.

Parker, W., Bruno, D., Holzknecht, Z.E., and Platt, J.L., 1994, Characterization and affinity isolation of xenoreactive human natural antibodies. *J. Immunol.* **153**:3791–803.

Ravindran, S., Satapathy, A.K., and Das, M.K., 1988, Naturally occurring anti-α-galactosyl antibodies in human *Plasmodium falciparum* infections: a possible role for autoantibodies in malaria. *Immunol. Lett.* **19**:137–142.

Sandrin, M., Vaughan, H.A., Dabkowski, P.L., and McKenzie, I.F.C., 1993, Anti-pig IgM antibodies in human serum react predominantly with Galα1-3Gal epitopes. *Proc. Natl. Acad. Sci. USA* 90:11391–11395.

Sandrin, M.S., Vaughan, H.A., Xing, P-X., and McKenzie, I.F.C., 1997, Natural human anti-Galα(1,3)Gal antibodies react with human mucine peptides. *Glycoconjugates J.* **14**:97–105.

Satake, M., Kawagishi, N., Rydberg, L., Samuelsson, B.E., Tibell, A., Groth, C.G., and Möller, E., 1994, Limited specificity of xenoantibodies in diabetic patients transplanted with fetal porcine islet cell clusters. Main antibody reactivity against α-linked galactose-containing epitopes. *Xenotransplantation* 1:89–101.

Shibata, S., Peters, B.P., Roberts, D.D., Goldstein, T.J., and Liotta, L.A., 1982, Isolation of laminin by affinity chromatography on immobilized *Griffonia simplicifolia* I lectin. *FEBS Lett.* 142:194–197.

Shinkel, T.A., Chen, C.G., Salvaris, E., Henion, T.R., Barlow, H., Galili, U., Pearse, M.J., and d'Apice, A.J., 1997, Changes in cell surface glycosylation in α-galactosyltransferase knock-out and α1,2-fucosyltransferase transgenic mice. *Transplantation* 95:574–579.

Socha, W.W., and Ruffie, J., 1983, Blood groups of primates: Theory, practice, evolutionary meaning. Liss. New York, 39–51.

Spiro, R.G., and Bhoyroo, V.D., 1984, Occurance of α-D-galactosyl residues in the thyroglobulin from several species. Localization in the saccharide chains of the complex carbohydrate units. *J. Biol. Chem.* 259:9858–9866.

Springer, G.F., 1971, Blood-group and Forssman antigenic determinants shared between microbes and mammalian cells. *Prog. Allergy* 15:9–77.

Springer, G.F., and Horton, R.E., 1969, Blood group isoantibody stimulation in man by feeding blood group-active bacteria. *J. Clin. Invest* 48:1280–1291.

Stellner, K., Saito, H., and Hakomori, S., 1973, Determination of aminosugar linkage in glycolipids by methylation. Aminosugar linkage of ceramide pentasaccharides of rabbit erythrocytes and of Forssman antigen. *Arch. Biochem. Biophys.* 133:464–472.

Stone, K.R., Ayala, G., Goldstein, J., Hurst, R., Walgenbach, A., and Galili, U., 1998, Porcine cartilage transplants in cynomolgus monkey: III. Transplantation of α-galactosidase treated porcine cartilage. *Transplantation* 65:1577–1583.

Stone, K.R., Walgenbach, A.W., Abrams, T., Nelson, J., Gellett, N., and Galili, U., 1997, Porcine and bovine cartilage transplants in cynomolgus monkey: I. A model for chronic xenograft rejection. *Tranplantation* 63:640–645.

Suzuki, E., and Naiki, M., 1984, Heterophile antibodies to rabbit erythrocytes in human sera and identification of the antigen as a glycolipid. *J. Biochem. (Tokyo)* 95:103–108.

Teneberg, S., Lönnroth, I.,Torres Lopez, J.F., Galili, U., Olwegard Halvarsson, M., Angstrom, J., and Karlsson, K-A., 1996, Molecular mimicry in the recognition of glycosphingolipids by Galα3Galβ4GlcNAcβ-binding *Clostridium difficile* toxin A, human natural anti-α-galactosyl IgG and the monoclonal antibody Gal-13: characterization of a binding-active human glycosphingolipid, non-identical with the animal receptor. *Glycobiology* 6:599–609.

Thall, A.D., Maly, P., and Lowe, J.B., 1995, Oocyte Galα1-3Gal epitopes implicated in sperm adhesion to the zona pellucida glycoprotein ZP3 are not required for fertilization in the mouse. *J. Biol. Chem.* 270:21437–21442.

Thall, A.D., Murphy, H.S., and Lowe, J.B., 1996, α-galactosyltransferase deficient mice produce naturally occurring cytotoxic anti-Gal antibodies. *Transplant. Proc.* 28:556–557.

Towbin, H., Rosenfelder, G., Weislander, J., Avila, J.L., Rojas, M., Szarfman, A., Esser, K., Nowack, H., and Timple, R., 1987, Circulating antibodies to mouse laminin in Chagas disease, American cutaneous Leishmaniasis and normal individuals recognize terminal galactosyl (α1-3) galactose epitopes. *J. Exp. Med.* 166:419–432.

Wang, L., Anaraki, F., Henion, T.R., and Galili, U., 1995a, Variations in activity of the human natural anti-Gal antibody in young and elderly populations. *J. Gerontol. (Med. Sci.)* 50A:M227-M233.

Wang, L., Radic, M.Z., and Galili, U., 1995b, Human anti-Gal heavy chain genes: Preferential use of $V_H3$ and the presence of somatic mutations. *J. Immunol.* 155:1276–1285.

Watkins, W.M., 1966, Blood group substances. *Science* 152:172–181.

Wiener, A.S., 1951, Origin of naturally occurring hemagglutinins and hemolysins: A review. *J. Immunol.* 66:287–295.

Weislander, J., Mannsono, O., Kallin, E., Gabrielli, A., Nowack, H., and Timple, R., 1990, Specificity of human antibodies against Galα1-3Gal carbohydrate epitope and distinction from natural antibodies reacting with Galα1–2Gal or Galα1-4Gal. *Glycoconjugate J.* 7:85–94.

Xu, Y., Lorf, T., Sablinski, T., Gianello, P., Bailin, M., Monroy, R., Kozlowski, T., Cooper, D.K.C., and Sachs, D.H., 1998, Removal of anti-porcine natural antibodies from human and nonhuman primate plasma in vitro and in vivo by a Galα1-3Galβ1-4Glc-R immunoaffinity column. *Transplantation* 65:172–179.

*Chapter 5*

# α-Gal Epitopes in Animal Tissue Glycoproteins and Glycolipids

Lennart Rydberg, Jan Holgersson, Bo E. Samuelsson, and Michael E. Breimer

## 1. CARBOHYDRATE STRUCTURAL COMPLEXITY

The great structural complexity of carbohydrate chains originates in the many ways the monosacchrides can be linked to each other. Binding position (e.g., 1,3 as opposed to 1,4), binding anomericity (i.e., α or β), monosaccharide structure (e.g., pentose vs hexose), ring size (i.e., furanose vs pyranose), and stereochemistry all contribute to structural complexity. Furthermore, the carbohydrate chain may be branched or non-branched and the monosaccharide units may be chemically modified through, for example sulfation or phosphorylation. For each linkage established between two monosaccharides, there is an enzyme, a glycosyltransferase, catalyzing the elongation of the growing carbohydrate chain (Natsuka and Lowe, 1994). Sometimes, there are even different glycosyltransferases catalyzing the

**Lennart Rydberg and Bo E. Samuelsson** Institute of Laboratory Medicine, Department of Clinical Chemistry and Transfusion Medicine, Sahlgrenska universitetssjukhuset, SE 413 45 Göteborg, Sweden. **Jan Holgersson** Division of Clinical Immunology, Karolinska Institute, Huddinge University Hospital, SE 141 86 Huddinge, Sweden. **Michael E. Breimer** Institute for Surgical Sciences, Department of Surgery, University of Göteborg, Sahlgrenska universitetssjukhuset, SE 413 45 Göteborg, Sweden.

*Subcellular Biochemistry, Volume 32: α-Gal and Anti-Gal,* edited by Galili and Avila.
Kluwer Academic / Plenum Publishers, New York, 1999.

ases catalyzing the formation of the same linkage. Carbohydrate antigens are therefore "secondary" gene products as they are produced by the joint action of a series of "primary" gene products, the glycosyltransferases. The structural diversity found in the primary sequence of a few monosaccharides is immense compared to the relatively limited structural variability obtained in the primary sequence of the same number of amino acids. If three different amino acids are combined in every possible way, six different polypeptide structures are obtained. If the same theoretical calculation is done with three different monosaccharides, one ends up with more than 1.000 different trisaccharide structures (Samuelsson and Breimer, 1987). However, nature does not, to our knowledge, make use of all these possible structures.

## 2. CELL SURFACE GLYCOCONJUGATES

A specific carbohydrate chain (i.e., an oligo- or polysaccharide) can be found in different contexts on the cell surface. It may be lipid-linked, as in

**Glycosphingolipid core saccharides**

| | |
|---|---|
| Galβ1-3GlcNAcβ1-3Galβ1-4Glcβ1-1Cer | lacto series; type 1 |
| Galβ1-4GlcNAcβ1-3Galβ1-4Glcβ1-1Cer | neolacto series; type 2 |
| GalNAcβ1-3Galα1-4Galβ1-4Glcβ1-1Cer | globo series; type 4 |
| GalNAcβ1-3Galα1-3Galβ1-4Glcβ1-1Cer | isoglobo series |
| Galβ1-3GalNAcβ1-4Galβ1-4Glcβ1-1Cer | ganglio series |
| Galβ1-3GalNAcβ1-3Galβ1-4Glcβ1-1Cer | isoganglio series |

**Blood group determinants**

Fucα1-2Galβ1-R — blood group H

Galα1-3Gal(Fucα1-2)β1-R — blood group B

GalNAcα1-3Gal(Fucα1-2)β1-R — blood group A

Galβ1-3GlcNAc(Fucα1-4)β1-R — Le^a determinant

Galβ1-4GlcNAc(Fucα1-3)β1-R — X hapten

Fucα1-2Galβ1-3GlcNAc(Fucα1-4)β1-R — Le^b determinant

Fucα1-2Galβ1-4GlcNAc(Fucα1-3)β1-R — Y hapten

**Figure 1.** Structures of glycosphingolipid core saccharides, blood group ABH, Lewis and X/Y determinants (Holgerson *et al.*, 1992).

glycosphingolipids or glyceroglycolipids, or protein-bound, as in glycoproteins. In glycosphingolipids, the ceramide lipophilic part is comprised of an amino alcohol, sphingosine, to which a fatty acid is amide linked and this part anchors the glycolipid in the outer leaflet of the membrane (Holgerson *et al.*, 1992). Glycolipids can be either non-charged, or negatively charged due to the presence of phosphate, sulfate or negatively charged monosaccharides such as sialic acid. Glycolipids containing sialic acid are called gangliosides. Six major families of glycosphingolipids have been identified which are shown in Fig. 1 together with the blood group ABH, Lewis and X/Y determinants (Holgersson *et al.*, 1992).

Glycoproteins have their carbohydrate side chains either N- or O-linked as defined by the type of linkage between the carbohydrate chain and the peptide backbone (Kornfeld and Kornfeld, 1985; Iwase, 1988; Jentoft, 1990). In N-linked oligosaccharides the innermost sugar residue, *N*-acetylglucosamine, is bound to the amide nitrogen in the side chain of an asparagine residue located in the -AsnXSer/Thr- concensus sequence (Kornfeld and Kornfeld, 1985), whereas in O-linked oligosaccharides, *N*-acetylgalactosamine is bound to the hydroxyl oxygen of serine or threonine (Iwase, 1988; Jentoft, 1990). N-linked oligosaccharides contain a common pentasaccharide core structure (Fig. 2), which is extended to give three main types of N-linked oligosaccharides: the complex, the hybrid, and the high-mannose type (Fig. 2) (Kornfeld and Kornfeld, 1985). O-linked oligosaccharides on the other hand have been shown to have four major, common core

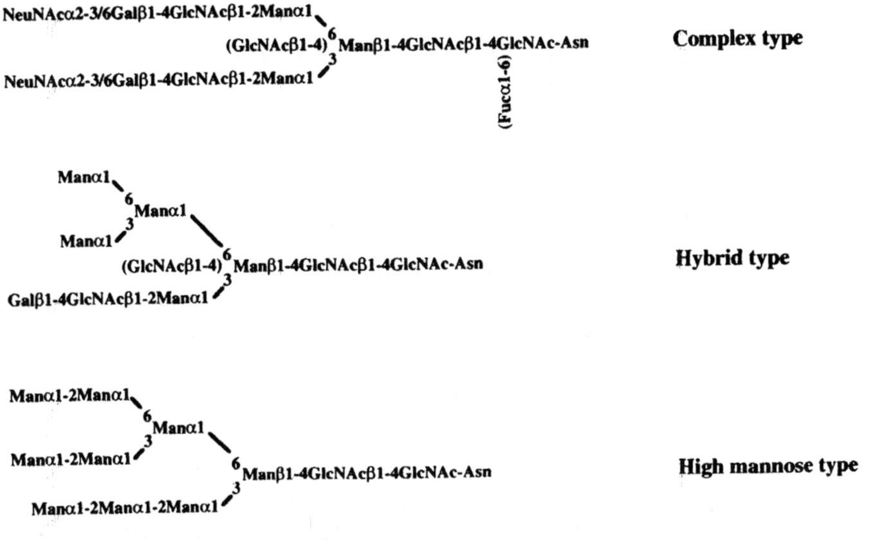


**Figure 2.** Structures of the three main types of N-linked oligosaccharides. The two Asn-linked *N*-acetylglucosamines and the following three mannose residues constititue a pentasaccharide core common to all N-linked oligosaccharides. Residues within brackets may or may not be present. Redrawn from (Kornfeld and Kornfeld, 1985).

**Galβ1-3GalNAcα-Ser/Thr** **Core 1**

**GlcNAcβ1**\
**6**
**Galβ1-3GalNAcα-Ser/Thr** **Core 2**

**GlcNAcβ1-3GalNAcα-Ser/Thr** **Core 3**

**GlcNAcβ1**\
**6**
**GlcNAcβ1-3GalNAcα-Ser/Thr** **Core 4**

**Figure 3.** Structures of the four main core saccharides of O-linked oligosaccharides (Iwase, 1988; Jentoft, 1990).

structures (Fig. 3) (Iwase, 1988; Jentoft, 1990). Another class of glycoproteins, the proteoglycans, are named so because the glycan part is by mass the dominating part (cp. glycoproteins) (Höök *et al.*, 1984). The glycan part is comprised of repeating disaccharide units with a varying degree of sulfation (Höök *et al.*, 1984). It can be linked to serine via xylose, representing yet another type of carbohydrate/peptide linkage. Recently, O-linked *N*-acetylglucosamines (GlcNAc) have been found in several cytoplasmic and nucleoplasmic glycoproteins, and its importance as a regulatory signaling mechanism comparable to, and sometimes mutually exclusive with, phosphorylation has been realized (Hayes and Hart, 1994; Hart, 1997). In addition to the above mentioned carbohydrate/peptide linkages, examples of new ones are constantly being added to the growing list (Hayes and Hart, 1994).

## 3. CELL-, TISSUE-, INDIVIDUAL-, AND SPECIES-SPECIFIC GLYCOSYLATION PATTERNS

Each cell type has its own, specific glycosylation machinery, i.e.,. a cell-specific set of glycosyltransferases which through their enzymatic activity give the cell a particular carbohydrate phenotype. These phenotypes can often be recognized by the immune system, which is why many cell markers of tumor cells and various hematopoietic cells are carbohydrates (Hakomori, 1990). The CA19-9 and CA50 carbohydrate epitopes (Sialyl-Le$^{a}$, NeuAcα2-3Galβ1–3(Fucα1-4)GlcNAβ1-3Galβ1-4Glcβ1-1Cer) are important in clinical medicine as diagnostic tools for gastrointestinal adenocarcinomas, especially pancreatic cancer. CD15 (**C**luster of **D**ifferentiation, X antigen see Fig.1) and CD65 (NeuNAcα2-3Galβ1-4GlcNAcβ1-3Galβ1-4(Fucα1-3) GlcNAcβ1-) are epitopes which are markers for neutrophilic granulocytes, and CD57 ($O_3$S3GlcAβ1-3Galβ1-4GlcNAcβ1-) is found on human NK cells. Although, the list of these epitopes can be made very long, the physiological relevance of these epitopes has been established only in a few cases. A number of factors such as, for example, cell/cell and cell/matrix interactions, gender and hormonal influences (Hay and Gray, 1970; Gray, 1971; Wang *et al.*, 1989; Wang *et al.*, 1990; Hakomori, 1994; Dharmesh and Baeniziger, 1993), cell growth rate and growth phase, and state of activation (Postigo *et al.*, 1994) influence the glycosylation status of a certain cell type. The reason for, and the control of, changes in the glycosylation status of the cell is poorly understood. Specific carbohydrate patterns in tissues and organs can also be recognized, reflecting the cell-specific carbohydrate phenotypes of the cell types constituting a particular tissue ( Clausen and Hakomori, 1989; Holgersson *et al.*, 1992; Oriol *et al.*, 1992). It has been long known that the gastrointestinal epithelium is rich in fucose-containing glycoconjugates, some of which has been utilized as attachment sites by microbes (Holgersson *et al.*, 1992; Falk *et al.*, 1993; Borén *et al.*, 1994). Cell surface carbohydrates are also receptors for viruses and toxins (Karlsson, 1995). The ABO and Lewis blood group systems are carbohydrate antigens exhibiting important inter-individual differences (Clausen and Hakomori, 1989; Ulfvin *et al.*, 1993), which have to be considered prior to blood transfusions and organ transplantations (Breimer and Samuelsson, 1986). In this review, we will focus on a species-specific carbohydrate epitope, Galα1,3Gal termed here the α-Gal epitope (Galili, 1993), thought to be involved in autoimmunity and as an important xenoantigen causing hyperacute rejection of pig organs transplanted to humans and non-human primates (this volume) (Good *et al.*, 1992; Welsh *et al.*, 1991; Sandrin *et al.*, 1993; Neethling *et al.*, 1994).

## 4. THE α-GAL EPITOPE AND ITS TISSUE EXPRESSION

Superficially, the α-Gal epitope can be looked upon as a blood group B epitope, Galα1,3(Fucα1,2)Galβ1-R, lacking the fucose residue. However, the two determinants are biosynthesized by two distinct α-1,3 galactosyltransferases and thus genetically unrelated. Even though a large fraction of the human anti-blood group B antibody population can be explained by anti-Gal antibodies (Galili *et al.*, 1987), the two determinants have been shown to be conformationally distinct due to the presence or absence of fucose (Teneberg *et al.*, 1996). The immune system is capable of recognizing this difference, mounting an antibody response which is specific for either of the two structures. Man, apes and Old World monkeys lack α-Gal, but have antibodies against this epitope due to an immune response possibly against intestinal bacteria carrying α -Gal determinants in their capsular lipopolysaccharide (Galili *et al.*,1988; Hamadeh *et al.*, 1992). In contrast, New World monkeys and other animals express α-Gal and consequently lack anti-Gal antibodies (Galili *et al.*, 1988). The presence of α-Gal in donor animals and of anti-Gal in human recipients constitute the most immediate obstacle to successful clinical pig to human xenotransplantation (Galili, 1993).

The tissue distribution of α-Gal epitopes in various species has been studied using *Griffonia (Bandeiraea) simplicifolia* isolectin $B_4$ and anti-Gal antibodies (Peters and Goldstein, 1979; Galili *et al.*, 1988; Oriol *et al.*, 1993; Alvarado *et al.*, 1995). By light microscopy, using these reagents, the α-Gal epitope in mice was shown to be confined to basement membranes, endothelium and certain epithelial cells (Peters and Goldstein, 1979). The distribution of this epitope in pig, a possible future donor species for human xenotransplantations, was similar to the distribution in mice with the highest expression levels seen in vascular endothelium, kidney proximal tubules, bronchioli, and alveoli of the tissues tested (Oriol *et al.*, 1993). Immunelectronmicroscopy using *Griffonia simplicifolia* lectin and anti-Gal antibodies, has revealed a more complex distribution of the α-Gal epitope in pig kidney than earlier found by light microscopy. In the proximal convoluted tubule, the S1 and S2 segments contain large amounts of the antigen while the S3 and S4 segments do not. In the glomeruli, the endothelial cells and parietal epithelial cells were positive for the α-Gal antigen while the podocytes and mesangial cells were negative. In the vascular tree, the α-Gal antigen was found in the endothelial cells and subendothelium while the deeper layers of the blood vessel wall were negative. Among the interstitial cells, part of the fibroblast-like cells, probably dendritic cells, and all macrophage-like cells were positive while the lipid-laden cells and pericytes were negative (Strokan *et al.*, 1998b).

## 5. α-GAL-TERMINATED GLYCOLIPIDS

In 1911 it was reported (von Dungern and Hirszfeld, 1911) that rabbit erythrocytes could absorb anti-B human isoagglutinin indicating that animal erythrocytes can contain blood group B material. In the late fifties, Yamakawa et al. reported that rabbit erythrocytes contained a globoside type glycolipid containing a glucosamine (Yamakawa *et al.*, 1956; Yamakawa *et al.*, 1960) but too little material was obtained to characterize its structure. Some years later they could present this glycolipid as a novel pentaglycosylceramide with the structure Gal1-3Gal1-4GlcNAc1-3Gal1-4Glc-Cer (Eto *et al.*, 1968) and postulated that the terminal galactose had an α anomeric configuration due to its ability to inhibit the hemagglutination of human blood group B red blood cells by the corresponding isoantibodies. The inhibition was slightly weaker compared to human blood group B glycolipids (Eto *et al.*, 1968). The complete structure of the rabbit erythrocyte pentaglycosylceramide, including the anomeric configuration of the glycosidic linkages, was later reported (Stellner *et al.*, 1973). Since then, this component has been identified in several animal tissues and cells as listed in Table I. Some of these compounds have been completely structurally characterized while in some tissues, the amount of material isolated allowed only a partial structural characterization. Hanfland et. al. showed in a series of papers (Hanfland *et al.*, 1981;

**Table I**
**Galα1-3Gal Terminating Glycolipids Isolated from Tissus and Cells of Animal Origin**

| No sugar residues | Chemical structure | Tissue | Ref* |
|---|---|---|---|
| *Neutral glycolipids* | | | |
| 3 | Galα1-3Galβ1-4Glcβ1-1Cer | Rat small intestine | Breimer *et al.*, 1982 |
| | | Dog intestine | Galili *et al.*, 1987 |
| 4 | Galα1-3Galα1-4Galβ1-4Glcβ1-1Cer | Rat small intestine | Breimer *et al.*, 1982 Ångström *et al.*, 1982 |
| | | Rat PC12h cells | Ariga *et al.*, 1988 |
| 5 | Galα1-3Galβ1-4GlcNAcβ1-3Galβ1-4Glcβ1-1Cer | Rabbit erythrocytes | Yamakawa, 1956, Yamakawa, 1960, Eto *et al.*, 1968, Stellner *et al.*, 1973, Suzuki *et al.*, 1984 |
| | | Rabbit kidney | Hendricks *et al.*, 1990 |
| | | Bovine erythrocytes | Uemura *et al.*, 1978 |
| | | Bovine kidney | Hendricks *et al.*, 1990 |
| | | Pig kidney | Hendricks *et al.*, 1990, Jalali-Araghi *et al.*, 1994, Samuelsson *et al.*, 1994, Bäcker *et al.*,1998, |
| | | Porcine aortic endothelial cells | Bouhours *et al.*, 1996 |
| | | Pig aorta | Hallberg *et al.*, 1998a, Hallberg *et al.*, 1998b |
| | | Pig intestine, spleen, liver, heart, salivary gland | Bäcker *et al.*, 1998 |
| | | Sheep kidney | Hendricks *et al.*, 1990, Strokan *et al.*, 1998a |

(*continued*)

**Table I**
***(Continued)***

| No sugar residues | Chemical structure | Tissue | Ref* |
|---|---|---|---|
| | | Sheep small/large intestine, pancreas | Strokan *et al.*, 1998a |
| | | Canine MDCK I cells | Hansson *et al.*, 1988 |
| 6 | Galα1-3Galβ1-4GlcNAcβ1-3Galβ1-4Glcβ1-1Cer<br>Fucα1-3 | Porcine aortic endothelial cells | Bouhours *et al.*, 1996 |
| | | Pig aorta | Hallberg *et al.*, 1998a |
| | | Pig kidney | Bouhours *et al.*, 1997, Bäcker *et al.*, 1998 |
| 5–8 | Galα1-3(Galα1-3)$_{1\text{-}4}$Galα1-4Galβ1-4Glcβ1-1Cer | Rat small intestine | Breimer *et al.*, 1982, Ångström *et al.*, 1982 |
| 7 | Galα1-3Galβ1-4GlcNAcβ1-3Galβ1-4GlcNAcβ1-3Galβ1-4Glcβ1-1Cer | Bovine erythrocytes | Chien *et al.*, 1979 |
| | | Rabbit erythrocytes | Dabrowski *et al.*, 1984, Egge *et al.*, 1985 |
| | | Sheep kidney | Hendricks *et al.*, 1990, Strokan *et al.*, 1998a, |
| | | Porcine aortic endothelial cells | Bouhours *et al.*, 1996 |
| 8 | Galα1-3Galβ1-4GlcNAcβ1-3Galβ1-4GlcNAcβ1-3Galβ1-4Glcβ1-1Cer<br>Fucα1-3 | Pig kidney | Bouhours *et al.*, 1997 |
| 10 | Galα1-3Galβ1-4GlcNAcβ1-3<br>Galβ1-4GlcNAcβ1-3Galβ1-4Glcβ1-1Cer<br>Galα1-3Galβ1-4GlcNAcβ1-6 | Rabbit erythrocytes | Hanfland *et al.*, 1981 |

| | | | |
|---|---|---|---|
| 12 | Fucα1-3<br>Galα1-3Galβ1-4GlcNAcβ1-6<br>Galβ1-4GlcNAcβ1-3Galβ1-4Glcβ1-Cer<br>Galα1-3Galβ1-4GlcNAcβ1-3<br>Fucα1-3 | Pig kidney | Bouhours *et al.*, 1997 |
| 15 | Galα1-3Galβ1-4GlcNAcβ1-3<br>Galβ1-4GlcNAcβ1-3<br>Galα1-3Galβ1-4GlcNAcβ1-6 Galβ1-4GlcNAcβ1-3Galβ1-4Glcβ1-1Cer<br>Galα1-3Galβ1-4GlcNAc1-6 | Rabbit erythrocytes | Dabrowski *et al.*, 1984, Egge *et al.*, 1985 |
| 25–30 | (Galα1-3Galβ1-4GlcNAc)$_n$- (-3Galβ1-4GlcNAc)$_{n+x}$-Cer** | Rabbit erythrocytes | Honma *et al.*, 1981, Breimer *et al.*, 1983 |
| 40 | (Galα1-3Galβ1-4GlcNAc)$_n$- (-3Galβ1-4GlcNAc)$_{n+x}$-Cer** | Rabbit erythrocytes | Dabrowski *et al.*, 1988 |
| *Gangliosides* | | | |
| 10 | Galα1-3Galβ1-4GlcNAcβ1-6<br>Galβ1-4GlcNAcβ1-3Galβ1-4Glcβ1-1Cer<br>NeuAcα2-3Galβ1-4GlcNAcβ1-3 | Bovine erythrocytes | Watanabe *et al.*, 1979 |

*Underlined references contains a complete chemical structural characterization of the glycolipid. In the remaining references, a partial characterization or tentative structure have been assigned.
**Complex branched structures.

Dabrowski *et al.*, 1984; Egge *et al.*, 1985; Dabrowski *et al.*, 1988) that rabbit red blood cells contained a series of complex Galα1-3 terminating glycolipids with 7, 10, 15, 30 and 40 sugar residues containing repetitive internal Galβ1-4GlcNAc units and branched saccharide chains. A series of glycolipids with up to 8 sugar residues having repetitive Galα1-3 saccharides linked to globotriaosylceramide were found in the rat small intestine (Ångström *et al.*, 1982). In the pentahexosylceramide fraction from rat intestine a novel fucolipid was found having the probable structure of Fucα1-2Galα1-3Galα1-4Galβ1-4Glcβ1-1Cer (Ångström *et al.*, 1982). If this compound is the result of a specific α1-2fucosyltransferase is not known. Recently, glycolipids with a Gal residue linked α1-3 to the blood group X (type 2 chain isomer of $Le^a$, Fig. 1) histo-blood group antigen has been identified in pig kidney (Bouhours *et al.*, 1997; Bäcker *et al.*, 1998), aorta (Hallberg *et al.*, 1998b) and porcine aortic endothelial cells (Bouhours *et al.*, 1996). Gangliosides having terminal α–Gal residues have been found also in bovine erythrocytes (Watanabe *et al.*, 1979; Hendricks *et al.*, 1990).

## 6. α-GAL-TERMINATED N-LINKED OLIGOSACCHARIDES

N-linked oligosaccharides carrying α-Gal determinants are mainly of the complex, bi-, tri- and tetraantennary types having an *N*-acetyllactosamine unit capped by α-Gal, and with or without bisecting *N*-acetylglucosamines or fucose-substituted core *N*-acetylglucosamines (Spiro and Bhoyroo, 1984; Kagawa *et al.*, 1988; Knibbs *et al.*, 1989; Krotkiewski *et al.*, 1990; Lipniunas *et al.*, 1993; Hironaka *et al.*, 1993). The branches in a multiantennary complex type oligosaccharide may all be terminated by α-Gal or may have a varying number of branches terminated by NeuNAcα2-3/6Gal (Spiro and Bhoyroo, 1984; Kagawa *et al.*, 1988; Knibbs *et al.*, 1989; Krotkiewski *et al.*, 1990; Lipniunas *et al.*, 1993; Hironaka *et al.*, 1993).

## 7. α-GAL-TERMINATED O-LINKED OLIGOSACCHARIDES

The O-linked oligosaccharides of a major glycoprotein, GP-3, of calf thyroid cells (Edge and Spiro, 1985) have been structurally characterized in detail (Edge and Spiro, 1997). α-Gal-terminated structures were found among both non-sialylated and sialylated compounds. The most abundant α -Gal containing structure of GP-3 was Galα3Galβ4GlcNAcβ(Galβ3)6GalNAcol accounting for 22 %, followed by Galα3Galβ4GlcNAcβ(NeuAcα3Galβ3)6GalNAcol constituting 9 %, and Galα3Galβ(Galβ3)6GalNAcol accounting for 5 % of GP-3 O-linked oligosaccharides (Edge and Spiro, 1997). The latter structure is interesting because the terminal αGal is bound to a galactose that is β-linked to C-6 of the serine/threonin-linked *N*-acetylgalactosamine and not, as usual, to C-4 of the *N*-acetylglucosamine in an *N*-acetyllactosamine unit. This raises two questions: are there more than one α 1,3 GT and are there human anti-Gal antibodies that can distinguish between different α -Gal-terminated structures having different core saccharide chains? If the answer to the latter question is yes, future anti-pig antibody absorbers need to be structurally more diverse and not solely based upon Galα3Galβ-R or Galα3Galβ4GlcNAcβ-R di- and trisaccharides, respectively (see below) (Liu *et al.*, 1997).

## 8. α-GAL ON RECOMBINANT PROTEINS AND PROTEINS FOR THERAPEUTIC USE IN HUMANS

Many secreted glycoproteins, for example laminin (Knibbs *et al.*, 1989; Shibata *et al.*, 1982; Galili *et al.*, 1998), thyroglobulin (Spiro and Bhoyroo, 1984; Thall and Galili, 1990), fibrinogen (Thall and Galili, 1990), factor VIII (Hironaka *et al.*, 1993) and immunoglobulins (Krotkiewski *et al.*, 1989; Krotkiewski *et al.*, 1990; Thall and Galili, 1990; Lipniunas *et al.*, 1993; Endo *et al.*, 1995) from different species carry α-Gal determinants. The glycosylation pattern of secreted proteins reflects the ability of the cell type in which they were produced to synthesize a particular carbohydrate determinant (Kagawa *et al.*, 1988; Liu *et al.*, 1997). Human anti-Gal antibodies may influence the half life in the circulation of recombinant proteins carrying α-Gal through immune complex formation and rapid clearance by the reticulo-endothelial system (Kagawa *et al.*, 1988; Hironaka *et al.*, 1993; Thall and Galili, 1990; Borrebaeck et al., 1993). Thus, host cells used for production of recombinant proteins to be used as therapeutics in humans, should be selected such that they don't have the ability to synthesize terminal α-Gal.

In contrast, we have taken advantage of the ability of host cells to determine the structure of oligosaccharide side chains on secreted proteins in order to make a protein which can be used in a pretransplant situation to remove anti-pig antibodies from the human circulation. COS cells that were cotransfected with vectors encoding a mucin/immunoglobulin chimeric protein and the porcine α 1,3 galactosyltransferase produced a heavily α-Gal substituted mucin/Ig protein which was easily affinity-purified from supernatants of transfected cells (Liu *et al.*, 1997). This glycoprotein was shown to be a very efficient absorber of human anti-pig antibodies *in vitro* (Liu *et al.*, 1997).

## 9. α-GAL IMMUNOGENICITY

Many carbohydrate structures function as strong immunogenic antigens which is also true for the α-Gal epitope. Natural antibodies are present in the serum

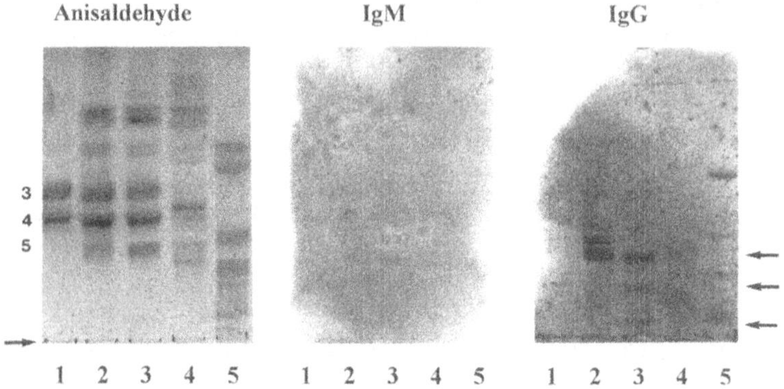


**Figure 4.** Thin-layer chromatography and immunostaining of total neutral glycolipid fractions isolated from pig aorta (lane 1), blood group O pig kidneys (lane 2), the contra-lateral reference kidney from the same animal as the xenoperfused kidney (lane 3), rabbit small intestine (lane 4) and the total acidic fraction from the contra-lateral reference kidney (lane 5). The two plates to the right are immunostained with serum collected from patient no 1 in (Breimer *et al.*, 1996; Rydberg *et.al.*, 1996) which had a pig kidney connected to his circulation for 65 minutes. The serum sample was collected two days before the perfusion experiment. Secondary antibodies used were anti-human IgM and IgG respectively. In the left chromatogram the plate is stained by a chemical reagent. Conditions for analysis as in (Rydberg *et al.*, 1996). The arrows to the left indicate the position of the α-Gal pentaglycosylceramide as well as compounds with 7–8 and 10–12 sugar residues.

of individuals lacking this epitope (Galili *et al.*, 1988) in a similar manner as formulated by Landsteiner in his law about the reciprocal presence of antigens/antibodies in the ABO histo-blood group system (Landsteiner, 1945). Natural α -Gal antibodies do not only recognize the α-Gal epitope in the pentasaccharide glycosphingolipid but also in longer glycosphingolipid compounds. This is shown in Fig. 4 where a serum sample form a renal patient on dialysis was tested in a chromatogram binding assay against glycolipid compounds prepared from pig organs. In addition to the binding of the 5 sugar compound, binding is also seen to compounds with Rf values corresponding to glycolipids with 7–8 and 10–12 sugar residues.

In the same way as immune antibodies can be induced by antigenic challenge in the ABO system, anti-Gal immune antibodies can be induced by antigenic challenge with the α-Gal epitope in individuals lacking this epitope. Thus, in two patients subjected to short term (65 minutes and 15 minutes respectively) extracorporeal ("ex vivo") pig kidney perfusions (Breimer *et al.*, 1996) we could identify a strong α-Gal humoral immune response comprised of all immunoglobulin classes, including the IgG1 and IgG2 subclasses, in the first individual, and a mild humoral immune response, consisting of anti-Gal antibodies of the IgA class, mainly the IgA1 subclass, in the second patient (Rydberg *et al.*, 1996). The im-

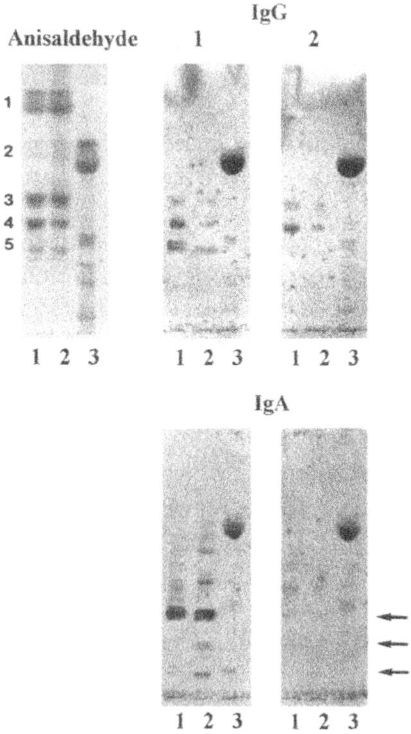


**Figure 5.** Thin-layer chromatography and immunostaining of total neutral glycolipid fractions isolated from blood group O pig kidneys (lane 1), the contra-lateral reference kidney from the same animal as the xenoperfused kidney (lane 2) and the total acidic fraction from the contra-lateral reference kidney (lane 3). The four plates to the right are immunostained with serum collected from patient no 2 in (Breimer *et al.*, 1996) which had a pig kidney connected to his circulation for 15 minutes. The serum sample was collected 3 weeks after the perfusion experiment. In plates 1, the native serum was added to the plates and in plates 2 the serum was preincubated with soluble Galα1-3Gal disaccharides prior to the immunostaining. Secondary antibodies used were anti-human IgG and IgA respectively. Note that α-Gal disaccharides completely block the binding of the antibody to the glycolipid components with 5, 7–8 and 10–12 sugar residues seen in lane 2. Conditions for analysis as in (Rydberg *et al.*, 1996). The arrows to the left indicates the position of the α-Gal pentaglycosylceramide as well as compounds with 7–8 and 10–12 sugar residues.

mune response in the second patient is illustrated in the chromatogram binding assay in Fig.5 showing intense binding of IgA antibodies to glycolipid compounds with Rf value corresponding to the α-Gal terminated pentasaccharide structure. Binding to glycolipids with more complex structures is also seen similar to the binding in the non-immune serum sample. Binding to these glycospingolipids could be completely inhibited by addition of soluble α-Gal disaccharides to the patient serum sample (Fig. 5).

## 10. SUMMARY AND CONCLUSIONS

α-Gal terminated saccharides are present on the cell surface both as glycolipids and glycoproteins in all mammals except Old World monkeys and humans. The structural diversity among identified saccharides terminated by this epitope in animal tissues is steadily increasing. The majority of these saccharides have the α-Gal linked to lactosamine but other core saccharides exist. The α-Gal terminated saccharides are recognized by the immune system as a specific antigen and antibodies directed to the α-Gal, which do not cross-react with the classic blood group B trisaccharide, are found in man and Old World monkeys. Similar to other complex carbohydrate cell surface antigens, the α-Gal epitope is heterogenously distributed in different organs and in different cells within an organ. It is present on the vascular entothelium and it is the primary target for human naturally occurring antibodies following pig to primate/man xenotransplantation leading to hyperacute rejection of the graft. Important for the future will be to further struturally charcterize this antigen system, its cellular/subcellular distribution, and to identify possible of additional glycosyltransferases, related to the already described α1,3galactosyltransferase that may explain the structural diversity. Such information will be of importance in the studies of, for example, the pathogenesis of autoimmune diseases and for the production of genetically modified pigs to prevent xenograft rejection.

ACKNOWLEDGEMENT. This work was supported by The Swedish Medical Research Council (grant no 6521, 11574 and 11612), EU Shared Cost Biotechnology Programme PL962242, the EU Biomed II grant (no BMH4-97-2596), IngaBritt and Arne Lundberg Foundation for Medical Research, The Lars-Erik Gelin Memorial Foundation, The Swedish Cancer Society (grant no 3645 and 3884), The Vardal Foundation (A1996 072) and The Swedish Society of Medicine.

## REFERENCES

Alvarado, C.G., Cotterell, A.H., McCurry, K.R., Collins, B.H., Magee, J.C., Berthold, J., Logan, J.S. and Platt, J.L., 1995, Variation in the level of xenoantigen expression in porcine organs, *Transplantation* **59:**1589–1596.

Ångström, J., Breimer, M.E., Falk, K.-E., Hansson, G., Karlsson, K.-A., Leffler, H. and Pascher, I., 1982, Structural Characterization of glycolipids of rat small intestine having one to eigth hexoses in a linear sequence. *Arch. Biochem. Biophys.* **213:**708–725.

Ariga, T., Yu, R.K., Scarsdale, J.N., Suzuki, M., Kuroda, Y., Kitagawa, H. and Miyatake, T., 1988, Accumulation of a globo-series glycolipid having Galα1–3Gal in PC12h pheochromocytoma cells, *Biochemistry* **27:**5335–5340.

Borén, T., Falk P., Roth, K.A., Larson, G. and Normark, S., 1994, Attachment of *Helicobacter pylori* to human gastric epithelium mediated by blood group antigens, *Science* **262,**:892–1895.

Borrebaeck, C.A.K., Malmborg, A.-C. and Ohlin, M., 1993, Does endogeneous glycosylation prevent the use of mouse monoclonal antibodies as cancer therapeutics? *Immunol. Today,* **14:**477–479.

Bouhours, D., Liaigre, J., Naulet, J., Maume, D. and Bouhours, J.-F., 1997, A novel glycosphingolipid expressed in pig kidney: Galα1–3Lewis$^x$ hexaglycosylceramide, *Glycoconjugate J.* **14:**29–38.

Bouhours, D., Pourcel, C. and Bouhours, J.-F., 1996, Simultaneous expression by porcine aorta endothelial cells of glycosphingolipids bearing the major epitope for human xenoreative antibodies (Galα1–3Gal), blood group H determinant and N-glycolylneuraminic acid, *Glycoconjugate J.* **13:**1–7.

Breimer, M.E. and Samuelsson, B.E., 1986, The specific distribution of glycolipid-based blood group A antigens in human kidney related to $A_1/A_2$, Lewis, and secretor status of single individuals. A possible molecular explanation for the successful transplantation of $A_2$ kidneys into O recipients, *Transplantation* **42**:88–91.

Breimer, M.E., Hansson, G.C., Karlsson, K.-A. and Leffler, H., 1982, Glycosphingolipids of rat tissues, *J. Biol. Chem.* **257**:557–568.

Breimer, M.E., Hansson, G.C., Karlsson, K.-A., Larson, G., Leffler, H. and Pimlott, W.,1983, Sequencing of large oligosaccarides by direct inlet mass spectrometry. Application to cell surface glycolipids, *Int. J. Mass Spectrom. Ion Phys.* **48**:113–116.

Breimer, M.E., Björk, S., Svalander, C.T. Bengtsson, A., Rydberg, L., Lie-Svensson, K. Attman, P.O., Aurell, M. And Samuelsson, B.E., 1996, Extracorporeal ("ex vivo") connection of pig kidneys to dialysis patients. I. Clinical data and studies of platelet destruction. Xenotransplantation **3**:328–339.

Bäcker, A.E., Holgersson, J., Samuelsson, B.E. and Karlsson, H., 1998, Identification of extended, Galα1,3Gal-terminated glycosphingolipids in porcine tissues, *Glycobiology*, **6**:000–000.

Chien, J.-L., Li, S.-C. and Li, Y.-T., 1979, Isolation and characterization of a heptaglycosylceramide from bovine erythrocyte membranes, *J.Lipid Res.* **20**:669–673.

Clausen, H. and Hakomori, S., 1989, ABH and related histo-blood group antigens: immunochemical differences in carrier isotypes and their distribution, *Vox. Sang.* **56**:1–20.

Dabrowski, J., Dabrowski, U., Bermel, W., Kordowicz, M. and Hanfland, P., 1988, Structure elucidation of the blood group B like and blood group I active octaantennary ceramide tetracontasaccharide from rabbit erythrocyte membranes by two-dimensional $^1$H-NMR Spectroscopy at 600 MHz, *Biochemistry* **27**:5149–5155.

Dabrowski, U., Hanfland, P., Egge, H., Kuhn, S. and Dabrowski, J., 1984, Immunochemistry of I/i-Active Oligo-and Polyglycosylceramides from rabbit erythrocyte membranes, *J. Biol. Chem.* **259**:7648–7651.

Dharmesh, S.M. and Baenziger, J.U., 1993, Estrogen modulates expression of the glycosyltransferases that synthesize sulfated oligosaccharides on lutropin, *Proc. Natl. Acad. Sci. USA* **90**:11127–11131.

Edge, A.S.B. and Spiro, R.G., 1985, Thyroid cell surface glycoproteins. Nature and disposition of carbohydrate units and evaluation of their blood group I activity, *J. Biol. Chem.* **260**: 15332–15338.

Edge, A.S.B. and Spiro, R.G., 1997, Structure of the O-linked oligosaccharides from a major thyroid cell surface glycoprotein, *Arch. Biochem. Biophys.* **343**:73–80.

Egge, H., Kordowicz, M., Peter-Katalinic, J. and Hanfland, P., 1985, Immunochemistry of I/i-active oligo- and polyglycosylceramides from rabbit erythrocyte membranes, *J. Biol. Chem.* **260**:4927–4935.

Endo, T., Wright, A., Morrison, S.L. and Kobata, A., 1995, Glycosylation of the variable region of immunoglobulin G - site specific maturation of the sugar chains, *Mol. Immunol.* **32**:931–940.

Eto, T., Ichikawa, Y., Nishimura, K., Ando, S. and Yamakawa, T., 1968, Chemistry of lipid of the posthemolytic residue or stroma of erythrocytes, *J.Biochem.* **64**:205–213.

Falk, P., Roth, K.A., Borén, T., Westblom, T.U., Gordon, J.I. and Normark, S., 1993, An *in vitro* adherence assay reveals that *Helicobacter pylori* exhibits cell lineage-specific tropism in the human gastric epithelium, *Proc. Natl. Acad. Sci. USA* **90**:2035–2039.

Galili, U., 1993, Interaction of the natural anti-gal antibody with α-galactosyl epitopes: a major obstacle for xenotransplantation in humans, *Immunol. Today* **14**:480–482.

Galili, U., LaTemple, D.C. and Radic, M.Z., 1998, A sensitive assay for measuring α-Gal epitope expression on cells by a monoclonal anti-Gal antibody, *Transplantation* **65**:1129–1132.

Galili, U., Buchler, J., Shohet, S.B. and Macher, B.A., 1987, The human natural anti-Gal IgG. III. The subtlety of immune tolerance in man as demonstrated by crossreactivity between natural anti-Gal and anti-B antibodies, J. Exp. Med. **165**:693–704.

Galili, U., Mandrell, R.E., Hamadeh, R.M., Shohet, S.B. and Griffiss, J.M., 1988, Interaction between human natural anti-α-galactosyl immunoglobulin G and bacteria of the human flora, *Infect. Immun.* **56**:1730–1737.

Galili, U., Basbaum, C.B., Shohet, S.B., Buehler, J. and Macher, B.A., 1987, Identification of Erythrocyte Galα1–3Gal glycosphingolipids with a mouse monoclonal antibody, Gal-13, *J. Biol. Chem.* **262**:4683–4688.

Galili, U., Shohet, S.B. Kobrin, E., Stults, C.L. and Macher, B.A., 1988, Man, apes and old world monkeys differ from other mammals in the expression of α-galactosyl epitopes on nucleated cells, *J. Biol. Chem.* **263**:17755–17762.

Good, A.H., Cooper, D.K.C., Malcolm, A.J., Ippolito, R.M., Koren, E., Neethling, F.A., Ye, Y., Zudhi, N. and Lamontagne, L.R., 1992, Identification of carbohydrate structures that bind human anti-porcine antibodies: Implications for discordant xenografting in humans, *Transplantation Proc.* **24**:559–562.

Gray, G.M., 1971, The effect of testosterone on the biosynthesis of the neutral glycosphingolipids in the C57/BL mouse kidney, *Biochim. Biophys. Acta* **239**:494–500.

Hakomori, S., 1990, Biochemical basis of tumor-associated carbohydrate Antigens. Current trends, future perspectives, and clinical applications. *Immunology and Allergy Clinics of North America*, **10**:781–802.

Hakomori, S., 1994, Role of carbohydrates in cell adhesion and recognition, In: *Complex carbohydrates in drug research*, Alfred Benzon Symposium 36 (K.Bock and H.Clausen, eds.), pp. 337–349, Munksgaard, Copenhagen.

Hallberg, E.C., Holgersson, J. and Samuelsson, B.E., 1998a, Glycosphingolipid expression in pig aorta: identification of possible target antigens for human natural antibodies, *Glycobiology* **8**:000–000

Hallberg, E.C., Strokan, V., Cairns, T.D.H., Breimer, M.E. and Samuelsson, B.E., 1998b Chemical and immune electron microscopical studies of the expression of the Galα1-determinant in the pig aorta, *Xenotransplantation*, in press.

Hamadeh, R.M., Jarvis, G.A., Galili, U., Mandrell, R.E., Zhou, P. and Griffiss, J.M., 1992, Human natural anti-Gal IgG regulates alternative complement pathway activation on bacterial surfaces, *J. Clin. Invest.*,**89**:1223–1235.

Hanfland, P., Egge, H., Dabrowski, U., Kuhn, S. Roelcke, D. and Dabrowski, J., 1981, Isolation and characterization of an I-Active ceramide decasaccharide from rabbit erythrocyte membranes, *Biochemistry* **20**:5310–5319.

Hansson, G.C., 1988, A blood group B-like pentaglycosylceramide is the major complex glycospingolipid of the Madin-Darby canine kidney epithelial cell line I (MDCK I), *Biochim.Biophys. Acta* **967**:87–91.

Hart, G.W., 1997, Dynamic O-linked glycosylation of nuclear and cytoskeletal proteins, *Annu. Rev. Biochem.* **66**:315–335.

Hay, J.B. and Gray, G.M., 1970, The effect of testosterone on the glycosphingolipid composition of mouse kidney, *Biochim. Biophys. Acta* **202**:566–568.

Hayes, B.K. and Hart, G.W., 1994 Novel forms of protein glycosylation, *Curr. Opin. Struct. Biol.* **4**: 692–696.

Hendricks, S.Å., Pingren, H., Stults, C.L.M. and Macher, B.A., 1990, Regulation of the expression of Galα1–3Galβ1–4GlcNAc glycosphingolipids in kidney, *J. Biol. Chem.* **265**:17621–17626.

Hironaka, T., Furukawa, K., Esmon, P.C., Yokota, T., Brown, J.E., Sawada, S., Fournel, M.A., Kato M., Minaga T. and Kobata A., 1993, Structural study of the sugar chains of porcine factor VIII - tissue- and species-specific glycosylation of factor VIII, *Arch. Biochem. Biophys.* **307**:316–330.

Holgersson, J., Breimer, M.E. and Samuelsson, B.E., 1992, Basic biochemistry of cell surface carbohydrates and aspects of the tissue distribution of histo-blood group ABH and related glycosphingolipids. *APMIS Suppl. 27* **100**:18–27.

Honma, K., Manabe, H., Tomita, M. and Hamada, A., 1981, Isolation and partial structural characterization of macroglycolipid from rabbit erythrocyte membranes, *J. Biochem.* **90:**1187–1196.

Höök, M., Kjellén, L., Johansson, S. and Robinson, J., 1984, Cell-surface glycosaminoglycans, *Annu. Rev. Biochem.* **53:**847–869.

Iwase, H., 1988, Variety and microheterogeneity in the carbohydrate chains of glycoproteins, *Int. J. Biochem.* **20:**479–491.

Jalali-Araghi, K. and Macher, B.A., 1994, Characterization of porcine kidney neutral glycosphingolipids: identification of a carbohydrate antigen recognized by human natural antibodies, *Glycoconjugate J.* **11:**266–271.

Jentoft, N., 1990, Why are proteins O-glycosylated? *Trends Biochem. Sci.* **15:**291–294.

Kagawa, Y., Takasaki, S., Utsumi, J., Hosoi, K., Shimizu, H., Kochibe, N. and Kobata, A., 1988, Comparative study of the asparagine-linked sugar chains of natural human interferon-β1 and recombinant human interferon-β1 produced by three different mammalian cells, *J. Biol. Chem.* **263:**17508–17515.

Karlsson, K.-A., 1995, Microbial recognition of target-cell glycoconjugates, *Curr. Opin. Struct. Biol.* **5:**622–635.

Knibbs, R.N., Perini, F. and Goldstein, I.J., 1989, Structure of the major Concanavalin A reactive oligosaccharides of the extracellular matrix component laminin, *Biochemistry* **28:**6379–6392.

Kornfeld, R. and Kornfeld, S., 1985, Assembly of asparagine-linked oligosaccharides, *Annu. Rev. Biochem.* **54:**631–664.

Krotkiewski, H., Grönberg, G., Krotkiewska, B., Nilsson, B. and Svensson, S., 1989, Heterogeneity in the sugar moieties of glycoproteins: Structures of the oligosaccharides of a mouse monoclonal antibody OKT3, in: *Xth International Symposium on Glycoconjugates* (N. Sharon, H. Lis, D. Duksin and I. Kahane, eds.), Jerusalem, Israel.

Krotkiewski, H., Grönberg, G., Krotkiewska, B., Nilsson, B. and Svensson, S., 1990, The carbohydrate structures of a mouse monoclonal IgG antibody OKT3, *J. Biol. Chem.* **265:**20195–20201.

Landsteiner, K., 1945, *The specificity of serological reactions*, 2nd Ed., Harvard University Press, Cambridge.

Lipniunas, P., Grönberg, G., Krotkiewski, H., Angel A.S. and Nilsson B., 1993, Investigation of the structural heterogeneity in the carbohydrate portion of a mouse monoclonal immunoglobulin A antibody, *Arch. Biochem. Biophys.* **300:**335–345.

Liu, J., Qian, Y. and Holgersson, J., 1997, Removal of xenoreactive human anti-pig antibodies by absorption on recombinant mucin-containing glycoproteins carrying the Galα1,3Gal epitope. *Transplantation*, **63,** 1673–1682.

Natsuka, S. and Lowe, J.B., 1994, Enzymes involved in mammalian oligosaccharide biosynthesis, *Curr. Opin. Struct. Biol.* **4:**683–691.

Neethling, F.A., Koren,, E., Ye Y., Richards S.V., Kujundzic, M., Oriol, R. and Cooper, D.K.C., 1994, Protection of pig kidney (PK15) cells from the cytotoxic effect of anti-pig antibodies by α-galactosyl oligosaccharides, *Transplantation* **57:**959–963.

Oriol, R., Mollicone, R., Coullin, P., Dalix, A.-M. and Candelier, J.-J., 1992, Genetic regulation of the expression of ABH and Lewis antigens in tissues, APMIS Suppl.27, **100:** 28–38.

Oriol, R., Ye, Y., Koren, E. and Cooper, D.K., 1993, Carbohydrate antigens of pig tissues reacting with human natural antibodies as potential targets for hyperacute vascular rejection in pig-to-man organ xenotransplantation, *Transplantation* **56:**433–42.

Peters, B.P. and Goldstein, I.J., 1979, The use of fluorescein-conjugated Bandeiraea simplicifolia $B_4$-isolectin as a histochemical reagent for the detection of α-D-galactopyranosyl groups, *Exp. Cell Res.* **120:**321–334

Postigo, A.A., Marazuela, M., Sánchez-Madrid, F. and de Landázuri M.O., 1994, B lymphocyte binding to E- and P-selectins is mediated through the de novo expression of carbohydrates on *in vitro* and *in vivo* activated human B cells. *J. Clin. Invest.* **94:**1585–1596.

Rydberg, L., Björck, S., Hallberg, E., Magnusson, S. Strokan, V., Svalander, C.T., Samuelsson, B.E. and Breimer, M.E., 1996, Extracorporeal ("ex vivo") connection of pig kidneys to humans. II. The anti-pig' antibody response, *Xenotransplantation* **3**:340–353.

Samuelsson, B.E. and Breimer, M.E., 1997, ABH antigens: Some basic aspects, *Transplantation Proc.* **19**:4401–4407.

Samuelsson, B.E., Rydberg, L., Breimer, M.E., Bäcker, A., Gustavsson, M., Holgersson, J., Karlsson, E., Uyterwaal, A.-C., Cairns, T. and Welsh, K., 1994, Natural Antibodies and Human Xenotransplantation, *Immunol. Rev.* **141**:151–168.

Sandrin, M.S., Vaughan, H.A., Dabkowski, P.L. and McKenzie, I.F.C., 1993, Anti-pig IgM antibodies in human serum react predominantly with gal(α1–3)gal epitopes, *Proc. Natl. Acad. Sci. USA* **90**:11391–11395.

Shibata, S., Peters, B.P., Roberts, D.D., Goldstein, I.J. and Liotta, L.A., 1982, Isolation of laminin by affinity chromatography on immobilized *Griffonia simplicifolia* I lectin, *FEBS Lett.* **142**:194–198.

Spiro, R.G. and Bhoyroo, V.D., 1984, Occurence of a-D-galactosyl residues in the thyroglobulins from several species. Localization in the saccharide chains of the complex carbohydrate units. *J. Biol. Chem.* **259**:9858–9866.

Stellner, K., Saito, H. and Hakomori, S.I., 1973, Determination of aminosugar linkage in glycolipids by methylation. Aminosugar linkage of ceramide pentasaccharides of rabbit erythrocytes and of Forssman antigen, *Arch. Biochem. Biophys.* **155**:464–472.

Strokan, V., Rydberg, L., Hallberg, E.C., Mölne, J. And Breimer M.E., 1998a, Characterisation of human natural anti-sheep xenoantibodies, *Xenotransplantation* **5**:111–121.

Strokan, V.,Mölne, J., Svalander, C. And Breimer, M.E., 1998b, Hetergeneous expressioin of Galα1–3Gal xenoantigen in pig kidney. A lectin- and immunogold ecelctron microscopic study, Transplantation, accepted.

Suzuki, E. and Naiki, M, 1984, Heterophile antibodies to rabbit erythrocytes in human sera and identification of the antigen as a glycolipid, *J. Biochem.* **95**:103–108.

Teneberg, S., Lönnroth, I., Torres López, J.F., Galili, U., Ölwegård Halvarsson, M., Ångström, J. and Karlsson, K.-A., 1996, Molecular mimicry in the recognition of glycosphingolipids by Galα3Galβ4GlcNAcβ-binding *Clostridium difficile* toxin A, human natural anti α-galatosyl IgG and the monoclonal antibody Gal-13: characterization of a binding-active human glycosphingolipid, non-identical with the animal receptor, *Glycobiology* **6**:599–609.

Thall, A. and Galili, U., 1990, Distribution of Galα1–3Galβ1–4GlcNAc residues on secreted mammalian glycoproteins (thyroglobulin, fibrinogen, and immunoglobulin G) as measured by a sensitive solid-phase radioimmunoassay, *Biochemistry*, **29**:3959–3965.

Uemura, K., Yuzuwa, M. and Taketomi, T., 1978, Characterization of major glycolipids in bovine erythrocyte membrane, *J. Biochem.* **83**:463–471.

Ulfvin, A., Bäcker, A.E., Clausen, H., Hakomori, S., Rydberg, L., Samuelsson, B.E. and Breimer, M.E., 1993, Expression of glycolipid blood group antigens in single human kidneys. Change in antigen expression of rejected ABO incompatible kidney grafts, *Kidney Int.* **44**:1289–1297.

von Dungern, E and Hirszfeld, L., 1911, Ueber Gruppenspezifishe Strukturen des Blutes, *Z. Immunitätsforsch.* **8**:526–562.

Wang X., O'Hanlon T.P. and Lau J.T.Y.,1989, Regulation of β-galactoside α2,6-sialyltransferase gene expression by dexamethasone. *J. Biol. Chem.* **264**:1854–1859.

Wang, X.C., Smith, T.J. and Lau, J.T.Y., 1990, Transcriptional regulation of the liver β-galactoside α2,6-sialyltransferase by glucocorticoids. *J. Biol. Chem.* **265**:17849–17853.

Watanabe, K., Hakamori, S., Childs, R.A. and Feizi, T., 1979, Characterization of a blood Group I-active ganglioside, *J Biol Chem.* **254**:3221–3228.

Welsh, K.I., Taube, D.H., Thick, M., Palmer, A., Stevens, N. and Binns, R.,1991, Human antibodies to pig determinants and their association with hyperacute rejection of xenografts, Chapter 31 in:

*Xenotransplantation* (D.K.C Cooper, K. Reemtsma and D.J.G. White, eds.), pp.501–510, Springer-Verlag, Heidelberg,

Yamakawa, T., Matsumoto, M. and Suzuki, S., 1956, The chemistry of the lipids of posthemolytic residue or stroma of erythrocytes. VIII. The nature of hexosamine and fatty acids of blood cells sphingolipids, *J.Biochem.* **43**:63–72.

Yamakawa, T., Irie, R. and Iwanaga, M., 1960, The chemistry of lipid of posthemolytic residue or stroma of erythrocytes. IX. Silicic acid chromatography of mammalian stroma glycolipids, *J. Biochem* **48**:490–507.

*Chapter 6*

# The *Griffonia simplicifolia* I - $B_4$ Isolectin
# A Probe for α-D-Galactosyl End Groups

Irwin J. Goldstein and Harry G. Winter

## 1. INTRODUCTION

Lectins are a heterogeneous group of proteins with the common property of recognizing carbohydrate molecules, often in a very specific manner. Although first discovered in plant seeds, they also have been found in other vegetative tissues of the plant world (leaves, stems, bark, roots, tubers, etc.). In recent years, lectins have been isolated from microorganisms, insects, crustacea, rodents, and most recently from various tissues of primates where they appear to be involved in cellular adhesion, clearance of glycoproteins from the circulating system, cell-cell recognition, etc.

Because of the relative ease of isolation and purification, and their wide range of sugar binding specificity, plant lectins have proven to be valuable reagents in biomedical research. They have been employed as probes for the detection and preliminary characterization of complex carbohydrates (glycoproteins,

---

**Irwin J. Goldstein and Harry G. Winter** Department of Biological Chemistry, University of Michigan, Ann Arbor, Michigan 48109-0606.

*Subcellular Biochemistry, Volume 32: α-Gal and Anti-Gal*, edited by Galili and Avila.
Kluwer Academic / Plenum Publishers, New York, 1999.

glycolipids, polysaccharides) in solution and on the surface of cells; for the sorting of cells based on differences of membrane carbohydrates; in their immobilized form as matrices for the affinity chromatographic isolation of glycoproteins and the resolution of mixtures of sugar nucleotides. (For reviews see Liener *et al.*, 1986; Lis and Sharon, 1998).

In this chapter, we concentrate on the isolation, purification, characterization and application to biomedical research of a single system of isolectins present in *Griffonia simplicifolia* seeds.

## 2. THE *GRIFFONIA SIMPLICIFOLIA* FAMILY OF LECTINS

*Griffonia simplicifolia* Baill. (formerly known as *Bandeiraea simplicifolia* Benth.) is a leguminous shrub which grows in the rain forests of northwest Africa. The seeds of this plant contain a series of lectins of differing carbohydrate binding specificity. These include the *Griffonia simplicifolia* I (GS I) lectin composed of five isolectins which recognize Gal and GalNAc (the subject of this chapter); a GlcNAc–binding lectin (GS II); and GS IV which recognizes the Lewis b blood group antigen.

## 3. ISOLATION AND PURIFICATION OF THE *GRIFFONIA SIMPLICIFOLIA* I LECTIN (GS I)

Mäkelä and Mäkelä (1956) first reported potent anti-blood group B activity in *Griffonia simplicifolia* seed extracts, an observation confirmed by Mäkelä (1957) and Mäkelä *et al.* (1959). This activity was inhibited by galactose and galactose-containing oligosaccharides and type B blood group substance (Mäkelä *et al.*, 1959). Horejsî and Kocourek (1973) isolated a lectin from *G. simplicifolia* seeds using an α–galactosyl–substrate matrix as an affinity adsorbent but provided little characterization of the protein obtained. Employing a melibionate–BioGel affinity column, Hayes and Goldstein (1974) isolated a lectin that appeared homogeneous by polyacrylamide gel electrophoresis, immunoelectrophoresis, gel filtration and sedimentation analysis. The lectin was a tetrameric glycoprotein of mol wt. 114,000 containing 9.0% by weight carbohydrate (mannose, L–fucose, xylose, and N–acetylglucosamine). Amino acid analysis of the lectin revealed a high content of hydroxylic and acidic amino acids and a striking paucity of sulfur containing amino acids (one cysteine/monomeric unit), a common feature of legume lectins. The finding of 0.4 residues of methionine (Hayes and Goldstein, 1974) presaged the existence of isolectins (Murphy and Goldstein, 1977). The lectin requires $Ca^{++}$ for activity (Hayes and Goldstein, 1974).

Equilibrium dialysis and cell binding studies on the GS I lectin were conducted by Hayes and Goldstein (1975). It was determined that there was one binding site per subunit for methyl α-D-galactopyranoside with an intrinsic association constant of $8.6 \times 10^4\ M^{-1}$ at 2°C and $3.3 \times 10^4\ M^{-1}$ at 20°C. The receptor sites per human B erythrocyte varied from 0.72 to $1.34 \times 10^5$ with an apparent association constant of 1.1 to $2.9 \times 10^7\ M^{-1}$.

Chemical modification studies indicated the importance of cysteine residues and carboxyl groups (Glu, Asp) for the carbohydrate binding activity of the lectin whereas free amino groups do not appear to be involved (Lönngren and Goldstein, 1976). Similar to other lectins, GS I was shown to contain 30–40% β-structure by CD spectroscopy (Lönngren *et al.*, 1976).

## 4. SEPARATION AND PROPERTIES OF THE GS I ISOLECTINS

Upon further investigation, the *Griffonia simplicifolia* I lectin was found to consist of a mixture of five isolectins (Murphy and Goldstein, 1977). They are tetrameric structures composed of varying proportions of two different subunits, A and B and are designated GS I -$A_4$, -$A_3B$, -$A_2B_2$, -$AB_3$, and -$B_4$ (Murphy and Goldstein, 1977). The A subunit is specific for α-D-GalNAc end groups and binds readily to the blood type A determinant, as well as to terminal α-D-Gal groups found on type B cells. The B subunits are specific for α-D-Gal end groups and bind very specifically to type B erythrocytes. The A and B subunits differ in two further fundamental respects: subunit B has a mass of 1000 daltons greater than subunit A; and subunit B has a single methionyl residue whereas A has none (see Table I).

A structural comparison between the A and B units of the GS I lectin was undertaken to determine the extent of homology between the subunits (Lamb and Goldstein, 1984). The first 25 *N*-terminal amino acids of both A and B subunits were determined following the enzymatic removal of *N*-terminal pyroglutamate blocking groups with pyroglutamate aminopeptidase. Although 21 amino acids were common to both subunits, there were four unique amino acids in the N–terminal sequence of the A and B subunits. Residues 8, 9, 17, and 19 were asparagine, leucine, lysine and asparagine in subunit A, and threonine, phenylalanine, glutamic acid and serine in subunit B. Although complete amino acid sequences have not been determined, tryptic peptide mapping using reverse-phase HPLC showed that approximately 30% of the tryptic peptides were common to both subunits. Thus, although the *N*- and *C*-terminal regions of A and B are similar, each subunit possesses unique sequences. In contrast, homology of the *N*-terminal sequences with other leguminous lectins, while evident, was much less (Young and Oomen, 1992), even with the lectin from the most closely-related species (*Bauhinia purpurea*), which is also Gal/GalNAc-specific (Kusai *et al.*, 1991).

The five GS I isolectins were resolved by successive chromatography on two affinity matrices, namely melibionate–BioGel and insolubilized hog mucin A+ H (or human cyst A) substance. The GS I-$A_4$ and -$A_3B$ isolectins were purified on the

**Table I**
**Physical-Chemical Properties of the GS I-A and -B Subunits***

| Property | | Subunit A | Subunit B |
|---|---|---|---|
| Molecular mass | | 32 kDa | 33 kDa |
| | | 1 5 10 15 20 25 | |
| *N*-terminal amino acids | A PyroGln | S D S V S F N L P N F W S D V K D N I I F Q G D A | |
| | B PyroGln | S D S V S F T F P N F W S D V E D S I I F Q G D A | |
| *C*-terminal amino acids | | R (F, V) L T S | |
| Methionyl residues | | 0 | 1 |
| Cysteinyl residues | | 1 | 1 |
| Carbohydrate analysis** | | 6.9% | 7.2% |
| | | Man(8); Xyl (2); Fuc (2.0) | |
| $K_{assoc}$ methyl α-D-galactoside | | $(A_4)$ $1.45 \times 10^4$ $M^{-1}$ | $(B_4)$ $2.06 \times 10^4$ $M^{-1}$ |
| $K_{assoc}$ methyl α-D *N*-acetylgalactosaminide | | $(A_4)$ $1.87 \times 10^5$ $M^{-1}$ | $(B_4)$ not detectable |

*From Murphy and Goldstein, 1979 and Lamb and Goldstein 1984.
**Neutral carbohydrates by phenol-$H_2SO_4$ assay.

melibionate-BioGel column, and the $A_2B_2$, $AB_3$ and $B_4$ separated on hog mucin A+H substance by elution with methyl α-D-galactoside (Murphy and Goldstein, 1977). An improved procedure for the isolation of the five isolectins involves differential elution from a *p*-aminophenyl β-D-galactopyranosyl-succinylamino-hexylaminyl-Sepharose-4B column (Delmotte and Goldstein, 1980). *N*-acetyl galactosamine displaced the $A_4$ isolectin. A step-wise or gradient increase of methyl α-D-galactoside concentration displaced the remaining four isolectins in order of increasing content of B subunit. Guaran beads, prepared by cross-linking guar gum with epichlorohyrin, also proved to be an efficient immunosorbent for the isolation of the mixture of the five GS I isolectins (Lönngren *et al.*, 1976). Subtypes of the A and B subunits were demonstrated by isoelectric focusing in 8 M urea and the content of subunit subtypes was shown to vary from seed to seed (Murphy and Goldstein, 1979). Two-dimensional agar gel diffusion studies revealed that, in addition to common structural features, each subunit contains its own distinct antigenic determinants (Murphy and Goldstein, 1979).

Equilibrium dialysis and fluorescence enhancement measurements, employing $^{14}C$-labeled methyl α-D-glucopyranoside and methyl 2-acetamido-2-deoxy-α-D-galactopyranoside, were conducted on three of the GS I isolectins ($A_4$, $A_2B_2$, and $B_4$) (Goldstein *et al.*, 1981). The A and B subunits appear to have approximately the same $K_{assoc}$ for methyl α-D-Gal*p*: $1.45 \times 10^4$, $1.98 \times 10^4$ and $2.06 \times 10^4\ M^{-1}$ for $A_4$, $A_2B_2$ and $B_4$, respectively. Each A and B subunit has a single binding site for Me α D-Gal*p*. It was further determined by equilibrium dialysis against Me α-GalNAc*p*, that $A_4$ had 3.3 (theory 4) sites and $K_{assoc} = 1.87 \times 10^5\ M^{-1}$; $A_2B_2$, 1.9 sites (theory 2) and $K_{assoc}$ $1.19 \times 10^5\ M^{-1}$ whereas $B_4$ has no detectable binding sites. However, four very weak Me α D-GalNAc binding sites for $B_4$ were detected by fluorescence enhancement measurements ($K_{assoc} = 1.26 \times 10^2\ M^{-1}$). Thus the A subunit has an affinity for methyl α-D-GalNAc*p* three orders of magnitude greater than the B subunit (Goldstein *et al.*, 1981).

Carbohydrate binding studies on the GS I lectin have been carried out using the techniques of quantitative precipitation and hapten inhibition analysis, and equilibrium dialysis and fluorescent enhancement approaches (see above). These studies were conducted both on the native *mixture* of the GS I isolectins before it became evident that the lectin consisted of a mixture of five isolectins, as well as on the individual A and B subunits (Murphy and Goldstein, 1977; 1979; Wood *et al.*, 1979).

Both GS I-$A_4$ and -$B_4$ formed precipitates with biopolymers containing multiple, non-reducing α-D-Gal*p* end groups; these include guaran (a galactomannan) and type B blood group substance (Murphy and Goldstein, 1979). Biopolymers containing β-galactosyl end groups (e.g. larch arabinogalactan) did not precipitate with either isolectin.

A number of biopolymers containing multiple, nonreducing α-D-GalNAc*p* end groups were also subjected to quantitative precipitation analysis with GS I-$A_4$

and $B_4$. Type A blood group substance, a streptococcus type C cell wall polysaccharide and asialo bovine submaxillary mucin precipitated strongly with $A_4$, whereas only asialo bovine submaxillary mucin precipitated a small amount of $B_4$. These precipitation studies demonstrated the very strong preference of the $B_4$ isolectin for α-D-Gal*p* end groups.

Inhibition by mono- and oligosaccharides and their derivatives was studied by quantitative hapten inhibition of the precipitation reaction between GS I-$A_4$ and -$B_4$ and guaran (Murphy and Goldstein, 1979) and blood group A substance (for GS I-$A_4$) and B substance (for GS I-$B_4$) (Wood *et al.*, 1979).

The best inhibitors of the GS I-$B_4$-guaran system were Me α-D-Gal*p* and *p*-nitrophenyl α- and β-D-Gal*p*. Comparison of Me α-D-Gal*p* with its β-anomer and a series of oligosaccharides containing nonreducing α- and β-D-galactopyranosyl end units revealed that the α-anomer is bound significantly more strongly than the corresponding β-anomer. For example, melibiose (Galα1-6Glc) is bound almost 100-fold more strongly than lactose (Murphy and Goldstein, 1979). Similarly, Galα1-3Gal is an excellent ligand for the $B_4$ isolectin (Wood *et al.*, 1979). Substitution at the O-2 position of D-galactose has a substantial effect on its binding to $B_4$. D-Galactose is 41 times more inhibitory than 2-deoxy-D-galactose, and methyl α-D-Gal*p* is almost 100-fold more potent an inhibitor than Me α-D-GalNAc*p*. Substitution by *O*-methyl, *O*-acetyl or *O*-benzyl at the C-2 hydroxyl groups of Me α-D-Gal*p* drastically reduce their affinity for GS I-$B_4$ (Kaifu *et al.*, 1985). The C-6 hydroxymethyl group also is an important binding locus for interaction with $B_4$: D-fucose (6-deoxy-D-galactose) is approximately 30 times less potent as an inhibitor for the $B_4$-guaran precipitation reaction than D-Gal (Murphy and Goldstein, 1979).

On the other hand, the GS I-$A_4$ isolectin displays potent binding activity for D-GalNAc end groups: Me α-D-GalNAc is bound 3 orders of magnitude more strongly than Me α-D-Gal*p*. Oligosaccharides with terminal nonreducing α-D-GalNAc groups are excellent ligands for GS I-$A_4$ but do not bind to the $B_4$ isolectin. The environment of the saccharide binding sites for GS I-$A_4$ and -$B_4$ are significantly different as revealed by fluorescence quenching studies using methyllumbelliferyl glycosides (de Boeck *et al.*, 1981).

## 5. USES AND APPLICATIONS OF THE GS I ISOLECTINS

### 5.1. Affinity Chromatographic Applications

The use of the mixture of the GS I isolectin and the GS I-$B_4$ isolectin, in their immobilized form, have been employed to isolate polysaccharides and glycoproteins with terminal, nonreducing α-D-galactopyranosyl groups. Immobilization has generally been to cyanogen bromide-activated Sepharose 4B (Ross *et al.*, 1976). Galactomannans from *Cassia alata*, *Cyamopsis tetragonolobus* (guar gum), and yeast (*Torulopsis gropengiesseri*) bound to the mixture of immobilized GS I isolectins and were displaced by D-galactose (Ross *et al.*, 1976).

In similar fashion numerous glycoproteins (and glycopeptides) with α-D-galactosyl end-groups have been purified. These include murine laminin (Shibata *et al.*, 1982) and α-D-galactosyl-terminated oligosaccharides from laminin (Knibbs *et al.*, 1989).

Calf thyroid plasma membranes contain glycoproteins with α-D-galactosyl end groups. Using immobilized GS I and the GS I-$B_4$ isolectin, glycoproteins, glycopeptides and oligosaccharides derived from these glycoproteins have been isolated and characterized from both *N*- and *O*-linked chains (Okada and Spiro, 1980; Edge and Spiro, 1985; 1997). Also isolated on immobilized GS I columns were a family of α-D-galactopyranosyl-terminated glycopeptides from Ehrlich ascites tumor cell membranes (Eckhardt and Goldstein, 1983); a trypsin-released glycopeptide, with α-D-galactosyl end-groups, present on activated murine macrophages, but not on resident populations (Maddox *et al.*, 1982a; Petryniak *et al.*, 1992); glycoproteins containing α-D-galactosyl-terminated glycoproteins from human breast carcinoma cell line MCF-7 not present, or differentially expressed, on human mammary epithelial cells (Petryniak *et al.*, 1991); and bovine and porcine fibrinogen molecules containing nonreducing terminal α-D-galactosyl groups (Thall and Galili, 1990). Conversely, it was shown that bovine prothrombin failed to form a precipitate with GS I-$B_4$ or to bind to the same lectin immobilized on Sepharose 4B (Eckhardt *et al.*, 1980) indicating it does not contain α-D-galactosyl groups as previously reported (Nelsestuen and Suttie, 1972). The one step purification of murine IgD on immobilized GS I was reported by Oppenheim *et al.* (1990).

It has also been possible to resolve mixtures of isomers, nucleotide sugars and oligosaccharides on immobilized GS I and GS I-$B_4$. For example, when a mixture of UDP-glucose and UDP-galactose was loaded onto a GS I-Sepharose column, a clean separation of the two nucleotide sugar was obtained: the UDP-glucose emerged with the void volume whereas the UDP-galactose was sufficiently retarded to completely resolve the two sugar nucleotides (Blake and Goldstein, 1980; 1982). In similar fashion, a GS I-$B_4$-Sepharose column was employed to separate a mixture of UDP-GalNAc and UDP-Gal: the UDP-GalNAc was slightly retarded, but separated completely from the UDP-Gal which was significantly retarded (Blake and Goldstein, 1980; 1982). The disaccharide alditol, [$^3$H] lactitol, was separated from [$^3$H] melibiitol on a GS I-Sepharose column by taking advantage of the strong preference of the GS I lectin for α-D-galactosyl end groups resulting in the strong retardation of the melibiitol (α-D-gal-(1-6)-D-glucitol) (Blake and Goldstein, 1980; 1982).

## 5.2. Use of GS I-$B_4$ as a Probe for α-D-Galactosyl End-Groups

Because of its pronounced specificity for non-reducing, terminal α-D-galactosyl groups, the *Griffonia simplicifolia* $B_4$ isolectin (GS I-$B_4$) has been one of the most widely used and useful reagents for the detection of these groups in biological materials. The isolectin has been labeled with fluorescein, ferritin, colloidal gold, biotin, and radioisotopes.

Ehrlich ascites tumor cells were strongly agglutinated by GS I-$B_4$ (0.4 μg/ml) (Eckhardt and Goldstein, 1983). Incubation of the cells with GS I-$B_4$ labeled with either fluorescein isothiocyanate or ferritin followed by examination with the light and electron microscope revealed a random distribution of α-D-Gal*p* units over the entire cell surface (Eckhardt and Goldstein, 1983). Cell binding studies with [$^3$H] propionate-labeled GS I-$B_4$ demonstrated a minimum of $1.8 \times 10^3$ α-D-Gal*p* sites per Ehrlich cell (Eckhardt and Goldstein, 1983).

In similar fashion, GS I-$B_4$ labeled with colloidal gold was used to explore the subcellular distribution of α-D-gal units in Ehrlich tumor cells grown as ascites or monolayer. The microvillar, but not the smooth plasma membranes, were labeled by GS I-$B_4$. In the Golgi apparatus, labeling was observed to commence in the trans cisternal vacuoles, indirect evidence for the occurrence of Galα(1,3)Gal units in Ehrlich cell membrane glycoproteins (Roth and Goldstein, 1995).

In a comprehensive study of the occurrence of α-D-galactosyl end-groups in cryostat sections of the mouse, rat, rabbit and TA3 murine mammary carcinoma, Peters and Goldstein (1979) reported that fluorescein-labeled GS I-$B_4$ (FITC-$B_4$) bound to the basement membrane in a variety of tissues including kidney cortex (peritubular and glomerular structures); capillaries of cardiac and skeletal muscle and brain; intercellular substance of the cornea and cells of the stratum basale of the skin. The binding of FITC-$B_4$ to tissue sections was abolished by pretreatment with coffee bean α-galactosidase but not by either *A. niger* or *E. coli* β-galactosidase. Of importance is the fact that a second probe of kidney basement membrane, namely, antibody to that structure, exhibited the same peritubular and glomerular fluorescence localization as that observed with FITC-$B_4$. It is suggested that one of the specific molecular structures with which FITC-$B_4$ is interacting is laminin, a glycoprotein shown to contain α-galactosyl end-groups (Shibata *et al.*, 1982; Knibbs *et al.*, 1989). Indeed, Rao *et al.* (1983) showed by a solid phase assay using GS I-$B_4$ that laminin contains α-D-galactosyl end groups.

The Galα-1,3 Gal epitope (α-gal epitope) present on porcine organs is the major target of human xenotransplants by natural anti-Gal antibodies. The α-1,3 galactosyltransferase (α-1,3 GT) which is responsible for the synthesis of the α-gal epitope is located within the Golgi compartment, in accordance with the pattern of staining of pig organs by FITC-$B_4$ (Oriol *et al.*, 1993). Localization of the α-gal epitope as a product of four isoforms of porcine α1,3 GT in the Golgi apparatus was conducted using FITC-$B_4$ (Vanhove *et al.*, 1997). Similarly, Cho and

Cummings (1997) showed the expression of α-gal epitopes on the surface of mouse teratocarcinoma F9 cells by flow cytometry in the presence of FITC-$B_4$.

A rather large series of studies have dealt with the problem of preventing immunorejection of organs from animals to humans. As indicated above, it involves the presence of the α-gal epitope on xenographs. Two possible genetic approaches to eliminate the α-gal epitope involve "knocking out" the α-1,3 GT gene by homologous recombination, and competition for the enzyme's *N*-acetyllactosamine substrate by transgenically expressed α1,2-fucosyltransferase (H-transferase). These "solutions" have been explored by several groups. The GS I-$B_4$ isolectin has played an important role in assessing the results of these experiments. For example, Shinkel *et al.* (1997) showed an 80% reduction in FITC-$B_4$ staining in "knock out" mice and 70% reduction in the α-gal epitope expression in mice expressing H-transferase (Koite *et al.*, 1997). Similarly Tearl *et al.* (1996) used the avidin-biotin-peroxidase technique for labeling $B_4$ in their "knock out" mouse model. On the other hand, Gorelik et al. (1995) employed biotinylated GS I-$B_4$ to confirm the existence of α-D-galactosyl groups on the cell surface of mice tranfected with α-1,3 GT.

A further technique for circumventing the immune rejection of xenogenic porcine endothelial cells involved removal of terminal α-galactosyl units with coffee bean α-galactosidase as evaluated by flow cytometry with FITC-$B_4$ (Watier *et al.*, 1996).

Other examples of the use of GS I lectin as a probe for the presence of α-D-galactosyl end-groups include: the binding of FITC-$B_4$ to the basal cells of the newborn rat epidermis (Brabec *et al.*, 1980); the staining of human breast carcinomas with peroxidase-labeled $B_4$ (Ito *et al.*, 1996) thus confirming the presence of α-D-galactosyl groups on these tumors (c.f. Castronovo *et al.*, 1989; and Petryniak *et al.*, 1991); the localization of a glycolipid(s) bearing α-D-galactosyl groups on endothelial cells in human kidney and possibly tubular epithelia (Faraggiana *et al.*, 1989) using peroxidase-labeled GS I. It is known that patients with Fabry's disease display high contents of "Fabry's glycolipid" (Galα1-4 Galβ1-4 Glcβ1-sphingosine) (Faraggiana *et al.*, 1982). Model glycolipids containing melibionate groups (e.g., *N*-dodecylmelibionate) formed specific precipitates with GS I-$B_4$ (Williams *et al.*, 1979).

Oocyte α-gal epitopes have been implicated in sperm adhesion to the zona pellucida glycoproteins of mice. Murine oocytes show strong staining with FITC-$B_4$ whereas mice deficient in α1,3GT do not (Thall *et al.*, 1995). It was shown that female α1,3GT (–/–) mice yield oocytes devoid of the α-gal epitope but are fully fertile (Thall *et al.*, 1995).

The content of blood group B antigen in human tissues was estimated by microphotomeric quantitation of GS I-$B_4$ staining (Nagaike *et al.*, 1997). The lectin was conjugated with horseradish peroxide and used with or without prior treatment of the tissue with coffee bean α-galactosidase.

### 5.3. Biological Effects of GS I Lectin

The demonstration that the GS I-$B_4$ isolectin binds to the α-gal epitope present on the surface of many types of cells opened the door to its possible involvement in the modulation of biological processes. Indeed, it was shown that consistent with its ability to agglutinate Ehrlich ascites tumor cells (Eckhardt and Goldstein, 1983a) the GS I lectin is cytotoxic to these cells *in vivo* (Eckhardt *et al.*, 1982). Mice given intraperitoneal injections of $10^7$ Ehrlich cells followed by 100 μg of GS I daily showed significantly enhanced survival rates compared with mice given only $10^7$ Ehrlich cells. It was speculated that "lectin-dependent, macrophage-mediated tumor cell lysis may be the mechanism whereby the GS I lectin exerts its cytotoxic effect" (Eckhardt *et al.*, 1982). This hypothesis was supported when it was shown that the lectin was not toxic to Ehrlich tumor cells directly, nor to nonspecifically initiated complement-mediated lysis of these target cells. Rather it was found that the lectin appears to be able to provide a linkage between effector (activated) macrophages (Maddox *et al.*, 1982b) and Ehrlich tumor cells resulting in the triggering of a cytolytic event (Maddox *et al.*, 1982a; 1982b).

The observation that GS I-$B_4$ bound to stimulated, but not resident murine macrophages (Maddox, 1982a) prompted a study of the binding of $B_4$ to murine peritoneal cells (Petryniak *et al.*, 1986). Intraperitoneal injection of thioglycolate resulted in the production of a population of *activated* murine macrophages which showed an increase in agglutinability by $B_4$ (12–50 μg/mL vs. > 200 μg ml) over resident macrophages. These results indicate the expression of cell surface α-D-galactosyl-containing glycoproteins and/or glycolipids. There also was a significant increase in the binding of $B_4$ (2.4 to 20-fold) to stimulated as opposed to resident macrophages (Petryniak *et al.*, 1986).

The finding that GS I-$B_4$ was cytotoxic to Swiss 3T3 mouse cells stimulated its use for the generation of mutants resistant to the cytotoxic effects of the lectin. After mutagenesis with ethyl methanesulfonate, and selection with GS I-$B_4$, a variant clonal cell line resistant to both this isolectin and the α-D- and β-D-galactose-binding toxic lectin abrin was isolated (Stanley *et al.*, 1979). Thus this variant 3T3 clonal cell line displayed an altered membrane phenotype from its wild parental cell line. A second lectin, GS II, derived from the same *G. simplicifolia* seeds, but specific for nonreducing terminal *N*-acetyl-D-glucosaminyl groups, was cytotoxic to the variant, but not the parental cells. These results suggest altered glycoprotein biosynthesis resulting in the exposure of subterminal *N*-acetyl-D-glucosaminyl moieties on the surface of the variant cell line.

A unique application of the GS I-$B_4$ isolectin to the purification of type II alveolar epithelial cells was reported by Simon *et al.* (1986). It was found that the $B_4$ isolectin bound to alveolar macrophages due to the presence of α-D-galactosyl groups on the surface of these cells, substantiated by the liberation of galactose upon treating the cells with coffee bean α-galactosidase but not by β-galactosidase. Type II alveolar cells do not possess terminal α-D-gal end groups. When proteinase-derived suspensions of pulmonary cells are incubated with GS I lectin, the macrophages agglutinate and can be removed by filtration through nylon mesh. The filtrate contained a cellular suspension of the type II alveolar epithelial cells that were 94 ±2% pure (Simon *et al.*, 1986). In similar fashion, Bankston *et al.* (1991) reported that biotinylated GS I bound to the endothelial cells of alveolar capillaries and to interstitial macrophages of rat lung, but not to the type II alveolar cells in adult rats. GS I-$B_4$ specifically labeled the intracellular canaliculi and the apical surface on the fundic gland parietal cells (Ihida *et al.*, 1990).

### 5.4. Comparison of the Carbohydrate Binding Specificity of Human Natural Anti-Gal Antibody with the *Griffonia simplicifolia* I-$B_4$ Isolectin

Among the natural human antibodies which occur in human serum is an IgG protein which recognizes the Galα(1-3) Gal epitope (α-gal epitope) (Galili *et al.*, 1985). As much as 1% of circulating IgG is directed against this structure. Interestingly, the α-gal epitope is extensively expressed on cells from New World monkeys but not found on cells of Old World monkeys, apes, and humans (Galili *et al.*, 1987; 1988; 1993). The possible origin of this population of anti-Gal antibodies and their properties are described elsewhere in this book. Here, we compare the carbohydrate binding properties of the natural, human anti-Gal antibodies with those of GS I-$B_4$.

Both anti-Gal and GS I-$B_4$ recognize terminal, non-reducing α-D-galactopyranosyl groups. However, the specificity of the anti-Gal antibodies extends to include a subterminal galactosyl residue linked α-1,3(Galα1,3Gal). The α1,4 isomer is a noninhibitor of anti-Gal-mediated agglutination of rabbit red blood cells, and the α1,6 linkage as found in melibiose [Galα(1-6)Glc] was a much poorer inhibitor than for GS I-$B_4$ (Galili *et al.*, 1985). On the other hand, the GS I-$B_4$ isolectin appears to lack the extended specificity of the α-gal antibodies. It was shown by hapten inhibition assay that GS I-$B_4$ had the same binding affinity for the α-gal epitope as it did for methyl α-D-galactopyranoside (Wood *et al.*, 1979). Melibiose [Galα(1-6)Glc] and Galα(1-4)Galα-OMe were only one half as active as the α-1,3-isomer (cf. Wu *et al.*, 1995). Therefore, it is difficult to conclude from these data that the GS I-$B_4$ isolectin has an extended binding site.

## REFERENCES

Bankston, P. W., Porter, G. A., Milici, A. J., and Palade, G. E., 1991, Differential and specific labeling of epithelial and vascular endothelial cells of the rat lung by *Lycopersicon esculentum* and *Griffonia simplicifolia* I lectins, *Eur. J. Cell Biol.* **54**: 187–195.

Blake, D. A. and Goldstein, I. J., 1980, Resolution of nucleotide sugars and oligosaccharides by lectin affinity chromatography, *Anal. Biochem.* **102**: 103–109.

Blake, D. A. and Goldstein, I. J., 1982, Resolution of carbohydrates by lectin affinity chromatography, *Meth. Enzymol.* **83**: 127–132.

Brabec, R. K., Peters, B. P., Bernstein, I. A., Gray, R. H., and Goldstein, I. J., 1980, Differential lectin binding to cellular membranes in the epidermis of the newborn rat, *Proc. Natl. Acad. Sci. USA* **77**: 477–479.

Castronovo, V., Colin, C., Parent, B., Foidart, J.-M., Lambotte, R., and Mahieu, P., 1989, Possible role of human natural anti-Gal antibodies in the natural antitumor defense system, *J. Natl. Cancer Inst.* **81**: 212–216.

Cho, S. K., and Cummings, R. D., 1997, A soluble form of α-1,3 galactosyltransferase within cells to galactosylate glycoproteins, *J. Biol. Chem.* **272**: 13622–13628.

de Boeck, H., Loontiens, F. G., Delmotte, E. M., and de Bruyne, C. K., 1981, The use of 4-methylumbelliferyl glycosides in binding studies with the lectins BS I-$A_4$, BS I-$B_4$ and BS II from *Bandeiraea (Griffonia) simplicifolia*, *FEBS Lett.* **126**: 227–230.

Delmotte, F. M. and Goldstein, I. J., 1980, Improved procedure for purification of the *Bandeiraea simplicifolia* I isolectins and *Bandeiraea* II lectin by affinity chromatography, *Eur. J. Biochem.* **112**: 219–223.

Eckhardt, A. E. and Goldstein, I. J. 1983a, Occurrences of α-D-galactosyl-containing glycoproteins on Ehrlich tumor cell membranes, *Biochemistry* **22**: 5280–5289.

Eckhardt, A. E. and Goldstein, I. J., 1983b, Isolation and characterization of a family of α-D-galactosyl-containing glycopeptides from Ehrlich ascites tumor cells, *Biochemistry* **22**: 5290–5297.

Eckhardt, A. E., Goldstein, I. J., and Nelsestuen, G L, 1985, Bovine prothrombin does not contain α-D-galactopyranosyl groups, *Arch. Biochem. Biophys.* **169**: 635–636.

Eckhardt, A. E., Malone, B. N., and Goldstein, I. J., 1982, Inhibition of Ehrlich ascites tumor cell growth by *Griffonia simplicifolia* I lectin *in vivo*, *Cancer Res.* **42**, 2977–2979.

Edge, A. S. B. and Spiro, R. G., 1985, Thyroid cell surface glycoproteins. Nature and deposition of carbohydrate units and evaluation of their blood group I activity, *J. Biol. Chem.* **260**: 1532–1538.

Edge, A. S. B. and Spiro, R. G., 1997, Structure of the O-linked oligosaccharides from a major thyroid cell surface glycoprotein, *Arch. Biochem. Biophys.* **343**(1): 73–80.

Faraggiana, T., Churg, J. Grishman, E., Strauss, L., Prado, A., Bishop, D. F., Schuchman, E., and Desnick, R. J., 1982, Light- and electron-microscope histochemistry of Fabry's Disease, *Am. J. Pathol.* **103**: 247–262.

Faraggiana, T., Crescenze, A., and Marinozzi, V., 1989, Presence of an α-galactolipid on the cell surfaces of endothelial cells of human kidney, *Histochem. J.* **21**: 235–240.

Galili, U., 1993, Interaction of the natural anti-Gal antibody with α-galactosyl epitopes: a major obstacle for xenotransplantation in humans, *Immunol. Today* **14**: 480–482.

Galili, U., Clark, M R., Shohet, S. B., Buehler, J., and Macher, B A, 1987, Evolutionary relationship between the natural anti-Gal antibody and the Galα1,3Gal epitope in primates, *Proc. Natl. Acad. Sci. USA* **84**: 1369–1373.

Galili, U., Macher, B A, Buehler, J. and Shohet, S. B., 1985, Human natural anti-α-galactosyl IgG. II. The specific recognition of α-(1 → 3)-linked galactose residues, *J. Exp. Med.* **162**: 573–582.

Galili, U., Shohet, S. B., Kobrin, E., Stults, C. L., and Macher, B. A., 1988, Man, apes, and old world monkeys differ from other mammals in the expression of α-galactosyl epitopes on nucleated cells, *J. Biol. Chem.* **263**: 17755–17762.

Goldstein, I. J., Blake, D. A., Ebisu, S., Williams, T. J. and Murphy, L. A., 1981, Carbohydrate binding studies on the *Bandeiraea simplicifolia* I isolectins. Lectins which are mono-, di-, tri-, and tetravalent for *N*-acetyl-D-galactosamine, *J. Biol. Chem.* **256**: 3890–3893.

Goldstein, I. J., Winter, H. C., and Poretz, R. D., 1997, Plant lectins: tools for the study of complex carbohydrates in: *Glycoproteins II*, J. Montreuil, J. F. G. Vliegenthart, and H. Schachter, eds., Elsevier, Amsterdam, pp. 403–474.

Gorelik, E., Duty, L., Anaraki, F., and Galili, U., 1995, Alteration of cell surface carbohydrates and inhibition of metastatic property of murine melanomas by α-1,3 galactotransferase gene transfection, *Cancer Res.* **55**: 4168–4173.

Hayes, C. E. and Goldstein, I. J., 1974, An α-D-galactosyl-binding lectin from *Bandeiraea simplicifolia* seeds. Isolation by affinity chromatography and characterization, *J. Biol. Chem.* **249**: 1904–1914.

Hayes, C. E. and Goldstein, I. J., 1975, Equilibrium dialysis and cell binding studies on *Bandeiraea simplicifolia* lectin, *J. Biol. Chem.* **250**: 6837–6840.

Horejsí, V. and Kocourek, J., 1973, Studies on phytohemagglutinins XII. O-Glycosyl polyacrylamide gels for affinity chromatography of phytohemagglutinins, *Biochim. Biophys. Acta.* **297**: 346–351.

Ihida, K., Tsuyama, S., and Murata, F., 1990, The use of *Griffonia simplicifolia* agglutinin-$IB_4$ in staining the developing rat fundic gland, *Acta. Histochem. Cytochem.* **23**: 475–486.

Ito, N., Imai, S., Haga, S. Nagaike, C., Morimura, Y., and Hatake, K., 1996, Localization of binding of *Ulex europeus I, Helix pomata* and *Griffonia simplicifolia* I-$B_4$ lectins and analysis of their backbone structures by several glycosidases and poly-*N*-acetyllactosamine specific lectins in human breast carcinomas, *Histochem. Cell Biol.* **106**: 331–339.

Ito, N., Nagaike, C., Morimura, Y., and Hatake, H., 1997, Estimation and comparison of the content of blood group B antigen in selected human tissues by microphotomeric quantitation of *Griffonia simplicifolia* agglutinin I-$B_4$ staining with or without prior α-galactosidase treatment, *Histol. Histopathol.* **12**: 415–424.

Kaifu, R., Plantefaber, L. C., and Goldstein, I. J., 1985, 2-substituted methyl α-D-galactopyranosides: synthesis and binding affinity for the A and B subunits of the *Griffonia simplicifolia* I isolectins, *Carbohydr. Res.* **140**: 37–49.

Kusai, K., Yamamoto, K., Konami, Y., and Osawa, T., 1991, cDNA cloning and expression of *Bauhinia purpurea* lectin, J. Biochem. (Tokyo) **109**: 899–903.

Koite, C., Katayama, A., Kadomatsu, K., Muramatsu, T., Hiraiwa, N., Kannagi, R., Nakashima, I., Yokoyama, I., Takagi, H., 1997, Direct gene replacement of the mouse α(1,3)-galactosyltransferase gene with human α(1,2)-fucosyltransferase gene: converting α-galactosyl epitopes into H antigens, *Xenotransplantation* **4**: 147–153.

Knibbs, R. N., Perini, F. and Goldstein, I. J., 1989, Structure of the major concanavalin A-reactive oligosaccharides of the extracellular matrix component laminin, *Biochemistry* **28**: 6379–6392.

Lamb, J. E. and Goldstein, I. J., 1984, A structural comparison of the A and B subunits of *Griffonia simplicifolia* I isolectins, *Arch. Biochem. Biophys.* **229**: 15–26.

Liener, I. E., Sharon, N., and Goldstein, I. J., eds., 1986, *The lectins: properties, function and application in biology and medicine*, pp. 1–600.

Lis, H., and Sharon, N., 1998, Lectins: carbohydrate specific proteins that mediate cellular recognition, *Chem. Rev.* in press.

Lönngren, J. and Goldstein, I. J., 1976, Chemical modification studies on the α-D-galactopyranosyl binding lectin from the seeds of *Bandeiraea simplicifolia*, *Biochim. Biophys. Acta.* **439**: 160–166.

Lönngren, J., Goldstein, I. J., and Bywater, R., 1976, Cross-linked guaran: a versatile immunosorbent for D-galactopyranosyl-binding lectins, *FEBS Lett.* **68**: 31–34.

Lönngren, J., Goldstein, I. J., and Zand, R., 1976, Circular dichroism studies on the α-D-galactopyranosyl binding lectin isolated from the seeds of *Bandeiraea simplicifolia*, *Biochemistry* **15**: 436–440.

Maddox, D. E., Shibata, S., and Goldstein, I. J., 1982a, Stimulated macrophages express a new glycoprotein receptor reactive with *Griffonia simplicifolia* I-$B_4$ isolectin, *Proc. Natl. Acad. Sci. USA* **79**: 166–170.

Maddox, D. E., Goldstein, I. J., and Lobuglio, A. E., 1982b, *Griffonia simplicifolia* I lectin mediates macrophage-induced cytotoxicity against Ehrlich ascites tumor, *Cell. Immunol.* **71**: 202–207.

Mäkelä, O., 1957, Studies on hemagglutinins of Leguminosae seeds, *Ann. Med. Exp. Biol. Fenn.* **35**: Suppl. 11, 1–133.

Mäkelä, O. and Mäkelä, P., 1956, Some new blood groups specific phytoagglutinins, *Ann. Med. Exp. Biol. Fenn.* **34**: 402–404.

Mäkelä, O., Mäkelä, P., and Krüpe, M., 1959, Zur Spezifität der Anti-B-Phythämagglutinine, *Z. Immunitasetsforsch. Exp. Ther.* **117**: 220–229.

Murphy, L. A. and Goldstein, I. J., 1977, Five α-D-galactopyranosyl-binding isolectins from *Bandeiraea simplicifolia* seeds, *J. Biol. Chem.* **252**: 4739–4742.

Murphy, L. A. and Goldstein, I. J., 1979, Physical-chemical characterization and carbohydrate-binding activity of the A and B subunits of the *Bandeiraea simplicifolia* I isolectins, *Biochemistry* **18**: 4999–5005.

Nelsestuen, G.L. and Sattie, J. W., 1972, The carbohydrate of bovine prothrombin. Partial structural determination demonstrating the presence of α-galactose residues, *J. Biol. Chem.* **247**: 6096–6102.

Okada, Y. and Spiro, R. G., 1980, Isolation and characterization of three major glycoproteins from thyroid plasma membranes, *J. Biol. Chem.* **255**: 8865–8872.

Oppenheim, J. D., Amin, A. R., and Thorbeck, C. J., 1990, A rapid one-step purification procedure for murine IgD based on the specific affinity of *Bandeiraea (Griffonia) simplicifolia* I for *N*-linked carbohydrates on IgD, *J. Immunol. Meth.* **130**: 243–250.

Peters, B.P. and Goldstein, I.J.,1979, The use of fluorescein-conjugated *Bandeiraea simplicifolia* $B_4$ isolectin as a histochemical reagent for the detection of α-D-galactosyl groups, *Exp. Cell Res.* **120**:321–334.

Petryniak, J., Varani, J., Ervin, P. R., and Goldstein, I. J., 1991, Differential expression of glycoproteins containing α-D-galactosyl groups on normal human breast epithelial cells and MCF-7 human breast carcinoma cells, *Cancer Lett.* **60**: 59–65.

Petryniak, J., Huard, T. K., and Goldstein, I. J., 1992, α-D-Galactose-bearing glycoproteins on the surface of stimulated murine peritoneal macrophages. Biochemical and immunochemical characterization of purified glycoproteins, *Eur. J. Biochem.* **206**: 197–207.

Rao, C. N., Goldstein, I. J., and Liotta, L. A., 1983, Lectin-binding domains on laminin, *Arch. Biochem. Biophys.* **227**: 118–124.

Ross, T. T., Hayes, C. E., and Goldstein, I. J., 1976, Carbohydrate-binding properties of an immobilized α-D-galactopyranosyl binding protein (lectin) from the seeds of *Bandeiraea simplicifolia*, *Carbohydr. Res.* **47**: 91–97.

Roth, J. and Goldstein, I. J., 1995, Subcellular distribution of terminal α-D and β-D galactosyl residues in Ehrlich tumor cells studied by lectin-gold techniques, *Glycoconjugate. J.* **12**: 142–149.

Shibata, S., Peters, B. P. Roberts, D. D., Goldstein, I. J. and Liotta, L A, 1982, Isolation of laminin by affinity chromatography on *Griffonia simplicifolia* I lectin, *FEBS Letts.* **42(2)**: 194–198.

Shinkel, T. A., Chen, C.-G., Solvaris, E., Henion, T. R., Barlow, H., Galili, U., Pearse, M. J., and D'Apice, A. J. F., 1997, Changes in cell surface glycosylation in α-1,3 galactosyl-transferase knockout and α-1,2-fucosyltransferase transgenic mice, *Transplantation* **64**: 197–204.

Tearle, R. G., Tange, M. J., Zannettino, Z. L., Katerelos, M., Shinkel, T. A., Van Denderen, B. J. W., Lonie, A. J., Lyons, I., Nottle, M. B., Cox, T., Becker, C., Peura, A. M., Wigley, P. L., Crawford, R. J., Robins, A. J., Pearse, M. J., and D'Apice, A. J. F., 1996, The α 1.3 galactosyltransferase knockout mouse: Implications for xenotransplantation, *Transplantation* **61**: 13–19.

Thall, A. D. and Galili, U., 1990, Distribution of Galα1→3Galβ1→4GlcNAc residues on secreted mammalian glycoproteins (thyroglobulin, fibrinogen, and immunoglobulin G) as measured by a sensitive solid-phase radioimmunoassay, *Biochemistry* **29**: 3959–3965.

Thall, A. D., Maly, P., and Lowe, J. B., 1995, Oocyte gal α 1,3 gal epitopes implicated in sperm adhesion to the zona pellucida glycoprotein ZP3 are not required for fertilization in the mouse, *J. Biol. Chem.* **15**: 21437–21440.

Vanhove, B., Goret, F., Soulillou, J.-P., and Pourcel, C., 1997, Porcine α-1,3 galactosyl-transferase: tissue specific and regulated expression of splicing isoforms, *Biochim. Biophys. Acta.* **1356**: 1–11.

Watier, H., Guillaumin, J.-M., Piller, F., Lacord, M., Thibault, G., Lebranchu, Y., Monsigny, M., and Bardos, P., 1996, Removal of terminal α-galactosyl residues from xenogenic porcine endothelial cells, *Transplantation* **62**: 105–113.

Williams, T. J., Plessas, N. R., and Goldstein, I. J. and Lönngren, J., 1979, A new class of model glycolipids: synthesis, characterization, and interaction with lectins, *Arch. Biochem. Biophys.* **195**: 145–151.

Wood, C., Kabat, E. A., Murphy L. A., and Goldstein, I. J., 1979, Immunochemical studies of the combining sites of the two isolectins $A_4$ and $B_4$ isolated from *Bandeiraea simplicifolia*, *Arch. Biochem. Biophys.* **198**: 1–11.

Wu, A. M., Song, S.-C., Wu, J. H., and Kabat, E. A., 1995, Affinity of *Bandeiraea (Griffonia) simplicifolia* lectin I, isolectin $B_4$ for Galα1→4Gal ligand, *Biochem. Biophys. Res. Commun.* **216**: 814–820.

Young, N. M., and Oomen, R. P., 1992, Analysis of sequence variation among legume lectins. A ring of hypervariable residues forms the perimeter of the carbohydrate-binding site, *J. Mol. Biol.* **228**: 924–934.

*Chapter 7*

# α-Gal Epitopes on Viral Glycoproteins

Russell P. Rother and Uri Galili

## 1. INTRODUCTION

The α-gal epitope (Galα1,3Galβ1,4GlcNAc-R) is a terminal glycosidic structure that is expressed on the surface of cells from most mammalian species other than humans, apes and Old World monkeys (Galili *et al.*, 1988a; Galili *et al.*, 1987). The terminal α-galactosyl unit of this epitope is added to nascent glycolipids and glycoproteins in the Golgi apparatus by α1,3galactosyltransferase (α1,3GT). In primates lacking the α-gal epitope, the α1,3GT gene is not transcribed, and non-sense mutations are present within the coding region of some species (Galili and Swanson, 1991; Larsen *et al.*, 1990a; Joziasse *et al.*, 1989). The presence of a functional α1,3GT in New World monkeys suggests that this gene was inactivated in ancestral Old World monkeys and apes after their divergence from New World monkeys (Galili *et al.*, 1988a). A detailed comparison of the α1,3GT pseudogene sequences in Old World monkeys and apes further suggests that this gene was inactivated after these two groups diverged from each other (Galili and Andrews, 1995; Galili and Swanson, 1991).

---

**Russell P. Rother** Molecular Development, Alexion Pharmaceuticals, New Haven, Connecticut 06511. **Uri Galili** Department of Microbiology and Immunology, MCP Hahnemann School of Medicine, Philadelphia, Pennsylvania 19129.

*Subcellular Biochemistry, Volume 32: α-Gal and Anti-Gal*, edited by Galili and Avila.
Kluwer Academic / Plenum Publishers, New York, 1999.

It is likely that the evolutionary events that resulted in the alteration of the carbohydrate make-up of Old World monkeys, apes and humans via suppression of α-gal epitope expression similarly altered the glycosylation of the viruses infecting them. For example, it is now known that virions produced in cells with α1,3GT activity express the α-gal epitope because the carbohydrate chains on the viral envelope glycoproteins are synthesized by the host glycosylation machinery. As described in Chapter 1, the putative exposure of ancestral Old World monkeys and apes to a pathogen (such as a virus) expressing the α-gal epitope might have resulted in the selection of primates with mutations in the α1,3GT gene causing the suppression of α-gal epitope synthesis.

The evolutionary inactivation of the α1,3GT gene has resulted in loss of immune tolerance to the α-gal epitope and allows humans, apes and Old World monkeys to produce high titers of natural antibody specific to this epitope (anti-Gal). Reciprocally, other mammalian species express the α-gal epitope and do not produce anti-Gal (Galili *et al.*, 1988a; Galili *et al.*, 1987). As much as 1% of the total circulating IgG (i.e., 30–100 μg/ml) in humans recognizes the α-gal epitope (Galili *et al.*, 1985; Galili *et al.*, 1984). In addition, significant levels of anti-Gal IgM and IgA have also been reported (Hamadeh *et al.*, 1995a; Sandrin *et al.*, 1993). A detailed description of the α-gal epitope and the anti-Gal antibody can be found in other chapters within this book.

As described below, there is now convincing evidence that anti-Gal is primarily responsible for the characteristic inactivation of various enveloped viruses in normal human serum (NHS) following viral propagation in cells expressing functional α1,3GT. This ability to inactivate viruses, taken together with the continued presence of anti-Gal in humans, apes and Old World monkeys, suggests that this antibody may have provided a unique type of innate immunity against viruses expressing the α-gal epitope, thereby creating a barrier against viral zoonosis. This anti-Gal-mediated destruction of viruses expressing the α-gal epitope also presents a barrier to current approaches in gene therapy, as retroviruses expressing this epitope are commonly used as vectors for gene transduction. Understanding these phenomena has helped to overcome this biological barrier and should improve the efficacy of *in vivo* procedures where such viral vectors are used to provide gene therapy.

## 2. EXPRESSION OF THE α-GAL EPITOPE ON ENVELOPED VIRUSES

### 2.1. Host Cell Glycosylation of Viral Envelope Glycoproteins

Asparagine (N)-linked carbohydrate chains are commonly found on envelope glycoproteins of viruses that bud from vertebrate cells where they are synthe-

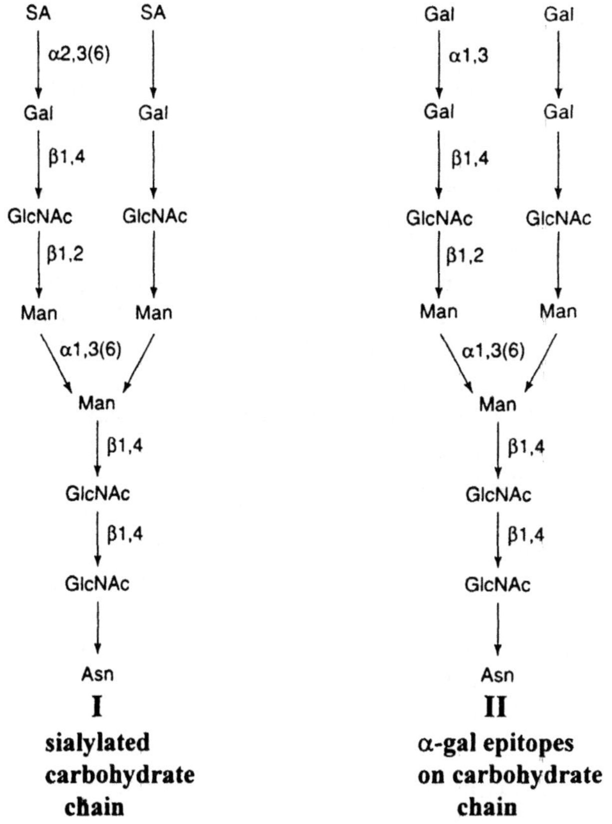


**Figure 1.** Characteristic complex carbohydrate chains on mammalian glycoproteins. The number of antennae may vary. Sialic acid (SA) and terminal α-galactosyl units (Gal) may be present on different antennae of the same carbohydrate chain or may be absent. Other carbohydrates (e.g., α1,2fucosyl of the blood group H) may cap the N-acetyllactosamine units (Galβ1,4GlcNAc) as well.

sized by host cell glycosyltransferases (Klenk, 1990; Rademacher *et al.*, 1988; Schlesinger and Schlesinger, 1987a; Schlesinger and Schlesinger, 1987b; Kornfeld and Kornfeld, 1985; Hsieh *et al.*, 1983; Klenk and Rott, 1980; Kornfeld and Kornfeld, 1976). These enzymes build the carbohydrate chain in a sequential manner analogous to an assembly line. Each carbohydrate unit is transferred by a corresponding glycosyltransferase from a nucleotide-sugar donor (e.g., UDP-Gal or CMP-SA) to the growing carbohydrate chain to form a final structure such as those depicted in Figure 1. The terminal portion of each carbohydrate chain is assembled within the various compartments of the Golgi on the viral glycoprotein molecules that are transported to the surface of the cell, where they assemble with membrane lipids to form the envelope of the virus.

Like glycoproteins of the host cell, viral envelope glycoproteins contain two types of N-linked carbohydrate chains. These include the high mannose type, where all the antennae have mannose structures, and the complex type, in which the antennae contain N-acetyllactosamine (Galβ1,4GlcNAc-R) units (Figure 1). Complex carbohydrate chains usually contain between two and four antennae, depending upon the activity of branching enzymes within the Golgi apparatus.

Relative concentrations and activities of glycosyltransferases may vary in different species, as well as in different cells within the same species. This differential enzyme expression makes it possible for the same virus to express varying types of carbohydrate chains when grown in different cell lines (Galili *et al.*, 1996; Repik *et al.*, 1994; Hsieh *et al.*, 1983; Burke and Keegstra, 1976; Stollar *et al.*, 1976; Strauss *et al.*, 1970). Sialylated N-linked complex carbohydrate chains (Figure 1, structure I), are common on viral glycoproteins because sialyltransferases are prevalent in the Golgi apparatus of mammalian cells. These transferases cap terminal N-acetyllactosamine residues of the growing carbohydrate chains with sialic acid (Rademacher *et al.*, 1988; Kornfeld and Kornfeld, 1985; Klenk and Rott, 1980). Similarly, in those species and cell types where it is expressed, α1,3GT can transfer galactose in an α1,3 linkage to such N-acetyllactosamine residues (Blanken and Ban den Eijnden, 1985; Betteridge and Watkins, 1983; Blake and Goldstein, 1981; Basu and Basu, 1973), forming the α-gal epitope on cellular glycoproteins as well as on the viral envelope glycoproteins (Figure 1, structure II) where it may interact with anti-Gal.

## 2.2. Natural Antibody Reactivity to Enveloped Viruses

Two decades have passed since the identification of naturally occurring antibodies in NHS that mediate the killing of enveloped viruses. The initial report showed that lymphocytic choriomeningitis virus (LCMV) is inactivated by naturally occurring antibodies through the classical complement pathway (Welsh, 1977). These sensitizing antibodies also recognize the host cell of the virus (e.g., the murine cell line L-929) suggesting that the virus acquires a cell-specific antigen during replication. The sensitivity of the virus to NHS is also highly dependent on the species of the host cell, as LCMV propagated in human HeLa cells or hamster BHK 21 cells, but not certain murine cells, resists serum killing. Similarly, vesicular stomatitis virus is highly sensitive to NHS following propagation in dog cells, but is completely resistant to serum killing following passage in HeLa cells (Thiry *et al.*, 1978).

By examining the ability of serum to immunoprecipitate retroviral glycoproteins, it was demonstrated that NHS contains antibodies to the envelope protein of various retroviruses and that antibody recognition is entirely dependent on the species of the host cell (Barbacid *et al.*, 1980; Snyder and Fleissner, 1980). These studies supported the inference that natural antibody reactivity to viral glycoproteins is directed toward cellular antigens rather than resulting from prior exposure to retroviruses. Most strikingly, it was observed that human natural antibody binding to envelope glycoproteins is primarily restricted to retroviruses propagated in non-primate mammalian cells. Furthermore, the binding of natural antibodies to murine leukemia virus envelope glycoprotein was found to be efficiently blocked by normal serum components from various mammalian species other than humans, apes and Old World monkeys. Utilizing glycosidases to remove terminal glycosidic structures from the viral envelope glycoproteins, the antigenic target of human anti-viral reactivity was characterized as a carbohydrate moiety (Lower *et al.*, 1981; Snyder and Fleissner, 1980). Taken together, these data parallel the pattern of human natural antibody reactivity to the α-gal epitope (Galili *et al.*, 1988a).

### 2.3. α-Gal Epitope Expression on Various Viral Envelopes

The association of a terminal glycosidic structure, now recognized as the α-gal epitope, with the surface glycoproteins of a virus was first demonstrated in the Friend leukemia virus (Geyer *et al.*, 1984). Using nuclear magnetic resonance spectroscopy and methylation analysis, it was shown that the N-glycosidically linked carbohydrates of the major envelope glycoprotein consist of bi-, tri- and tetraantennary oligosaccharitols of the complex type. Furthermore, it was demonstrated that the asialoglycans of the viral envelope contain N-acetyllactosamine repeating units which are, in some cases, substituted by nonreducing terminal Galα1,3 residues (thus forming the α-gal epitope).

#### 2.3.1. Eastern Equine Encephalitis Virus

The first reported experiment correlating the presence of the α-gal epitope on the surface of a virus with anti-Gal-mediated viral neutralization was performed using the eastern equine encephalitis (EEE) virus, a member of the *Alphavirus* genus within the Togaviridae family (Repik *et al.*, 1994). The EEE virus has two envelope glycoproteins, $E_1$ and $E_2$, which are predicted to have one and two putative N-linked glycosylation sites (Asn-X-Ser/Thr), respectively (Chang and Trent, 1987). Since the EEE virus, like other alphaviruses, is capable of infecting a large variety of mammalian cells including primate cells (Morris, 1988; Schlesinger and Schlesinger, 1987b; Strauss and Strauss, 1977), this virus was selected as a model for demonstrating the acquisition of the α-gal epitope.

The EEE virus was grown in mouse 3T3 fibroblasts and in the African green monkey cell line, Vero, and analyzed for expression of the α-gal epitope on the E1 and E2 envelope glycoproteins (Repik *et al.*, 1994). Mouse 3T3 fibroblasts express an abundance of the α-gal epitope on cell surface glycoproteins (Santer *et al.*, 1989). In contrast, African green monkey-derived Vero cells, like cells of other

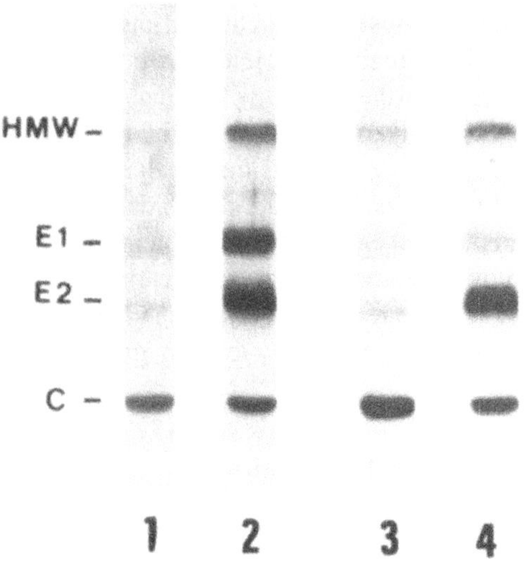


**Figure 2.** Interaction of $EEE_{Vero}$ and $EEE_{3T3}$ virus envelope proteins with anti-Gal and BS lectin in western blots. Proteins from $EEE_{Vero}$ (lane 1) and $EEE_{3T3}$ (lane 2) viruses immunostained with anti-Gal. Proteins from $EEE_{Vero}$ (lane 3) and $EEE_{3T3}$ (lane 4) viruses immunostained with BS lectin. The positions of the viral proteins are indicated on the left. The binding to the C protein is nonspecific. (Reproduced from Repik *et al.*, 1994, with permission.)

Old World monkeys, are completely lacking in α1,3GT activity (Joziasse *et al.*, 1989; Galili *et al.*, 1988a) and thus are totally devoid of the α-gal epitope (Galili *et al.*, 1988a). Whether or not the α-gal epitope is expressed on EEE viral glycoproteins following propagation in either Vero or 3T3 cells was determined by western blot analysis. As shown in Figure 2, purified anti-Gal antibody readily binds both E1 and E2 glycoproteins of EEE virus propagated in 3T3 cells ($EEE_{3T3}$), indicating that the α-gal epitope is expressed on these molecules. Anti-Gal also interacts with a high molecular weight (HMW) protein that represents aggregates of the envelope glycoprotein molecules. In contrast, anti-Gal does not bind to either $E_1$ or $E_2$ of EEE virus propagated in Vero cells ($EEE_{Vero}$) suggesting that these proteins lack the α-gal epitope. Similar results were obtained with the lectin *Bandeiraea simplicifolia* IB4 (BS lectin), which interacts specifically with the α-gal epitope (Wood *et al.*, 1979). This lectin binds to $E_2$ and the HMW protein of $EEE_{3T3}$, but not to the same molecules on $EEE_{Vero}$. The inability of BS lectin to bind to the $E_1$ glycoprotein of $EEE_{3T3}$ could reflect the paucity of the α-gal epitope on this molecule, estimated at one carbohydrate chain per molecule (Chang and Trent, 1987). The BS lectin tetramer must interact with at least two epitopes on the glycoprotein molecule in order to maintain binding during subsequent wash steps. Antibody and lectin bind nonspecifically to the capsid protein of EEE virus in both viral preparations. These data imply that $EEE_{3T3}$ contains carbohydrate chains linked to the $E_1$ and E2 glycoproteins that are of the complex type terminating in the α-gal epitope (Figure 1, structure II). Based on the interaction of EEE virus with anti-Gal in a radioimmunoassay, it has been estimated that each virion propa-

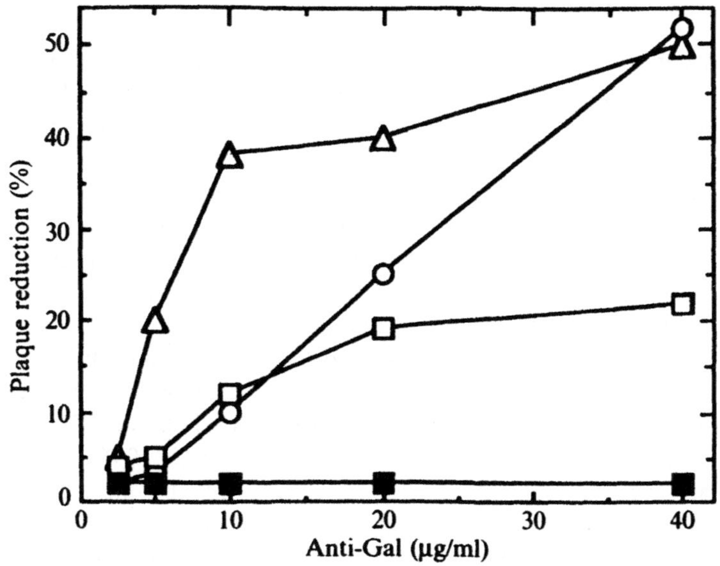


**Figure 3.** Plaque reduction neutralization of $EEE_{Vero}$ (■) and $EEE_{3T3}$ (Δ, O, □; three separate experiments) viruses by anti-Gal. Percentage plaque reduction curves obtained following incubation of viruses with increasing concentration of anti-Gal. (Reproduced from Repik *et al.*, 1994, with permission.)

gated in 3T3 cells contains approximately 80 α-gal residues (Repik *et al.,* 1994). In other viruses, this number can be much higher, depending on the number of N-linked carbohydrate chains per glycoprotein molecule and the relative activity of the various glycosyltransferases (including α1,3GT) within the host cell. In fact, the α-gal epitope has been shown to be the most prevalent epitope on the surface of the Friend murine leukemia virus (Geyer *et al.*, 1984).

The ability of anti-Gal to neutralize $EEE_{3T3}$ expressing the α-gal epitope was tested in a plaque reduction assay (Repik *et al.*, 1994). As shown in Figure 3, incubation of $EEE_{3T3}$ with anti-Gal results in a 20–50% decrease in plaque formation. Anti-Gal, however, had no effect on the ability of $EEE_{Vero}$ to infect Vero cells and form plaques. This finding is in accord with the observation that $EEE_{Vero}$ lacks the α-gal epitope and therefore does not bind anti-Gal. Taken together, these results demonstrate a direct correlation between the presence of a functional α1,3GT gene in the host cell and acquisition of the α-gal epitope by the virus. Furthermore, since anti-Gal is present in humans at high concentrations, these observations suggest that *in vivo* binding of the antibody to virions that express the α-gal epitope may aid in protection against viral infection by neutralizing the ability of the virus to penetrate into host cells.

### 2.3.2. Influenza Virus

A second example of differential expression of the α-gal epitope on a virus propagated in various host cells is that of influenza virus (Galili *et al.*, 1996). The major envelope glycoprotein of influenza virus is the hemagglutinin molecule (HA) which facilitates the binding of the virus to sialic acid residues on host cells and the subsequent infection of the cells (Air and Laver, 1990). The HA molecule has seven N-linked carbohydrate chains, most of which are of the complex type (Kiel *et al.*, 1985). Unlike many N-linked carbohydrate chains, those associated with the HA of influenza virus lacks terminal sialic acid residues. This is attributed to the activity of a virally encoded neuraminidase.

Type N1H1 influenza virus A/Puerto Rico/8/34 (termed PR8 virus) was propagated in embryonated chicken eggs or in the mammalian cell lines Madin Darby bovine kidney (MDBK) and Madin Darby canine kidney (MDCK) and analyzed for α-gal epitope expression by ELISA (Galili *et al.*, 1996). Chickens and other avian species do not express α1,3GT, whereas both bovine and canine species have a functional α1,3GT and produce the α-gal epitope (Galili *et al.*, 1988a; Galili *et al.*, 1987). As shown in Figure 4A, the α-gal epitope-specific BS lectin does not bind the PR8 virus propagated in eggs ($PR8_{egg}$), whereas BS lectin readily binds to the PR8 virus propagated in either MDCK cells ($PR8_{MDCK}$) or MDBK cells ($PR8_{MDBK}$). Purified anti-Gal also binds $PR8_{MDCK}$ and $PR8_{MDBK}$ but not $PR8_{egg}$ (Figure 4B). Similarly, western blot analysis shows that anti-Gal binds to 53 and 28 kDa polypeptides of $PR8_{MDCK}$ and $PR8_{MDBK}$ (Figure 4C). The 53 kDa polypeptide corresponds to the $HA_1$ fragment of the HA molecule, whereas the 27 kDa polypeptide represents the $HA_2$ fragment. The two fragments are linked to each other by a disulfide bond in the intact HA molecules (Air and Laver, 1990; Schulze, 1970). No binding of anti-Gal to HA proteins of $PR8_{egg}$ is observed. Taken together, these data show that the α-gal epitope is expressed on influenza virus propagated in bovine and canine cells but not on the same virus propagated in chicken cells, further supporting the association between α1,3GT activity in the host cell and the expression of the α-gal epitope on the propagated virus.

### 2.3.3. Murine Retroviruses

Virolysis of murine leukemia virus occurs in human, ape and Old World monkey sera but not New World monkey sera, while the addition of purified human anti-Gal renders New World monkey serum capable of inactivating virus (Rother *et al.*, 1995a). These data suggest that the α-gal epitope is present on the surface of the murine leukemia retrovirus and that this epitope is the primary factor responsible for the characteristic inactivation of murine retroviruses in NHS. As was demonstrated with EEE and influenza viruses, the α-gal epitope is associated with an envelope glycoprotein (in this case, the gp70 protein of both amphotropic and ecotropic murine retroviruses) (Rother *et al.*, 1995a). The role that the α-gal

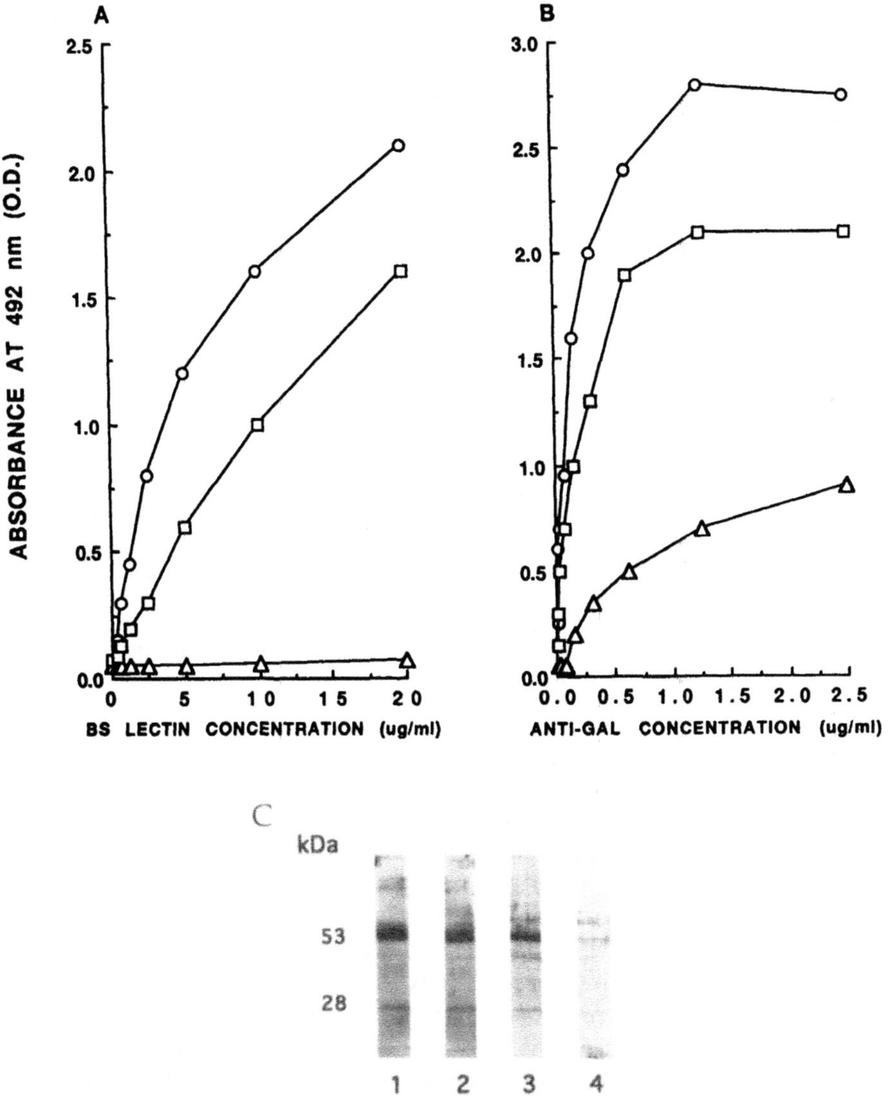


**Figure 4.** Binding of BS lectin (A) and anti-Gal (B) to PR8 virus propagated in various cells. ○ $PR8_{MDBK}$ virus; □ $PR8_{MDCK}$ virus; △ $PR8_{egg}$ virus. Data were determined in ELISA using different concentrations of biotinylated lectin or antibody. (C) Western blot analysis showing the binding of anti-Gal to HA1 (53 kDa band) and to HA2 (28 kDa band) of $PR8_{MDBK}$ virus (lane 1); inactivated $PR8_{MDBK}$ virus (lane 2); $PR8_{MDCK}$ (lane 3); and $PR8_{egg}$ virus (lane 4). (Reproduced from Galili *et al.*, 1996, with permission.)

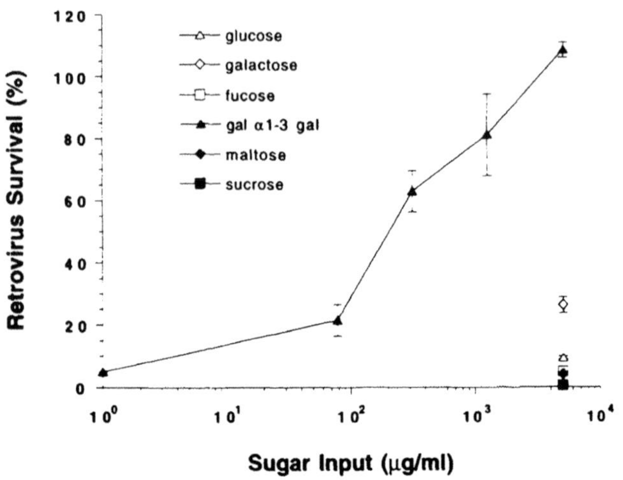


**Figure 5.** Inhibition of retrovirus inactivation in NHS using soluble galα1-3gal. Retroviral particles were incubated in human serum in the presence of soluble carbohydrates including D(+) glucose (glucose), D(+) galactose (galactose), α-L(-) fucose (fucose), galactose α1,3 galactose (galα1,3gal), maltose or sucrose. The curve represents the percentage of infectious particles remaining at various concentrations of galα1,3gal relative to virus survival in heat-inactivated human serum. Single points indicate retrovirus survival in the presence of various other carbohydrates at a concentration of 5 mg/ml. (Reproduced from The Journal of Experimental Medicine, 1995, Volume 182, Pages 1345–1355, by copyright permission of The Rockefeller University Press.)

epitope plays in the serum inactivation of amphotropic retrovirus is shown by the data set forth in Figure 5. Viral inactivation in NHS is effectively inhibited by the addition of synthetic galα1,3gal in a dose-dependent manner. This soluble disaccharide has been shown to specifically bind anti-Gal (Galili, 1993a) and effectively inhibit anti-Gal-mediated complement lysis of cells expressing the α-gal epitope (Neethling *et al.*, 1994). In addition, D(+) galactose, which has also been shown to partially block anti-Gal reactivity (Vaughan *et al.*, 1994; Sandrin *et al.*, 1993; Galili *et al.*, 1984), provides some protection from viral inactivation by anti-Gal. Conversely, addition of other disaccharides to NHS does not affect retrovirus survival. Murine retrovirus inactivation is also effectively abrogated by the selective removal of anti-Gal from NHS or by the down-regulation of α-gal epitope expression on the retroviral host cell (Rother *et al.*, 1995a).

The critical role of anti-Gal in murine retroviral inactivation is supported by data showing that murine retrovirus passaged through cells that are deficient in α-gal epitope expression (e.g. human cells and hamster BHK cells) are also insensitive to virolysis in NHS (Cosset *et al.*, 1995; Takeuchi *et al.*, 1994; personal observations). Further, expression of an exogenous α1,3GT gene in viral host cells of human origin more definitively demonstrates the association between the α-gal epitope and retrovirus sensitivity. Human HT1080 cells transfected with porcine α1,3GT express the α-gal epitope on the cell surface, rendering both the cells and the murine retroviruses that they produce sensitive to NHS (Takeuchi *et al.*, 1996).

### 2.3.4. Human Immunodeficiency Virus

Human HUT-78 cells transduced with recombinant α1,3GT express high levels of the α-gal epitope on the cell surface, resulting in increased susceptibility of these cells to NHS killing (Reed *et al.*, 1997). Human immunodeficiency virus type I (HIV-1) propagated through these modified cells acquires the α-gal epitope in association with the gp120 envelope protein. This carbohydrate modification of HIV-1 renders the virus highly sensitive to inactivation in NHS relative to virus harvested from wild-type HUT-78 cells (Figure 6, panel A). The inactivation of modified HIV-1 is mediated by anti-Gal, as preincubation of NHS with synthetic galα1,3gal blocks anti-viral activity (Figure 6, panel B).

These results shed light on the observations that, in contrast to animal retroviruses, HIV-1 and human T cell leukemia virus type I are insensitive to NHS (Banapour *et al.*, 1986; Hoshino *et al.*, 1984). Previous reports suggested that this species-restricted phenomenon is a result of the acquisition of human complement inhibitors by the viruses during propagation in the host cell (Marschang *et al.*, 1995; Saifuddin *et al.*, 1995; Spear *et al.*, 1995). However, given the data reviewed above showing that animal viruses are inactivated in NHS through an anti-Gal-mediated event, and considering that HIV-1 is equally sensitive to NHS following passage in human cells expressing the α-gal epitope, it is likely that the reported sensitivity of retroviruses to NHS is primarily due to the expression of the α-gal epitope on the virus as a result of the presence of a functional α1,3GT in the host cell.

### 2.3.5. Other Viruses

Demonstration of the acquisition of the α-gal epitope by viruses following passage in cells containing a functional α1,3GT gene has recently been extended to several other viruses (Welsh, *et al.*, 1998). LCMV passaged in human melanoma cells modified with recombinant α1,3GT is effectively inactivated in NHS. This result correlates with earlier work showing that, when harvested from certain cell types, LCMV is neutralized by naturally occurring serum antibodies (Welsh, 1977).

Somewhat surprising is the demonstration that Sindbis virus expressing the α-gal epitope following passage in the recombinant human cells is not inactivated in NHS (Welsh, *et al.*, 1998). In contrast, Sindbis virus modified via passage through α1,3GT competent murine cells is moderately sensitive to NHS, suggesting that the virus may acquire a species-restricted factor from the surface of the human cells that protects it from complement attack. These data suggest that the α-gal epitope is acquired by enveloped viruses in general during passage through

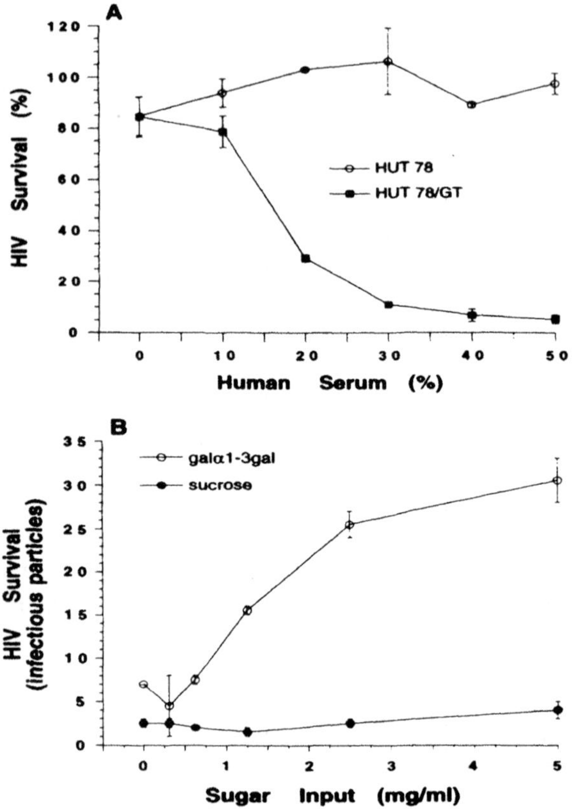


**Figure 6.** Inactivation of HIV expressing the α-gal epitope in NHS. (A) Virus harvested from HUT 78 control or HUT 78/GT was subjected to increasing concentrations of NHS and survival was determined. Curves represent the percentage of infectious particles remaining following incubation in NHS relative to survival of the same virus in heat-inactivated serum. (B) Virus from HUT 78/GT cells was subjected to NHS previously incubated with increasing concentrations of either galα1,3gal or sucrose and survival was determined. Curves represent the number of infectious particles remaining after challenge in serum. (Reproduced from Reed *et al.*, 1997, by copyright permission of The American Association of Immunologists.)

cells that possess a functional α1,3GT gene, but also indicate that the presence of the epitope is not sufficient to confer sensitivity to NHS in all cases.

## 2.4. Potential Role of Anti-Gal in Preventing Cross-Species Transfer of α-Gal Epitope Bearing Viruses

Complement-mediated inactivation of viruses in human serum triggered by anti-Gal represents a unique type of innate immunity. Antibodies are usually considered to be an element of acquired immunity elicited in response to a specific foreign antigen. In contrast, anti-Gal is pre-formed and recognizes a carbohydrate moiety that is present on a variety of viruses. This raises the possibility that anti-Gal provides a barrier to the horizontal transmission of viruses from species that express the α-gal epitope to those that lack it. In fact, it has been speculated that α1,3GT gene null mutations in ancestral Old World moneys and apes were selected following an epidemic potentiated by an infectious agent that either carried the α-gal epitope or used it as a receptor (Galili, 1993a; Galili and Andrews, 1995). Epitopes reactive with anti-Gal have been detected on several potential pathogens including enveloped viruses, bacteria and protozoa (Welsh, *et al.*, 1998; Galili *et al.*, 1996; Rother *et al.*, 1995a; Repik et al., 1994; Hamadeh *et al.*, 1992; Almeida *et al.*, 1991; Avila *et al.*, 1989; Galili *et al.*, 1988b).

A phenomenon suggesting a possible functional role for anti-Gal in providing protection against α-gal epitope bearing enveloped viruses has been observed in primates. It has been known for some time that murine retroviruses induce T cell neoplasms in rodents (Tsichlis and Lazo, 1991; Tsichlis, 1987; Famulari, 1983) and transform human cells *in vitro* (Teich *et al.*, 1975; Aaronson and Todaro, 1970; Boiron *et al.*, 1969; Jensen *et al.*, 1964). However, the pathogenic potential of these retroviruses in humans, apes and Old World monkeys has been controversial. In particular, murine leukemia virus infused into rhesus monkeys fails to cause disease (Cornetta *et al.*, 1990). In contrast, stem cells from rhesus monkeys transduced *ex vivo* with murine leukemia virus induce fulminant lymphomas when transplanted back into the donor (Donahue *et al.*, 1992). Consideration of the potential role of the α-gal epitope in immunity provides an explanation for these data. Retroviruses derived from murine cells express the α-gal epitope and are rapidly inactivated in human, or Old World monkey serum. On the other hand, *ex vivo* transduction of stem cells derived from these primates in the absence of serum complement would allow survival of retroviruses expressing the α-gal epitope and their subsequent replication. Retroviruses liberated from these cells would not express the α-gal epitope and, therefore, escape anti-Gal-mediated immune destruction following transplantation back into primates.

The occurrence of human zoonotic diseases caused by enveloped viruses such as influenza, rabies and vesicular stomatitis suggests that anti-Gal does not provide a strong impediment to cross-species transfer of all enveloped viruses. However, the existence of zoonotic diseases caused by specific enveloped viruses does not rule out a role for anti-Gal in innate immunity. Effective elimination of viruses expressing the α-gal epitope is likely to be dependent on several factors including the number of α-gal epitopes per virion, route of transmission and viral dose and the possible presence of other factors on the viral envelope that may modify the immune response to the α-gal epitope. For example, viruses introduced through aerosol transmission will not initially encounter high-titer anti-Gal IgG. Although IgA class secretory anti-Gal may be encountered during this mode of transmission, IgA is a poor activator of complement and its binding has been shown to actually protect some bacteria from complement lysis (Hamadeh *et al.*, 1995b). In addition, if inactivation of a particular virus by anti-Gal is complement-dependent, concentration of complement at the site of viral entry may play a critical role, as may the presence of complement inhibitory proteins on the viral surface. Finally, the size of the initial dose of virus could determine the difference between pathogenesis or viral clearance. If a single virion escapes anti-Gal-mediated inactivation and is amplified in a human cell, it will no longer express the α-gal epitope and could more readily spread to other cells. Likewise, once a virus successfully infects one individual, transmission to other humans will not be subject to inactivation through this mechanism.

The difficulty in defining the role that anti-Gal may play in preventing the inter-species transmission of viruses to species that express this antibody has been increased by a lack of animal models (Welsh, *et al.*, 1998). Recently, α1,3GT gene knock-out mice have been generated as a small animal model for anti-Gal production (Thall *et al.*, 1995). These mice produce high titers of anti-Gal IgG and IgM in response to repeat injections of rabbit red blood cell membranes, which are a rich source of the α-gal epitope (Welsh, *et al.*, 1998; Galili *et al.*, 1998; LaTemple and Galili, 1998).

To test the hypothesis that anti-Gal-mediated complement inactivation of viruses passed through cells expressing the α-gal epitope may be involved in innate immunity to virus infections *in vivo*, the replication of LCMV in α1,3GT gene knock-out mice expressing anti-Gal has been examined (Welsh, *et al.*, 1998). Although it was anticipated that LCMV expressing the α-gal epitope would replicate poorly in the anti-Gal expressing mice, the results fail to support this hypothesis. However, preincubation of α-gal epitope-expressing LCMV in freshly drawn mouse serum containing anti-Gal successfully reduces the ability of the virus to infect Vero cells *in vitro*. By contrast, under the same conditions anti-Gal fails to inhibit LCMV infection of mouse peritoneal macrophage cells *in vitro*. Macrophages, which have receptors for mouse antibody and complement, are a primary target for LCMV infection *in vivo* (Welsh, *et al.*, 1976). These data suggest that anti-Gal in mouse serum may not induce complement lysis of the modified virus. Rather, anti-Gal may sterically hinder the entry of LCMV into Vero cells, while anti-Gal and/or complement deposition may promote LCMV infection of macrophages, which express receptors for these molecules.

According to this hypothesis, the absence of viral neutralization *in vivo* may reflect a deficiency in complement activity in the α1,3GT gene knock-out mice. In fact, laboratory mice are known to be low in complement activity compared with wild mice, guinea pigs, rabbits and humans (Welsh, *et al.*, 1976). Accordingly, infectious LCMV/antibody/complement complexes can be isolated from the serum of persistently infected mice. These findings suggest that the α1,3GT gene knock-out mice may not be a good model for assessing the role of anti-Gal in protecting humans from interspecies transmission of viruses and indicate that, to provide a more accurate model, *in vivo* studies may have to be performed in Old World monkeys.

## 3. RETROVIRAL-MEDIATED GENE THERAPY AND THE α-GAL EPITOPE

### 3.1. Inactivation of Retroviral Vector Particles in Human Serum

The innate immune barrier provided by anti-Gal against enveloped viruses that express the α-gal epitope, may have important effects on the practice of gene therapy. Retroviral-mediated gene transfer accounts for the vast majority of approved human gene transfer trials to date (Anderson, 1992; Miller, 1992). The majority of vector systems used today in retroviral-mediated gene transfer experiments are derived from murine leukemia virus (Jolly, 1994; Cornetta *et al.*, 1991; Miller and Rosman, 1989). There are several advantages of retroviral vectors over other vector systems, including high efficiency of gene transfer, stable expression of transferred genes, capacity to transfer large genes and lack of cellular cytotoxicity. However, most retroviral-based human gene therapy trials involve transduction of autologous cells *ex vivo* and subsequent transplantation into patients. Direct *in vivo* retroviral-mediated gene transfer has been limited due to the inability to transduce nondividing cells, the lack of targetable vectors and the concern of retroviral vector safety. Although progress continues in overcoming these obstacles, the survival of retroviral vector particles in human serum will be a prerequisite for successful *in vivo* gene transfer in many cases.

It was initially proposed that murine retrovirus inactivation in NHS involves the direct activation of the classical complement system (Cooper *et al.*, 1976; Welsh, *et al.*, 1975). This hypothetical mechanism of classical complement activation is unusual in that it is antibody independent. Antibody-dependent activation of classical complement requires the activation of C1 through the interaction of C1q with the Fc region of antibodies deposited on a membrane surface (Muller-Eberhard, 1988). Further studies showed that the C1 component of complement can directly bind the viral envelope protein p15E resulting in complement activation and efficient inactivation of virus in NHS (Bartholomew *et al.*, 1978; Cooper *et al.*, 1976). This long-standing hypothetical mechanism of antibody-independent inactivation of murine retroviruses in NHS has recently been called into question by several studies that have identified anti-Gal as the primary mediator of retroviral inactivation (Rother and Squinto, 1996; Takeuchi *et al.*, 1996; Rother *et al.*, 1995a). In this model, (described above) natural antibody specific for the (host cell derived) α-gal epitope is primarily responsible for complement activation and lysis of murine-derived retroviruses in human, ape and Old World monkey sera.

As was demonstrated for murine retroviruses, murine-derived retroviral vector particles are also quickly inactivated following *in vitro* exposure to NHS. The role that complement plays in the inactivation of retroviral vector particles has been examined using the vector particle LXSN harvested from the amphotropic packaging cell line PA317, a 3T3 fibroblast derivative (Rother *et al.*, 1995b). The inactivation of this retroviral vector particle is entirely complement dependent, as LXSN survives exposure to human serum deficient in various complement components or serum depleted of complement by cobra venom factor treatment. In addition, functionally blocking anti-complement monoclonal antibodies are potent inhibitors of complement-mediated retroviral vector particle inactivation in human serum or heparinized blood. Retroviral vector particle inactivation apparently results through virolysis, as serum deficient in the terminal complement component C9 (the final component sequentially assembled on membrane surfaces to form the lytic membrane attack complex) does not dramatically reduce virus titer.

### 3.2. Retroviral Vector Producer Cells and the α-Gal Epitope

As mentioned above, anti-Gal is believed to be the primary antibody molecule responsible for hyperacute rejection of non-primate xenogeneic organ transplants in humans, apes and Old World monkeys (Galili, 1993b; Dalmasso *et al.*, 1992). Following transplantation of vascularized xenografts, anti-Gal binds to the α-gal epitope on vascular endothelium and activates the recipient's complement through the classical pathway (Dalmasso *et al.*, 1992). Activation of the complement cascade leads to endothelial cell activation and/or lysis, platelet aggregation, collapse of the vascular bed and cessation of graft circulation within minutes following transplantation (Platt *et al.*, 1990).

Similar to xenogeneic organ transplant rejection, cells that express the α-gal epitope are also rapidly killed *in vitro* in NHS, and carbohydrates known to specifically block anti-Gal have been shown to successfully protect these cells (Neethling *et al.*, 1994; Vaughan *et al.*, 1994; Sandrin *et al.*, 1993). Considering that mouse cells express an abundance of the α-gal epitope on their cell surfaces (Santer *et al.*, 1989), these data suggest that murine-derived retroviral vector producer cells (RVPC) will also be killed in serum containing anti-Gal. RVPC are currently used to generate high titer replication incompetent vector particles for gene therapy applications. The survival of RVPC in humans has recently become of interest based on data suggesting that the transplantation of these cells may increase the efficiency of *in vivo* gene therapy.

Current *in vivo* retroviral-mediated gene transfer procedures struggle with the difficulty of achieving high transduction efficiencies (Cardoso *et al.*, 1993; Rettinger *et al.*, 1993; Culver *et al.*, 1992; Ferry *et al.*, 1991; Hatzoglou *et al.*, 1990). However, studies performed *in vitro* have shown that transduction efficiencies are improved by co-culturing target cells with RVPC (Lemischka *et al.*, 1986; Williams *et al.*, 1984). The same strategy has been utilized in animal models where transplantation of RVPC into or near established tumors yields transduction efficiencies superior to the introduction of retroviral particles alone (Ram *et al.*, 1994a; Ram *et al.*, 1994b; Ram *et al.*, 1994c; Ram *et al.*, 1993a; Takamiya *et al.*,

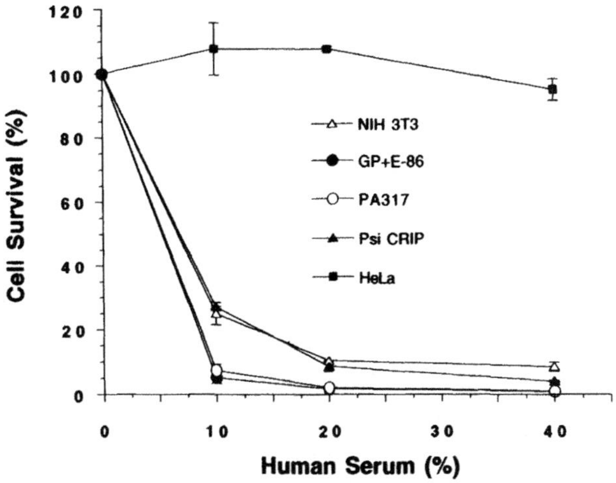


**Figure 7.** RVPC survival in NHS. Murine-derived RVPC were incubated in NHS at various concentrations and cell survival was determined. Curves represent serum survival of GP+E-86 (ecotropic), PA317 (amphotropic) and Psi CRIP (amphotropic) RVPC relative to survival of the same cells in the absence of serum. Serum survival of NIH 3T3 cells from which these RVPC were derived and human HeLa cells are also shown. (Reproduced from Rollins *et al.*, 1996, with permission.)

1993; Culver *et al.*, 1992;). Several clinical protocols have now been submitted or accepted regarding the transplantation of RVPC into humans for the treatment of solid tumors (Oldfield and Ram, 1995; Culver, 1994; Raffel *et al.*, 1994; Oldfield *et al.*, 1993).

Preclinical studies to assess RVPC survival *in vivo* have primarily been performed in non-primate mammalian species (Moorman *et al.*, 1994; Ram *et al.*, 1994b; Ram *et al.*, 1993b). However, amphotropic retroviral vector packaging cell lines and RVPC are effectively killed in NHS (Rother *et al.*, 1995a; Russell *et al.*, 1995). This observation has been extended by the assessment of the survival of RVPC lines in sera from various species (Rollins *et al.*, 1996). Three RVPC lines including GP+E-86, PA317 and Psi CRIP are effectively killed in NHS, as are NIH 3T3 cells from which the RVPC were derived (Figure 7). Survival of the human HeLa cell line is not affected by exposure to NHS. In contrast, the murine-derived RVPC lines survive exposure to dog, rabbit, rat and mouse sera, none of which produce anti-Gal (Rollins *et al.*, 1996). These results confirm that murine-derived RVPC do not survive exposure to NHS and that survival of RVPC in non-primate mammalian species may be a poor predictor of the survival of the same cells in humans. Furthermore, the blockade of anti-Gal or complement in NHS protects RVPC, indicating that the killing of these cells is due to complement activation through anti-Gal reactivity with the α-gal epitope expressed on the cell surface.

### 3.3. Strategies to Overcome the Serum Sensitivity of Retroviral Vector Particles

While murine-derived retroviral vector particles injected into the bloodstream of primates are quickly inactivated (Cornetta *et al.*, 1990), they survive *in vivo* in non-primate mammals (Rettinger *et al.*, 1994; Cardoso *et al.*, 1993; Rettinger *et al.*, 1993; Culver *et al.*, 1992; Ferry *et al.*, 1991; Hatzoglou *et al.*, 1990). Taken together with the observations discussed above regarding the role of anti-Gal and complement in viral inactivation, these data suggest that in the absence of complement activation, retroviral vector particles may survive in humans and other primates. Therefore, strategies have emerged to circumvent the inactivation of retroviral vector particles in NHS. These strategies include neutralizing either components of the complement cascade or the α-gal epitope, or both.

Complement blockade using soluble inhibitors may provide a method for improving retroviral-mediated gene transfer outcomes *in vivo* in primates. As shown in Figure 8, there is an inverse relationship between the level of complement activity in human serum and vector survival in that serum. These data also indicate that blockade of complement component C5 using a monoclonal antibody effectively protects retrovirus in NHS. Such anti-C5 monoclonal antibodies are currently being tested in clinical trials, albeit for other indications, and have been administered to human patients without ill effects. Similar levels of retrovirus pro-

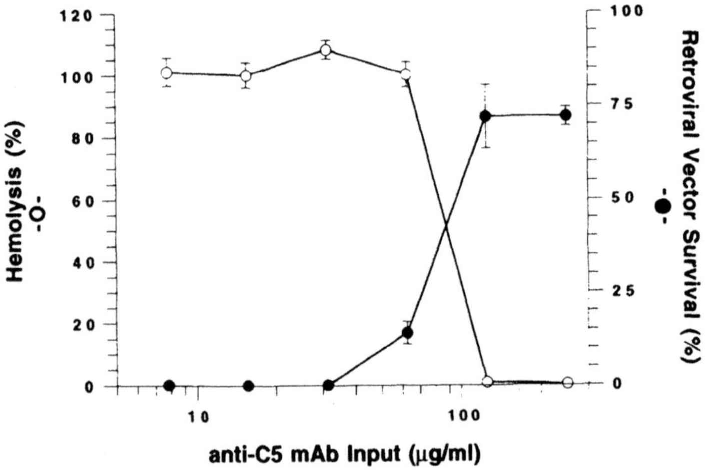


**Figure 8.** Retroviral vector particle survival correlates inversely with complement activity. NHS was incubated in the presence of various concentrations of anti-C5 monoclonal antibody and the residual complement activity was determined in an erythrocyte hemolytic assay. Retroviral vector particle survival was simultaneously assessed by preincubating vector with the anti-C5 treated serum before titering. (Reproduced from Rother *et al.*, 1995b, with permission.)

tection in NHS are attainable using functionally blocking antibodies that target other complement components, as well as the complement depleting reagent cobra venom factor (Rother *et al.*, 1995b).

The ability to protect retroviral vector particles from inactivation by NHS through blockade of terminal complement components does not address the fate of these particles in whole blood. Deposition of the activated complement fragments C3b and C4b on the surface of these particles could lead to their inactivation through binding to complement receptors present on blood cells. Additionally, the binding of the vector particles to viral receptors potentially found on blood cells could quickly deplete the titer of virus in whole blood. This does not appear to be the case, however, as retroviral vector particles survive in human whole blood in the presence of a functionally blocking anti-C5 monoclonal antibody (Rother *et al.*, 1995b). These results indicate that human blood cells do not significantly affect retroviral vector particle survival in the absence of complement-mediated virolysis.

As discussed above, RVPC are also effectively killed in NHS by the activation of complement (Rollins *et al.*, 1996; Rother *et al.*, 1995a; Russell *et al.*, 1995). Similar to that demonstrated for retroviral vector particles, preincubation of NHS with a functionally blocking anti-C5 monoclonal antibody effectively protects RVPC from complement-mediated killing (Rollins *et al.*, 1996). These results suggest that inhibition of complement may increase transduction efficiencies in retroviral-mediated gene transfer experiments involving the transplantation of RVPC into humans. Further, complement inhibition will not only allow prolonged survival of RVPC *in vivo*, but will simultaneously protect the retroviral vector particles as they are released from the cell. Upon cessation of complement inhibition, however, complement-mediated virolysis and cytolysis may destroy the introduced RVPC and virus, thus providing a convenient means of terminating therapy.

The observation that retroviral inactivation is mediated through recognition of the α-gal epitope has also prompted attempts to prevent incorporation of this carbohydrate structure into viral envelopes. One strategy that has recently been employed to generate serum resistant retroviral vector particles involves the manipulation of RVPC (Rother *et al.*, 1995a). This strategy is based on observations showing that competition between glycosyltransferases for a common acceptor can determine which terminal glycosidic residues are added to the substrate. The N-acetyllactosaminyl group (Figure 9) is the acceptor for sugars from at least five glycosyltransferases in mammalian cells, including α1,3GT and α1,2fucosyltransferase (H-transferase) (Smith *et al.*, 1990). H-transferase generates the terminal glycosidic structure Fucα1,2Galβ1,4GlcNAc-R (H-antigen), which is universally compatible in human blood transfusions (Larsen *et al.*, 1990b). Expression of recombinant H-transferase in cells containing a functional α1,3GT gene results in elevated expression of H-antigen modified membrane proteins (Figure 9, reaction B) with a concomitant reduction in α-gal epitope expression (Figure 9, reaction A) (Rother and Squinto, 1996; Sandrin *et al.*, 1995).

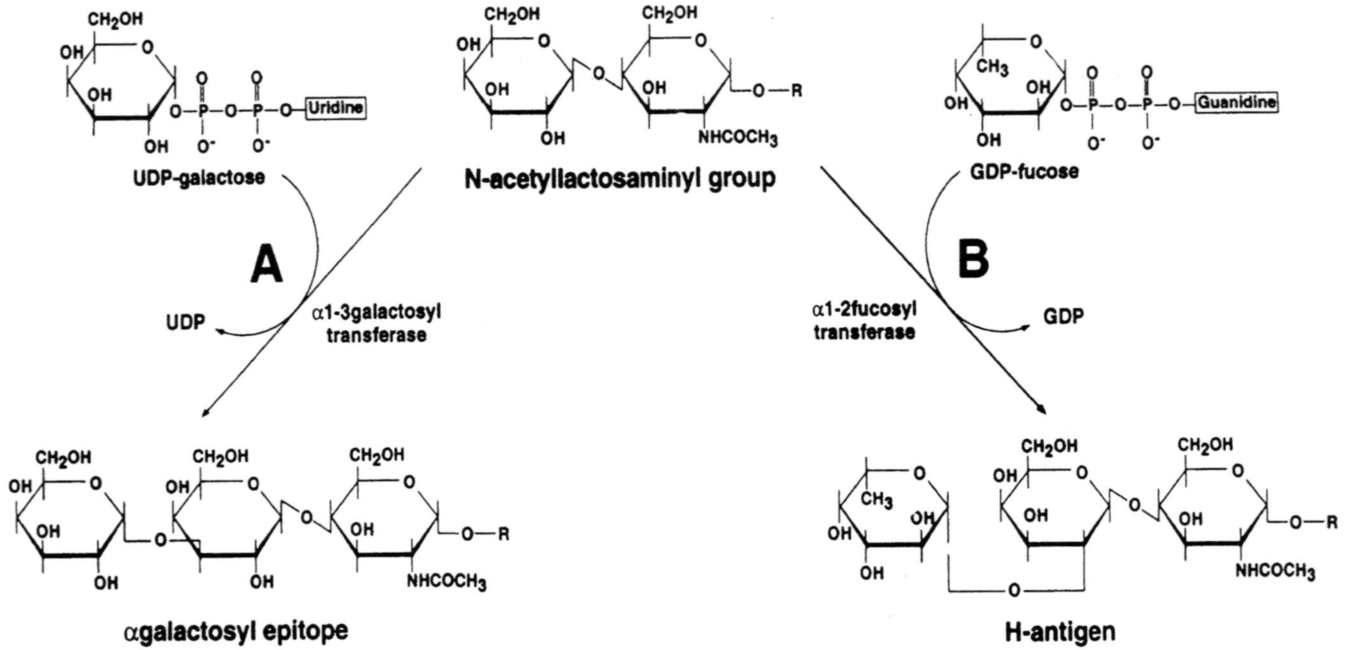


**Figure 9.** Terminal enzymatic reactions leading to the generation of the α-gal epitope or H-antigen. (A) Transfer of galactose from UDP-galactose to the N-acetyllactosaminyl group in an α 1,3 linkage. (B) Transfer of fucose from GDP-fucose to the N-acetyllactosaminyl group in an α1,2 linkage. The bolded arrow in reaction B signifies the dominance of α1,2fucosyltransferase over α1,3GT for the common N-acetyllactosaminyl acceptor group. (Reproduced from Rother and Squinto, 1996, with permission.)

To investigate the effect of α-gal epitope down-regulation on the serum sensitivity of retrovirus, H-transferase was expressed in RVPC. Unmodified RVPC express high levels of the α-gal epitope, while expression of H-antigen in these cells is low (Figure 10, panel A). Conversely, H-transferase modified RVPC show an increase in H-antigen expression while α-gal epitope expression is reduced by more than 90% (Figure 10, panel B). Similarly, binding of purified anti-Gal to H-transferase modified RVPC is greatly reduced (Figure 10, panel C). These results show that expression of H-transferase in RVPC drastically reduces α-gal epitope expression on the cell surface.

In an attempt to correlate α-gal epitope expression with NHS killing, the sensitivity of both H-transferase modified RVPC and retroviral vector particles produced by these cells was investigated. Modified RVPC show a marked increase in survival relative to unmodified cells following exposure to NHS (Figure 11, panel A). These results indicate that the level of α-gal epitope expression inversely correlates with the survival of RVPC in NHS. Concomitantly, retroviral vector particles obtained from the H-transferase modified RVPC survive exposure to NHS while vector particles from unmodified cells are effectively inactivated (Figure 11, panel B). These data indicate that down-regulation of α-gal epitope expression on RVPC results in the release of retroviral vector particles that are resistant to inactivation by human serum complement.

Cells that are deficient in α-gal epitope expression are also good candidates for the generation of cell lines that produce complement resistant retrovirus. Such cells include those of humans, apes and Old World monkeys. Human RVPC have been reported to produce retrovirus that is resistant to inactivation in NHS (Cosset *et al.*, 1995). In addition, RVPC can be generated from certain non-primate cells that are deficient in α-gal epitope expression. For example, baby hamster kidney cells and cells obtained from α1,3GT gene knock-out mice do not express detectable levels of the α-gal epitope (Welsh, *et al.*, 1998; Goochee *et al.*, 1991), and both cell lines have been used to generate retroviral vector particles that are complement resistant (personal observations). Finally, the recently developed lentiviral vector systems (Naldini *et al.*, 1996), which utilize human producer cells, should also provide particles that do not activate complement via anti-Gal.

## 4. CONCLUSION

There is now an established correlation between the presence of a functional α1,3GT gene in the host cell and the expression of the α-gal epitope on enveloped viruses that are propagated through this cell. Furthermore, the α-gal epitope has been demonstrated in association with the viral envelope glycoproteins on the surface of viruses from several viral groups. Most striking is the demonstration that,

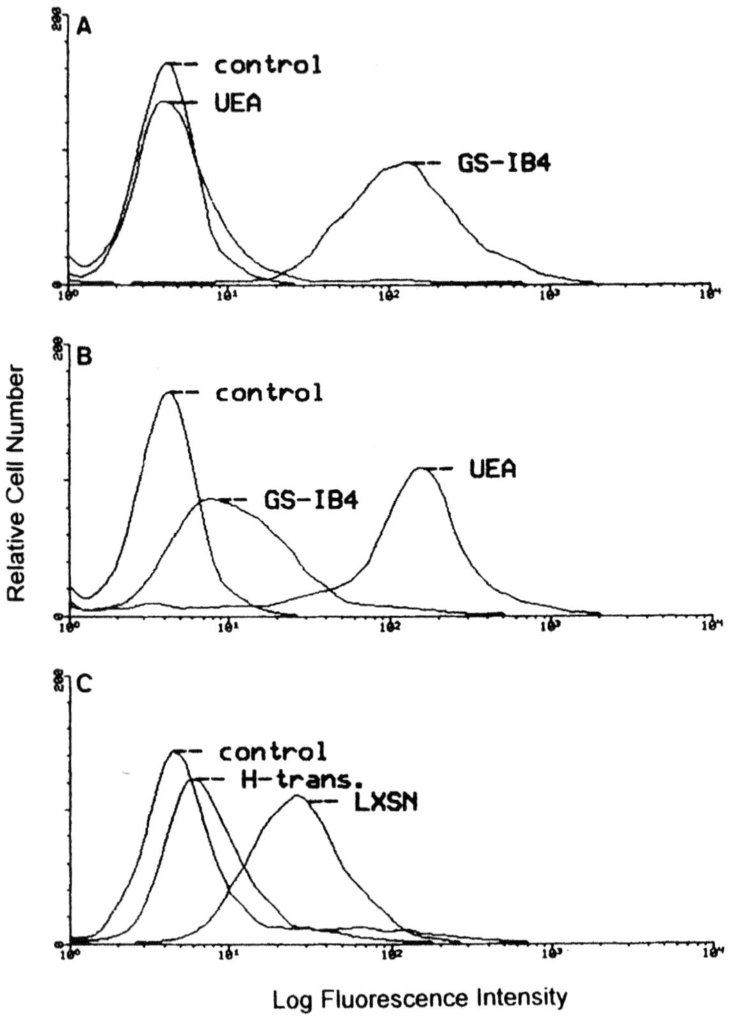


**Figure 10.** α-Gal epitope expression in H-transferase modified RVPC. The RVPC PA317 was modified with either H-transferase or vector alone (unmodified). RVPC were reacted with BS lectin, UEA lectin or anti-Gal and analyzed by FACS. UEA lectin staining (UEA) indicates expression levels of the H-transferase product (H-antigen) while BS lectin staining (GS-IB4) shows α-gal epitope expression. Panel A and B show lectin binding to unmodified RVPC or RVPC modified with H-transferase, respectively. Unstained cells are indicated in each panel (control). Panel C shows the reactivity of purified anti-Gal with unmodified RVPC (LXSN) or RVPC modified with H-transferase (H-trans.). H-transferase modified cells incubated with secondary antibody alone are also shown (control). (Reproduced from The Journal of Experimental Medicine, 1995, Volume 182, Pages 1345–1355, by copyright permission of The Rockefeller University Press.)

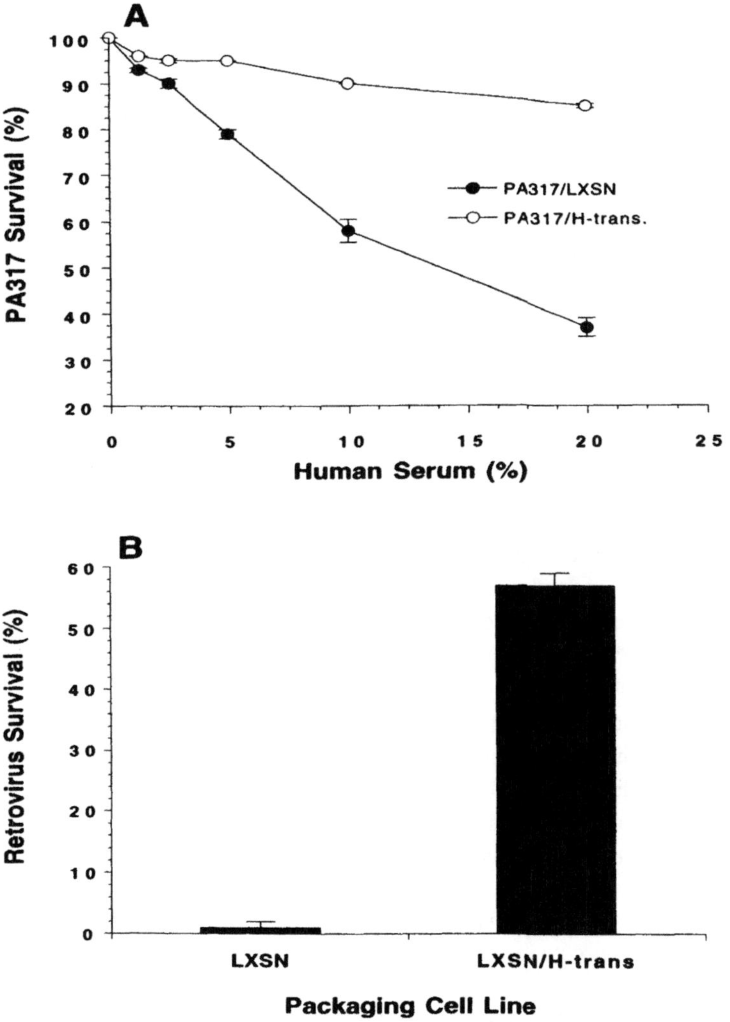


**Figure 11.** NHS sensitivity of H-transferase modified RVPC and retrovirus following α-gal epitope down-regulation. (A) RVPC modified with either H-transferase (PA317/H-transferase) or unmodified RVPC (PA317/LXSN) assayed for survival in NHS. Curves represent RVPC survival at increasing concentrations of serum. (B) Retroviral vector particles liberated from RVPC modified with H-transferase (PA317/H-trans.) or unmodified RVPC (PA317/LXSN) were subjected to NHS and retrovirus survival was determined. Bars represent the percentage of infectious vector particles remaining relative to input virus. (Reproduced from The Journal of Experimental Medicine, 1995, Volume 182, Pages 1345–1355, by copyright permission of The Rockefeller University Press.)

in most cases, there is a direct correlation between the presence of this epitope and the susceptibility of viruses to inactivation in NHS.

It is interesting to speculate that the presence of high titer anti-Gal in humans may be an important impediment to the cross-species transmission of enveloped viruses originating from mammalian species other than humans, apes or Old World monkeys. The dramatic evolutionary division between mammalian species that express either the anti-Gal antibody or the α-gal epitope strongly suggest that, at least at one point in primate evolution, the presence of anti-Gal and/or the absence of the α-gal epitope provided a strong survival advantage to primates lacking a functional α1,3GT gene. However, experimentation is still required to delineate the precise role that anti-Gal plays in the human immune response against pathogens that express this epitope. As discussed in this chapter, animal modeling will likely need to be performed in Old World monkeys. In addition, various routes of administration and different doses of α-gal epitope bearing viruses will be required to accurately determine a potential protective role for anti-Gal.

Although beyond the scope of this chapter, the presence of the α-gal epitope on the surface of enveloped viruses may prove useful in the field of viral vaccines. The virally-expressed α-gal epitope may provide an adjuvant effect in species that produce anti-Gal. Indeed, incorporation of this epitope on the surface of the influenza virus results in enhanced presentation of antigenic determinants to T cells (Galili *et al.*, 1996). The benefits of this vaccine strategy remain to be demonstrated in an animal model.

While much attention has been devoted to eliciting strong immune responses against human pathogens, in the case of certain gene therapy applications, immunity is undesirable. Recent evidence now identifies the α-gal epitope as the primary determinant on the surface of retroviral vector particles that targets them for elimination in NHS. Consequently, several strategies have been devised to reduce or eliminate the α-gal epitope on the surface of retroviral vector particles. This technology should bring the field of gene therapy one step closer to efficiently delivering genes *in vivo*.

ACKNOWLEDGMENTS. The authors thank Seth Fidel and Susan Faas for critical review of this article.

## REFERENCES

Aaronson, S.A. and Todaro, G.J., 1970. Transformation and virus growth by murine sarcoma viruses in human cells, *Nature* **225**:458–459.

Air, G.M. and Laver, W.G., 1990, Influenza Viruses, in: *Immunochemistry of Viruses. II. The Basis for Serodiagnosis, and Vaccines* (M.H.V. van Regemorted, and A.R. Neurath, eds.), Elsevier Science Publications, Amsterdam, pp. 171–216.

Almeida, I.C., Milani, S.R., Gorin, A.J., and Travoassos, L.R., 1991, Complement-mediated lysis of *Trypanosoma cruzi* tryptomastigotes by human anti α-galactosyl antibodies, *J. Immunol.* **146**:2394–2400.

Anderson, W.A., 1992, Human gene therapy, *Science* **256**:808–813.

Avila, J.L., Rojas, M., and Galili, U., 1989, Immunogenic Galα1–3Gal carbohydrate epitopes are present on pathogenic American Trypanosoma and Leishmania, *J. Immunol.* **142**:2828–2834.

Banapour, B., Sernatinger, J., and Levy, J.A., 1986, The AIDS-associated retrovirus is not sensitive to lysis or inactivation by human serum, *Virol.* **152**:268–271.

Barbacid, M., Bolognesi, D., and Aaronson, S.A., 1980, Humans have antibodies capable of recognizing oncoviral glycoproteins: demonstration that these antibodies are formed in response to cellular modification of glycoproteins rather than as consequence of exposure to virus, *Proc. Natl. Acad. Sci. USA* **77**:1617–1621.

Bartholomew, R.M., Esser, A.F., and Muller-Eberhard, H.J., 1978, Lysis of oncornaviruses by human serum: isolation of the viral complement (C1) receptor and identification as p15E, *J. Exp. Med.* **147**:844–53.

Basu, M. and Basu, S., 1973, Enzymatic synthesis of a blood group-related pentaglycosyl ceramide by an α-galactosyltransferase from rabbit bone marrow, *J. Biol. Chem.* **248**:1700–1706.

Betteridge, A. and Watkins, W.M., 1983, Two α-2-D-galactosyltransferases in rabbit stomach mucosa with different acceptor substrate specificities, *Eur. J. Biochem.* **132**:29–35.

Blake, D.D. and Goldstein, I.J., 1981, An α-D-galactosyltransferase in Ehrlich ascites tumor cells. Biosynthesis and characterization of a trisaccharide [α-D-galactose (1–3)-N-acetyllactosamine], *J. Biol. Chem.* **256**:5387–5393.

Blanken, W.M. and Ban den Eijnden, D.H., 1985, Biosynthesis of terminal Galα1–3Galβ1–4GlcNAc-R oligosaccharide sequence on glycoconjugates: purification and acceptor specificity of a UDP-Gal: N-acetyllactosaminide α1–3 galactosyltransferase, *J. Biol. Chem.* **260**:12927–12934.

Boiron, R.R., Bernard, C., and Chuat, J.C., 1969, Replication of mouse sarcoma virus Moloney strain (MSV-N) in human cells, *Proc. Amer. Assoc. Cancer Res.* **10**:8.

Burke, D.J. and Keegstra, K., 1976, Purification and composition of the proteins from Sindbis virus grown in chick and BHK cells, *J. Virol.* **20**:676–686.

Cardoso, J.E., Branchereau, S., Jeyaraj, P.R., Houssin, D., Danos, O., and Heard, J.-M., 1993, *In situ* retrovirus-mediated gene transfer into dog liver, *Hum. Gene Ther.* **4**:411–418.

Chang, G.-J.J. and Trent, D.W., 1987, Nucleotide sequence of the genome region encoding the 26S mRNA of eastern equine encephalomyelitis virus and the deduced amino acid sequence of the viral structural proteins, *J. Gen. Virol.* **68**:2129–2142.

Cooper, N.R., Jensen, F.C., Welsh, R.M., Jr., and Oldstone, M.B.A., 1976, Lysis of RNA tumor viruses by human serum: direct antibody-independent triggering of the classical complement pathway, *J. Exp. Med.* **144**:970–984.

Cornetta, K., Moen, R.C., Culver, K., Morgan, R.A., McLachlin, J.R., Sturm, S., Selegue, J., London, W., Blaese, R.M., and Anderson, W.F., 1990, Amphotropic murine leukemia retrovirus is not an acute pathogen for primates, *Hum. Gene Ther.* **1**:15–30.

Cornetta, K., Morgan, R.A., and Anderson, W.F., 1991, Safety issues related to retroviral-mediated gene transfer in humans, *Hum. Gene Ther.* **2**:5–14.

Cosset, F.-C., Takeuchi, Y., Battini, J.-L., Weiss, R.A., and Collins, M.K.L., 1995, High-titer packaging cells producing recombinant retroviruses resistant to human serum, *J. Virol.* **69**:7430–7436.

Culver, K.W., 1994, Clinical applications of gene therapy for cancer, *Clin. Chem.* **40**:510–512.

Culver, K.W., Ram, Z., Wallbridge, S., Ishii, H., Oldfield, E.H., and Blaese, R.M., 1992, *In vivo* gene transfer with retroviral vector-producer cells for treatment of experimental brain tumors, *Science* **256**:1550–1552.

Dalmasso, A.P., Vercellotti, G.M., Fischel, R.J., Bolman, R.M., Bach, F.H., and Platt, J.L., 1992, Mechanism of complement activation in the hyperacute rejection of porcine organs transplanted into primate recipients, *Am. J. Pathol.* **140:**1157–1168.

Donahue, R.E., Kessler, S.W., Bodine, D., Goodman, S., Agricola, B., Byrne, E., Raffeld, M., Moen, R., Bacher, J., Zsebo, K.M., and Nienhuis, A.W., 1992, Helper virus induced T cell Lymphoma in nonhuman primates after retroviral mediated gene transfer, *J. Exp. Med.* **176:**1125–1135.

Famulari, N.G., 1983, Murine leukemia viruses with recombinant env genes: a discussion of their role in leukemogenesis, *Curr. Top. Microbiol. Immunol.* **103:**103–108.

Ferry, N., Duplessis, O., Houssin, D., Danos, O., and Heard, J.-M., 1991, Retroviral-mediated gene transfer into hepatocytes *in vivo*, *Proc. Natl. Acad. Sci. USA* **88:**8377–8381.

Galili, U., 1993a, Evolution and pathophysiology of the human natural anti-α-galactosyl IgG (anti-Gal) antibody, *Springer Semin. Immunopathol.* **15:**155–171.

Galili, U., 1993b, Interaction of the natural anti-Gal antibody with α-galactosyl epitopes: a major obstacle for xenotransplantation in humans, *Immunol. Today* **14:**480–482.

Galili, U. and Andrews, P., 1995, Suppression of α-galactosyl epitopes synthesis and production of the natural anti-gal antibody: a major evolutionary event in ancestral Old World primates, *J. Hum. Evol.* **29:**433–443.

Galili, U., and Swanson, K., 1991, Gene sequences suggest inactivation of α1,3galactosyltransferase in catarrhines after the divergence of apes from monkeys, *Proc. Natl. Acad. Sci. USA* **88:**7401–7404.

Galili, U., Rachmilewitz, E.A., Peleg, A., and Flechner, I., 1984, A unique natural human IgG antibody with anti-α-galactosyl specificity, *J. Exp. Med.* **160:**1519–1531.

Galili, U., Macher, B.A., Buehler, J., and Shohet, S.B., 1985, Human natural anti-α-galactosyl IgG. II. The specific recognition of α(1–3)-linked galactose residues, *J. Exp. Med.* **162:**573–582.

Galili, U., Clark, M.R., Shohet, S.B., Buehler, J., and Macher, B.A., 1987, Evolutionary relationship between the natural anti-Gal antibody and the Galα1–3Gal epitope in primates, *Proc. Natl. Acad. Sci. USA* **84:**1369–1373.

Galili, U., Shohet, S.B., Kobrin, E., Stults, C.L.M., and Macher, B.A., 1988a, Man, apes, and Old World monkeys differ from other mammals in the expression of α-galactosyl epitopes on nucleated cells, *J. Biol. Chem.* **263:**17755–17762.

Galili, U., Mandrell, R.E., Hamadeh, R.M., Shohet, S.B., and Griffis, J.M., 1988b, Interaction between human natural anti-αgalactosyl immunoglobulin G and bacteria of the human flora, *Infect. Immun.* **56:**1730–1737.

Galili, U., Repik, P.M., Anaraki, F., Mozdzanowska, K., Washko, G., and Gerhard, W., 1996, Enhancement of antigen presentation of influenza virus hemagglutinin by the natural human anti-Gal antibody, *Vaccine* **256:**160–178.

Galili, U., La Temple, D.C., and Radic, M.Z., 1998, A sensitive assay for measuring α-gal epitope expression by a monoclonal anti-Gal antibody, *Transplantation* **65:**1129–1132.

Geyer, R., Geyer, H., Stirm, S., Hensmann, G., Schneider, J., Dabrowski, U., and Dabrowski, J., 1984, Major oligosaccharides in the glycoprotein of Friend murine leukemia virus: structure elucidation by one- and two-dimensional proton nuclear magnetic resonance and methylation analysis, *Biochem.* **23:**5628–5637.

Goochee, C.F., Gramer, M.J., Andersen, D.C., Baher, J.B., and Rasmussen, J.R., 1991, The oligosaccharides of glycoproteins: bioprocess factors affecting oligosaccharide structure and their effect on glycoprotein properties, *Biotechnology* **9:**1347–1355.

Hamadeh, R.M., Jarvis, G.A., Galili, U., Mandrell, R.E., Zhou, P., and Griffiss, J.M., 1992, Human natural anti-Gal IgG regulates alternative complement pathway activation on bacterial surfaces, *J. Clin. Invest.* **89:**1223–1235.

Hamadeh, R.M., Galili, U., Zhou, P., and Griffis, J.M., 1995a, Anti-α-galactosyl immunoglobulin A (IgA), IgG and IgM in human secretions, *Clin. Diagnos. Lab. Immunol.* **2:**125–131.

Hamadeh, R.M., Estabrook, M.M., Zhou, P., Jarvis, G.A., and Griffiss, J.M., 1995b, Anti-gal binds to pili of *Neisseria meningitidis*: the immunoglobulin A isotype blocks complement-mediated killing, *Infect. Immun.* **63**:4900–4906.

Hatzoglou, M., Lamers, W., Bosch, F., Wynshaw-Boris, A., Clapp, D.W., and Hanson, R.W., 1990, Hepatic gene transfer in animals using retroviruses containing the promoter from the gene for phosphoenolpyruvate carboxykinase, *J. Biol. Chem.* **265**:17285–17293.

Hoshino, H., Tanaka, H., Miwa, M., and Okada, H., 1984, Human T-cell leukemia virus is not lysed by human serum, *Nature* **310**:324–325.

Hsieh, P., Rosner, M.R., and Robbins, P.W., 1983, Host-dependent variation of asparagine-linked oligosaccharides at individual glycosylation sites of Sindbis virus glycoproteins, *J. Biol. Chem.* **258**:2548–2554.

Jensen, F.C., Girardi, A.J., Gilden, R.V., and Koprowski, H., 1964, Infection of human and simian tissue cultures with Rous sarcoma virus, *Proc. Natl. Acad. Sci. USA* **52**:53–57.

Jolly, D., 1994, Viral vector systems for gene therapy, *Cancer Gene Ther.* **1**:51–64.

Joziasse, D.H., Shaper, J.H., Van den Eijnden, D.H., Van Tunen, A.J., and Shaper, N.L., 1989, Bovine α1–3-galactosyltransferase: isolation and characterization of a cDNA clone, *J. Biol. Chem.* **264**:14290–14297.

Kiel, W., Geyer, R., Dabrowski, J., Dabrowski, U., Niemann, H., Strim, S., and Klenk, H.-D., 1985, Carbohydrates of influenza virus. Structure elucidation of the individual glycans of the hemagglutinin by two-dimensional H NMR, and methylation analysis, *EMBO J.* **4**:2711–2720.

Klenk, H.-D., 1990, II. Influence of Glycosylation on Antigenicity of Viral Proteins, in: *Immunochemistry of Viruses* (M.H.V. van Regemortel, and A.R. Neurath, eds.), Elsevier Science Publications, New York, pp. 25–37.

Klenk, H.-D. and Rott, R., 1980, Cotranslational and posttranslational processing of viral glycoproteins, *Curr. Top. Microbiol. Immunol.* **90**:19–48.

Kornfeld, R. and Kornfeld, S., 1976, Comparative aspects of glycosylation structure, *Ann. Rev. Biochem.* **45**:217–237.

Kornfeld, R. and Kornfeld, S., 1985, Assembly of asparagine-linked oligosaccharides, *Ann. Rev. Biochem.* **54**:631–664.

Larsen, R.D., Rivera-Marrero, C.A., Ernst, L.K., Cummings, R.D., and Lowe, J.B., 1990a, Frameshift and nonsense mutations in a human genomic sequence homologous to a murine UDP-Gal β-D-Gal(1,4)-D-GlcNAc α(1,3)galactosyltransferase cDNA, *J. Biol. Chem.* **265**:7055–7062.

Larsen, R.D., Ernst, L.K., Nair, R.P., and Lowe, J.B., 1990b, Molecular cloning, sequence, and expression of a human GDP-L-fucose:β-D-galactoside 2-α-L-fucosyltransferase cDNA that can form the H blood group antigen, *Proc. Natl. Acad. Sci. USA* **87**:6674–6678.

La Temple, D.C., and Galili, U., 1998, Adult and neonatal anti-Gal response in knock-out mice for α1,3galactosyltransferase, *Xenotransplantation* (in press).

Lemischka, I.R., Raulet, D.H., and Mulligan, R.C., 1986, Developmental potential and dynamic behavior of hematopoietic stem cells, *Clin. Diagnos. Lab. Immunol.* **45**:917–927.

Lower, J., Davidson, E.A., Teich, N.M., Weiss, R.A., Joseph, A.P., and Kurth, R., 1981, Hetrophil human antibodies recognize oncovirus envelope antigens: epidemiological parameters and immunological specificity of the reaction, *Virol.* **109**:409–417.

Marschang, P., Sodroski, J., Wurzner, R., and Dierich, M.P., 1995, Decay-accelerating factor (CD55) protects human immunodeficiency virus type 1 from inactivation by human complement, *Eur. J. Immunol.* **25**:285–290.

Miller, A.D., 1992, Human gene therapy comes of age, *Nature* **357**:455–460.

Miller, A.D. and Rosman, G.J., 1989, Improved retroviral vectors for gene transfer and expression, *Biotechniques* **7**:980–990.

Moorman, D.W., Butler, D.A., Stanley, J.D., Lamsam, J.L., Ackermann, M.R., Jacobson, C.D., and Culver, K.W., 1994, Survival and toxicity of xenogeneic murine retroviral vector producer cells in liver, *J. Surg. Oncol.* **57**:152–156.

Morris, C.D., 1988, Eastern Equine Encephalomyelitis, in: *The Arboviruses: Epidemiology and Ecology* (T.P. Month, ed.), CRC Press, Boca Raton, pp. 1–20.

Muller-Eberhard, H.J., 1988, Molecular organization and function of the complement system, *Ann. Rev. Biochem.* **57**:321–347.

Naldini, L., Blomer, U., Gallay, P., Ory, D., Mulligan, R., Gage, F.H., Verma, I.M., and Trono, D., 1996, *In vivo* gene delivery and stable transduction of nondividing cells by a lentiviral vector, *Science* **272**:263–267.

Neethling, F.A., Koren, E., Ye, Y., Richards, S.V., Kujundzic, M., Oriol, R., and Cooper, D.K.C., 1994, Protection of pig kidney (PK15) cells from the cytotoxic effect of anti-pig antibodies by α-galactosyl oligosaccharides, *Transplantation* **57**:959–963.

Oldfield, E.H. and Ram, Z., 1995, Intrathecal gene therapy for the treatment of leptomeningeal carcinomatosis, *Hum. Gene Ther.* **6**:55–85.

Oldfield, E.H., Ram, Z., Culver, K.W., Blaese, R.M., and DeVroom, H.L., 1993, Gene therapy for the treatment of brain tumors using intra-tumoral transduction with the thymidine kinase gene and intravenous ganciclovir, *Hum. Gene Ther.* **4**:39–69.

Platt, J.L., Vercellotti, G.M., Dalmasso, A.P., Matas, A.J., Bolman, R.M., Najarian, J.S., and Bach, F.H., 1990, Transplantation of discordant xenografts: a review of progress, *Immunol. Today* **11**:456–457.

Rademacher, T.W., Parekh, R.B., and Dwek, R.A., 1988, Glycobiology, *Ann. Rev. Biochem.* **57**:785–838.

Raffel, C., Culver, K., Kohn, D., Nelson, M., Siegel, s., Gillis, F., Link, C.J., and Villablanca, J.G., 1994, Gene therapy for the treatment of recurrent pediatric malignant astrocytomas with *in vivo* tumor transduction with the herpes simplex thymidine kinase gene/ganciclovir system, *Hum. Gene Ther.* **5**:863–890.

Ram, Z., Culver, K.W., Walbridge, S., Blaese, R.M., and Oldfield, E.H., 1993a, *In situ* retroviral-mediated gene transfer for the treatment of brain tumors in rats, *Cancer Res.* **53**:83–88.

Ram, Z., Culver, K.W., Walbridge, S., Frank, J.A., Blaese, R.M., and Oldfield, E.H., 1993b, Toxicity studies of retroviral-mediated gene transfer for the treatment of brain tumors, *J. Neurosurg.* **79**:400–407.

Ram, Z., Walbridge, S., Heiss, J.D., Culver, K.W., Blaese, R.M., and Oldfield, E.H., 1994a, *In vivo* transfer of the human interleukin-2 gene: negative tumoricidal results in experimental brain tumors, *J. Neurosurg.* **80**:535–540.

Ram, Z., Walbridge, S., Oshiro, E.M., Viola, J.J., Chiang, Y., Mueller, S.N., Blaese, R.M., and Oldfield, E.H., 1994b, Intrathecal gene therapy for malignant leptomeningeal neoplasia, *Cancer Res.* **54**:2141–2145.

Ram, Z., Walbridge, S., Shawker, T., Culver, K.W., Blaese, R.M., and Oldfield, E.H., 1994c, The effect of thymidine kinase transduction and ganciclovir therapy on tumor vasculature and growth of 9L gliomas in rats, *J. Neurosurg.* **81**:256–260.

Reed, D.J., Lin, X., Thomas, T.D., Birks, C.W., Tang, J., and Rother, R.P., 1997, Alteration of glycosylation renders HIV sensitive to inactivation by normal human serum, *J. Immunol.* **159**:4356–4361.

Repik, P.M., Strizki, J.M., and Galili, U., 1994, Differential host-dependent expression of α-galactosyl epitopes on viral glycoproteins: a study of eastern equine encephalitis virus as a model, *J. Gen. Virol.* **75**:1177–1181.

Rettinger, S.D., Ponder, K.P., Saylors, R.L., Dennedy, S.C., Hafenrichter, D.G., and Flye, M.W., 1993, *In vivo* hepatocyte transduction with retrovirus during in-flow occlusion, *J. Surg. Res.* **54**:418–425.

Rettinger, S.D., Kennedy, S.C., Wu, X., Saylors, R.L., Hafenrichter, D.G., Flye, M.W., and Ponder, K.P., 1994, Liver-directed gene therapy: quantitative evaluation of promoter elements by using *in vivo* retroviral transduction, *Proc. Natl. Acad. Sci. USA* **91**:1460–1464.

Rollins, S.A., Birks, C.W., Setter, E., Squinto, S.P., and Rother, R.P., 1996, Retroviral vector producer cell killing in human serum is mediated by natural antibody and complement: strategies for evading the humoral immune system, *Hum. Gene Ther.* **7**:619–626.

Rother, R.P. and Squinto, S.P., 1996, The α-galactosyl epitope: a sugar coating that makes viruses and cells unpalatable, *Cell* **86**:185–188.

Rother, R.P., Fodor, W.L., Springhorn, J.P., Birks, C.W., Setter, E., Sandrin, M.S., Squinto, S.P., and Rollins, S.A., 1995a, A novel mechanism of retrovirus inactivation in human serum mediated by anti-αgalactosyl natural antibody, *J. Exp. Med.* **182**:1345–1355.

Rother, R.P., Squinto, S.P., Mason, J.M., and Rollins, S.A., 1995b, Protection of retroviral vector particles in human blood through complement inhibition, *Hum. Gene Ther.* **6**:429–435.

Russell, D.W., Berger, M.S., and Miller, A.D., 1995, The effects of human serum and cerebrospinal fluid on retroviral vectors and packaging cell lines, *Hum. Gene Ther.* **6**:635–641.

Saifuddin, M., Parker, C.J., Peeples, M.E., Gorny, M.K., Zolla-Pazner, S., Ghassemi, M., Rooney, I.A., Atkinson, J.P., and Spear, G.T., 1995, Role of virion-associated glycosylphosphatidylinositol-linked proteins CD55 and CD59 in complement resistance of cell line-derived and primary isolates of HIV-1, *J. Exp. Med.* **182**:501–509.

Sandrin, M.S., Vaughan, H.A., Dabkowski, P.L., and McKenzie, I.F.C., 1993, Anti-pig IgM antibodies in human serum react predominantly with gal(α1–3)gal epitopes, *Proc. Natl. Acad. Sci. USA* **90**:11391–11395.

Sandrin, M.S., Fodor, W.L., Mouhtouris, E., Osman, N., Cohney, S., Rollins, S.A., Guilmette, E.R., Setter, E., Squinto, S.P., and McKenzie, I.F.C., 1995, Enzymatic remodelling of the carbohydrate surface of a xenogenic cell substantially reduces human antibody binding and complement-mediated cytolysis, *Nature Med.* **1**:1261–1267.

Santer, U.V., DeSantis, R., Hard, K.J., van Kuik, J.A., Vliegenthart, J.F.G., Won, B., and Glick, M.C., 1989, N-linked oligosaccharide changes with oncogenic transformation require sialylation of multiantennae, *Eur. J. Biochem.* **181**:249–260.

Schlesinger, M.J. and Schlesinger, S., 1987a, Formation and assembly of alphavirus glycoproteins, in: *The Togaviridae and Flaviviridae* (S. Schlesinger, and M.J. Schlesinger, eds.), Plenum Press, New York, pp. 121–148.

Schlesinger, M.J. and Schlesinger, S., 1987b, Domains of virus glycoproteins, *Advances in Virus Research* **33**:1–44.

Schulze, I.T., 1970, The structure of influenza virus. I. The polypeptides of the virion, *Virol.* **41**:890–904.

Smith, D.F., Larsen, R.D., Mattox, S., Lowe, J.B., and Cummings, R.D., 1990, Transfer and expression of a murine UDP-Gal:β-D-Gal-α1,3-galactosyltransferase gene in transfected Chinese hamster ovary cells, *J. Biol. Chem.* **265**:6225–6234.

Snyder, H.W., Jr. and Fleissner, E., 1980, Specificity of human antibodies to oncovirus glycoproteins: recognition of antigen by natural antibodies directed against carbohydrate structures, *Proc. Natl. Acad. Sci. USA* **77**:1622–1626.

Spear, G.T., Lurain, N.S., Parker, C.J., Ghassemi, M., Payne, G.H., and Saifuddin, M., 1995, Host cell-derived complement control proteins CD55 and CD59 are incorporated into the virions of two unrelated enveloped viruses, *J. Immunol.* **155**:4376–4381.

Stollar, V., Stollar, B.D., Koo, K., Harrap, K.A., and Schlesinger, W.R., 1976, Sialic acid contents of Sindbis virus from vertebrate and mosquito cells: equivalence of biological and immunological viral properties, *Virol.* **69**:104–115.

Strauss, J.H., Burge, B.W., and Darnell, J.E., 1970, Carbohydrate content of the membrane protein of Sindbis virus, *J. Mol. Biol.* **47**:437–448.

Strauss, J.H. and Strauss, E.G., 1977, Togaviruses, in: *The Molecular Biology of Animal Viruses* (D.P. Nayak, ed.), Marcel Dekker, New York, pp. 111–166.

Takamiya, Y., Short, M.P., Moolten, F.L., Fleet, C., Mineta, T., Breakefield, X.O., and Martuza, R.L., 1993, An experimental model of retrovirus gene therapy for malignant brain tumors, *J. Neurosurg.* **79:**104–110.

Takeuchi, Y., Cosset, F.-L., Lachmann, P.J., Okada, H., Weiss, R.A., and Collins, M.K.L., 1994, Type C retrovirus inactivation by human complement is determined by both the viral genome and the producer cell, *J. Virol.* **68:**8001–8007.

Takeuchi, Y., Porter, C.D., Strahan, K.M., Preece, A.F., Gustafsson, K., Cosset, J.-L., Weiss, R.A., and Collins, M.K.L., 1996, Sensitization of cells and retroviruses to human serum by (α1–3) galactosyltransferase, *Nature* **379:**85–88.

Teich, N.M., Weiss, R.A., Salahuddin, S.Z., Gallagher, R.E., Gillespie, D.H., and Gallo, R.C., 1975, Infective transmission and characterization of a C-type virus released by cultured human myeloid leukemia cells, *Nature* **256:**551–555.

Thall, A.D., Maly, P., and Lowe, J.B., 1995, Oocyte Galα1,3Gal epitopes implicated in sperm adhesion to the zona pellucida glycoprotein ZP3 are not required for fertilization in the mouse, *J. Biol. Chem.* **270:**21437–21440.

Thiry, L., Cogniaux-Le Clerc, J., Content, J., and Tack, L., 1978, Factors which influence inactivation of vesicular stomatitis virus by fresh human serum, *Virol.* **87:**384–393.

Tsichlis, P.N., 1987, Oncogenesis by Moloney murine leukemia virus, *Anticancer Res.* **7:**171–180.

Tsichlis, P.N. and Lazo, P.A., 1991, Virus-host interactions and the pathogenesis of murine and human oncogene retroviruses, *Curr. Top. Microbiol. Immunol.* **171:**95–171.

Vaughan, H.A., Loveland, B.E., and Sandrin, M.S., 1994, Galα(1,3)gal is the major xenoepitope expressed on pig endothelial cells recognized by naturally occurring cytotoxic human antibodies, *Transplantation* **58:**879–882.

Welsh, R.M., Jr., 1977, Host cell modification of lymphocytic choriomeningitis virus and Newcastle disease virus altering viral inactivation by human complement, *J. Immunol.* **118:**348–354.

Welsh, R.M., Jr., Cooper, N.R., Jensen, F.C., and Oldstone, M.B.A., 1975, Human serum lyses RNA tumour viruses, *Nature* **257:**612–614.

Welsh, R.M., Jr., Lampert, P.W., Burner, P.A., and Oldstone, M.B.A., 1976, Antibody-complement interactions with purified lymphocytic choriomeningitis virus, *Virol.* **73:**59–71.

Welsh, R.M., Jr., O'Donnell, C.L., Reed, D.J., and Rother, R.P., 1998, Evaluation of the galα1–3gal epitope as a host modification factor eliciting natural humoral immunity to enveloped viruses, *J. Virol.* (in press).

Williams, D.A., Lemischka, I.R., Nathan, D.G., and Mulligan, R.C., 1984, Introduction of a new genetic material into pluripotent haematopoietic stem cells of the mouse, *Nature* **310:**476–480.

Wood, C., Kabat, E.A., Murphy, L.A., and Goldstein, I.J., 1979, Immunochemical studies on the combining sites of two isolectins $A_4$ and $B_4$ isolated from *Bandeiraea simplicifolia*, *Arch. Biochem. Biophys.* **198:**1–9.

*Chapter 8*

# α-Galactosyl-Bearing Epitopes as Potent Immunogens in Chagas' Disease and Leishmaniasis

José Luis Avila

## 1. INTRODUCTION

Anti-α-galactosyl (anti-Gal) is a human natural antibody which constitutes as much as 1% of circulating IgG and which interacts specifically with the carbohydrate epitope Galα1-3Galβ1-4GlcNac-R (termed the α-gal epitope). This natural antibody is unique among known human natural antibodies because of its unusually high concentration in the serum (30–100 μg/ ml)( Avila *et al.*, 1989; Davin *et al.*, 1987; Galili *et al.*, 1984), and is presence in all humans (Galili *et al.*, 1984). The only other mammals producing anti-Gal are Old World monkeys and apes (Galili *et al.*, 1987b).

Anti-Gal is a natural antibody which may be of physiologic significance since it is continuously produced in large amounts by a high proportion of B lymphocytes in all humans.

The ubiquitous presence of anti-Gal in high concentrations in human serum throughout life (Galili *et al.*, 1984) raised the question of the source for such a

---

**José Luis Avila** Instituto de Biomedicina, Caracas 1010A, Venezuela.

*Subcellular Biochemistry, Volume 32: α-Gal and Anti-Gal*, edited by Galili and Avila.
Kluwer Academic / Plenum Publishers, New York, 1999.

chronic immunogenic stimulus. Studies on other natural carbohydrates antibodies, i.e., anti-blood group A and anti-blood B antibodies suggested that bacteria within the normal intestinal flora provide the constant antigenic stimulation for the production of these antibodies (Springer and Horton, 1969).

This review discusses the possible role of anti-Gal against *Trypanosoma* and *Leishmania* infection.

## 2. AMERICAN TRYPANOSOMIASIS OR CHAGAS' DISEASE

Chagas' disease is the most common form of chronic myocarditis in the world (Lancet, 1965). On the basis of published data from various parts of Latin America, it is estimated that at least 20 million individuals are infected with *Trypanosoma cruzi*, and that over 65 million live in endemic areas where the risk of this infection exists (WHO, 1995).

### 2.1. Acute Chagas' Disease

The acute stage is characterized by general malaise with a variety of clinical manifestations. The symptoms can be very mild and atypical. Consequently the disease is often not recognized at this stage, indeed, and diagnosed in only 1% to 2% of all patients, whereas in the remaining cases pass unnoticed. General symptoms are fever, enlarged liver and spleen, generalized oedema, and swollen lymph-nodes. Up to 30% of cases show electrocardiographic or radiological abnormalities due to acute myocarditis of different degrees.

### 2.2. Indeterminate Stage

The indeterminate stage begins 8–10 weeks after the acute stage, whether there have been clinical manifestations or not. This stage may last several years or persist indefinitely. It is characterized by the absence of clinical symptoms, the subjects being fully capable of normal physical activity and showing normal electrocardiograms and chest X-rays. However, serological tests for Chagas disease remain positive, and parasitemia can be recognized in 20–60% of cases by xenodiagnosis(this technique is performed by feeding third-instar nymphs of *Triatoma infestans* or *Rhodnius prolixus* with patient's blood. Thirty and 60 days after the blood-meal, their faeces and intestines are examined under the microscope for trypomastigotes or epimastigotes of *T. cruzi*).

### 2.3. Chronic Chagas' Disease

It is estimated that up to 30% of persons with the indeterminate form of the infection will suffer from cardiac, digestive, or neurological damage, 10–20 years after having contracted the disease. The cardiac form is the most studied, the best known, and the easiest to diagnose. The clinical manifestations depend on the degree of myocardial damage, the presence of arrhythmias and the degree of heart failure. Carrasco *et al* (1982) according to the clinical and electrocardiographic characteristics and the results of the left cineventriculogram proposed to include chagasic patients in three groups with the following characteristics: group I was subdivided into groups Ia and Ib (group Ia= normal electrocardiogram (EKG) and normal cineventriculogram; group Ib = normal EKG and abnormal left cineventriculogram); group II = abnormal EKG, but not signs of congestive failure; group III = abnormal EKG and congestive failure.

In the digestive form any portion of the digestive tract can be involved but the most commonly affected segments are the oesophagus and the colon.

Mild to profound involvement of the central, peripheral or autonomic nervous system have been reported in chronic stage.

### 2.4. Pathogenesis

In spite of the importance of the disease, little is known of its pathogenesis. The myocardial fibrosis develops in association with a chronic inflammatory reaction (Andrade *et al.*, 1989; Andrade and Grimaud, 1986; Andrade, 1985). Sequential studies of mice with experimentally induced chagasic myocarditis have shown that the inflammatory reaction is initially associated with insterstitial deposits of fibronectin and subsequently with increased amounts of interstitial connective tissue proteins, including type III and IV collagen and laminin (Andrade *et al.*, 1989). These observations are important because "anti-laminin" antibodies (latter demonstrated to be exclusively directed against the α-gal epitope abundantly present in murine laminin [Towbin *et al*, 1987]) were found to increase in the sera of patients, monkeys and mice infected with *T. cruzi* (Sanchez *et al.*, 1993; Unterkircher *et al.*, 1993; Avila, 1992; Laguens *et al.*, 1991; Gazzinelli *et al.*, 1991a; Towbin *et al.*, 1987; Milei and Storino, 1987; Avila *et al.*, 1984; Szarfman *et al.*, 1982). Ultrastructural studies of endomyocardial biopsy specimens from patients with the chronic form of the disease have revealed severe thickening of the basement membranes of the myocytes, endothelial cells and vascular smooth muscle cells (Milei *et al.*, 1992; Palacios-Prü *et al.*, 1989; Ferrans *et al.* 1988). More recently, Sanchez *et al.*(1993) have found in 10 patients with chagasic cardiomyopathy, marked thickening of the basement membranes on staining with the periodic acid silver-methenamine (PAM) method, specific for glycoconjugates.

## 3. HUMAN LEISHMANIASIS

The true incidence and prevalence of leishmaniasis is uncertain because many cases are undiagnosed or unreported in areas where the infection is endemic. In 1993, the World Health Organization estimated that 350 million people world-

**Table I**
**Correlation of *Leishmania* Species, Geographical Distribution and Disease**

| Agent | Geographical distribution | Clinical presentations |
|---|---|---|
| *L. braziliensis* | Argentina, Belize, Bolivia,Brazil, Colombia, Costa Rica, Guatemala, Honduras, Panama Paraguay, Perú, Venezuela | CL, MCL |
| *L. mexicana* | Belize, Colombia, Costa Rica, Dominican Republic, Guatemala, Mexico, Panama, Venezuela, USA | CL, VL, DCL |
| *L. amazonensis* | Bolivia, Brazil, Colombia, Costa Rica, Ecuador, Guyana, Panama, Perú, Venezuela | CL, MCL, VL, DCL |
| *L. tropica* | Israel, Jordan, Iran, Iraq, Libya, Azerbaijan, Tiukmen, Uzbek, Syria, Afghanistan, India, Greece | CL |
| *L. major* | NW China, Pakistan, Kazakhstan, Afghanistan, India, Saudi Arabia, Iran, Iraq, Kuwait, Jordan, Libya, Israel, Algeria, Tunisia, Ethiopia, United Arab Republic, Senegal, Sudan, Kenya, Mali, | CL, VL |
| *L. aethiopica* | Ethiopia, Kenya | CL, DCL, MCL |
| *L. donovani* | Africa, Asia | VL |
| *L. infantum* | Mediterranean basin | VL |
| *L. chagasi* | Central and South America | VL |

CL: cutaneous leishmaniasis; DCL: diffuse cutaneous leishmaniasis; MCL: mucocutaneous leishmaniasis; VL: visceral leishmaniasis.

wide were at risk for infection. The incidence of cutaneous leishmaniasis has been estimated to be 1.0–1.5 million cases per year, and the incidence of visceral disease has been estimated to be 500,000 cases per year (PAHO, 1994).

Leishmaniases are caused by protozoa of the genus *Leishmania* (Table I), family Trypanosomatidae, order Kinetoplastida. Histologically the lesions are characterized by a diffuse inflammatory infiltrate composed of macrophages, plasma cells and lymphocytes, which may or may not exhibit amastigotes. A wide spectrum of clinical manifestations can be observed in tegumentary leishmaniasis, varying from the localized cutaneous and mucocutaneous leishmaniasis (CL and MCL), representing the responsive pole, to diffuse cutaneous leishmaniasis (DCL), the unresponsive pole. Between these polar forms a borderline or intermediate disease has been characterized.

*Leishmania* are generally divided into parasites which cause tegumentary disease (*L. braziliensis*, *L. amazonensis* and *L. mexicana* in the New World and *L. tropica* and *L. aethiopica* in the Old World) (Table 1) and those agents of visceral disease (*L. donovani* and *L. chagasi*).

### 3.1. New World Cutaneous Leishmaniasis

The cutaneous lesions begin at the site of parasite entrance as a small papule which develops into a nodule that ulcerates in the center. Lesions can take aspects of papules, nodules, ulcers, tubercles or infiltrated plaques. Less frequently, vegetation and verrucous lesions may be observed. The most frequent aspect observed in CL cases is an ulcer with elevated borders and a sharp crater. *L. braziliensis* is responsible for the more severe forms of CL, lesions develop rapidly and show a tendency to heal slowly without treatment. Rarely (approximately 1% of cases), disseminated CL can be observed with several lesions in a patient. The lesions appear as acneiform elements, papules and small ulcers. Disseminated CL is completely different from diffuse CL, an entity in which many lesions are also observed but they are nodular and non-ulcerated.

### 3.2. Mucocutaneous Leishmaniasis

Mucosal involvement, probably due to hematogenous spread from the primary lesion, occurs in approximately 3% of patients infected by *L. braziliensis*, and is related to multiple or extensive skin lesions above the belt and inadequate therapy of the primary lesion. Initial involvement of the nasal mucosa can spread to the hard and soft palate, uvula, pharynx, gums and upper lip. Septal perforation is observed in up to 42% of the MCL patients. A great enlargement of the upper lip and nose can result from infiltration and confer a tumoral aspect.

### 3.3. Diffuse and Borderline Cutaneous Leishmaniasis

The initial lesion of DCL resemble those of CL without ulceration. The lesions are erythematous and appear as papules, nodules, tuberous lesions, infiltrated plaques and diffuse infiltration of a part of the body. Even in the DCL pole there are subtle differences in presentation, with patients exhibiting the classical aspects and other exhibiting small ulcerations in a few lesions, despite the great predominance of nodular non-ulcerated lesions.

### 3.4. Human Visceral Leishmaniasis

Visceral leishmaniasis (Kala azar) is typically caused by *L. donovani* in India and Africa and by two closely related species, *L. infantum* in the Mediterranean littoral and *L. chagasi* in Latin America. Cases tend to occur sporadically in rural areas where the organisms are endemic. The majority of cases occur in children < 10 years of age. It is now known that only in a minority of cases of infections with *Leishmania* progress to classic visceral leishmaniasis. The remainder of these infections are asymptomatic or are associated with mild symptoms that eventually resolve spontaneously when protective immune responses develop. The clinical manifestations of visceral leishmaniasis seem to be similar throughout the world.

Kala azar is characterized by fever, malaise, weight loss, hepatomegaly and splenomegaly. Severe cachexia develops over time. Measles, diarrhea, bacterial pneumonia, tuberculosis, and other secondary infections are common late in the course of visceral leishmaniasis and frequently contribute to death (Pearson and Queiroz-Sousa, 1996).

## 4. α-GAL EPITOPE ON SEVERAL *TRYPANOSOMA CRUZI* MACROMOLECULES

Almeida *et al.* (1994, 1993) using McConville technique for extraction of glycoinositolphospholipids (GIPLs) (McConville and Bacic, 1989) and lipophosphoglycan (LPG) from *Leishmania* (reviewed in Avila 1993; McConville and Ferguson, 1993) extracted GIPL and LPGs from *T. cruzi* trypomastigotes and obtained three different fractions called F1-F3. The more hydrophobic fraction F1 was separated by thin layer chromatography and immunostained with chagasic sera. Reactive components migrated slowly in this system, in a region corresponding to the migration of *L. major* GIPLs, which also react with chagasic sera (Avila, 1993; Avila *et al.*, 1991). Although a strong reaction with F1 components was obtained with a serum from a patient with a positive hemoculture the reaction was rather poor with purified chagasic anti-Gal. In contrast, fractions F2 and F3 contained most of the components reacting with chagasic anti-Gal present in lysates of trypomastigotes. When separated by SDS-PAGE, fraction F2 formed two bands at Mr 74 kD and 95.6 kD that could be distinguished by their reactivity with anti-Gal. Fraction F3 formed a broad band at Mr 120–200 kD (main peak at 150 kD) that reacted with both chagasic anti-Gal and the 3C9 monoclonal antibody (recognizing sialyl epitopes of the Ssp-3 glycoconjugate from trypomastigotes). Interestingly, adsorption of chagasic serum *T. cruzi* lytic antibodies by F2 and F3 fractions fixed on polyvinylidene difluoride (PVDF) paper removed 70% of the initial lytic activity. These adsorbed antibodies could be eluted from the paper with citric acid, and were shown to promote *T. cruzi* lysis.

Using ELISA technique, reactions of the F1 fraction with chagasic sera showed low affinity and cross-reactivity with cutaneous leishmaniasis was detected, as has previously been demonstrated (Avila *et al.*, 1991). Fractions F2 and F3 contained antigens that displayed better sensitivity and specificity. Fractions F2 and F3 demonstrated positive reactions with chagasic and treated not-cured chagasic sera and did not react with treated-cured chagasic, visceral and mucocutaneous leishmaniasis and normal human sera (Almeida *et al.*, 1994; 1993).

An Octyl Sepharose sample of purified F2/F3 fractions was found to be rich in carbohydrate (Gal, Man, GlcNac, sialic acid, Glc and Xyl) and certain amino acids (Thr, Glx, Ser, Pro, Asx, Gly and Ala). In addition it contained myo-inositol and ethanolamine in a molar ratio of approximately 1:1. The finding of the latter components, together with lipid in the form of monoalkylglycerol (1-0-hexadecylglycerol) suggested the presence of glycosylphos-phatidylinositol(GPI)-anchored glycoconjugates. Labelling with galactose oxidase/$NaB_3H_4$

and reductive β-elimination showed that the F2/F3 material contained: 1) a small (4.5 glucose units [GU]) chagasic anti-Gal-positive oligosaccharide that was not found in a sialylated form; 2) a group of oligosaccharitols in the 5–12 GU size range, some of which terminated in α-Gal and were chagasic anti-Gal positive, and some of which (mostly from the chagasic anti-Gal negative fraction) were found in a sialylated form; 3) a group of chagasic anti-Gal positive and chagasic anti-Gal negative acidic structures that were neuraminidase-resistant, which was not further characterized.

The 4.5 GU oligosaccharide, obtained by hydrazinolysis followed by $NaB_3H_4$ reduction and Bio-Gel P-4 purification, was subjected to strong acid hydrolysis and re-N-acetylation, plus coffee-bean α-galactosidase treatment. The oligosaccharide was identified as Galα1-3Galβ1-4GlcNAcol(N-acetylglucosaminitol).

Binding studies with immobilized purified F2/F3 fractions in chemiluminiscence ELISA showed that the affinity of chagasic anti-Gal for these glycoconjugates was about 100-fold that of normal human serum anti-Gal (Figure 1). As expected, treatment of the F2/F3 glycoconjugates with α-galactosidase abolished almost all of the immunoreactivity. Similarly, only chagasic anti-Gal bound significantly on several αGal-containing structures from the GPI anchor of *T. brucei* VSG (Zamse *et al.*, 1991) and the glycoinositol phospholipids (GIPL) of *Leishmania major* (Avila, 1993; Avila *et al.*, 1991). These data are summarized in Table 2. Previous results suggest that chagasic anti-Gal can bind to structures terminating in Galα1-2Gal, Galα1-3Gal, Galα1-6Gal and Galα1-3Galf (where Galf is galactofuranose) (Almeida *et al.*, 1994, 1993; Avila, 1992; Avila *et al.*,1992,1991,1989) because it has little specificity for the penultimate units or longer oligo-saccharides linked to the terminal non-reducing α-Gal residues.

In other experiments, Almeida *et al* (1994) found that most cell-derived trypomastigotes react strongly with chagasic anti-Gal, but not with normal human serum anti-Gal. Treatment with phosphatidylinositol-specific phospholipase C (PI-PLC) almost completely abolished binding of chagasic anti-Gal to the parasites. These results show that most of the antigens recognized by chagasic anti-Gal on the trypomastigote cell surface are PI-PLC-sensitive and are therefore probably GPI-anchored. Consistent with the above result, partially purified F2/F3 glycoconjugates digested with PI-PLC were rendered hydrophilic and could no longer bind to the PVDF membrane. Taken together, these data demonstrate that the majority of the chagasic anti-Gal reactive molecules were GPI-anchored glycoconjugates. A GPI-anchored trypomastigote-specific glycoprotein named Tc-85 also have been described in *T. cruzi* (Couto *et al.*, 1993) and its N-linked carbohydrate chain has the α-gal epitope (Couto *et al.*, 1990).

Avila *et al.* (1991) reported the presence of several GIPLs in *T. cruzi*. Some of them comigrated with purified *L. major* GIPL- 1, GIPL-2 and GIPL-3 in high-performance thin layer chromatography. Several other GIPLs have been described

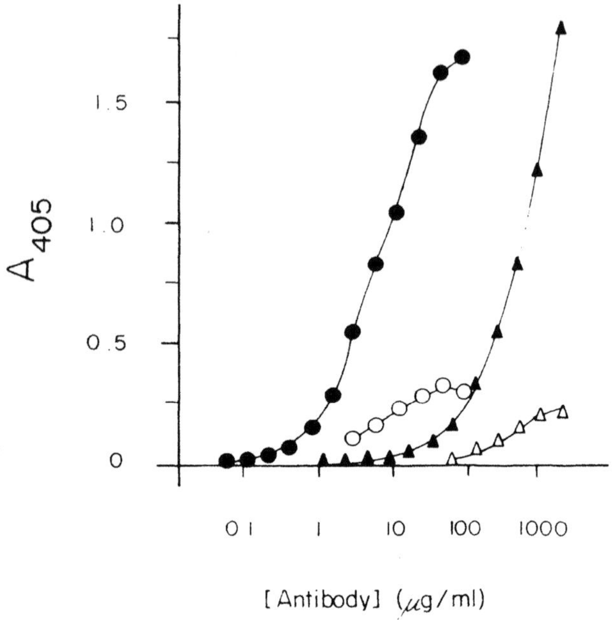


**Figure 1.** Microtitre plates coated with F2/3 were probed with a range of chagasic anti-Gal (●) and normal human serum (▲) antibody concentrations. The intensity of the reactions was determined by chemiluminescent ELISA. The open symbols represent the same experiment following prior treatment of the coated plates with coffee-bean α-galactosidase. From Almeida *et al* (1994), with permission.

in *T. cruzi*, series 1 oligosaccharides having only terminal $Gal_f$(β1-3)Man units (Carreira *et al.*, 1996). These may correspond to GIPLs reported by Avila *et al.* (1991), that show high chromatographic mobility. According to Carreira *et al.* (1996) these GIPLs have 2–3 phosphate groups whereas *L. major* GIPLs have one phosphate.

## 5. α-GAL EPITOPE ON SEVERAL *LEISHMANIA* MACROMOLECULES

The glycoinositolphospholipids (GIPLs) are the major glycolipids synthesized by *Leishmania* parasites. Three distinct types of GIPLs have been identified, which are expressed in markedly different levels in different species or developmental stages (McConville and Ferguson (1993)(for a chemical description see Table II). The α-gal epitope is present in GIPL-2 from *L. major* (McConville *et al.*, 1990; McConville and Bacic, 1989) and from *L. mexicana* (McConville *et al.*, 1993). The several GIPLs may coat a significant proportion of the plasma membrane, thus acting as a shield against lysosomal hydrolases and as receptors for the interaction with host cells (Winter *et al.*, 1994).

**Table II**
**Binding of Chagasic and Normal Human Serum Anti-Gal to Different α-Gal-Containing Epitopes**

| | | Anti-Gal binding | |
|---|---|---|---|
| Epitope | Origin | ChHS | NHS |
| Galpα1,3Galpβ1,4GlcNAcβ1-0-R | Synsorb 115 | + | – |
| Galpα1,3Galpβ1,4GlcNAcβ1-0-R | CPH | + | – |
| Galpα1,3Galpβ1,4GlcNAcol | F2/F3 *T. cruzi* | + | – |
| Galpα1,2Galpα1,6Galpα1,3[Man$_3$AHM] | VSG 117 | + | – |
| Galpα1,2Galpα1,6(Galpα1,2)Galpα1,3[Man$_3$AHM] VSG 117 | | + | – |
| Galpα1,6(Galpα1,2)Galpα1,3[Man$_3$AHM] | VSG 117 | + | – |
| Galpα1,3Galfβ1,3ManGlcNH$_2$-PI | GIPL-2 | + | – |
| Galpα1,6Galpα1,3Galβ1,3ManGlcNH$_2$-PI | GIPL-3 | + | – |

ChHS: chagasic human serum; NHS: normal human serum; CPH; ceramide pentahexoside from rabbit erythrocytes; VSG, variant surface glycoprotein from *Trypanosoma* brucei; GIPL-2 and GIPL-3, glycoinositolphospholipid from *L. major*; AHM, anhydromannitol; PI, phosphatidylinositol.
Taken from Almeida *et al* (1994) with permission.

## 6. MECHANISMS OF PRODUCTION OF ELEVATED ANTI-GAL IN KINETOPLASTIDAE INFECTIONS

Towbin *et al.* (1987) proposed that members of the *Leishmania* and *Trypanosoma* possess highly immunogenic α-gal epitopes, which induced elevated anti-Gal antibody production.

This hypothesis was studied using three different approaches:

### 6.1. Biochemical Isolation and Characterization of α-Gal Epitope-Bearing Molecules

See above Sections 4 and 5.

### 6.2. Ligand Analysis of α-Gal Epitope Expression on *Trypanosoma* and *Leishmania*

This was assessed by the use of two antibodies and two lectins, all of which interact specifically with this epitope: a) the natural human anti-Gal, b) the monoclonal antibody Gal-13. This mouse IgG mab was found to be highly specific for glycolipids with terminal Gal(α1-3)Galβ1-4GlcNac-R residues (Galili *et al*, 1987a). Gal-13 does not bind, however, to these residues on N-glycosylated glycoproteins such as the Galα1-3Galβ1-4GlcNac residues on mouse laminin or bovine thyroglobulin (Galili, 1987a), c) The specific lectin *Bandeiraea simplicifolia* IB$_4$ (BS). This lectin was found to bind to the α-gal epitope of

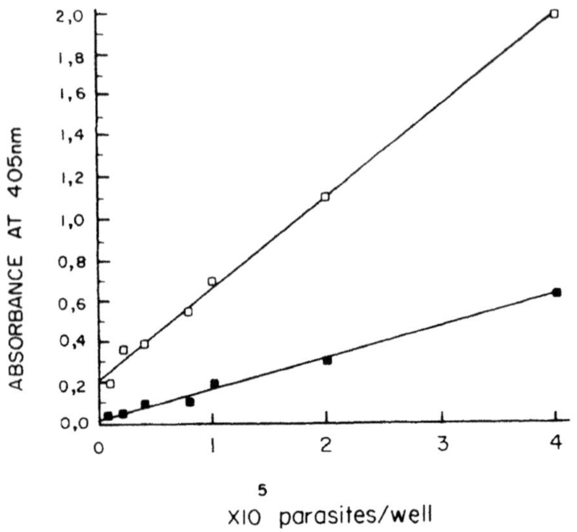


**Figure 2.** Binding of purified human anti-Gal to *T. cruzi* epimastigotes. Several parasite concentrations (0 to 4 × $10^5$ parasites/well) were exposed to freshly purified human anti-Gal antibody (10 μg/ml) for 1 h at 37 °C. □, anti-Gal binding to *T. cruzi* epimastigotes; ■, anti-Gal binding to 10 mM periodate-treated *T. cruzi* epimastigotes. From Avila *et al* (1989), with permission.

glycoconjugates and with less affinity to the Gal(α1-3(Fucα1-2)Gal residues which represent the blood group B antigen (Egge *et al.*, 1985; Wood *et al.*, 1979). Figure 2 shows the dose-dependent binding of anti-Gal to *T. cruzi* epimastigotes. Periodate treatment of the parasites greatly decreased the binding of anti-Gal, further demonstrating the carbohydrate nature of the anti-Gal binding sites on the parasite. Figure 3 shows the binding of Gal-13 to fixed *T. cruzi* epimastigotes and American *Leishmania* promastigotes. The strong binding of Gal-13 to the parasite surface suggested that a substantial proportion of the α-gal epitope on these parasites is expressed as epitopes on glycolipids.

To further corroborate the presence of terminal α-gal-bearing glycoconjugates in *T. cruzi* and American *Leishmania*, Avila *et al.*(1989) used BS lectin. Table III and Figure 4 shows BS lectin binding to *T. cruzi* epimastigotes and American *Leishmania* promastigotes. Periodate treatment of fixed parasites strongly blocked lectin binding, confirming the carbohydrate nature of the lectin binding sites on *T. cruzi* epimastigotes and *Leishmania* promastigotes.

To study the occurrence of glycolipids with α-gal epitope in the parasites' membrane, crude total lipids fraction was assessed for interaction with the three α-gal binding molecules. Figure 5 shows the dose-dependent binding of purified human anti-Gal, of mouse mAb Gal-13 and of BS lectin to increasing amounts of *Leishmania* crude lipids. These data confirm that at least part of the α-gal epitopes on the parasite are on glycolipid molecules.

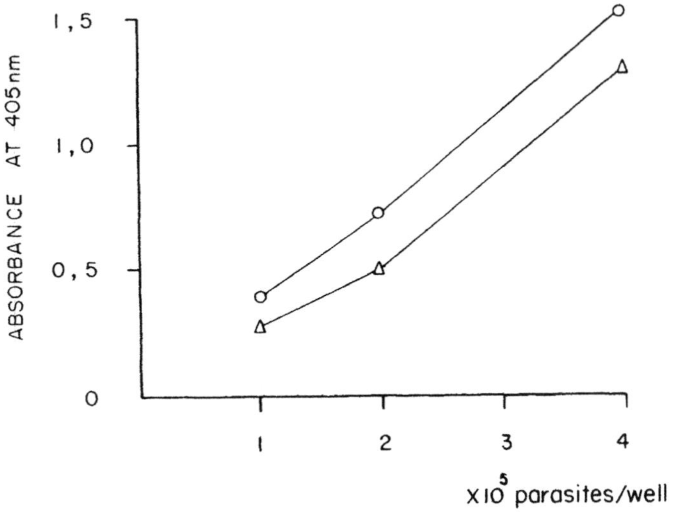


**Figure 3.** Binding of the mAb Gal-13 to fixed *T. cruzi* epimastigotes and American *Leishmania* promastigotes as a function of parasite concentration. (○). *T. cruzi* epimastigotes; (Δ) *Leishmania* promastigotes. From Avila *et al* (1989), with permission.

**Table III**
**Binding of *Bandeiraea simplicifolia* $IB_4$ Lectin to American *Leishmania* and *T. cruzi* Epimastigotes and Trypomastigotes under Different Experimental Conditions**

| | % of control[a] | | |
|---|---|---|---|
| | | *T. cruzi*[b] | |
| Experimental conditions | *Leishmania*[b] | Epimastigotes | Trypomastigotes |
| 20 mM PMSF-10 mM EDTA | 100 | 100 | 100 |
| 1 % formaldehyde | 94 ± 4 | 90 ± 2 | 93 ± 6 |
| 1 % glutaraldehyde | 65 ± 6 | 60 ± 9 | 74 ± 8 |
| 0.01 % sodium azide | 83 ± 5 | 78 ± 7 | 69 ± 6 |
| 0.5 M galactose[c] | 18 ± 3 | 21 ± 4 | 19 ± 4 |
| 0.5 M glucose[c] | 94 ± 5 | 91 ± 7 | 87 ± 6 |
| 10 mM periodate[c] | 23 ± 3 | 20 ± 6 | 16 ± 3 |

[a]Mean ± standard deviation of six different experiments.
[b]Data obtained with MP, HM, LR, JAP, MHOM/4147 and MHOM/ M2903 strains of American *Leishmania* and Marin-1, Bertoldo, FL, Ma, Ya and Y strains of *T. cruzi*.
[c]Experiments were carried out in PMSF-EDTA fixed parasites.
Taken from Avila *et al* (1989) with permission.

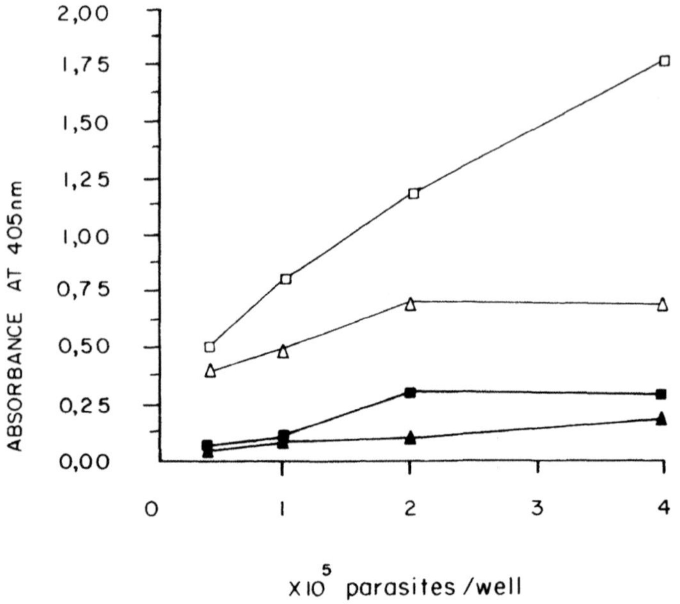


**Figure 4.** Binding of *Bandeiraea simplicifolia* $IB_4$ (BS) lectin to fixed *T. cruzi* epimastigotes and American *Leishmania* promastigotes as a function of parasite concentration. □, *T. cruzi*; Δ, American *Leishmania*; (■, ▲), binding in the presence of 0.5 M Galactose. From Avila *et al* (1989), with permission.

## 6.3. Immunolocalization of the α-Gal Epitope by Electron and Immunofluorescence Microscopy

Bretaña *et al* (1986) using immunogold technique studied the immunocytochemical localization of a "laminin-like protein" in the plasma membrane of *T. cruzi* and American *Leishmania*. They reported the immunoreactivity as located in *T. cruzi* trypomastigotes on the external surface of the plasma membrane, close to the sites where the flagellar veil attaches to the plasma membrane. Laminin immunoreactivity was rapidly lost when trypomastigotes were cultured in liquid medium and no reactivity was found in fresh epimastigotes. Promastigotes and amastigotes of American *Leishmania* spp. also showed a specific localization of laminin immunoreactivity, this being limited to the lips of the flagellar pocket and to the parasitic side exactly opposite to the flagellar exit. These results confirmed the presence of a "laminin-like molecule(s)" in both trypanosomatids. Once established that human anti-laminin antibodies are indeed reacting with α-gal epitopes (as demonstrated using mouse laminin by Towbin *et al.*, 1987) we used more specific methods to detect the presence of the α-gal epitope on Trypanosomatidae. Thus, Bretaña *et al* (1992) using three colloidal

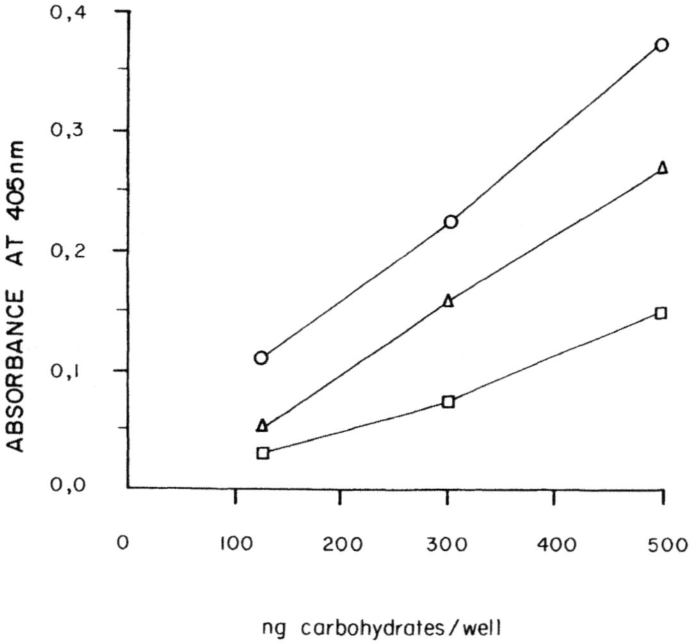


**Figure 5.** Binding of (□), BS lectin; (Δ), Gal-13; and (○) human anti-Gal to American *Leishmania* lipid fraction, as a function of carbohydrate concentration. From Avila *et al* (1989), with permission.

gold-linked specific markers: mAb Gal-13, rabbit anti-nidogen antibodies and a lectin (*B. simplicifolia* $IB_4$) found terminal α-gal epitopes on the *T. cruzi* external surface of trypomastigotes but not in intact epimastigotes (although disrupted epimastigotes strongly stained), in the lips of the flagellar pocket, and on the parasitic side exactly opposite to the flagellar pocket in amastigote and promastigote forms of American *Leishmania*. These results resemble those obtained before using anti-laminin antibodies in both trypanosomatids (Bretaña *et al*, 1986).

In other studies, Gazzinelli *et al* (1991b) using the fluorescent-labeled lectin *Euonymus europaeus* (EE)(specific for the α-gal epitope) and acute chagasic serum studied the distribution of α-gal epitope in mono- and heteroxenic trypanosomatids. They found that this epitope is present in all trypanosomatids. Although more EE receptors seem to be present in some monoxenic trypanosomatids and *Phytomonas* than in *Trypanosoma* species, the differences are not as evident. EE recognizes a wide range of glycoconjugates in all trypanosomatids. In *T. cruzi* this lectin recognizes a major glycoprotein of apparent Mr of 60 to 90 kDa. In contrast, acute chagasic serum recognizes many glycoconjugates of the several trypanosomatids tested. Most of these cross-reacting antigens are absorbed when acute chagasic serum is treated with mouse laminin or melibiose.

Using antibody-mediated agglutination and immunofluorescence, Almeida *et al* (1991) demonstrated anti-Gal binding on live liver infusion-tryptose-cultured metacyclics and Hela cell-derived trypomastigotes of two different *T. cruzi* strains. In both cases the minimum antibody concentration needed for parasite agglutination was higher for normal human serum anti-Gal than for chagasic anti-Gal. The observed anti-Gal binding to living trypomastigotes was followed by lysis of the parasites when complement was added.

Finally, more recently Souto-Padron *et al.* (1994), using *T. cruzi* epimastigotes, trypomastigotes and amastigotes, an immunogold labeling method and human chagasic anti-Gal, have confirmed Bretaña *et al.*(1992) in that epimastigotes were poorly labeled, whereas extracellular trypomastigotes and amastigotes bound heterogeneously the antibody with many cells being intensely labeled at the cell surface. Generally, a strong surface reaction was accompanied by a weak cytoplasmic labeling. Conversely, a more intense intracellular or vacuolar labeling was observed in parasites with low surface labeling, suggesting different stages of processing of the carbohydrate epitope in different parasites. The Golgi complex of trypomastigotes was strongly labeled. Intracellular parasites were labeled at the parasite cell surface or within vacuolar structures.

### 6.4. The α-Gal Epitope in Other Protozoa

The presence of terminal α-gal epitope has been demonstrated in other protozoa. In the asexual blood stages of *Plasmodium falciparum* by: a) the α-galactosidase sensitivity of particular parasite antigens; b) by the ability of *B. simplicifolia* lectin to bind to the parasite (Ramasamy and Reese, 1986) and c) because anti-Gal inhibit *P. falciparum* growth in culture (Ramasamy and Rajakaruna, 1997).

The α-gal epitope was also reported on the type II and type III variant surface glycoproteins of *Trypanosoma* brucei (Zamse *et al.*, 1991).

## 7. ANTI-GAL RESPONSE IN *T. CRUZI*-INFECTED HUMAN SUBJECTS

### 7.1. In Acute Infection

Szarfman *et al* (1982) were the first in demonstrating, using mouse laminin as antigen, that the sera from both acute and chronic chagasic patients contained antibodies to laminin and that the highest titers were present in the acute phase. Patients with acute *T. cruzi* infection had high titers of IgM and IgG antibodies, which react with laminin. Similar findings have been later demonstrated by Gazzinelli and Brener (1988). They found about 6- and 12-fold more immunoreactivity in acute chagasic than in chronic chagasic patients and in controls respectively. Grauert *et al.* (1993) have studied the kinetics of anti-laminin antibody response in an acute case of human Chagas' disease. Anti-laminin IgG and IgM antibodies peaked on day 30, antibody reaching levels higher than 200% of basal values. Acute chagasic sera presented high quantities of anti-Gal but low levels of antibodies against total *T. cruzi* antigens, suggesting that the α-gal epitope is a major *T. cruzi* antigen occurring in this stage of the disease.

### 7.2. In Chronic Patients

As mentioned above, Szarfman *et al.* (1982) using mouse laminin as antigen, were the first to demonstrate that the sera from nine chronic chagasic patients contained elevated levels of antibodies to laminin. Subsequently, Avila *et al.* (1984) studying a larger group (37 patients) found 84% of chronic chagasic patients had anti-laminin antibody levels higher than normal. These antibodies were mainly of the IgG class. Similar results were later reported by Unterkircher *et al.*, (1993) (80% of 80 patients with confirmed Chagas' disease), Milei *et al.* (1993) (58% of 47 patients), Gazzinelli *et al.* (1988) and Avila *et al.* (1987) (73% of 92 chronic chagasic patients).

That anti-murine laminin antibodies are similar to anti-Gal was reported by Towbin *et al.* (1987), after the report of Avila *et al.* (1986) that chagasic patients also have increased anti-mouse nidogen antibody levels, nidogen being a structurally different protein that laminin (Paulsson *et al.*, 1987).

Avila *et al.* (1989) have reported that when measured by hemagglutination assay with rabbit red cells, anti-Gal in healthy individuals usually yields a titer of 1/640 to 1/1280 (Figure 6A). In sera passed through a Galα1-3Galβ1-4GlcNac-R adsorbent (Synsorb 115), most (>85%) of the anti-Gal reactivity is retained on the column, as can be seen by a more than 10-fold decrease in the hemagglutinating reactivity in the effluent serum (Figure 6C). The antibodies eluted from the adsorbent column by low pH displayed a titer of 1/320 to 1/640 in healthy individuals (Figure 6B).

A similar pattern of anti-Gal reactivity was observed in sera of five patients with various bacterial infections. Eighty per cent of these patients displayed normal anti-Gal activity. However, in the serum of patients with Chagas' disease and *Leishmania* infection, the mean anti-Gal titer was 10-fold and 16-fold higher than that of healthy individuals or infection controls respectively (Figure 6A). This marked elevation in hemagglutinating reactivity was due to an increase in anti-Gal reactivity because, as observed with sera from healthy individuals, passage on Galα1-3Galβ1-4GlcNac-R adsorbent removed more than 90% of the hemagglutinating reactivity in the sera of the chagasic and leishmanic patients (Figure 6C). The adsorbed anti-Gal of these patients, when eluted, displayed a 6- to 20-fold higher titers, respectively, as compared to that in normal individuals

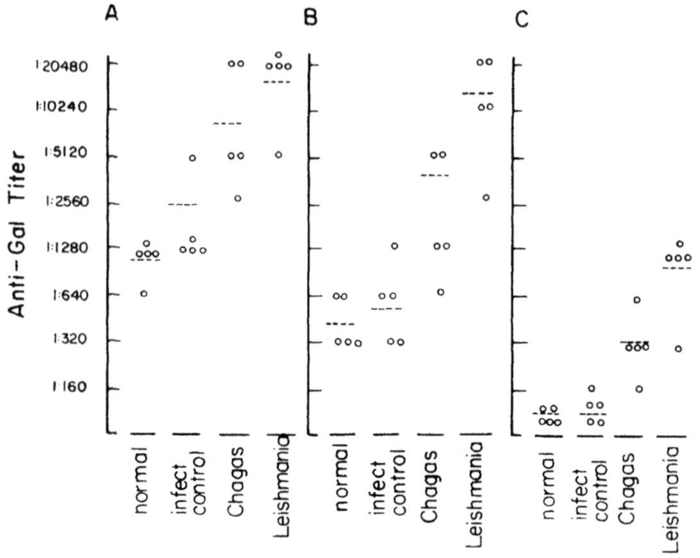


**Figure 6.** Anti-Gal titers in sera of patients with Kinetoplastida infections. Normal, healthy individuals; Infect. controls, patients with bacterial infections. Each circle represent one patient. A, Anti-Gal in serum. B, Anti-Gal eluted from Galα1-3Galβ1-4GlcNac-R adsorbent. C, anti-Gal in the serum effluent passed on the Galα1-3Galβ1-4GlcNac-R adsorbent. Dashed line represent the mean titer. From Avila *et al* (1989), with permission.

(Figure 6B). When IgG concentration was measured in the sera of healthy individuals, infection controls and chagasic or leishmaniasic patients, all the patient groups exhibited only a 2–3-fold increase in total IgG concentration when compared with that in healthy individuals (Table IV). This suggested that the severalfold increase in anti-Gal reactivity in chagasic patients exceeded the increase in IgG concentration and thus may have resulted from a specific immune response against the α-gal epitope in these patients. These data confirm previous observations on elevation in antibody reactivity to α-gal epitopes in chagasic patients (Avila *et al.*, 1988a,b; Towbin *et al.*, 1987).

In order to study the association between anti-Gal levels and the clinical stage, Avila (1992) measured this antibody activity in patients at different clinical stages of chagasic and non-chagasic dilatatory cardiomyopathies. No differences were found in anti-Gal activity or in the proportion of anti-Gal positive patients. However anti-Gal in the two groups was significantly elevated compared with control subjects suggesting that : a) there is no relationship between anti-Gal levels and clinical stage and b) the existence in tropical areas of another anti-Gal antibody-elevated cardiomyopathy (nutritional, toxic, viral, rickettsial ?) (Table V). Similar results, were also reported by Milei *et al.* (1993). These authors divided

**Table IV**
**IgG and Anti-Gal Concentrations in Sera of Various Patients**

| Antibody concentrations in subjects | IgG concentration (mg/ml) | Anti-Gal concentration (μg/ml) |
|---|---|---|
| Healthy individuals (n= 9) | | |
| Range | 7–11.0 | 48–91 |
| Mean[a] | 8.7 ± 1.0 | 74 ± 28 |
| Infection controls (n= 23) | | |
| Range | 10–40 | 56-96 |
| Mean[a] | 26 ± 1.2 | 71 ± 2.6 |
| Leishmanic patients (n= 31) | | |
| Range | 12-40 | 302-490 |
| Mean[a] | 24 ± 1.1 | 380 ± 9.2 |
| Chagasic patients (n= 31) | | |
| Range | 14–40 | 72–419 |
| Mean[a] | 28 ± 1.6 | 213 ± 16.6 |

[a]± standard error.
Taken from Avila *et al* (1989) with permission.

**Table V**
**Anti-*Trypanosoma cruzi* and Anti-Gal Levels in Control Subjects and in Chagasic and Non-Chagasic Cardiomyopathy Patients Classified According to Their Degree of Myocardial Damage**

| Groups | Anti-*T. cruzi* antibodies[a,b] | Anti-Gal antibodies[b,c] | % of anti-Gal positive |
|---|---|---|---|
| Control (106) | 343 ± 94 | 235 ± 151 | 8 |
| Chagasic Ia (21) | 1106 ± 359[d] | 549 ± 388[e] | 29 |
| Chagasic Ib (12) | 1170 ± 353[d] | 420 ± 236[f] | 30 |
| Chagasic II (56) | 1134 ± 355[d] | 825 ± 928[e] | 45 |
| Chagasic III (13) | 1244 ± 327[d] | 631 ± 535[e] | 36 |
| Non-chagasic I (12) | 247 ± 43 | 422 ± 202[f] | 25 |
| Non-chagasic II (12) | 285 ± 128 | 697 ± 658[e] | 38 |
| Non-chagasic III (16) | 268 ± 151 | 482 ± 386 | 5[g] |

[a]Expressed as $OD_{405}$ at a serum dilution of 1:800. A value of > 437 is considered as positive.
[b]Mean ± standard deviation ($10^{-3}$).
[c]Expressed as $OD_{405}$ at a serum dilution of 1:50. A value of > 613 is considered positive. Antigen used: purified rabbit erythrocyte ceramide pentasaccharide.
[d]$P < 0.001$ compared with control group.
[e]$P < 0.005$ compared with control group.
[f]$P < 0.010$ compared with control group.
[g]$P < 0.010$ compared with chagasic type III group.
Taken from Avila *et al* (1992) with permission.

their chronic chagasic patients into three groups: group I: asymptomatic patients, normal thorax X-ray and ECG; Group II: conduction disturbances and/or arrhythmia and Group III: cardiomegaly with heart failure and abnormal ECG. Abnormal levels of anti-laminin antibodies were detected in 46% of the patients from group I, in 58% from group II and in 50% from group III. Percentages of the three groups were statistically similar. Nevertheless, very high titers of antibodies were detected in 10% of the patients from group II and 37% from group III.

Studies in chronic chagasic patients, have indicated that the levels of anti-Gal do not differ within the different clinical forms of Chagas' disease (indeterminate, cardiac and digestive patients) (Gazzinelli, 1992).

Interestingly, in chronic Chagas' disease patients, anti-Gal antibodies are significantly elevated mainly in the IgG class and there is lesser immunoreactivity in IgA and IgM classes (Unterkircher *et al.*, 1993; Avila *et al.*, 1992, 1988a, 1986, 1984).

Based on the affinity of chagasic antibodies to laminin and *T. cruzi* antigens fixed to immunoadsorbents, Unterkircher *et al.*(1993) have reported that there are two populations of anti-laminin antibodies in chronic chagasic patients: one that seems to be specific for the α-gal epitope on murine laminin and on the parasite and the other cross-reacting with other antigens on the parasite as well as with other antigens, both populations having a dissociation constant of $7 \times 10^{-8}$–$10^{-7}$.

Of note, Milei *et al.* (1993) have reported higher titres of these antibodies in some chagasic patients who progressed in the severity of their disease (worsening of its clinical conditions). However, as emphasized by Kierszembaum (1986), it remains to be shown whether such antibodies develop subsequently to the production of tissues damage or are elicited by the parasite prior to the occurrence of tissue lesions.

Interesting, Goin *et al.*(1990) have reported that anti-Gal can mimic the negative inotropic action of cholinergic agonists. On the other hand, Milei and Storino (1987) could not reproduce myocardial lesions resembling chronic Chagas' cardiomyopathy in rabbits immunized with murine laminin. However we must point out that it is hard to compare the immunogenicity of α-gal epitope present in murine laminin with that elicited by α-gal epitopes present in *T. cruzi* trypomastigotes, which are mostly linked to glycolipid structures where other membranous components can possibly act as powerful adjuvants (Bretaña *et al.*, 1992; Avila *et al.*, 1989; Avila, Harner and Barrios (unpublished results)[see also Section 15.1]).

It has been suggested that the findings of anti-laminin antibodies are in accord with the observation that the surfaces of cardiac myocytes of patients with chagasic cardiomyopathy are the sites of deposition of immunoglobulins and complement (Molina *et al.*, 1984). it is possible that infected cardiac muscle cells may change their surface structure by inserting parasite-derived α-gal epitopes, or changing host carbohydrate components by new posttranslational modifications due to the intracellular residence of parasites. The autoimmunity observed in Chagas disease may therefore not be the result of cross-reacting structures present *per se*, but due to a modified structure arising in the *T. cruzi*-infected cells (Avila, 1994). Another alternative explanations, proposed by Avila *et al.* (1987), is that injury to cardiac basement membranes occurs in Chagas' disease, and leads to the release of basement membrane components (including laminin) which become antigenic, or that released *T. cruzi* metabolites may induce an inflammatory response. This interaction (analogous to xenograft rejection) may explain part of the extensive inflammatory response in the infected tissues in humans. These three explanations are not mutually exclusive.

It is worth mentioning that Wolff *et al.* (1989) have reported alterations in laminin distribution and antibodies to "laminin" (measured using mouse laminin and therefore clearly representing anti-Gal) in myocarditis (73% of the cases) and dilated cardiomyopathy (78% of the cases).

## 7.3. Anti-Gal(α1-2)Gal Antibodies in Chronic Chagas' Disease Patients

Since Gal(α1-2)Gal residues have been recently shown to be a structural unit of a phosphatidylinositol anchor for inserting variable surface glycoproteins of *T. brucei* (Ferguson *et al.*, 1988) and certain proteins of *T. cruzi* (Ledekremer *et al.*, 1990) into their cell surface, it was hypothesized that specific Gal(α1-2)Gal antibodies could arise in *T. cruzi*-infected human subjects (Avila *et al.*, 1992).

Significantly elevated Gal(α1-2)Gal antibody levels were indeed found in 66% of chronic chagasic cardiomyopathy patients (Table VI). In contrast, other acute or chronic Kinetoplastidae infections such as *T. rangeli*-infection, localized, mucocutaneous, diffuse cutaneous or visceral leishmaniasis did not induce significantly elevated Gal(α1-2)Gal antibody levels.

To investigate whether results in chronic Chagas' disease patients were associated with degree of myocardial damage, patients were subdivided into several groups (Carrasco *et al.*, 1982). Among chagasic groups there was no significant difference in anti-Gal(α1-2)Gal antibody levels or in percentage of patients that have elevated antibody levels. Nevertheless, Gal(α1-2)Gal antibody levels were significantly higher in all chagasic groups when compared with healthy individuals or with patients having other non-chagasic cardiomyopathies.

Analysis of previous results has suggested that in human healthy individuals, as well as in *T. cruzi*-infected subjects, there are at least three types or clones of terminal α-galactosyl-reactive antibodies: One is detected by ELISA with rabbit neutral glycosphingolipids-enriched for ceramide pentasaccharide (Avila *et al.*, 1988a), murine laminin (Unterkircher *et al.*, 1993; Avila *et al.*, 1984; Szarfman *et al.*, 1982) or murine nidogen (Avila *et al.*, 1986) as solid phase antigens. These antibodies are mainly IgG and highly specific for the α-gal epitope as they are almost completely adsorbed by rat, rabbit, and guinea pig red blood cells as well as for

**Table VI**
**Anti-Gal(α1-2)Gal Antibody Levels in Healthy Controls, in Patients with Localized Cutaneous (LCL), Mucocutaneous (MCL), and Diffuse Leishmaniasis (DCL), Kala azar, *T. rangeli*-Infection, Chronic Chagasic Cardiomyopathy, and Different Inflammatory Diseases**

| Study group (No. patients) | $OD_{405}$ $(10^{-3})$[a] | % Positive[b] | P[c] |
|---|---|---|---|
| Healthy controls (83) | 1,000 ± 535 | 0 | |
| LCL (47) | 2,379 ± 1,063 | 28 | > 0.050 |
| MCL (30) | 1,762 ± 1,224 | 20 | > 0.050 |
| DCL (22) | 2,011 ± 1,266 | 32 | > 0.050 |
| Kala azar (20) | 1,854 ± 951 | 10 | > 0.050 |
| *T. rangeli*-infected (28) | 2,422 ± 1,266 | 39 | > 0.050 |
| Chronic chagas' disease (81) | 3,656 ± 2,333 | 66 | < 0.005 |
| Inflammatory diseases (69) | 932 ± 370 | 0 | > 0.050 |

[a]Mean ± standard deviation in ELISA (serum dilution of 1:800).
[b]An OD of 2,338 represents the upper limit of normality, defined as the mean plus 2.5 standard deviations, of the healthy control group. The percentage of patients with levels higher than this is presented.
[c]Differences between $OD_{405}$ values determined in comparison with the control group.
Taken from Avila *et al.* (1992) with permission.

murine laminin or nidogen. All these red cells and glycoproteins have many Gal(α1-3)Gal-bearing residues (Towbin *et al.*, 1987). These antibodies are likely correspond to those described by Unterkircher *et al.* (1993) as exhibiting high affinity for the α-gal epitope.

The second type of antibodies is detected in ELISA using an American *Leishmania* glycoinositolphospholipid preparation bearing terminal Gal(α1-3)Man or $Gal_f$(α1-3)Man disaccharide as solid phase antigen (Avila and Rojas, 1990). These antibodies are mainly IgM, and also bound to Gal(α1-3)Gal-linked synthetic antigens, but did not interact with the same residues present on rabbit, rat and guinea-pig erythrocytes or in murine laminin or nidogen. These antibodies are equally blocked by low concentrations of Gal(α1-3)Man and Gal(β1-4)Man. They are strongly elevated in 89% and 84% of diffuse and localized cutaneous leishmaniasis patients (Avila and Rojas 1990).

The third type of terminal α-galactosyl-reactive antibodies are Gal(α1-2)Gal antibodies, which are not blocked by > 200 mM Gal(α1-3)Man or Gal(β1-4)Man and are elevated only in 38% and 28% of diffuse and localized cutaneous leishmaniasis patients. The fact that they are unadsorbed neither by rabbit, rat or guinea-pig red blood cells or murine laminin and nidogen or by pronase-treated human O erythrocytes, nor is their binding affected by methyl-α-Galactopyranoside, melibiose and stachyose, adds further evidence for the fact that these latter antibodies are indeed different from anti-Gal. They may correspond to the second type of α-gal epitope-reactive antibodies reported by Unterkircher *et al* (1993). Given the polyclonal nature of anti-Gal (Towbin *et al.*, 1987), our results suggest that among the three different antibody types we found to increase in chronic chagasic cardiomyopathy sera, a considerable proportion of antibody clones might have a different affinity for the protein-(Couto *et al.*, 1990) or lipid-linked α-gal epitopes (Avila *et al.*, 1991; 1989) present on *T. cruzi* membrane. Furthermore, individual differences in the overall affinity to α-gal epitopes on *T. cruzi* may determine the capacity of this family of antibodies to prevent infection and destroy the invading parasite. In supporting of this view, Almeida *et al* (1991) reported that chagasic anti-Gal had a lower affinity for Gal(α1-6)Glc bound to agarose than for the α-gal epitope, and was several times more effective in lysis of trypomastigotes than corresponding anti-Gal from normal human serum. This demonstrates a considerable diversity in the recognition of the α-gal epitope by natural antibodies.

Anti-Gal(α1-2)Gal antibodies also may have a biological significance since they may cross-react with α-gal epitope on the parasite, thus they may contribute to the protection against invading protozoa (Gazzinelli *et al.*, 1991a; Avila *et al.*, 1988b).

### 7.4. Lysis of *T. cruzi* Trypomastigotes by Human Anti-Gal Antibodies

Participation of antibodies in the resistance against *Trypanosoma cruzi* has been demonstrated at the chronic phase by protection subsequent to passive transfer of immune serum (Krettli and Brener, 1976) or immunoglobulins fractions to infected mice (Takehara *et al*, 1981). Specific anti-*T. cruzi* antibodies are involved in resistance through alternative and classical complement mediated lysis (Krettli and Brener, 1982; Krettli *et al.*, 1979) and antibody cell-mediated cytotoxicity (Lima-Martins *et al.*, 1985). A lytic activity against *T. cruzi* induced by antibody binding and which is complement independent has recently been detected, in acute chagasic sera, but not in chronic chagasic sera or normal human sera (Stefani *et al*, 1987). Araújo (1989) found that 17 out of 27 patients at the acute phase of Chagas disease displayed lytic activity.

#### 7.4.1. Direct Lysis in Acute Chagas' Disease

The fact that:

a. In acute Chagas disease there are high levels of anti-Gal antibodies (IgM and IgG) in contrast to low titres of specific anti-*T. cruzi* antibodies (Gazzinelli *et al*, 1988),
b. that lytic activity in acute Chagas disease has been associated with Ig fraction (Stefani *et al*, 1987), and

c. that highly purified anti-Gal antibodies from chronic chagasic sera induced infectivity inactivation (Milani and Travassos, 1988; Milani *et al.*, 1988), made Gazzinelli *et al.* (1991a) to propose that lytic activity of anti-Gal antibodies is responsible in the decline of the parasitemia from the acute to the chronic chagasic phase in Chagas' disease.

Thus pre-absorption of acute Chagas serum with mouse laminin, that contains α-gal epitopes (Towbin *et al*, 1987), or with melibiose (Galactosyl(α1-6)Glucopyranoside), or α-D-Galactopyranoside, completely abolished the direct lytic activity. Furthermore, the lytic effect of acute chagasic serum was also inhibited by the presence in the incubation medium of melibiose or α-D-Galactopyranoside with Gal(α1-3)Gal being 5-fold more active than Gal(α1-4)Gal (Gazzinelli *et al.* (1991a).

Investigation of the characteristics of the cell lysis reaction showed that the lysis of *T. cruzi* by chagasic human sera occurred a few minutes after the samples of the parasite and anti-Gal antibodies are mixed and that this interaction is temperature-dependent (optimal temperature being 37°C). In contrast, chronic chagasic sera and normal sera were unable to carry out direct lysis of parasites.

### 7.4.2. Complement-Mediated Lysis in Chronic Chagas' Disease

Anti-Gal from chronic chagasic patients lysed *T. cruzi* trypomastigotes in a complement-dependent reaction (Figure 7a). At the concentration of 300 μg/ml, (a concentration frequently found in chagasic sera [Avila *et al.*, 1989]) this antibody caused the lysis of more than 80% of the parasites. Pooled anti-Gal from chagasic sera and complement readily lysed the CL strain trypomastigotes, even at 1 μg/ml (Figure 7b). The complement-mediated lysis seemed to occur via the alternative pathway, inasmuch as the addition of EGTA did not prevent lysis. The specificity for α-gal-containing epitopes in the lysis reaction was confirmed by inhibition with α- but not β-galactopyranosides (Table VII).

To assess the importance of anti-Gal from chagasic serum in the total serum lytic power, these antibodies were adsorbed by Synsorb 115 (i.e. synthetic α-gal epitopes linked to silica beads) followed by rabbit erythrocyte adsorption. Purified chagasic anti-Gal at 300 μg/ml, whole chagasic serum, and normal human serum were tested for residual lytic activity. With ligand-free Synsorb (silica beads) as the adsorbent control, no significant decrease in the lytic activity was obtained. Following adsorption on Synsorb 115 and rabbit erythrocyte, the lytic activity of both the chagasic anti-Gal and the chagasic whole serum was reduced to 25 to 30% of its initial value (Gazzinelli *et al.*, 1991a). Almeida *et al.* (1991) concluded that most of the chagasic serum lytic activity is a result of the presence of antibodies recognizing α-gal epitope-containing structures (Figure 8). They further proposed that in individuals contracting Chagas' disease, the affinity of natural anti-Gal

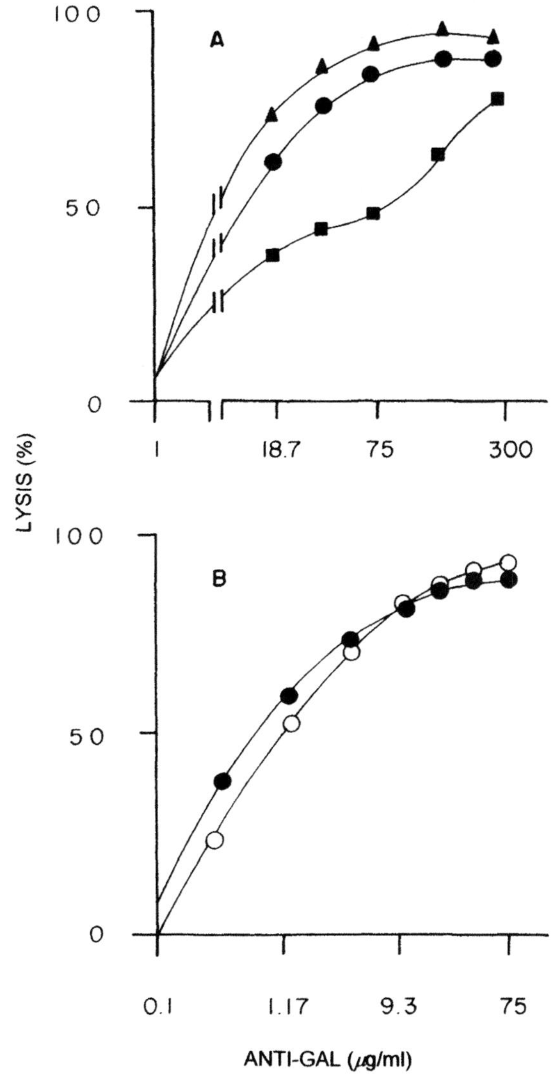


**Figure 7.** Complement-mediated lysis of metacyclic trypomastigotes by anti-Gal. A, Y strain trypomastigotes; black triangles, circles, and squares, three different anti-Gal preparations from pools pCh7; pCh-5 and pCh-6, respectively, of chagasic sera. B, Cl strain trypomastigotes; lytic response to chagasic anti-Gal in presence of EGTA (alternative complement pathway), black circles; without EGTA, open circles. From Almeida *et al* (1991), with permission.

**Table VII**
**Effect of Carbohydrates on the Immunolysis of Metacyclic Trypomastigotes of *Trypanosoma cruzi***

| | % lysis of parasites[b] | | |
|---|---|---|---|
| Carbohydrate[a] | Anti-Gal[c] | Chagasic serum[d] | Normal serum[d] |
| None | 96 (90)[e] | 92 (91) | 14 (12) |
| O-Methyl-α-D-Galp[f] | 18 (25) | 21 (20) | 6 (5) |
| O-Methyl-β-D-Galp | 87 (85) | 86 (89) | 12 (13) |
| Galactose | 24 (23) | 25 (22) | 8 (8) |
| N-Acetylgalactosamine | 32 (31) | 36 (32) | 11 (9) |
| Glucose | 91 (87) | 86 (87) | 15 (17) |
| Melibiose | 19 (26) | 18 (22) | 8 (10) |
| Lactose | 91 (86) | 88 (24) | 15 (11) |

[a]At final concentration of 100 mM.
[b]Values ± 5 % average the results of three experiments.
[c]At final concentration of 300 μg/ml, from chagasic serum.
[d]Chagasic serum and normal human serum at a final dilution of 1:2.5.
[e]Numbers in parentheses: complement-mediated lysis (%), alternative pathway (in presence of 10 mM EGTA).
[f]Galactopyranoside.
Taken from Almeida *et al* (1991) with permission.

clones may be too low to mount an effective complement-mediated lysis, thus, some of the parasites succeed in invading cells and thereby evade destruction by the immune system.

We mentioned above, that Gazinelli *et al* (1991b) have demostrated that monoxenic trypanosomatids are very rich in α-gal epitopes, and indeed Souza *et al* (1974) have demonstrated a partial protection of *T. cruzi* infection induced by immunization with monoxenic trypanosomatids. This increased resistance may be mediated by elevated anti-Gal induced by monoxenic parasites α-gal residues.

## 8. ANTI-GAL RESPONSE IN *T. RANGELI*-CHRONICALLY INFECTED HUMAN SUBJECTS

Avila *et al* (1987) detected elevated levels of anti-Gal in 82% of 38 *T. rangeli*-infected patients, who are free of any *T. cruzi* infection. This strong immune response is due probably to the presence of α-gal residues on *T. rangeli* as latter demonstrated by Gazzinelli *et al* (1991b). It is noteworthy that 33% of apparently hemoflagellate-free subjects living in a *T. rangeli*-endemic area and who have normal levels of antibody against *T. cruzi* and *T. rangeli* antigens were found to have elevated anti-Gal levels. This may be the result either of a cryptic or a very low parasitemia *T. rangeli* infection, undetected even after three consecutive separate hemocultures.

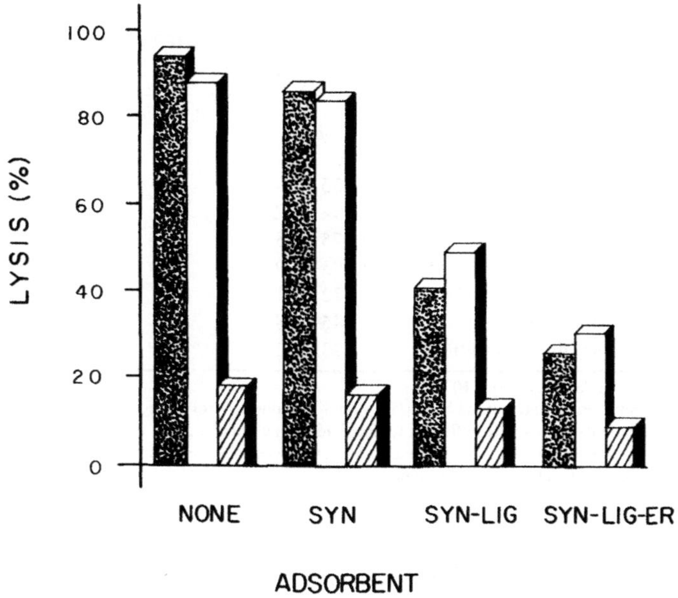


**Figure 8.** Removal of lytic antibodies by Synsorb 115 followed by rabbit erythrocyte absorption. Chagasic anti-Gal at 300 μg/ml (▒), whole chagasic serum (□) and normal human serum (▨), were mixed with 2 vol dry ligand-free Synsorb (SYN) as well as with Synsorb 115 (SYN-LIG) for 2 h at 37°C. Supernatants were assayed for their lytic activity with metacyclic trypomastigotes (Y strain). They were then further absorbed with 2 vol of a washed pellet of rabbit erythrocytes (SYN-LIG + ER) and tested for lytic activity. From Almeida *et al* (1991), with permission.

## 9. ANTI-GAL IN CUTANEOUS AND VISCERAL LEISHMANIASIC PATIENTS

Anti-Gal response to *Leishmania* infection have been studied in cutaneous, mucocutaneous and visceral leishmaniasis (Towbin *et al.*, 1987; Avila *et al.*, 1984; 1988b). Eighty-one percent of American cutaneous leishmaniasis patients (i.e., patients having ulcers or recently closed ulcers) and 34% of the patients with treated and healed ulcers had high "anti-laminin" antibody levels. Further studies of our group revealed that anti-Gal accounted for most of the anti-laminin antibody reactivity in these patients, because these antibodies interact with the α-gal epitope on the N-glycosylated carbohydrate chains of mouse laminin (Towbin *et al*, 1987). Additional studies showed no correlation between the time course of the leishmaniasis and the anti-Gal antibody levels. We found that mucocutaneous leishmaniasis patients had the highest anti-Gal levels whereas patients with active localized cutaneous leishmaniasis and those with a recently closed cutaneous

**Table VIII**
**Anti-Gal Antibody Levels in Healthy Controls and in Study Group of Patients**

| Study group (no. of patients) | $A_{405}$[a] | Positivity (%) | P value[b] |
|---|---|---|---|
| Controls (169) | 235 ± 151 | 8 | |
| Cutaneous ACL (91) | 1,594 ± 1,208 | 67 | < 0.01 |
| MCL (6) | 1,975 ± 657 | 87 | < 0.01 |
| Recently closed ACL ulcer[c] (54) | 1,493 ± 993 | 80 | < 0.01 |
| Chronic Chagas' disease (26) | 972 ± 812 | 50 | < 0.05 |
| *T. rangeli* infection (31) | 946 ± 582 | 74 | < 0.05 |
| Inflammatory diseases (69) | 293 ± 126 | 4 | > 0.1 |

[a]Mean ± standard deviation ($10^{-3}$).
[b]Determined in comparison with anti-Gal values found in control subjects.
[c]These patients were studied after ACL ulcer had been clinically cured for 2 to 3 months.
ACL= American cutaneous leishmaniasis; MCL= Mucocutaneous leishmaniasis.
Taken from Avila *et al* (1988a) with permission.

leishmaniasis ulcer have significantly elevated levels when compared with control patients (Table VIII).

Both IgG and IgM classes of anti-Gal were elevated in acute leishmaniasis patients whereas anti-Gal IgA levels was low (Avila *et al.*, 1984). In a subsequent study, Avila *et al* (1988b) detected elevated anti-Gal levels in 76% of children with active visceral leishmaniasis and in 42% of clinically cured patients with this disease who had been treated about 5 years previously with meglumine antimonate. In that studies rabbit erythrocyte ceramide pentasaccharide-rich neutral glycolipids were used as solid phase antigen in ELISA.

**Table IX**
**Anti-Gal Antibody Levels in Healthy Control Children and in Patients with Active, Cured or Treated Presumptive Leishmaniasis Visceral**

| Study group (no. of patients) | A ($10^{-3}$)[a] | Positivity (%)[b] | Value[c] |
|---|---|---|---|
| Control (62) | 235 ± 151 | | |
| Active Kala Azar (10) | 956 ± 262 | 76 | < 0.001 |
| Treated Kala Azar (38)[d] | 603 ± 311 | 42 | < 0.050 |
| Treated presumptive Kala azar (25)[d] | 547 ± 293 | 56 | < 0.050 |

[a]Mean ± standard deviation.
[b]An $A_{405}$ of 0.613 represents the upper limits of normal values, defined as the mean ± 2.5 standard deviation units of the control group. A reading of 0.613 or greater was therefore considered positive.
[c]Determined in comparison with anti-Gal levels found in control subjects.
[d]The posttreatment period varied between 0.1 and 13 years, being 5.2 ± 3.3 years for patients with Kala Azar and 4.7 ± 3.0 years for patients with presumptive Kala azar.
Taken from Avila *et al* (1988b) with permission.

The long-term persistence of elevated anti-Gal was also found in 56% of children living in the same geographic zone who, at the time of the initial clinical examination, had fever and evident splenomegaly with hyperglobulinemia but had a negative bone marrow aspirate for leishmanial bodies. Five years after antimonate treatment, these clinically cured children with presumptive Kala azar were studied serologically. Their mean anti-Gal values were slightly lower than those in patients with active Kala azar but were still abnormal (Table IX). Anti-Gal was primarily of the IgG and IgM classes in patients with active Kala azar and in antimonate-treated patients with clinically cured Kala azar. The possibility of the remnant living parasites or the persistence of inserted α-gal epitopes in parasitized macrophages was proposed as a mechanism to explain the long-term persistence of abnormal anti-Gal in patients apparently cured of Kala azar (Avila *et al.*, 1988b).

## 10. IS THERE AUTOIMMUNITY MEDIATED BY ANTI-GAL IN HUMAN CHAGAS' DISEASE ?

It has long been speculated that many autoimmune diseases are the result of infectious agents. Examples of autoimmune diseases believed to have an infectious "trigger" include rheumatic heart disease (group A streptococcus)(Van der Rijn *et al.*, 1977), ankylosing spondylitis and Reiter's syndrome (*Klebsiella pneumoniae*)(Schwimmbeck *et al.*, 1987), celiac disease (adenovirus), non-rheumatoid arthritides (Oldstone, 1987) and myasthenia gravis (herpes simplex virus)(Schwimmbeck *et al.*, 1989). For example, six consecutive amino acids (QTDRED) are identical between the hypervariable domain of HLA-B27 *and K. pneumoniae* nitrogenase. Sera from a significant proportion of HLA-B27 individuals with Reiter's syndrome (53%) or ankylosing spondylitis (29%), but not from appropriate controls, reacted with a synthetic peptide containing the homologous region of HLA-B27 and *K. pneumoniae* nitrogenase. These observations suggest that Reiter's syndrome and ankylosing spondylitis are autoimmune diseases, with HLA-B27 serving as the autoantigen. Induction would be caused by a microbe(s) encoding a protein with sequence homology to HLA-B27 (variable region) and the disease would presumably be related to an unusual concentration of HLA antigen in certain tissues-like joints. In fact, joint tissue from 92% ankylosing spondylitis patients studied heavily expressed HLA-B27 sequences (QTDRED). In contrast, these sequences were not observed either in synovial tissue of individuals with non-arthropathies or in skin biopsy from non-ankylosing spondylitis but HLA-B27 positive individuals.

Another possibility is that a pathogen with the capacity to persist in its host may continuously or cyclically express its antigens. Although expression of a pathogen genome may be restricted so that no infectious multiplication exists, pro-

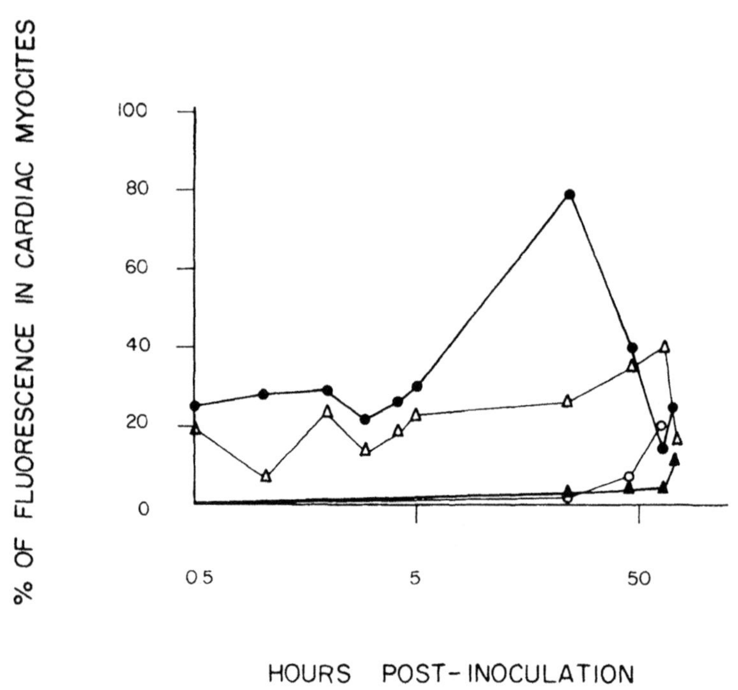


**Figure 9.** Indirect immunofluorescence staining of fetal murine cardiac myocites infected with *T. cruzi* trypomastigotes. Kinetics of expression of parasitic antigens on the plasma membrane (Δ), cytoplasm (●) and contractile fibers (▲). (○) Indicates percent of *T. cruzi*-infected cells.

duction of a pathogen antigen in common with that of the host might continue. This would allow initiation of an immune response and/or autoimmunity, either one leading to cyclic, chronic, or progressive disease. Moreover, some viruses and other microbes contain chemical structural components that mimic normal host "self" proteins or terminal oligosaccharide (molecular mimicry).

*Trypanosoma cruzi* is implicated in autoimmunity on the basis of three findings. First, autoimmune responses are *de novo* concomitant with infection. This point is strengthened by the second finding that, in experimental animals, both acute and persistent *T. cruzi* infection can induce autoimmune responses and cause autoimmune disease (Cunha-Neto *et al.*, 1995; Avila, 1994; Petry and Eisen, 1989; Schmuñis, 1987). Third, by evaluating molecular mimicry in human Chagas' disease, several cross-reacting antigens have been found including: Fl-160 antigen (Van Voorhis *et al.*, 1991), cardiac myosin (Cunha-Neto *et al.*, 1995), cholesterol sulphate-like structure (Avila *et al.*, 1996) and asialoganglioside and monosialoganglioside (Avila *et al.*, 1998). This may be related to Chagas' disease pathogenesis where in the chronic stage, parasite numbers decline to virtual insignificance whereas the disease symptoms become most apparent. This paradox may be explained by the parasite-triggered autoimmunity, as originally proposed by Köberle (1968).

Autoimmunity in Chagas' disease may be associated with the α-gal epitope. Salma and Avila (1991) have recently shown that *T. cruzi*-infected murine myocites and nervous tissue cells abundantly express the α-gal epitope. This was demonstrated using polyvalent goat anti-human Ig-FITC conjugate and purified human chagasic anti-Gal. Figure 9 shows the percentage of *T. cruzi*-infected murine myocites positive by immunofluorescence. The reaction, located on plasma membrane, increased steadily reaching maximal value at 72 hours after infection, thereafter decreasing. Cytoplasmic staining reached maximum value at 24 hours after infection. Fluorescence in contractile fibers steadily increased up to 60 hours after infection. Under our working conditions, no immunofluorescence was found in non-infected (control) murine myocites or nervous tissue cells as chagasic human serum had been previously absorbed with respective uninfected murine cells.

Figure 10 shows the percentage of *T. cruzi*-infected murine nervous tissue cells positive by immunofluorescence. This localized in plasma membrane reaching a maximum at 24–48 hours after infection. Cytoplasm showed maximal fluorescence 48 hours post-infection, thereafter steadily decreasing. In contrast, immunofluorescence on neurofilaments slowly increased reaching maximal value at 60 hours post-infection.

A clear expression of α-gal epitopes have been reported, using immunogold labeling for electron microscopy, for Vero cells infected with *T. cruzi* trypomastigotes (Souto-Padron *et al*, 1994). Thus, it is possible that infected cells (macrophages, muscle cells, nervous cells) may change their surface structure by inserting parasite-derived α-gal epitopes, or changing host carbohydrate components by new posttranslational modifications due to the intracellular residence of parasites. The autoimmunity observed in Chagas disease may therefore not be the result of cross-reacting structures present *per se*, but due to a modified structure arising in the *T. cruzi*-infected cells (Avila, 1994). Antibodies involved in the initiation of this process might be the high affinity anti-Gal from chronic chagasic patients, rather than the natural anti-Gal from healthy individuals.

Whereas the epitope recognized by anti-Gal may be a simple disaccharide, the conditions allowing antibodies to bind to that epitope are complex. Platt (1996) have proposed that expression of the α-gal epitope by itself may not be sufficient to allow substantial binding of complement fixing in xenoreactive antibodies. Rather, natural antibodies seem to bind preferentially to certain glycoproteins bearing that epitope.

One important factor in antibody binding appears to be the clustering of epitopes allowing multivalent interactions between antibody molecules and the cell surface. The clustering of α-gal epitope have been clearly demonstrated on *T. cruzi* and *Leishmania* plasma membrane outer face (Souto-Padron *et al.*, 1994;

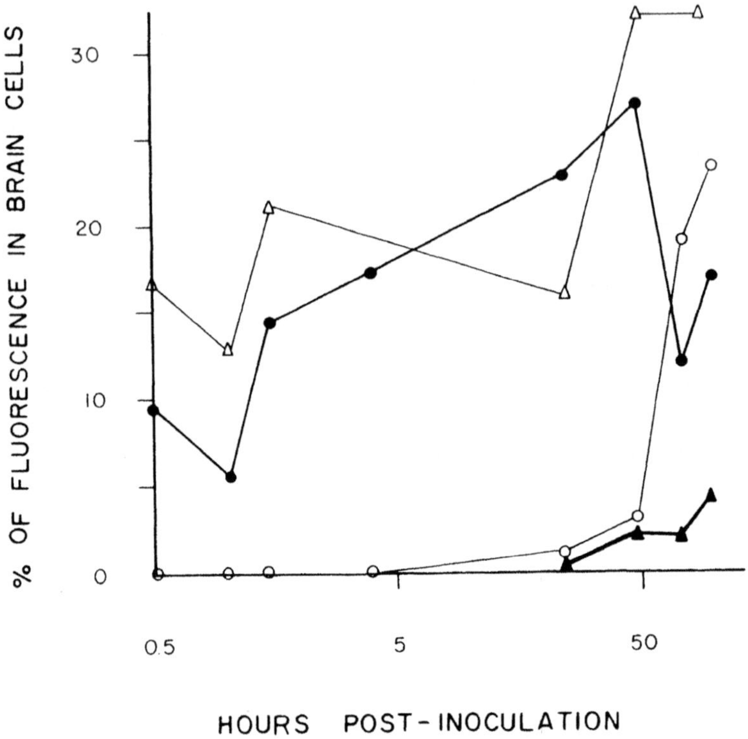


**Figure 10.** Indirect immunofluorescence staining of fetal murine brain cells infected with *T. cruzi* trypomastigotes. Kinetics of expression of parasitic antigens on the plasma membrane (Δ), cytoplasm (●) and neurofilaments (▲). (O) Indicates percent of *T. cruzi*-infected cells.

Bretaña *et al.*, 1992). It remains to be determined whether this α-gal epitope clustering occurs in *T. cruzi*-infected cells.

The optimal clustering of epitopes is not random, however, as the binding of xenoreactive anti-Gal to purified proteins containing similar numbers of α-gal epitope varies over approximately 1000-fold (Platt and Holzknecht, 1994). Further evidence that the manner in which α-gal epitope is expressed, rather than the total number of epitopes, determine the extent of antibody binding, emerged from studies on variation in antigen expression by the population of potential organ donors (Geller *et al.*, 1994). Analysis of the binding level of xenoreactive antibodies to cells from populations of pigs suggests that there is an up to 10-fold range of antibody binding suggesting variation in antigen expression (Alvarado *et al.*, 1995; Geller *et al.*, 1994). The nature of the core structures bearing the α-gal epitope modifications recognized by xenoreactive natural anti-Gal has several significant implications for the development of xenograft reaction. Thus, the core structures determine to a certain extent the three-dimensional array of the epitopes and may thus determine the amount of antibody binding to a target cell (Cotterell *et al.*, 1995). Accordingly, interaction of *T. cruzi*-infected cells with high affinity chagasic anti-Gal will lead to cell death only if the α-gal epitope express optimal conditions for a successful antigen-antibody interaction. If so, this would lead to a local inflammatory reaction mediated by granulocytes, macrophages, and killer lymphoid cells interacting with the Fc portion of the bound anti-Gal. These cytotoxic cells would subsequently exert their lytic activity toward the cells having the α-gal epitope in an antibody-dependent, cell mediated cytolysis (ADCC) process, as demonstrated recently by Takeuchi *et al* (1996) in human cells transfected with a porcine cDNA encoding α1-3Galactosyltransferase. Mononuclear cellular infiltrates (consisting of macrophages, lymphocytes and plasma cells)are the usual morphological findings in acute (Palacios-Prü *et al.*, 1989) as well as in chronic chagasic cardiomyopathy (consisting of macrophages and lymphocytes)(Milei *et al.*, 1992).

## 11. THE UDP-GAL:β-GAL-α1-3GAL-GALACTOSYLTRANSFERASE KNOCKOUT MICE

In 1995, Thall *et al* generated mice deficient in a gene (α1,3GT) encoding the UDP-Gal:β-Gal-α1-3galgalactosyl-transferase enzyme responsible for α-gal synthesis and expression. These mice develop normally and exhibit no gross phenotypic abnormalities. The α-gal epitope is absent from the vascular endothelium and other tissues in α1,3GT(–/–) adult mice.By contrast, α1,3GT(–/–) mice, like humans, develop naturally occurring anti-Gal normally absent in wild type mice. We hypothesized that these mice could be a good experimental model for testing not only the importance of the anti-Gal in experimental acute Chagas' disease and in cutaneous leishmaniasis but also the humoral immune response to the α-gal epitope. Thus, we have compared the parasitemia curve, and anti-Gal levels of adult α1,3GT(–/–) and of wild type mice that were experimentally infected with *T. cruzi*. We also monitored the development of *Leishmania* infection in rear footpads.

### 11.1. Acute *T. cruzi* Infection

Adult female α1,3GT(–/–) and wild type control mice were used in these experiments. Non-infected α1,3GT(–/–) mice have about 1.5-fold higher anti-Gal levels compared with wild type mice. Both groups were infected with *T. cruzi* trypomastigotes and the parasitemia curve and serum anti-Gal levels followed by about 30 days. In α1,3GT(–/–) mice parasitemia began to be detected at about eight days post-inoculation and was paralleled by an increase in anti-Gal levels. Interestingly, about day 13 post-inoculation there was a small fall in both para-

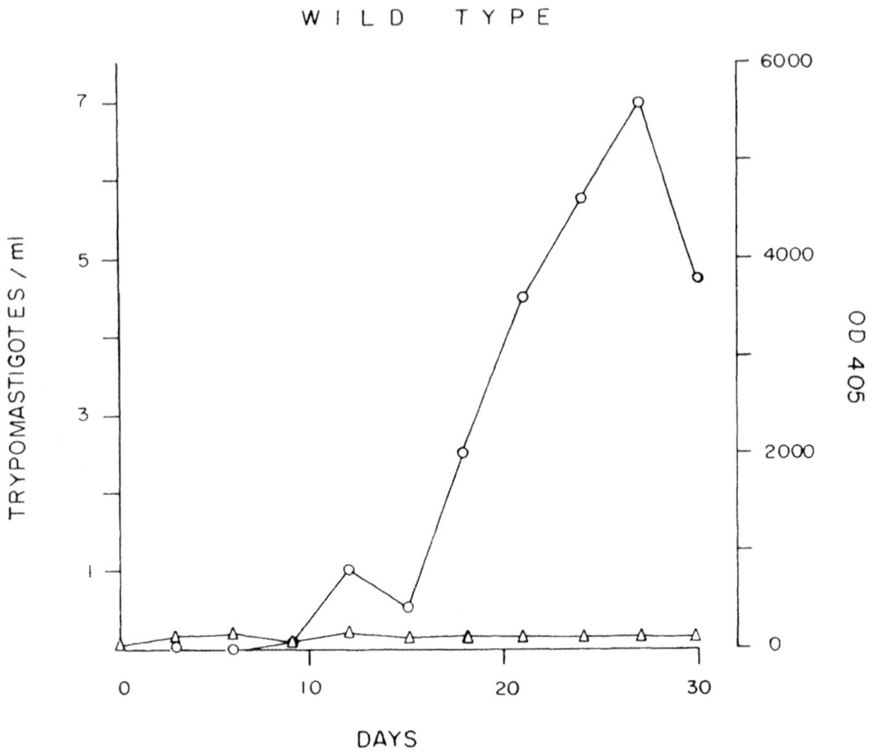


**Figure 11.** Kinetics of parasitemia (O) and anti-Gal response (Δ) of α1,3GT (–/–) and wild type mice intraperitoneally inoculated with 1 × $10^2$ or 5 × $10^2$ *T. cruzi* trypomastigotes of the Be strain.

sitemia and anti-Gal levels, thereafter parasitemia increased about 6-fold while anti-Gal levels maintained in a plateau (Figure 11). This represented an increase of 54-fold for α1,3GT(–/–) mice when compared with same uninfected mice. Anti-Gal levels were not modified in wild type mice. Noteworthy, parasitemia curve was similar to that found for α1,3GT(–/–) mice. However, parasitemia levels in α1,3GT(–/–) mice was significantly lower ($P < 0.01$) than in wild type mice suggesting that anti-Gal is capable of partially controlling *T. cruzi* infection. The rapid response in α1,3GT(–/–) mice in anti-Gal levels to *T. cruzi* inoculation suggest that indeed parasitic α-gal epitopes are highly immunogenic. Wild type mice were immunotolerant to α-gal epitope.

## 11.2. Cutaneous Leishmaniasis Infection

No difference in the time course of *Leishmania* nodule growth nor in the appearance of anti-Gal levels was found between α1,3GT(–/–) and wild type control

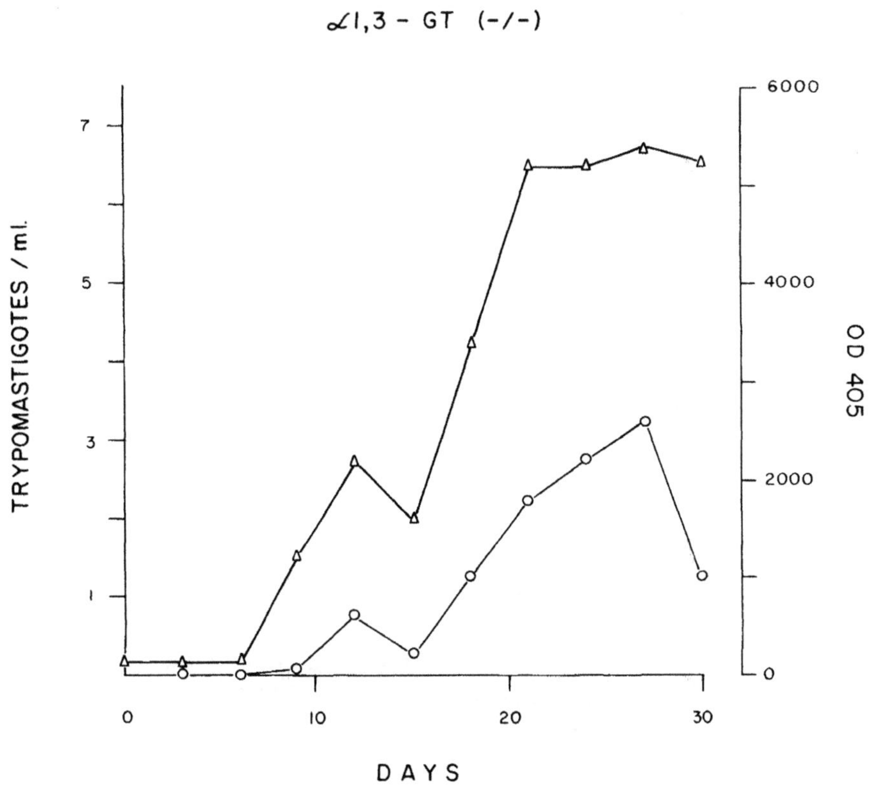


**Figure 11.** (*Continued*)

mice (Figure 12). It should be noted however that *Leishmania* acted as a powerful immunogen as in α1,3GT(–/–) mice anti-Gal levels began to increase only at day 12 post- inoculation with 100 or 500 amastigotes, remaining in a plateau thereafter. This delayed response may be due to the initial slow development of the disease. The sharp increase in nodule size observed 120 days post-inoculation was not followed by a significant ($P > 0.05$) modification in anti-Gal levels.

*Leishmania*-infected wild type animals did not modified its anti-Gal values. Similar to the observations with *T. cruzi* it is evident that anti-Gal response by itself is not capable of controlling cutaneous *Leishmania* infection.

## 12. CONCLUSIONS

Anti-Gal is a human natural antibody which interacts specifically with the mammalian carbohydrate structure Galα1-3Galβ1-4GlcNac-R, termed, the α-gal

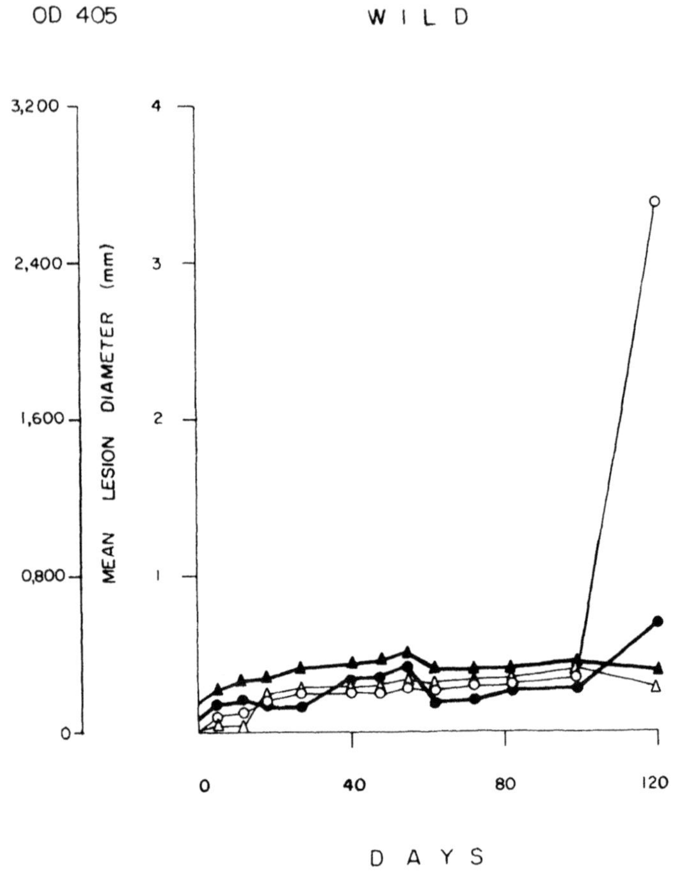


**Figure 12.** Development of leishmanial lesions in foot-pads (O,●) and anti-Gal response (Δ,▲) of α1,3GT (–/–) mouse and wild type control mice. Mice were inoculated with $1 \times 10^2$ (O) or $5 \times 10^2$ (●) of GML473 isolate of *Leishmania mexicana mexicana*.

epitope. This antibody constitutes approximately 1% of circulating IgG in human serum and is produced, upon stimulation, by 1% of circulating B lymphocytes. The antigenic source for the constant production of anti-Gal seems to be the α-gal-like epitopes found on many bacteria of the gastrointestinal flora. The physiological role of this antibody is not clear as yet.

Several biochemical and morphological reports have given clear evidence of the presence in *Trypanosoma* and in *Leishmania* of α-gal-rich structures on glycolipids and glycoproteins.

Patients with acute Chagas disease have about nine-fold higher anti-Gal levels compared with chronic chagasic patients. In this latter condition no relationship was found between clinical evolution and anti-Gal levels. Anti-Gal readily

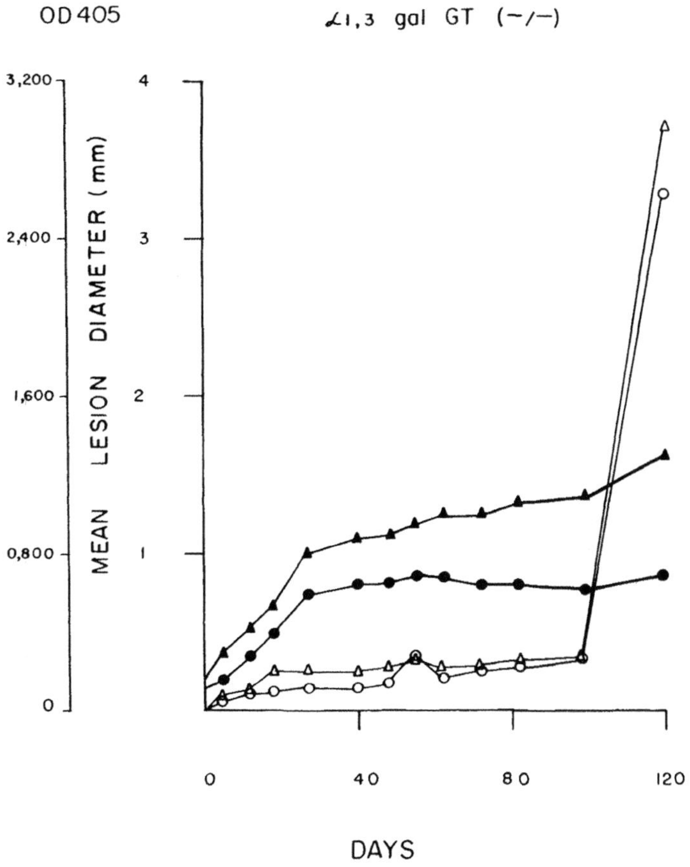


**Figure 12.** (*Continued*)

binds to *T. cruzi* and can induce either direct lysis or complement-mediated lysis of the parasite. Thus, anti-Gal may contribute to natural immunity against *T. cruzi*.

Eighty-two percent of *T. rangeli*-infected patients displayed significantly elevated anti-Gal levels. Noteworthy, 33% of apparently hemoflagellate-free subjects living in a *T. rangeli*-endemic area and having control antibody levels against *T. cruzi* and *T. rangeli* antigens showed elevated anti-Gal levels. This suggests the possibility either of a cryptic or a very low parasitemia *T. rangeli* infection, undetected even after three consecutive separate hemocultures.

Cutaneous and visceral leishmaniasis patients also have elevated anti-Gal levels. In Kala azar antigenic stimulation persists even years after apparent clinical cure suggesting that given the strong immunogenic capacity of α-gal epitopes present on *Leishmania* parasites, even a few remaining parasites are able to maintain a longstanding immunological stimulation. No information is still available on the importance of anti-Gal on *Leishmania* promastigote or amastigote lysis or in the development of lesions.

Using adult female mice deficient in a gene encoding the UDP-Gal:β-Gal-α1-3galgalactosyltransferase (enzyme responsible for α-gal synthesis and expression)(α1,3GT) [α1,3GT(–/–)] and wild type mice experimentally infected with *T. cruzi* trypomastigotes, we found a significant difference in the parasitemia levels: α1,3GT(–/–) mice having significantly lower parasitemia than wild type mice. The contrary was true for anti-Gal levels, thus while α1,3GT(–/–) mice have a 54-fold increase in anti-Gal levels, wild type mice maintained basal levels. These results suggests that anti-Gal has an important role in the control of parasitemia and that *T. cruzi* is a highly immunogenic parasite. This was not the case in Leishmania-infected mice, whose foot-pad lesion diameter (*Leishmania*) between both groups of mice was close suggesting a minor role for anti-Gal in the control of experimental cutaneous *Leishmania* infection. *Leishmania*-infected α1,3GT(–/–) mice showed a significant increase in anti-Gal levels, but very discrete compared with *T. cruzi*-infected mice, due perhaps to the fact that this infection was limited only to one rear foot-pad. Again wild type mice did not increase anti-Gal levels. Finally, the eventual role of anti-Gal in the autoimmunity existing in Chagas' disease is discussed. Thus, the possibility exist that *T. cruzi*-infected cells (macrophages, muscle cells, nervous cells) due to the intracellular residence of parasites, may change their surface structure by inserting parasite-derived α-gal epitopes or changing host components by new posttranslational modifications. The autoimmunity observed in Chagas disease may therefore not be due to cross-reacting structures present *per se*, but to a modified structure arising in the *T. cruzi*-infected cells.

ACKNOWLEDGMENTS. This work is a product of fruitful collaborations with many colleagues. In particular, I thank Miguel Rojas, Antonio Bretaña, Nunciada Salma, Alvaro Acosta-Serrano, María Argelia Polegre, Angela Avila, Gladys Velazquez-Avila, Rando García, Ivo Tremont-Lukats, Jaime Wilder, Esther Arbona, Leonardo García, Marisol Zambrano, Olga L. Wittig, Egidio Romano, Manuel Rieber, Rupert Timpl, H. Von der Mark, Harry Towbin, Uri Galili, Michael A.J. Ferguson, Malcom McConville, Hugo Carrasco, Jacinto Convit, G.A Maekelt and Felipa Merecuana. I thank María Leboso for graphic art assistance. This work was supported by Consejo Nacional de Investigaciones Científicas y Tecnológicas (CONICIT) and World Bank.

## REFERENCES

Almeida, I.C., Milani, S.R., Gorin, P.A.J., and Travassos, L.R., 1991, Complement mediated lysis of *Trypanosoma cruzi* trypomastigotes by human anti-α-galactosyl antibodies. *J. Immunol.* **146**: 2394–2400.

Almeida, I.C., Krautz, G.M., Krettli, A.U., and Travassos, L.R., 1993, Glycoconjugates of *Trypanosoma cruzi*: A 74 Kd antigen of trypomastigotes specifically reacts with lytic anti-α-galactosyl antibodies from patients with chronic Chagas disease. *J. Clin. Laborat. Anal.* **7**: 307–316.

Almeida, I.C., Ferguson M.A., Schenkman, S., and Travassos, L., 1994, Lytic anti-α-galactosyl antibodies from patients with chronic Chagas' disease recognize novel O-linked oligosaccharides on mucin-like glycosylphosphatidylinositol-anchored glycoproteins of *Trypanosoma cruzi*. *Biochem. J.* **304**: 793–802.

Alvarado, C.G., Cotterell, A.H., McCurry, K.R., Collins, B.H., Magee, J.C., Berthold, J., Logan, J.S., and Platt, J.L., 1995, Variation in the level of xenoantigen expression in porcine organs. *Transplantation* **59**: 1589–1594.

Andrade, Z., 1985, A patologia da doença de Chagas no homen. *Ann. Soc. Belg. Med. Trop.* **65** (Suppl 1): 15–30.

Andrade, S.G., and Grimaud, J.A., 1986, Chronic murine myocarditis due to *Trypanosoma cruzi*- an structural study and immunochemical characterization of cardiac interstitial matrix. *Mem. Inst. Oswaldo Cruz* **81**: 29–41.

Andrade, S.G., Grimaud, J.A., and Stocker-Guerret, S., 1989, Sequential changes of the connective matrix components of the myocardium (fibronectin and laminin) and evaluation of cardiac fibrosis in mice infected with *Trypanosoma cruzi*. *Am. J. Trop. Med. Hyg.* **40**: 151–60.

Araujo, M.A., 1989, Lise de *Trypanosoma cruzi* por soros de pacientes na fase aguda da doenca de Chagas. Ph.D. Thesis, *University of Sao Paulo*.

Avila, J.L., 1992, Molecular mimicry between *Trypanosoma cruzi* and host nervous tissues. *Acta. Cient. Venez.* **43**: 330–340.

Avila, J.L., 1993, Glycosyl phosphatidylinositols in *Leishmania* and *Trypanosoma*. *Trends in Comparat. Biochem. Physiol.* **1**: 881–906.

Avila, J.L., 1994, Molecular mimicry between *Trypanosoma* cruzi and host nervous tissues. In: *Chagas'disease and the nervous system*. Pan American Health Organization. Sci. Publ. N° 547, Washington, pp. 268–292.

Avila, J.L., and Rojas, M., 1990, A galactosyl(α1–3)mannose epitope on phospholipids of *Leishmania mexicana and L. braziliensis* is recognized by trypanosomatid-infected human sera. *J. Clin. Microbiol.* **28**: 1530–1537.

Avila, J.L., Rojas, M., and Reiber, M., 1984, Antibodies to laminin in American cutaneous leishmaniasis. *Infect. Immun.* **43**: 402–406.

Avila, J.L., Rojas, M., and Towbin, H., 1988a, Serological activity against galactosyl-α(1-3)galactose in sera from patients with several kinetoplastida infections. *J. Clin. Microbiol.* **26**: 126–132.

Avila, J.L., Rojas, M., and Garcia, L., 1988b, Persistence of elevated levels of galactosyl-α(1–3)galactose antibodies in sera from patients cured of visceral leishmaniasis. *J. Clin.Microbiol.* **26**: 1842–1847.

Avila, J.L., Rojas, M., and Galili, U., 1989, Immunogenic Galα1-3Gal carbohydrate epitopes are present on pathogenic American *Trypanosoma* and *Leishmania*. *J. Immunol.* **142**: 2828–2834.

Avila, J.L., Rojas, M., and Acosta, A., 1991, Glycoinositol phospholipids from American *Leishmania* and *Trypanosoma spp.*: Partial characterization of the glycan cores and the human humoral immune response to them. *J. Clin. Microbiol.* **29**: 2305–2312.

Avila, J.L., Rojas, M., and Velasquez-Avila, G., 1992, Characterization of a natural human antibody with anti-galactosyl(α1–2)galactose specificity that is present at high titers in chronic *Trypanosoma cruzi* infection. *Am. J. Trop. Med. Hyg.* **47**: 413–421.

Avila, J.L., Rojas, M., and Avila, A., 1996, Cholesterol sulphate-reactive autoantibodies are specifically increased in chronic chagasic human patients. *Clin. Exp. Immunol.* **103**: 40–46.

Avila, J.L., Rojas, M., and Avila, A., 1998, Increase in asialoganglioside- and monosialoganglioside-reactive antibodies in chronic Chagas' disease patients. *Am J. Trop. Med. Hyg.* **58**: 338–342.

Avila, J.L., Rojas, M., Velazquez-Avila, G., Von Der Mark, H., and Timpl., R., 1986, Antibodies to basement membrane protein nidogen in Chagas' disease and American cutaneous leishmaniasis. *J. Clin. Microbiol.* **24**: 775–778.

Avila, J.L., Rojas, M., Velazquez-Avila, G., and Rieber, M., 1987, Antibodies to laminin in *Trypanosoma rangeli*-infected subjects. *Parasitol. Res.* **73**: 178–179.

Bretaña, A., Avila, J.L., Arias-Flores, M., Contreras, M., and Tapia F.J., 1986, *Trypanosoma cruzi* and American *Leishmania* spp.: immunocytochemical localization of a laminin-like protein in the plasma membrane. *Exp. Parasitol.* **61**: 168–175.

Bretaña, A., Avila, J.L., Contreras-Bretaña, M., and Tapia, F., 1992, American *Leishmania* spp. and *Trypanosoma cruzi*: Galactosyl$\alpha$(1–3)galactose epitope localization by colloidal gold immuno-cytochemistry and lectin cytochemistry. *Exp. Parasitol.* **74**: 27–37.

Carrasco, H.A., Barboza, J.S., Inglessis, G., Fuenmayor, A., and Molina, C., 1982, Left ventricular cineangiography in Chagas' disease: detection of early myocardial damage. *Am. Heart J.* **104**: 595–602.

Carreira, J.C., Jones, C., Wait, R., Previato, J.O., and Mendonca-Previato, L., 1996, Structural variation in the glycoinositol phospholipids of differente strains of *Trypanosoma cruzi*. *Glycoconjugate J.* **13**: 955–966.

Cotterell, A.H., Collins, B.H., Parker, W., Harland, R.C., and Platt, J.L.,1995, The humoral immune response in humans following cross-perfusion of porcine organs. *Transplantation* **60**: 861–864.

Couto, A.S., Goncalves, M.F., Colli, W., and Lederkremer, R.M., 1990, The N-linked carbohydrate chain of the 85 kilodalton glycoprotein from *Trypanosoma cruzi* trypomastigotes contains sialyl, fucosyl and galactosyl($\alpha$1–3)galactose units. *Mol. Biochem. Parasitol.* **39**: 101–108.

Couto, A.S., Lederkremer, R.M. de, Colli, W., and Alves, M.J., 1993, The glycosylphosphatidylinositol anchor of the trypomastigote-specific Tc-85 glycoprotein *from Trypanosoma cruzi*. Metabolic-labeling and structural studies. *Eur. J. Biochem.* **217**: 597–602.

Cunha-Neto, E., Duranti, M., Gruber, A., Zingales, B., De Messias, I., Stolf, N., Bellotti, G.,Patarroyo, M. E., Pilleggi, F., and Kalil, J., 1995, Autoimmunity in Chagas disease cardiopathy: Biological relevance of a cardiac myosin-specific epitope crossreactive to an immunodominant *Trypanosoma cruzi* antigen. *Proc. Natl. Acad. Sci. USA.* **92**: 3541–3545.

Davin, J.C., Malaise, M., Foidart, J.M., and Mahieu, P., 1987, Anti-$\alpha$-galactosyl antibodies and immune complexes in children with Henoch-Schonlein purpura of IgA nephropathy. *Kidney Int.* **31**: 1132–1139.

Editorial, 1965, New light on Chagas' disease. *Lancet.* **1**: 1150–1151.

Egge, H., Kordowicz, M., Peter-Katalinic, J., and Hanfland, P., 1985, Immunochemistry of I/i-active oligo- and polyglycosylceramides from rabbit erythrocyte membranes. *J. Biol. Chem.* **260**: 4927–4935.

Ferguson, M.A.J., Homans, S.W., Dwek, R.A., and Rademacher, T.W., 1988, Glycosylphosphatidylinositol moiety that anchors *Trypanosoma* brucei variant surface glycoprotein to the membrane. *Science* **239**: 753–759.

Ferrans, V.J., Milei, J., Tomita, Y., and Storino, R., 1988, Basement membrane thickening in cardiac myocytes and capillaries in chronic Chagas' disease. *Am. J. Cardiol.* **61**: 1137–1140.

Galili, U., Rachmilewitz, E.A., Peleg, A., and Flechner, I., 1984, A unique natural human IgG antibody with anti-$\alpha$-galactosyl specifity. *J. Exp. Med.* **160**: 1519–1531.

Galili, U., Basbaum, C.B., Shohet, S.B., Buehler, J., and Macher, B.A., 1987a, Identification of erythrocyte Gal$\alpha$1–3Gal glycosphingolipids with a mouse monoclonal antibody Gal-13. *J. Biol. Chem.* **262**: 4683–4688.

Galili, U., Clark, M.R., Shohet, S.B., Buehler, J., and Macher, B.A., 1987b, Evolutionary relationship between the anti-Gal antibody and the Gal$\alpha$1–3Gal epitope in primates. *Proc. Natl. Acad. Sci. USA* **84**: 1369–1373.

Gazzinelli, R.T., 1992, Natural anti-Gal antibodies prevent, rather than cause, autoimmunity in human Chagas' disease. *Res. Immunol.* (36th Forum in Immunology) **142**: 164–167.

Gazzinelli, G., and Brener, Z., 1988, Anti-laminin and specific antibodies in acute Chagas disease. *Trans. R. Soc. Trop. Med. Hyg.* **82**: 574–576.

Gazzinelli, R.T., Galvao, L.M.C., Cardoso, J.R., Cancado, J.R., Kreettli, A.U., Brener, Z., and Gazzinelli, G., 1988, Anti-*Trypanosoma cruzi* and antilaminin antibodies in chagasic patients after specific treatment. *J. Clin. Microbiol.* **26**: 1795–1800.

Gazzinelli, R.T., Pereira, M.E.S., Romanha, A., Gazzinelli, G., and Brener, Z., 1991a, Direct lysis of *Trypanosoma cruzi*: a novel effector mechanism of protection mediated by human anti-gal antibodies. *Parasite Immunol.* **13**: 345–356.

Gazzinelli, R.T., Romanha, A.J., Fontes, R., Chiari, E., Gazzinelli, G., and Brener, Z., 1991b, Distribution of carbohydrates recognized by the lectins *Euonymus europaeus* and Concanavalin A in monoxenic and heteroxenic trypanosomatids. *J. Protozool.* **38**: 320–325.

Geller, R.L., Rubinstein, P., and Platt, J.L., 1994, Variation in expression of porcine xenogeneic antigens. *Transplantation* **58**: 272- 274.

Grauert, M.R., Houdayer, M., and Hontebeyrie-Joskowciz, M., 1993, *Trypanosoma cruzi* infection enhances polyreactive antibody response in an acute case of human Chagas' disease. *Clin. Exp. Immunol.* **93**: 85–92.

Goin, J.C., Borda, E., Denduchis, B., Milei, J., Storino, R., and Sterin-Borda, L., 1990, Anticuerpos circulantes en pacientes chagásicos con actividad colinérgica que regulan la función cardíaca. *Medicina (Bs.As)* **50**: 385–386.

Kierszenbaum, F., 1986, Autoimmunity in Chagas' disease. The debate goes on. Reply. *Parasitol Today* 2: 22–23.

Koberle, F., 1968, Chagas' disease and Chagas' syndromes: The pathology of American trypanosomiasis. *Adv. Parasitol.* **6**: 63–116.

Krettli, A.U., and Brener, Z., 1976, Protective effects of specific antibodies in *Trypanosoma cruzi* infection. *J. Immunol.* **116**: 755–760.

Krettli, A.U., and Brener, Z., 1982, Resistance against *Trypanosoma cruzi* associated to anti-living trypomastigotes antibodies. *J. Immunol.* **128**: 2009–2012.

Krettli, A.U., Weisz-Carrington, P., and Nussenzweig, R.S., 1979, Membrane-bound antibodies to bloodstream *Trypanosoma cruzi* in mice: strain differences in susceptibility to complement-mediated lysis. *Clin. Exp. Immunol.* **37**: 416–423.

Laguens, R.P., Meckert, P.C., and Chambo, J., 1991, Origin and significance of anti-heart and anti-skeletal muscle autoanti-bodies in Chagas' disease. *Res. Immunol.* **142**: 160–163.

Lederkremer, R.M., Lima, C., Ramírez, M.I., and Casal, O.L., 1990, Structural features of the lipopeptidophosphoglycan from *Trypanosoma cruzi* common with the glycophosphatidylinositol anchors. *Eur. J. Biochem.* **192**: 337–345.

Lima-Martins, M.V.C., Sánchez, G., Krettli, A.U., and Brener, Z., 1985, Antibody-dependent cell cytotoxicity *against Trypanosoma cruzi* is only mediated by protective antibodies. *Parasite Immunol.* **7**: 367–370.

McConville, M.J., and Bacic, A., 1989, A family of glycoinositol phospholipids from *Leishmania major*. Isolation, characterization and antigenicity. *J. Biol. Chem.* **24**: 757–766.

McConville, M.J., Homans, S.W., Thomas-Oates, J.E., Dell, A., and Bacic, A., 1990, Structures of the glycoinositol phospholipids from *Leishmania major*. A family of novel galactofuranose-containing glycolipids. *J. Biol. Chem.* **265**: 7385–7394.

McConville, M.J., Collidge, T.A.C., Ferguson, M.A.J., and Schneider, P., 1993, The glycoinositol phospholipids of *Leishmania mexicana* promastigotes. Evidence for the presence of three distinct pathways of glycolipid biosynthesis. *J. Biol. Chem.* **268**: 15595–15604.

McConville, M.J., and Ferguson, M.A.J., 1993, The structure, biosynthesis and function of glycosylated phosphatidylinositols in the parasitic protozoa and higher eukaryotes. *Biochem. J.* **294**: 305–324.

Milani, S.R., and Travassos, L.R., 1988, Anti-α-galactosyl antibodies in chagasic patients. Possible biological significance. *Brazil. J. Med. Biol. Res.* **21**: 1275–1286.

Milani, S.R., Oliveira, T.G., and Travassos, L.R., 1988, Protective effect of lytic human antibodies binding to carbohydrate epitopes against trypomastigote infection in mice. *Memorias do Instituto Oswaldo Cruz* **82** (Suppl): 142.

Milei, J., and Storino, R., 1987, Laminin in chagasic auto-immunity? (Letter to Editor). *Parasitol Today* **3**: 119.

Milei, J., Storino, R., Fernández-Alonso, G., Beigelman, R., Vanzulli, S., and Ferrans, V.J., 1992, Endomyocardial biopsies in chronic chagasic cardiomyopathy. *Cardiology* **80**: 424–437.

Milei, J., Sánchez, J., Storino, R., Yu, Z.X., Denduchis, B., and Ferrans, V.J., 1993, Antibodies to laminin and immunohistochemical localization of laminin in chronic chagasic cardiomyopathy: a review. *Mol. Cell. Biochem.* **129**: 161–170.

Molina, H.A., Milei, J., and Storino, R., 1984, Chronic Chagas myocardiopathy. Demonstration of in vivo immunoglobulins in heart structures by the immunoperoxidase technique. *Cardiology* **71**: 297–306.

Oldstone, M.B.A., 1987, Molecular mimicry and autoimmune diseases. *Cell* **50**: 819–820.

PAHO, Division of Communicable Disease Prevention and Control, Communicable Disease Program, HPC/HCT, 1994, *Leishmaniasis in the Americas. Epidemiol Bull.* **15**: 8–11.

Palacios-Pru, E., Carrasco, H., Scorza, C., and Espinoza, R., 1989, Ultrastructural characteristics of different stages of human chagasic myocarditis. *Am. J. Trop. Med. Hyg.* **41**: 29–40.

Paulsson, M., Aumailley, M., Deutzmann, R., Timpl, R., Beck, K., and Engel, J., 1987, Laminin-nidogen complex. Extraction with chelating agents and structural characterization. *Eur. J. Biochem.* **166**: 11–19.

Pearson, R.D., and de Queiroz-Sousa, A., 1996, Clinical spectrum of leishmaniasis. *Clin. Infect. Dis.* **22**: 1–13.

Petry, K., and Eisen, H., 1989, Chagas's disease: a model for the study of autoimmune diseases. *Parasitol Today* **5**: 111–116.

Platt, J.L., 1996, The immunological barriers to xenotransplantation. *Crit. Rev. Immunol.* **16**: 331–358.

Platt, J.L., and Holzknecht, Z.E., 1994, Porcine platelet antigens recognized by human xenoreactive natural antibodies. *Transplantation* **57**: 327–329.

Ramasamy, R., and Reese, R.T., 1986, Terminal galactose residues and the antigenicity of *Plasmodium falciparum* glycoproteins. *Mol. Biochem. Parasitol.* **19**: 91–101.

Ramasamy, R., and Rajakaruna, R., 1997, Association of malaria with inactivation of α-1,3-galactosyl transferase in catarrhines. *Biochim. Biophys. Acta* **1360**: 241–246.

Salma, N., and Avila, J.L., 1991, Interaction between various strains of *T. cruzi* and cardiac primary cultures. *Acta Cient. Venez.* **42**: 185.

Sánchez, J.A., Milei, J., Yu, Z.X., Storino, R., Wenthold, R. Jr., and Ferrans, V.J., 1993, Immunohistochemical localization of laminin in the heart of patients with chronic chagasic cardiomyopathy: Relationship to thickening of basement membranes. *Amer. Heart J.* **126**: 1392–1401.

Schwimmbeck, P.L., Yu, D.T.Y., and Oldstone, M.B.A., 1987, Autoantibodies to HLA-B27 in the sera of HLA-27 patients with ankylosing spondylitis and Reiter's syndrome: Molecular mimicry with *Klebsiella pneumoniae* as potential mechanism of autoimmune disease. *J. Exp. Med.* **166**: 173–181.

Schwimmbeck, P.L., Dyrberg, T., Drachman, D., and Oldstone, M.B. A., 1989, Molecular mimicry and myasthenia gravis: Uncovering a novel site of the acetylcholine receptor that has biologic activity and reacts immunochemically with herpes simplex virus. *J. Clin. Invest.* **84**: 1174–1180.

Schmuñis, G.A., 1987, Autoimmunity in Chagas' disease. *Mem. Inst. Oswaldo Cruz* **82**: 288–310.

Souto-Padrón, T., Almeida, I.C., De Souza, W., and Travassos, L.R., 1994, Distribution of α-galactosyl-containing epitopes on *Trypanosoma cruzi* trypomastigote and amastigote forms from infected Vero cells detected by chagasic antibodies. *J. Euk. Microbiol.* **41**: 47–54.

Souza, M.C., Reis, A.P., Días da Silva, W., and Brener, Z., 1974, Mechanism of acquired immunity induced by *Leptomonas pessoai* against *Trypanosoma cruzi* in mice. *J. Protozool.* **21**: 579–584.

Springer, G.F., and Horton, R.F., 1969, Blood group isoantibody stimulation in man by feeding blood group-active bacteria. *J. Clin. Invest.* **48**: 1280–1291.

Stefani, M.M.A., Oliveira, E.B., Luquetti, A.O., and Días Da Silva, W., 1987, Complement independent lysis of *Trypanosoma cruzi* Y strain by sera from acute phase chagasic patients. C-reactive protein and protein involvement. *Memorias do Instituto Oswaldo Cruz* **82** (Suppl): 140.

Szarfman, A., Terranova, V.P., Rennard, S.I., Foidart, J.M., De Fatima Lima, M., Scheinman, J.I., and Martin, G.R., 1982, Antibodies to laminin in Chagas' disease. *J. Exp. Med.* **155**: 1161–1171.

Takehara, H.A., Perini, A., Da Silva, M.H., and Mota, I., 1981, *Trypanosoma cruzi*: role of different antibody classes in protection against infection in the mouse. *Exp. Parasitol.* **52**: 137–146.

Takeuchi, Y., Porter, C.D., Strahan, K.M., Preece, A.F., Gustafsson, K., Cosset, F-L., Weiss, R.A., and Colins, M.K.L., 1996, Sensitization of cells and retroviruses to human serum by (α-1–3)galactosyltransferase. *Nature* **379**: 85–88.

Thall, A.D., Maly, P., and Lowe, J.B., 1995, Oocyte gal-α-1,3gal epitopes implicated in sperm adhesion to the zona pellucida glycoprotein ZP3 are not required for fertilization in the mouse. *J. Biol. Chem.* **270**: 21437–21440.

Towbin, H., Rosenfelder, G., Wieslander, J.., Avila, J.L., Rojas, M., Szarfman, K., Esser, K., Nowack, H., and Timpl, R., 1987, Circulating antibodies to mouse laminin in Chagas disease, American cutaneous leishmaniasis, and normal individuals recognize terminal galactosyl(α-1–3)galactose epitopes. *J. Exp. Med.* **166**: 419–432.

Unterkircher, C., Avrameas, S., and Ternynck, T., 1993, Autoantibodies in the sera of *Trypanosoma cruzi*-infected individuals with or without clinical Chagas' disease. *J. Clin. Laborat. Anal.* **7**: 60–69.

Van der Rijn, I., Zabriskie, J.B., and McCarty, M., 1977, Group A streptococcal antigens cross-reactive with myocardium: purification of heart-reactive antibody and isolation and characterization of the streptococcal antigen. *J. Exp. Med.* **146**: 579–599.

Van Voorhis, W.C., Schlekewy, L., and Trong, H.L., 1991, Molecular mimicry by *Trypanosoma cruzi*: The Fl-160 epitope that mimics mammalian nerve can be mapped to a 12-amino acid peptide. *Proc. Natl. Acad. Sci.USA.* **88**: 5993–5997.

WHO, 1995, Twelfth Programme Report of the UNDP/World Bank/WHO *Special Programme for research and training in tropical diseases.* In WHO-Tropical Diseases Research, World Health Organization, Geneva, p. 125.

Winter, G., Fuchs, M., McConville, M.J., Stierhof, Y-D., and Overath, P., 1994, Surface antigens of *Leishmania mexicana* amastigotes: characterization of glycoinositol phospholipids and a macrophage-derived glycosphingolipid. *J. Cell. Sci.* **107**: 2471–2482.

Wolff, P.G., Kuhl, U., and Schultheiss, H-P., 1989, Laminin distribution and autoantibodies to laminin in dilated cardiomyopathy and myocarditis. *Am. Heart J.* **117**: 1303–1309.

Wood, C., Kabat, E.A., Murphy, L.A., and Goldstein, I.J., 1979, Inmunochemical studies on the combining sites of two isolectins $A_4$ and $B_4$ isolated from *Bandeiraea simplicifolia*. *Arch. Biochem. Biophys.* **198**: 1–8.

Zamze, S.E., Ashford, D.A., Wooten, E.W., Rademacher, T.W., and Dwek, R.A., 1991, Structural characterization of the asparagine- linked oligosaccharides from *Trypanosoma brucei* type II and III variant surface glycoproteins. *J. Biol. Chem.* **266**: 20244–20261.

*Chapter 9*

# Enterotoxin A of *Clostridium difficile* and α-Gal Epitopes

Charalabos Pothoulakis

## 1. INTRODUCTION

*Clostridium difficile*, a noninvasive Gram positive toxigenic bacterium, is the principal agent of antibiotic-associated diarrhea and colitis, one of the most important complications of antibiotic therapy (Kelly *et al.*, 1994). *C. difficile* diarrhea and colitis affects millions of patients in hospitals and nursing homes around the world and represents the most frequent form of infectious colitis in hospitalized patients. Thus, up to a quarter of hospitalized patients can be infected with this anaerobic pathogen and many nosocomial outbreaks of *C. difficile* infection have been described (McFarland *et al.*, 1989). One of the characteristic features of *C. difficile* colitis is the acute inflammatory infiltrate in the colonic mucosa which is associated with severe destruction of colonic epithelial cells (Kelly *et al.*, 1994). *C. difficile* infection is a toxin-mediated disease. Studies in animal models coupled with receptor binding experiments documented that *C. difficile*-toxin A induced diarrhea and intestinal inflammation is mediated by binding of *C. difficile* toxin A to a carbohydrate receptor containing α-gal epitopes.

---

**Charalabos Pothoulakis** Division of Gastroenterology, Beth Israel Deaconess Medical Center, Harvard Medical School, Boston, Massachusetts 02215. email: cpothoul@bidmc.harvard.edu

*Subcellular Biochemistry, Volume 32: α-Gal and Anti-Gal*, edited by Galili and Avila.
Kluwer Academic / Plenum Publishers, New York, 1999.

### 1.1. *Clostridium difficile* Toxins

*C. difficile* mediates its effects in animal and human intestine by release of two large potent protein exotoxins, toxin A and toxin B. Cloning and nucleotide sequencing of the toxin genes indicate that toxins A and B are encoded by two closely positioned chromosomal genes (Johnson *et al.*, 1990). Sequence analysis predicts molecular weights of 308 and 270kDa for toxin A and B respectively and amino acid analysis revealed a 44.8% amino acid homology and similar spacing of cystein residues, suggesting that their secondary structures are similar (Johnson *et al.*, 1990; Dove *et al.*, 1990). An interesting feature of the toxin genes is the presence of a series of tandem peptide repeats of 20–30 amino acids each located at the carboxyl terminal one-third of the toxins (von Eichel-Streiber, 1993). Interestingly, the C-terminal repeats of toxin A have striking sequence similarity with glucosyltransferase enzymes of *Streptococcus mutans* and other Streptococci strains (von Eichel-Streiber and Sauerborn, 1990). These parts of the toxin A and B molecules contain the receptor binding domain and mediate toxin binding to small intestinal and colonocyte receptors as well as on surface receptors in non-intestinal cells. These structural similarities probably account for the similar biologic activities of these toxins in cells in culture and human colonocytes (Pothoulakis, 1996).

### 1.2. Biologic Actions of *Clostridium difficile* Toxins

The biologic actions of *C. difficile* toxins are quite diverse, indicating the presence of functional toxin receptors in many different cell types. The main toxin effect on cultured cells is rounding caused by disorganization of actin filament formation (Pothoulakis *et al.*, 1986). Studies with cells in culture indicate that after receptor binding toxins A and B are internalized into the cytosol by endocytosis via coated pits (Fiorentini and Thelestam, 1991). The catalytic activity of toxins A and B in cells is identical. Once into the cytosol, these toxins catalyze the transfer of glucose from UDP-glucose to Threonine at position 37 of the small GTP binding proteins Rho, Rac and Cdc-42 (Just *et al.*, 1994; Just et al., 1995; Just *et al.*, 1995b). As a result, these small GTP binding proteins are unable to regulate actin polymerization, leading to actin filament breakdown and cell rounding (Just *et al.*, 1994; Dillon *et al.*, 1995). Disaggregation of filamentous actin also causes dysfunction of tight junctions in the colon as evidenced by studies in human T84 colonocyte monolayers, and human colonic mucosal strips (Hecht *et al.*, 1988; Riegler *et al.*, 1995). In addition, toxin A, but not toxin B, possess weak agglutinating activity against rabbit erythrocytes (Krivan *et al.*, 1986) and is also chemoattractant for neutrophils *in vitro* (Pothoulakis *et al.*, 1988). Furthermore, both toxins are lethal when injected parenterally into experimental animals and death appears to be caused by neuromuscular paralysis leading to apnea (Arnon *et al.*, 1984).

The major action of *C. difficile* occurs in the intestine where toxin A elicits intestinal fluid secretion and inflammation. Administration of toxin A into animal intestine elicits fluid secretion and mucosal damage characterized by a prominent neutrophil infiltration in the intestinal mucosa which is evident 1–4 hr after toxin A injection into closed intestinal loops (Triadafilopoulos *et al.*, 1989). Although the precise pathways leading to the enterotoxic effects of toxin A are not completely understood, it appears that binding of toxin A to its intestinal brush border receptor leads to a signal cascade which involves interactions between inflammatory mediators released from the enterocyte with sensory nerves and immune and inflammatory cells of the intestinal lamina propria, including macrophages and mast cells (Pothoulakis *et al.*, 1994; Castagliuolo *et al.*, 1994; Castagliuolo *et al.*, 1997) . This leads to the release of proinflammatory cytokines and neuropeptides and activation and recruitment of neutrophils (Pothoulakis *et al.*, 1988; Castagliuolo *et al.*, 1988). Finally, neutrophil-derived inflammatory cytokines, including prostaglandins and leukotrienes act on epithelial cells causing epithelial cell damage and necrosis (Triadafilopoulos *et al.*, 1989; Pothoulakis *et al.*, 1998). Toxin B is inactive in animal intestine, probably reflecting the absence of intestinal surface receptors for this toxin (Triadafilopoulos *et al.*, 1987; Lyerly *et al.*, 1985). Both toxins, however, cause epithelial cell damage and an increase in colonic permeability in native human colon, which is related to disaggregation of actin-containing filaments in the peri-juctional actinomycin ring (Hecht *et al.*, 1988; Riegler *et al.*, 1995). These results indicate that both toxins A and B possess receptors in the human intestine and both are important in the pathophysiology of human *C. difficile* infection.

## 2. TOXIN RECEPTORS

### 2.1. The Animal Toxin A Receptor Contains the α-Gal Epitope

Specific receptors for toxin A have been identified on rabbit, rat, hamster, and human enterocytes as well as in non-intestinal cell types such as rabbit erythrocytes, pancreatic acinar cells, neutrophils, and rat basophilic leukemia cells (reviewed in Pothoulakis, 1996). Studies from various laboratories indicate that the toxin A receptor is a glycoconjugate containing the α-gal epitope in the receptor binding domain. Using an enzyme-linked immunosorbent binding assay Krivan and colleagues were first to report that toxin A possess receptors on rabbit erythrocytes and hamster brush border membranes (Krivan *et al.*, 1986). Toxin A receptor binding was higher at 4° C and was dramatically inhibited by pretreatment of cell membranes and rabbit erythrocytes with α-galactosidase and the lectin *Bandereia simplicifolia* (BS-1) which binds to glycoconjugates bearing terminal α-gal epitopes (Krivan *et al.*, 1986). Toxin A-mediated cytotoxicity was also blocked by exposure of cells to the lectin BS-1 (Wikins and Tucker, 1989), indicating that this carbohydrate binding site is a functional receptor for toxin A. Interestingly, the toxin A binding domain itself, in the absence of holotoxin, causes agglutination when added to rabbit erythrocytes (Lyerly *et al.*, 1990). Since BS-1 also agglutinates rabbit erythrocytes by binding to α-gal epitopes, these results indicate that interaction of the toxin's binding domain with α-gal epitopes on its erythrocyte receptor can produce a lectin-like effect. Along the same lines bovine thyroglobulin, which contains large quantities of α-gal epitopes, has been successfully used for purification of toxin A (Krivan and Wilkins, 1987) and a BS-1 affinity column has been used for purification of the toxin A receptor on rabbit ileum (Pothoulakis *et al.*, 1996b).

One of the interesting lines of research on *Clostridium difficile* is the identification and characterization of the toxin A receptor in animal and human intestine. Early studies from our laboratory showed that the toxin A receptor in rabbit ileum is a G-protein coupled glycoprotein containing N-acetyl-glucosamine and α-gal epitopes on the toxin binding site (Pothoulakis *et al.*, 1991). Subsequent studies demonstrated that the brush border disaccharidase complex sucrase isomaltase is the toxin A receptor for toxin A in rabbit ileum. Toxin A bound directly to purified rabbit sucrase-isomaltase (Pothoulakis *et al.*, 1996b) with a similar affinity previously reported for toxin A binding to rabbit brush border membranes (Pothoulakis *et al.*, 1991). Moreover, pretreatment of rabbits with an antibody directed against purified sucrase-isomaltase inhibited intestinal secretion and inflammation in response to ileal toxin A administration (Pothoulakis *et al.*, 1996b). Thus, sucrase-isomaltase appears to be a biologically important receptor for toxin A in rabbit ileum. Interestingly, binding of toxin A to sucrase-isomaltase involves α-gal epitopes localized on carbohydrate domains of this glycoprotein since toxin A-sucrase-isomaltase binding is dramatically inhibited by pretreatment of purified sucrase-isomaltase with α-galactosidase (Pothoulakis *et al.*, 1996b). It should be noted, however, that sucrase-isomaltase is distributed only on small intestinal villus epithelium and not on colonocytes or other non-intestinal cells known to respond to toxin A. Thus, other intestinal or non-intestinal cell surface receptors containing α-gal epitopes, either in form of glycoproteins, or glycolipids, should mediate binding and biologic effects of this potent toxin in different target cells. Clark and colleagues showed that the toxin A receptor in rabbit erythrocytes is a glycolipid containing α-gal epitopes (Clark *et al.*, 1987), while Rolfe and Song reported that the toxin A brush border ileal receptor in infant hamsters is a 164 kDa glycoprotein containing N-acetyl-glucosamine and α-gal epitopes (Rolfe and Song, 1993). This hamster receptor is clearly different from the 120 kDa sucrase and 140 kDa isomaltase which represents the toxin A receptor in rabbit ileum.

## 2.2. The Human Toxin A Receptor

Although the toxin A receptor in animal intestine and other cells of animal origin has been characterized, the identity of the human toxin A receptor is not clear. Since human cells, including colonocytes, do not produce α-gal epitopes, toxin A receptor(s) on colonocytes should contain different carbohydrate structures. Tucker and Wilkins tested the ability of toxin A to bind to carbohydrate structures present in human colonic mucosa. Using various purified glycoproteins and carbohydrates they found that toxin A binds to the antigens I (Galβ1-4GlcNAc-R), Lewis X (Galβ1-4, [Fucα1-3] GlcNAc-R) and Lewis Y (Fuc a1-2 Galβ1-4[Fucα1-3]GlcNAc-R), suggesting that they may represent receptors for toxin A in human intestine (Tucker and Wilkins, 1991). Using purified human duodenal and colonic epithelial cells and radiolabelled toxin A Smith et al showed that toxin A specifically binds to a glycoprotein epithelial cell receptor (Smith *et al.*, 1997). Binding of toxin A to these cells was inhibited by β-galactosidase, indicating that β-linked galactose is part of the carbohydrate moiety of the toxin A receptor and further supporting the possibility that Lewis X, Y, and I antigens are involved in toxin A binding to human intestinal epithelial cells (Smith *et al.*, 1997). Other investigators examined the ability of toxin A, human anti-Gal, and the mouse monoclonal antibody Gal-13, which interacts with α-gal epitopes, to bind to various glycosphingolipids using thin layer chromatography (Tenemberg *et al.*, 1996). These studies showed that all three ligands bound, as expected, to glycosphingolipids from rabbit erythrocytes with α-gal epitopes (Tenemberg *et al.*, 1996). Toxin A, human anti-gal, and Gal-13 also reacted with the $x_2$ glycosphingolipid isolated from human erythrocytes with the structure GalNAcβ1-3Galβ1-4GlcNAc-R, but not with human Lewis X, Y, or I blood group determinants (Tenemberg *et al.*, 1996).

### 2.3. Functional Importance of α-Gal epitopes in *Clostridium difficile* Infection

Several pieces of evidence indicate that binding of toxin A to α-gal epitopes on its receptor binding site is important for expression of enterotoxic and cytotoxic effects. For example, mouse teratocarcinoma cells which contain high levels of α-gal epitopes are very sensitive to the cytotoxic effects of toxin A (Tucker *et al.*, 1990). Pretreatment of rat colonic loops with the lectin BS-1 also inhibits toxin A-mediated enterotoxicity (Pothoulakis *et al.*, 1996a). Silicon based inert supports chemically linked with the specific toxin A binding trisaccharide Galα1-3 Galβ1-4GlcNAc (Synsorbs) neutralized toxin A activity from toxin-containing solutions and absorbed toxin A from stool samples obtained from patients with *C. difficile*-mediated pseudomembranous colitis (Heerze *et al.*, 1994). The ability of Synsorbs linked to the toxin A binding trisaccharide to bind radiolabelled toxin A *in vitro* and reduce the enterotoxic effects of toxin A *in vivo* was also examined. Results from these studies indicate that Synsorbs attached to α-gal epitopes bound avidly and specifically to radiolabelled toxin A (Castagliuolo *et al.*, 1996). Furthermore, oral administration of this compound to rats dramatically reduced toxin A-mediated ileal fluid secretion, intestinal permeability as well as intestinal inflam-

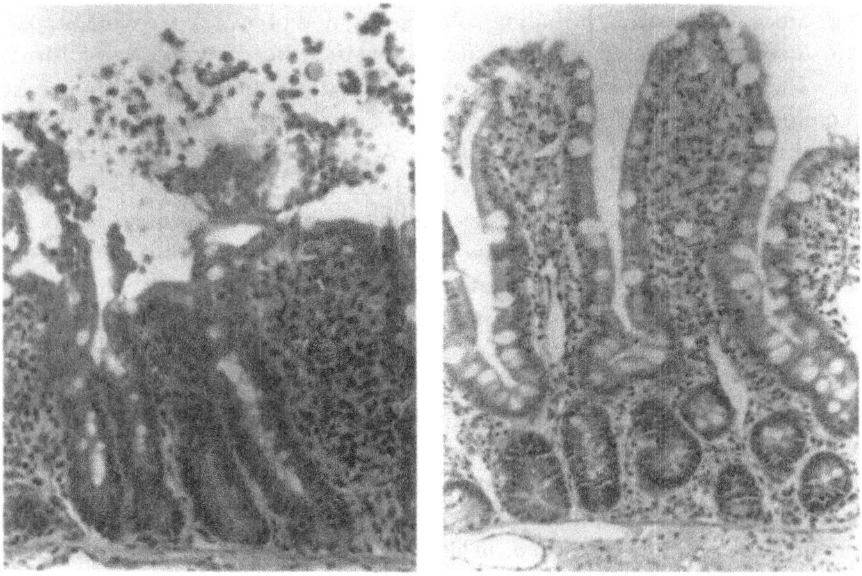

**Figure 1.** Silicon-based inert supports (Synsorbs) linked to the *C. difficile* toxin A binding trisaccharide Galα1-3Galβ1-4GlcNAc protects rats from toxin A-induced enteritis. Rats were pretreated with either a Synsorb linked to Galα1-3Galβ1-4GlcNAc (Synsorb 90) or a control Synsorb (Synsorb 83) before injection of toxin A into rat ileum. After 4 hr rats were killed and full thickness samples were fixed in formalin and sections were stained with H&E. Left panel: Ileal loop from a rat pretreated with the control Synsorb showing severe epithelial cell necrosis and loss of villus tips, marked congestion and edema and neutrophil infiltration of the lamina propria. Right panel: Ileal loop from a rat pretreated with Synsorb 90 showing a dramatic inhibition of toxin A-induced damage (From Pothoulakis *et al.* 1996. Gastroenterology., **111**:433., by permission).

mation and epithelial cell damage in response to toxin A (Castagliuolo *et al.*, 1996) (Fig. 1). These results demonstrate that a receptor decoy containing α-gal epitopes is able to sequester toxin A in the ileal lumen and inhibit its enterotoxic effects. This approach may be important in the management of *C. difficile* infection, and in particular in relapsing cases which are resistant to the conventional antibiotic therapy (Kelly *et al.*, 1994). Multicenter clinical trials with Synsorbs underway in the USA and Canada should give information on the potential therapeutic value of this compound in patients with *C. difficile* infection.

An interesting characteristic of *C. difficile* infection is that although up to 70% of newborn infants are colonized by this bacterium they do not develop diarrhea or colitis as adults do (Rolfe, 1988). Similarly, newborn rabbits have a diminished intestinal biologic response to *C. difficile* toxin A as compared to adults (Eglow *et al.*, 1992). Binding studies indicated that the appearance of specific toxin A receptor binding to rabbit ileal brush border is age-dependent with very little specific toxin A binding at age 2 and 5 days and gradually increased binding thereafter (Eglow *et al.*, 1992). Thus, the diminished intestinal effects of toxin A in newborn rabbits may be due to the absence of specific toxin receptors and may be related to the absence of intestinal disease in human infants infected with *C. difficile*. The reason(s) for this age-related reduction of toxin A binding sites in neonate intestine is not known. Previous studies have shown that microsomal galactosyltransferase activity in rat colon is age-dependent and increases gradually during neonatal life, reaching adult levels 2 weeks after birth (Rampal *et al.*, 1978). Thus, since the rabbit toxin A brush border receptor contains α-gal epitopes, the age-dependent appearance of toxin A binding sites may be related to the developmental regulation of microsomal galactosyltransferases in enterocytes. This age-dependent diminution in toxin A receptor binding appears to be species specific, since binding of toxin A to α-gal containing glycoprotein receptors on infant hamster brush border membranes is similar to brush border membranes obtained from adult hamsters (Rolfe, 1991).

## 2.4. Human Anti-Gal Mimics the Intestinal Effects of *C. difficile* Toxin A

It is well established that approximately 1% of human serum IgG is directed against glycoproteins or glycolipids containing α-gal epitopes (Galili *et al.*, 1984). While α-gal epitopes are expressed in mammals and New World monkeys, are absent in man and Old World primates (Galili *et al.*, 1988b) which lack α1,3-galactosyltransferase activity (Thall *et al.*, 1991). As discussed above, α-gal epitopes are also found in the rodent intestinal receptor for toxin A and appear to be important in the expression of enterotoxic and cytotoxic effects of this toxin. Dietary lectins are naturally occurring proteins and glycoproteins which bind to cell surface receptors containing specific carbohydrate sequences believe to serve as recognition markers. Interaction of lectins, such as wheat germ agglutinin (WGA) and concanavalin A (Con A), with specific carbohydrate-containing intestinal receptors can cause morphological as well as functional changes, including disarrangement of the actin cytoskeleton, shortening of the microvilli, and changes in permeability (Lorenzon and Olsen, 1982; Sjolander *et al.*, 1984; Sjolander *et al.*, 1986). Toxin A also functions like a lectin since binding of either the holotoxin or its binding domain to rabbit erythrocytes causes agglutination (Krivan *et al.*, 1986; Lyerly *et al.*, 1990), an effect also seen with the lectin BS-1 and human anti-Gal (Goldstein *et al.*, 1981; Avila *et al.*, 1989). Together, these findings support the hypothesis that binding of human anti-Gal to intestinal receptors containing α-gal epitopes may mimic some of the intestinal effects of toxin A. Indeed, injection of anti-Gal or toxin A to rat colonic loops *in vivo* stimulated fluid secretion, increased colonic permeability to mannitol, and increased colonic prostaglandin $E_2$ and rat mast cell protease II levels (Pothoulakis *et al.*, 1996b) (Fig. 2). Exposure of rat colon to the lectin BS-1, which also binds to α-gal epitopes, 15 min prior to toxin A

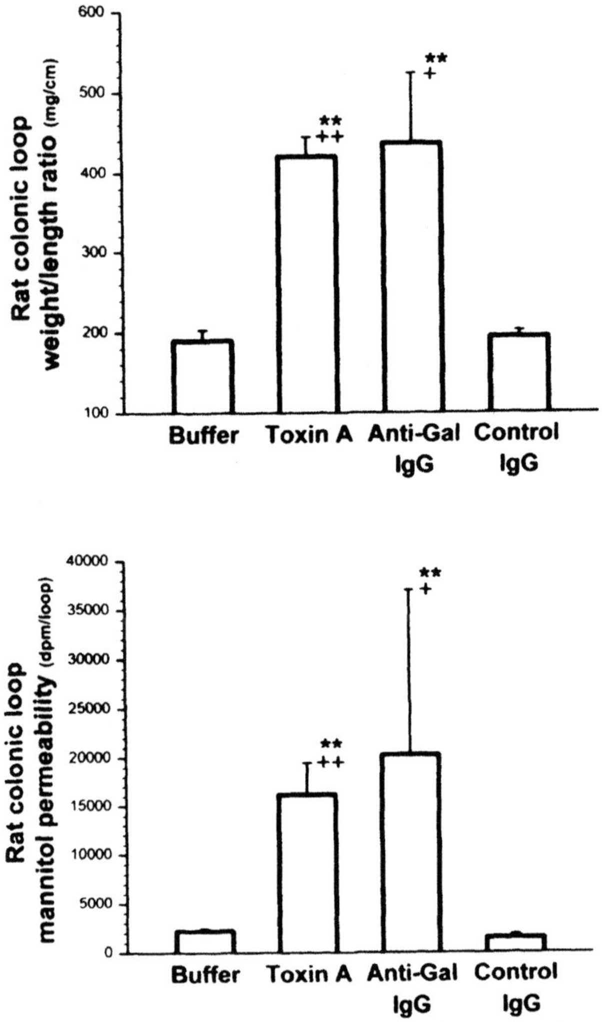


**Figure 2.** C. difficile toxin A and anti-Gal increase fluid secretion and mucosal permeability in rat colon. Rat colonic loops were administered with either buffer or buffer-containing purified toxin A (5 μg), or anti-Gal IgG (250 μg), or control human IgG (250 μg). Before injection, radiolabelled mannitol was injected into the inferior vena cava for intestinal permeability measurements. After 3 hr, animals were killed and fluid secretion (loop weight/length ratio) and mannitol permeability (scintillation counting of loop fluid samples) were measured. Both toxin A and anti-gal IgG caused a significant increase in colonic secretion (upper panel) and mucosal permeability to mannitol (From Pothoulakis et al., 1996. Gastroenterology., **110**:1704, by permission).

administration also inhibited toxin A-induced enterotoxicity (Pothoulakis *et al.*, 1996b). Exposure of rat colon to BS-1 itself for 3 hours also stimulated colonic secretion (Pothoulakis et al., 1996b). Binding studies revealed that both anti-Gal and toxin A bound specifically to rat colon brush border membranes and that pretreatment of brush border membranes with human anti-Gal reduced binding of toxin A to its colonic receptor (Pothoulakis *et al.*, 1996b). Preincubation of brush border membranes with toxin A or BS-1 inhibited anti-Gal receptor binding and anti-Gal itself bound specifically to partially purified rat colonic toxin A receptor (Pothoulakis *et al.*, 1996b). Taken together, these results suggest that toxin A, human anti-Gal, and BS-1 alter bowel function by binding to the same or different intestinal receptors containing α-gal epitopes.

Although human anti-Gal and BS-1 mimicked some of the intestinal effects of toxin A, none of these molecules cause colonic inflammation or epithelial cell destruction, a prominent characteristic of toxin A colitis. As discussed above, the mechanism of action of toxin A involves disorganization of the actin microfilament assembly due to inactivation of the rho family of proteins mediated by the catalytic part of the toxin A molecule localized on the N-terminal domain (Just *et al.*, 1995b). Although anti-Gal and BS-1 bind to receptors containing α-gal epitopes, their molecular structures are different than toxin A and they do not cause actin disorganization (Pothoulakis *et al.*, 1996b), an effect which appears to be important for expression of toxin A-mediated damage and inflammation. Another explanation for the similar, but not identical intestinal effects of toxin A and anti-Gal may be the different affinities by which these two molecules bind to their receptors. Scatchard plot analysis experiments indicated that, although anti-Gal and toxin A bound to a single class of rat colonic brush border receptors, the toxin A receptor binding affinity was much higher as compared to the affinity of anti-Gal for its brush border receptor (Pothoulakis *et al.*, 1996b).

## 3. CONCLUSIONS

The results showing that human anti-Gal alters colonic function in rodents by interaction with receptors containing α-gal epitopes open the possibility that a similar interaction may play a role in the pathophysiology of human diarrhea and inflammation, including inflammatory bowel disease (IBD). IBD is a complex tissue response to one or more environmental factors (microbial, dietary) in a genetically susceptible host (Fiocchi, 1998). In a preliminary report D'Alesandro et al found increased levels of circulating anti-Gal in the serum of patients with IBD as compared to normal individuals (D'Alessandro *et al.*, 1998). Increased circulating anti-Gal levels have been also identified in patients with certain autoimmune diseases such as Graves disease, Henoch-Schonlein purpura and IgA nepropathy (reviewed in Galili, 1989). Patients with Chagas disease, leismaniasis and encephalomyelitis subsequent to mycoplasma infection have also elevated anti-Gal levels (Towbin *et al.*, 1987; Kumada *et al.*, 1997). Although the mechanism of anti-Gal formation in man has not been elucidated, it may be the result of antigenic stimulus by antigens present on bacteria or protozoa, such as *E. coli*, Salmonella, and Klebsiella (Galili *et al.*, 1988a; Hamadeh *et al.*, 1996; Hamadeh *et al.*, 1992). The pathophysiologic importance of anti-Gal in IBD or possibly other forms of intestinal inflammation is not clear, since α-gal epitopes are absent in humans due to lack of α-1,3galactosyltransferase (Thall *et al.*, 1991). Although human anti-Gal does not react with normal human erythrocytes, it clearly reacts with erythrocytes from thalassemic and sickle cell anemia patients (Galili *et al.*, 1983; Galili *et al.*, 1986) as well as with human erythrocytes treated with proteases (Galili *et al.*, 1984). This raises the possibility that α-gal epitopes may be present on human cells in a cryptic form which can be uncovered by proteolytic enzymes that might become activated in certain disease states. Thus, during acute or chronic intestinal inflammation, protease enzymes released from phagocytes could digest intestinal cell surface glycoproteins, exposing receptors for anti-Gal. This antibody would bind to enterocytes and possibly trigger a receptor-mediated inflammatory response. In support of this scheme, human anti-Gal caused a striking increase in intestinal secretion and mucosa permeability equivalent to that seen with *C. difficile* toxin A.

## REFERENCES

Arnon, S. S., Mills, D. C., Day, P.A., Henrickson, R. V., Sullivan N. M., and Wilkins, T. D. 1984. Rapid death of infant rhesus monkeys injected with *Clostridium difficile* toxins A and B.: physiologic and pathophysiologic basis. J. Pediatr. **101**:34–40.

Avila, J. L., Rojas, M., and Galili, U. 1989. Galα1–3Gal carbohydrate epitopes are present on pathogenic American trypanosoma and leismania. J. Immunol. **142**:2828–2834.

Castagliuolo, I., Keates, A. C., Qiu, B., Kelly, C. P., Nikulasson, S., Leeman, S. E., and Pothoulakis, C. 1997. Substance P responses in dorsal root ganglia and intestinal macrophages during *Clostridium difficile* toxin A enteritis in rats. Proc. Natl. Acad. Sci. (USA). **94**:4788–4793.

Castagliuolo, I., LaMont, J. T., Letourneau, R., Kelly, C. P, O'Keane, J. C. Jaffer, A., Theoharides, T. C., and Pothoulakis, C. 1994. Neuronal involvement in the intestinal effects of *Clostridium difficile* toxin A and *Vibrio cholera* enterotoxin. Gastroenterology **107**:657–65.

Castagliuolo, I., LaMont, J. T, Qiu, B., and Pothoulakis, C. 1996. A receptor decoy inhibits the enterotoxic effects of *Clostridium difficile* toxin A in rat ileum. Gastroenterology. **111**:433–438.

Castagliuolo, I., Riegler, M., Nikulasson, S., Lu, B., Gerard, C., Gerard, N. P., and Pothoulakis, C. 1998. NK-1 receptor is required in *Clostridium difficile*-induced enteritis. J. Clin. Invest. **101**: 1547–1550.

Clark, G. F, Krivan, N. C., Wilkins, T. D., Smith, B. F. 1987. Toxin A from *Clostridium difficile* binds to rabbit erythrocyte glycolipids with terminal Galα1–3Galβ1–4GlcNAc sequences. Arch. Biochem. Biophys. **257**:217–229.

D'Alessandro, M., Mariani, P., Bachetoni, A., Lomanto, D., Mazzochi, P., and Speranza, V. 1998. Anti-Gal antibodies in patients with inflammatory bowel disease. Gastroenterology **114**:A958.

Dillon, S., Rubin, E., Yakubovich, M., Pothoulakis, C., LaMont, J. T., Feig, L. A., and Gilbert, R. J. 1995. Involvement of ras-related Rho proteins in the mechanism of action of *Clostridium difficile* toxin A and B. Infect. Immun. **63**:1421–26.

Dove, C. H., Wang, S. Z., Price, S. B., Phelps, C. J., Lyerly, D. M., Wilkins, T. D., and Johnson, J. L. 1990. Molecular characterization of the *Clostridium difficile* toxin A gene. Infect. Immun. **58**:480–488.

Eglow, R., Pothoulakis, C., Itzkowitz, S., Israel, E. J., O'Keane, C. J., Gong, D., Gao, N., Xu, Y. L., Walker, W. A., and LaMont, J. T. 1992. Diminished *Clostridium difficile* toxin A sensitivity in newborn rabbit is associated with decreased toxin A receptor. J. Clin. Invest. **90**:822–29.

Fiocchi, C. 1998. Inflammatory bowel disease: etiology and pathogenesis. Gastroenterology. **115**:182–205.

Fiorentini, C., and Thelestam, M. 1991. *Clostridium difficile* toxin A and its effects on cells. Toxicon. **29**:543–567.

Galili, U. 1989. Abnormal expression of α-galactosyl epitopes in am: a trigger for autoimmune processes? Lancet. **2**:358–360.

Galili, U., Clark, M. R., and Shohet. 1986. Excessive binding of the natural anti-α-galactosyl IgG to sickle red cells may contribute to extravascular cell destruction. J. Clin. Invest. **77**:27–33.

Galili, U., Korkesh, A., Kahane, I., and Rachmilewitz, E. A. 1983. Demonstration of a natural anti-galactosyl IgG antibody on thalassemic red blood cells. Blood. **61**:1258–1264.

Galili, U., Mandrell, R. E., Hamadeh, M., Shohet, S. B., and Griffis, M. 1988. Interaction between human natural anti-α-galactosyl immunoglobulin G and bacteria of the human flora. Infect. Immun. **56**:1730–1737.

Galili, U., Shohet, S. B., Cobrin, E., Stults, C. L. M., and Macher, B. A. 1988. Man, apes, and old world monkeys differ from other mammalians in the expression of α-galactrosyl epitopes on nucleated cells. J. Biol. Chem. **263**:17755–17762.

Galili, U., Rachmilewitz, E. A., Peleg, A., and Flechner, A. 1984. A unique natural human antibody with α-galactosyl specificity. J. Exp. Med. **160**:1519–1531.

Goldstein, I., Blake, D., Ebisu, S., Williams, T., and Murphy, L. 1981. Carbohydrate binding studies on the Bandeira simplicifolia I isolectins. J. Biol. Chem. **256**:3890–3897.

Hamadeh, R. M., Jarvis, G. A., Galili, U., Mandrell, R. E., Zhou, P., and Griffiss, J. M. 1992. Human natural anti-Gal IgG regulates alternative complement pathway activation on bacterial surfaces. J. Clin. Invest. **98**:1223–1235.

Hamadeh, R. M., Jarvis, G. A., Zhou, P., Cotleur, A. C., and Griffiss J. M. 1996. Bacterial enzymes can add galactose alpha 1,3 to human erythrocytes and creates a senescence-associated epitope. Infect. Immun. **64**:528–534.

Hecht, G., Pothoulakis, C., LaMont J. T., and Madara, J. L. 1988. *Clostridium difficile* toxin A perturbs cytoskeletal structure and junction permeability in cultured human epithelial cells. J. Clin. Invest. **82**:1516–1524.

Heerze, L. D., Kelm, M. A., Talbot, J. A., and Armstrong, G. D. 1994. Oligosaccharide sequences attached to an inert support (SYNSORB) as potential therapy for antibiotic-associated diarrhea and pseudomembranous colitis. J. Infect. Dis. **169**:1291–6.

Johnson, J. L., Phelps, C., Barroso, L., Roberts, M., Lyerly, D., and Wilkins, T. D. 1990. Cloning and expression of the toxin B gene of *Clostridium difficile*. Current Microbiology. 20:397–401.

Just, I., Fritz, G., Aktories, K., Giry, M., Poppoff, M. R., Boquet, P., Hegenbarth, S., and von Eichel-Streiber, C. 1994. *Clostridium difficile* toxin B acts on the GTP-binding protein Rho. J. Biol. Chem. **269**:10706–10712.

Just, I., Selzer, J., Wilm, M., von Eichel-Streiber, C., Mann, M., and Aktories, K. 1995. Glucosylation of *Rho* proteins by *Clostridium difficile* toxin B. Nature. **375**:500–503.

Just, I., Wilm, M., Selzer, J., Rex, G., von Eichel-Streiber, C., Mann, M., and Aktories, K. 1995. The enterotoxin from *Clostridium difficile* (ToxA) monoglucosylates the Rho proteins. J. Biol. Chem. **270**:13932–36.

Kelly, C. P., Pothoulakis, C., and LaMont, J. T. *Clostridium difficile* colitis. 1994. N. Engl. J. Med. **330**:257–262.

Krivan, H., Clark, C. F, Smith, D. F., and Wilkins, T. D. 1986. Cell surface binding site for *Clostridium difficile* enterotoxin: Evidence for a glycoconjugate containing the sequence Gal alpha 1–3Gal beta-4GlcNAc. Infect. Immun. **53**:573–581.

Krivan, H. C., and Wilkins, T. D. 1987. Purification of *Clostridium difficile* toxin A by affinity chromatography on immobilizes thyroglobulin. Infect. Immun. **55**:1873–77.

Lorenzon, V., and Olsen, W. A. 1982. *In vivo* responses of rat intestinal epithelium to intraluminal dietary lectins. Gastroenterology. 82:838–848.

Kumada, S., Kusaka, H., Okaniwa, M., Kobayashi, O., and Kusunoki, S. 1997. Encephalomyelitis subsequent to mycoplasma infection with elevated serum anti-Gal antibody. Pediatr. Neurol. **16**:241–244.

Lyerly, D. M., Johnson, J. L., Frey, S. M., and Wilkins, T. D. 1990. Vaccination against lethal *Clostridium difficile* enterocolitis with a nontoxic recombinant peptide of toxin A. Curr. Microbiol. **21**:29–32.

Lyerly, D. M., Saum, K. E. MacDonald, D., and Wilkins, T. D. 1985. Effect of toxins A and B given intragastrically to animals. Infect. Immun. **47**:349–352.

McFarland, L. V., Mulligan, M. E., Kwok, R. Y., and Stamm W. E. Nosocomial acquisition of *Clostridium difficile* infection. 1989. N. Engl. J. Med. **320**:204–210.

Pothoulakis, C. 1996. Pathogenesis of *Clostridium difficile*-associated diarrhoea. Eur. J. Gastroenterol. Hepatol. 8:1041–1047.

Pothoulakis, C., Barone, L. M., Ely, R., Faris, B., Clark, M. E., Franzblau, C., and LaMont JT. 1986. Purification and properties of *Clostridium difficile* cytotoxin B. J. Biol. Chem.**261**:1316–1321.

Pothoulakis, C., Castagliuolo, I., and LaMont, J. T.1998. Neurons and mast cells modulate secretory and inflammatory responses to enterotoxins. News Physiol. Sci. **13**:58–63.

Pothoulakis, C., Castagliuolo, I., LaMont, J. T., Jaffer, A., O'Keane, J. C., Snider, M., and Leeman, S. E. 1994. CP-96,345, a substance P antagonist, inhibits rat intestinal responses to toxin A but not cholera toxin. Proc. Natl. Acad. Sci. (USA). **91**:947–951.

Pothoulakis, C., Galili, U., Shen, C., Castagliuolo, I., Kelly, C. P., Nikulasson, S., Dudeja, P., Brasitus, T. A., and LaMont, J. T. 1996. Human anti-Gal binds to the same receptor and mimics the effects of *C. difficile* toxin A in rat colon. Gastroenterology **110**:1704–1712.

Pothoulakis, C., Gilbert, R. J, Cladaras, C., Castagliuolo, I., Semenza, G., Hitti, Y., Montcrief, J. S., Linevski, J., Kelly, C. P, Nikulasson, S., Desai, H. P., Wilkins, T. D., and LaMont, J. T. 1996. Rabbit sucrase-isomaltase contains a functional receptor for *Clostridium difficile* toxin A. J. Clin. Invest. **98**:641–649.

Pothoulakis, C., LaMont, J. T., Eglow, R., Gao, N., Rubins, J. B., Theoharides, T. C., and Dickey, B. F. 1991. Characterization of rabbit ileal receptors for *Clostridium difficile* toxin A. Evidence for a receptor-coupled G-protein. J. Clin. Invest. **88**:119–125.

Pothoulakis, C., Sullivan, R., Melnick, D., Triadafilopoulos, G., Gadenne, A. S., Meshulam, T., and LaMont, J. T. 1988. *Clostridium difficile* toxin A stimulates intracellular calcium release and chemotactic response in human granulocytes. J. Clin. Invest. **81**:1741–1745.

Rampal, P., LaMont, J. T., and Trier, J. C. 1978. Differentiation of glycoprotein synthesis in fetal rat colon. Am. J. Physiol. 235:E207-E212.

Riegler, M., Sedivy, R., Pothoulakis, C., Hamilton, G., Zacheri, J., Biscof, G., Cosetini, E., Feil, W., Schiessel, R., LaMont, J. T, and Wenzl, E. 1995. *Clostridium difficile* toxin B is more potent than toxin A in damaging human colonic epithelium *in vitro*. J. Clin. Invest. **95**:2004–2011.

Rolfe, R. D. 1988. Asymptomatic intestinal colonization by *Clostridium difficile*. *In*: Rolfe, R. D., and Finegold, S. M. eds. *Clostridium difficile*: its role in intestinal disease. Academic Press, New York, pp:201–25.

Rolfe, R. D. 1991. Binding kinetics of *Clostridium difficile* toxins A and B to intestinal brush border membranes from infant and adult hamsters. Infect. Immun. 59:1223–30.

Rolfe, D. F., and Song, W. 1993. Purification of a functional receptor for *Clostridium difficile* toxin A from intestinal brush border membranes of infant hamsters. Clin. Infect. Dis. **16**(Suppl 4):S219–27.

Smith, J. A., Cooke, D. L., Hyde, S., Borrielo, S. P., and Long, R. G. 1997. *Clostridium difficile* toxin A binding to human intestinal epithelial cells. J. Med. Microbiol. **46**:953–958.

Sjolander A., Magnusson, K-E., and Latcovic, S. 1984. The effect of concanavalin A and wheat germ agglutinin on the ultrastructure and permeability of the rat intestine. Int. Arch. Allergy Immunol. 75:230–236.

Sjolander, A., Magnusson, K. E., and Latkovic, S. 1986. Morphological changes of rat small intestine after short-time exposure to concanavalin A or wheat germ agglutinin. Cell Struct. Funct. **11**:285–293.

Teneberg, S., Lonnroth, I., Torres Loppez, J. F., Galili, U., Olwegard Halvarsson, M., Angsrom, J., and Karlsson K-A. 1996. Molecular mimicry of the recognition of glycosphingolipids by Galα3Galβ4GlcNAcβ-binding *Clostridium difficile* toxin A, human natural α-galactosyl IgG and the monoclonal antibody Gal-13: characterization of a binding-active human glycosphingolipid, non-identical with the animal receptor. Glycobiology. **6**:599–609.

Thall, A., Ettiene-Decerf, J., Winand, R., and Galili, U. 1991. The α-galactosyl epitope on mammalian thyroid cells. Acta Endocrinol. (Copenh). **124**:692–699.

Towbin, H., Rosenfelder, G., Wieslander, J., Avila, J. L., Rojas, M., Szarfman, A., Esser, K., Nowack, H., and Timpl, R. 1987. Circulating antibodies to mouse laminin in Chagas disease, American cutaneous leismaniasis, and normal individuals recognize terminal galactosyl(α1–3)-galactose epitopes. J. Exper. Med. **166**:419–432.

Triadafilopoulos, G., Pothoulakis, C., O'Brien, M., and LaMont, J. T: 1987. Differential effects of *Clostridium difficile* toxins A and B on rabbit ileum. Gastroenterology **93**:273–79.

Triadafilopoulos, G., Pothoulakis, C., Weiss, R., Giampaolo, C., and LaMont, J. T. 1989. Comparative study of *Clostridium difficile* toxin A and cholera toxin in rabbit ileum. Role of prostaglandins and leukotrienes. Gastroenterology **97**:1186–92.

Tucker, K. D., Carring, P. F., and Wilkins, T. D. 1990. Toxin A *of Clostridium difficile* is a potent cytotoxin. J. Clin. Microbiol. 28:869–871.

Tucker, K. D, and Wilkins, T. C. 1991. Toxin A of *Clostridium difficile* binds to the human carbohydrate antigens I, X, and Y. Infect. Immun. **59**:73–8.

von Eichel-Streiber, C. 1993. Molecular biology of *Clostridium difficile* toxins. *In*: Sebald, M. (ed). Genetics and Molecular Biology of Anaerobic Bacteria. Springer Verlag, New York, pp 264–289.

von Eichel-Streiber, C., and Sauerborn, M. 1990. *Clostridium difficile* toxin A carries a C-terminal repetitive structure homologous to the carbohydrate binding region of streptococcal glucosyltransferases. Gene. **96**:107–113.

Wilkins, T. D, and Tucker, K. D. 1989. *Clostridium difficile* toxin A (enterotoxin) uses Galα1–3Galβ1–4GlcNAc as a functional receptor. Microecol. Ther. **19**:225–227.

*Chapter 10*

# Anti-Gal, α-Gal Epitopes, and Xenotransplantation

Takaaki Kobayashi and David K. C. Cooper

## 1. INTRODUCTION

Significant progress in clinical transplantation has been achieved through the development of improved immunosuppressive therapy during the past two decades. Transplantation is clearly established as the treatment of choice for many patients suffering from end-stage organ failure. Although the number of organ donors becoming available has slowly increased, the relative shortage of organ donors caused by the extended indications for transplantation has become critical. Attention is now being focused on xenotransplantation—using animal organs, tissues or cells—as a potential and promising solution to this serious problem (Cooper, 1993a).

Clinical xenotransplantation, using kidneys, hearts and livers from nonhuman primates (chimpanzees, monkeys and baboons) was initially attempted in the 1960s. Successful long-term xenograft survival was never achieved (Auchincloss, 1988; Taniguchi and Cooper, 1996). Furthermore, the availability of chimpanzees

**Takaaki Kobayashi** Department of Surgery II, Nagoya University School of Medicine, Nagoya, Japan. **David K. C. Cooper** Transplantation Biology Research Center, Massachusetts General Hospital/Harvard Medical School, Boston, Massachusetts 02129.

*Subcellular Biochemistry, Volume 32: α-Gal and Anti-Gal*, edited by Galili and Avila.
Kluwer Academic / Plenum Publishers, New York, 1999.

and other nonhuman primates as potential organ donors for humans is inadequate. In many cases, suitability is also questionable because of factors such as slow breeding characteristics, inadequate size, the risk of transfer of infection, and ethical concerns. Attention has therefore been directed towards the study of xenograft rejection in more distantly related species combinations.

The concept that the susceptibility of xenografts to hyperacute rejection is dependent on the phylogenetic distance between donor and recipient was primarily suggested by the work of Perper and Najarian (1966a, 1966b). Their work suggested that the more distant the relationship between the two species, the more likely was hyperacute rejection, which occurred immediately (within minutes to hours) after reperfusion of the xenogeneic organ graft. Several *in vivo* models confirmed the critical role of preexisting (natural) antibody and complement in mediating hyperacute rejection. Preexisting antibody capable of mediating hyperacute rejection does not exist in allogeneic combinations unless blood group incompatibility or previous sensitization has occurred, while such antibody certainly exists in widely disparate (discordant) xenogeneic combinations regardless of previous exposure to xenogeneic antigens. More recent work has indicated that the susceptibility to hyperacute rejection is not primarily related to phylogenetic distance between donor and recipient but to the presence or absence of a single oligosaccharide, namely α-galactosyl (α-gal), on the vascular endothelial cells of the donor organ.

The pig is currently considered to be the most suitable donor of xenogeneic organs for humans. Advantages include short gestation period, large litter size, rapid growth, and similarities between pig and human organ size and physiology. Importantly, the pig is also amenable to modification by genetic engineering techniques. As approximately 90 million pigs are already killed annually as a food source in the US alone, the ethical considerations are significantly less than with nonhuman primates (Cooper et al., 1991; Ye et al., 1994a). However, since pig-to-human xenotransplantation is discordant, the major immunological hurdle of hyperacute rejection which is initiated by the binding of human preexisting anti-pig antibodies to pig tissues and subsequent activation of the complement system, has to be overcome.

## 2. DISCOVERY OF ANTI-GAL ANTIBODIES AND α-GAL EPITOPES

The existence of natural anti-Gal antibodies in humans was originally described by Galili et al. in 1984. They found that a significant percentage (approximately 1%) of circulating IgG in humans interacted with the carbohydrate epitope, Galα1-3Galβ1-4GlcNAc-R (α-gal) (Galili et al., 1985; Galili et al., 1993). Anti-Gal antibodies demonstrated a specific interaction with glycolipids containing a chemically-defined carbohydrate structure (Galili et al., 1985; Galili et al., 1987a) and with synthetic carbohydrates linked to proteins (Wieslander et al., 1990). An IgM class was also identified by Good et al. (1992) and an IgA class by Kujundzic et al. (1994).

Furthermore, Galili et al. showed that α-gal epitopes were present on red blood cells (1987b) and various nucleated cells (1988b) of New World monkeys and nonprimate mammals, but not in humans, apes and Old World monkeys. They have estimated that pig endothelial cells express approximately $10^7$ α-gal epitopes per cell (Galili et al. 1988b). As anti-Gal antibodies are detected only in humans and Old World nonhuman primates (Galili et al., 1987b), this indicated a unique reciprocal distribution pattern of anti-Gal antibodies and α-gal epitopes. By analyzing glycolipids of kidneys, α-gal epitopes were confirmed to be present only in New World monkeys and nonprimate mammals (Hendricks et al., 1990; Holgersson et al., 1990). At that time, Galili's major interest in this antigen-antibody system related to its potential role in autoimmune disorders in humans (Galili, 1991). Although the interaction that would take place if a pig organ was transplanted into a recipient with anti-Gal antibodies was clear to Galili and his colleagues, this fact was also determined independently by those interested in the use of the pig as an organ donor for humans.

## 3. SIGNIFICANCE OF THE α-GAL ANTIGEN-ANTIBODY SYSTEM IN XENOTRANSPLANTATION

### 3.1. Identification of Porcine Endothelial Epitopes Recognized by Human Anti-Pig Natural Antibodies

When renewed interest in xenotransplantation took place in the late 1980s, several groups began to examine the target epitopes (determinants) recognized by human anti-pig natural antibodies. Good et al. (1992) were the first to report that purified human natural anti-pig antibodies, which were obtained after the *ex vivo* perfusion of normal human plasma through pig hearts or kidneys, showed strong binding to synthetic oligosaccharides with an α-gal terminal residue. These included (i) αGal-R (α-galactose), (ii) αGal1-3βGal-R (B disaccharide), (iii) αGal1-3βGal1-4βGlcNAc-R (linear B type 2 trisaccharide), and (iv) αGal1-3βGal1-4βGlc-R (linear B type 6 trisaccharide) where R is $(CH_2)_8COOCH_3$ (Good et al., 1992; Cooper, 1992a; Cooper et al., 1993b; 1994).

### 3.2. Glycolipids

Samuelsson and his colleagues identified a terminal nonreducing αGal1-3βGal1-4GlcNAc epitope on glycolipids extracted from porcine aortic endothelial cells (Karlsson et al., 1993). Furthermore, they later analyzed the structure of the epitopes in pig organs by gas chromatography-mass spectrometry, and detected αGal1-3βGal1-4βGlcNAc1-3βGal1-4βGlc1-1 Ceramide, the kidney being the major source of this antigen (Samuelsson et al., 1994; Rydberg et al., 1994).

### 3.3. Glycoproteins

Xenoreactive antibody-binding epitopes were also identified in porcine glycoproteins of vascular endothelial cells, a kidney cell line (PK15), platelets, red blood cells, lymphocytes, spleen and heart. Western blots and/or immunoprecipitation techniques revealed that the molecular weights of these targets ranged from 26 to 250 kDA (Platt et al., 1990; Koren et al., 1992; Ratner et al., 1992; Tuso et al., 1993; Platt and Holzknecht, 1994; Satake et al., 1994; Thibaudeau et al., 1994; Cooke et al., 1995; Holzknecht et al., 1995; Vaughan et al., 1995; Aspeslet et al., 1996). Some of these demonstrated sensitivity to digestion by α-galactosidase, which removes terminal nonreducing α-gal epitopes, suggesting the participation of the α-gal carbohydrate structure.

### 3.4. Tissue Distribution and Pattern of Expression

The tissue distribution of α-gal epitopes in pigs was examined by immunohistochemistry using lectins, monoclonal antibodies and human natural antibodies. The α-gal epitope was found on all vascular endothelial cells, epithelial cells of renal proximal tubules, hepatocytes, lung epithelium, pancreatic ducts, and the deep layers of the epidermis (Oriol et al., 1993). Predigestion with α-galactosidase completely abolished reactivity with purified anti-Gal antibodies and with the *Griffonia simplicifolia* isolectin B4 (GS-I-B4), both of which specifically bind to αGal1-3Gal-R structures. A similar tissue distribution of α-gal epitopes was reported by McKenzie et al. (1994) and by Fournier et al. (1993). They reported that α-gal epitopes were also expressed on the apical pole of the epithelial cells of biliary ducts.

Oriol et al. (1994) also demonstrated that the precursor carbohydrate chain (βGal1-4βGlcNAc-R; N-acetyllactosamine) and NeuAc-substituted epitopes (αNeuAc 2-sialyl-N acetyllactosamine) were also found at the surface of pig vascular endothelium (Table I). However, it was considered unlikely that these two antigens are involved in hyperacute rejection because human vascular endothelial cells have the same lactosamine-ended precursor and sialylated chains as pigs. Instead of epitopes with a terminal α-gal, human vascular endothelial cells express the fucosylated polymorphic ABH histo-blood group epitopes (Table I) (Cooper et al., 1994; Oriol et al., 1994).

The existence of the α-gal epitope was considered to be monomorphic in the porcine species as it was found in every pig examined, involving 137 pigs belonging to 23 different breeds (Oriol et al., 1994). However, considerable variation in α-

**Table I**
**Structures of the Main Carbohydrate Epitopes Exposed at the Surface of Human and Porcine Vascular Endothelia[a]**

| Human | Pig |
|---|---|
| Galβ1-4GlcNAcβ1-R[b] | Galβ1-4GlcNAcβ1-R[b] |
| **ABH**-Galβ1-4GlcNAcβ1-R[c] | **Galα1**-3Galβ1-4GlcNAcβ1-R[d] |
| NeuAcα2-3Galβ1-4GlcNAcβ1-R[e] | NeuAcα2-3Galβ1-4GlcNAcβ1-R[e] |
| | NeuGcα2-3Galβ1-4GlcNAcβ1-R[f] |

Only the epitopes shown in bold type and underlined are different between the two species.
[a]Modified from Cooper, D.K.C., et al. (1994). R are glycolipid or glycoprotein carrier molecules anchored in the cell membrane.
[b]N acetyllactosamine.
[c]the A, B, H, or AB blood group antigen.
[d]the α-galactosyl antigen.
[e]N-acetylneuraminic acid.
[f]N-glycolylneuraminic acid (Bouhours, D., et al., 1997).

gal antigen expression in porcine organs was reported, suggesting a strategy to find and breed donor pigs exhibiting low expression of the antigen (Geller et al., 1994; Alvarado et al., 1995). To-date, this strategy has not been successful (Chae et al., 1998).

### 3.5. α1,3 Galactosyltransferase (α1,3GT)

The unique evolutionary distribution of α-gal epitopes in mammals is the result of different activities of the enzyme α1,3 galactosyltransferase (α1,3GT) which synthesizes the α-gal epitopes within the Golgi apparatus in the reaction: -

$$\beta\text{Gal1-4}\beta\text{GlcNAc-R+UDP-Gal} \rightarrow \alpha\text{Gal1-3}\beta\text{Gal1-4}\beta\text{GlcNAc-R+UDP}$$

where UDP-Gal is uridine diphosphate galactose (Galili, 1991).

This enzyme is active in New World monkeys and nonprimate mammals, but not in humans, apes and Old World monkeys (Galili et al., 1988b). Subsequently, the gene encoded for α1,3GT has been analyzed in the mouse (Larsen et al., 1989), cow (Joziasse et al., 1989), and pig (Sandrin et al., 1994). This gene is present in humans as a processed pseudogene (Joziasse et al., 1991) but is nonfunctional due to frameshift mutations (Larsen et al., 1990).

The significance of the α-gal epitopes to xenotransplantation was explored using the α1,3GT gene. Sandrin et al. demonstrated that the expression of α-gal epitopes on COS cells (a monkey kidney cell line) after transfection with a cDNA clone encoding α1,3GT enabled the cells to serve as targets for human natural anti-pig antibodies (Sandrin et al., 1993), resulting in cell death (Vaughan et al., 1994).

Furthermore, porcine α1,3GT cDNA was cloned and sequenced (Sandrin et al., 1994; Strahan et al., 1995), and the genomic DNA was analyzed (Katayama et al., 1998). Several splicing variants of porcine α1,3GT have been identified (Vanhove et al., 1997b), and a variant lacking the stem region has been found to be devoid of enzymatic activity (Katayama et al., 1998). This provides a possible new strategy for reducing α-gal antigen expression by the introduction of this genetic variant into pigs. The possibility of genetically engineering pigs that are deficient in the gene encoding for α1,3GT (α-gal-knockout pigs) has also been proposed (Cooper et al., 1993d; Sandrin et al., 1993) but the technology to do this is not available at present due to an inability to isolate porcine stem cells.

## 4. NATURE OF ANTI-GAL ANTIBODIES

### 4.1. Immunoglobulin Class

Koren and his colleagues and others determined that human xenoreactive (anti-Gal) antibodies belong to three immunoglobulin classes; IgG, IgM and IgA (Koren et al., 1992; 1993; Kujundzic et al., 1994; Schaapherder et al., 1994; Rieben et al., 1995; Goldberg et al., 1996; Kobayashi et al., 1996, 1997; Rydberg et al., 1996; Liu et al., 1997). With regard to IgG subclass, the predominant subclass is IgG2 (Ross et al., 1993; Schaapherder et al., 1994; Goldberg et al., 1996; Yu et al., 1996) or both IgG1 and IgG2 (Cooke et al., 1995; Rydberg et al., 1996: Watier et al., 1996), with a lesser amount of IgG3 (Schaapherder et al., 1994; Watier et al., 1996).

While antibodies of the IgG 2 subclass can only activate complement via the alternative pathway (Lucisano Valim et al., 1991), IgG1 and IgG3 are efficient in activating complement through the classical pathway (Bruggemann et al., 1987; Lucisano Valim et al., 1991). However, to cause cytotoxicity, complement activation via the classical pathway might require a much higher density of IgG1 or IgG3 antibody binding to the vascular endothelial cells when compared with IgM. Although these subclasses (IgG1, IgG2 and IgG3) were shown to recognize α-gal epitopes, treatment with α-galactosidase resulted in a greater reduction of binding of IgG2 to porcine endothelial cells than of IgG1, suggesting that a greater proportion of anti-pig IgG1 consists of non-Gal antibodies. This correlates with data that suggest that human IgG1 and IgG3 antibodies might recognize protein antigens on porcine endothelial cells, whereas IgG2 binds to carbohydrate antigenic determinants (Schaapherder et al., 1994).

Xenoreactive IgM antibody has been implicated as the predominant immunoglobulin involved both in cytotoxicity *in vitro* (Dalmasso et al, 1992; Sandrin et al., 1993; Tuso et al., 1993; Borche et al., 1994; Collins et al., 1994; Xu et al., 1994; Rydberg et al., 1995; Yu et al., 1996; Tachi et al., 1998) and in hyperacute rejection following pig organ transplantation into primates (Platt et al., 1991; Ross et al., 1993; Kaplon et al., 1995; Kroshus et al., 1996; Minanov et al., 1997). Although it has been suggested that anti-pig IgG does not play a role in hyperacute rejection and may even be protective (Yu et al., 1996), its relevance remains controversial. Anti-Gal IgG was cytotoxic to pig cells when rabbit complement was added *in vitro* (Koren et al., 1994; Kujundzic et al., 1994; Liu et al., 1997), but endogenous complement does not seem to be efficiently activated by IgG binding (Schaapherder et al., 1995). Anti-pig IgG antibodies almost certainly contribute to antibody-dependent cell-mediated cytotoxicity (ADCC) by NK cells via Fc receptors (Ross et al., 1993; Schaapherder et al., 1994; Watier et al., 1996) and are therefore a factor in the development of delayed xenograft rejection, which develops some days after transplantation if hyperacute rejection is avoided (see below). Anti-pig IgA may also be involved in rejection, as binding of dimeric IgA antibodies to porcine endothelial cells activates complement via the alternative pathway (Schaapherder et al., 1995).

It has been suggested that anti-Gal antibodies develop in neonates in response to the presence of certain gastrointestinal colonizing microorganisms, probably related to the body's defense mechanisms against infection (Galili et al., 1988a). α-Gal is known to be expressed on several microorganisms. Systemic or localized exposure to pig tissue leads to a marked increase in anti-Gal antibody levels (Galili et al., 1995; Galili et al., 1997; Kozlowski et al., 1998), but exposure to any structure bearing an α-gal terminal epitope, such as a monoclonal antibody or other product developed in mice or a polyclonal antibody raised in other α-gal-positive species, may have the same result. The intravenous infusion of cobra venom factor in baboons, which has been used for complement depletion to prevent hyperacute rejection, was demonstrated to promote the production of anti-Gal antibodies, presumably because cobra venom factor contains α-gal molecules (Taniguchi et al., 1996a).

## 4.2. Specificity, Avidity, and Heterogeneity

Xenoreactive natural IgM antibodies are polyreactive (Turman et al., 1991; Geller et al., 1993) but their binding, which shows high functional avidity (Kd: $10^{-8}$ M to $10^{-10}$ M), involves polyvalent and remarkably homogeneous interactions with antigens (Parker et al., 1994). Furthermore, some xenoreactive antibodies bind to porcine endothelial cells with a very low binding avidity, and play a pathogenetically important role *in vivo* (Vanhove and Bach, 1993).

McMorrow et al. have estimated that anti-Gal IgG antibody comprises 1 to 2.4% of total serum IgG, whereas anti-Gal IgM comprises 3.9% to 8.0% of total IgM (McMorrow et al., 1997b). In another report, xenoreactive IgM to porcine endothelial cells comprised 1% to 4% of total IgM (Parker et al., 1994). By using α-galactosidase digestion, it has been estimated that anti-Gal IgM accounts for approximately 70 to 90% of xenoreactive antibodies to porcine endothelial cells in humans (Collins et al., 1994; Parker et al., 1994; LaVecchio et al., 1995; Watier et al., 1996), while anti-Gal IgG accounts for about 50 to 80% (Cooke et al., 1995; LaVecchio et al., 1995; Watier et al., 1996). This suggests that the remaining antibodies are directed against epitopes that do not express a terminal α-gal residue. This would include possible neoantigens that are exposed after α-galactosidase digestion.

Anti-Gal antibodies exhibit a limited heterogeneity (polymorphism) (Good et al., 1992; Koren et al., 1992; 1996; Cooper et al., 1994; Goldberg et al., 1996; Parker et al., 1996a; McMorrow et al., 1997b) and it has been suggested that the majority are encoded by a limited number of conserved germline genes (McMorrow et al., 1997b) demonstrating limited clonal diversity (Parker et al., 1996a).

Considerable variability in the levels and cytotoxicity of xenoreactive or anti-Gal antibodies has been demonstrated (Kujundzic et al., 1994; Parker et al., 1994; Schaapherder et al., 1994; Cairns et al., 1995; Rydberg et al., 1995; McMorrow et al., 1997b; Tachi et al., 1998).

The relationship between the ABO histo-blood group antibodies and anti-Gal antibody is of interest. Higher levels of anti-Gal antibodies have been observed in humans with the blood groups A or O when compared to those with B or AB (McMorrow et al., 1997a). A similar observation has been made in baboons (Neethling and Cooper, 1998). This would appear to be related to the fact that anti-Gal antibodies in human non-B antigen-expressing individuals (A or O) include a subset that is reactive with the structurally-related B antigen; this subset is absent in B or AB subjects since the α-gal expressed on pig cells is similar to the human blood group B epitope. This observation supports the initial report on cross-reactivity between anti-Gal and anti-B antibodies by Galili et al. (1987a).

## 5. ATTEMPTS TO PREVENT HYPERACUTE REJECTION

### 5.1. *In Vitro* Studies

Various efforts have been made to inhibit the α-gal antigen-antibody reaction *in vitro*. α-Galactosidase has been used to digest the α-gal epitopes, and this confirmed α-gal to be the major determinant recognized by natural anti-pig antibodies (Oriol et al., 1993; Cairns et al., 1994; Collins et al., 1994; Parker et al., 1994; Satake et al., 1994; Cooke et al., 1995; Holzknecht et al., 1995). α-Galactosidase treatment reduces xenoreactive antibody binding and cytotoxicity to pig cells (LaVecchio et al., 1995; Watier et al., 1996). This enzyme has been shown to work reasonably well even at 4°C, which might allow its use during cold preservation of pig organs (Luo et al., 1997). However, its effectiveness is temporary, as 40% of the α-gal epitopes reappear within 48 hours (LaVecchio et al., 1995). The introduction of the gene for this enzyme into the donor animal or organ would appear to be a more promising alternative method of reducing α-gal expression (Cooper et al., 1996b; Sandrin et al., 1997).

Prevention of the α-gal antigen-antibody reaction has been achieved by the use of immunoaffinity columns of α-gal to adsorb the antibody from human serum (Good et al., 1992; Cooper, 1992a; Cooper et al., 1994). The cytotoxicity of human and baboon sera to pig cells is significantly decreased by the use of such immunoaffinity columns. The addition of an α-gal oligosaccharide to the serum also reduces or prevents cytotoxicity as the anti-Gal antibodies are inactivated by binding to the sugar and are no longer free to bind to the pig cells (Neethling et al., 1994). The specificity of α-gal to inhibit anti-Gal antibody cytotoxicity was clearly demonstrated by Neethling et al. (1994, 1996) (Table II). The inhibitory effect of synthetic α-gal oligosaccharides (disaccharide: Galα1-3Gal, trisaccharide: Galα1-3Galβ1-4GlcNAc (or Glc or Gal), tetrasaccharide: Galα1-3Galβ1-4Galα1-3Gal, and pentasaccharide: Galα1-3Galβ1-4GlcNAcβ1-3Galβ1-4Glc) on antibody binding and cytotoxicity to pig cells has been confirmed by numerous investigators (Cairns et al., 1995; Cooke et al., 1995; Rieben et al., 1995; Rydberg et al., 1995; Aspesiet et al., 1996; Galili and Matta, 1996; Goldberg et al., 1996; Neethling et al., 1996; Nagasaka et al., 1997). These α-gal oligosaccharides demonstrated at least 50% inhibition of antibody binding or cytotoxicity to pig cells at a concentration of 0.03 to 10 mM, although the results have varied between the reported studies, probably due to differences in the sera, targeted cells and synthetic

**Table II**
**Concentrations (in μM) of Various Oligosaccharides Needed to Obtain 50% Inhibition of Cytotoxicity of Unmodified Human or Baboon Serum on Pig Kidney (PK15) Cells**

| | Serum | |
|---|---|---|
| Oligosaccharide* | Human | Baboon |
| 1. ***Fucα1-2Galβ1-R** | >10000 | >10000 |
| 2. Gal**β1-R** | >10000 | >10000 |
| 3. Galα1-**2Gal1β1-R′** | 7000 | >10000 |
| 4. Galα1-3Gal | 386 (±149) | 301 (±44) |
| 5. Galα1-3Galβ1-4**Gal** | 163 (±73) | 141 (±60) |
| 6. Galα1-3Galβ1-4**Galα1-3Gal** | 54 (±31) | 119 (±30) |
| 7. Galα1-3Galβ1-4GlcNAc | 27 (±11) | 31 (±4) |

*Bold type indicates structural differences of the oligosaccharide with the major pig vascular endothelium glycolipid Galα1-3Galβ1-4GlcNAcβ1-3Galβ1-4Glcβ1-Cer.
R represents 1-3 or 1-4 linkages to Gal or to GlcNAc; R′ is $-O(CH_2)_3NHCOCF_3$. The results of the strong inhibitors (4-7) are expressed as mean ± SD (n=3).
(From Neethling et al., 1996.)

oligosaccharides used in these experiments. The inhibition resulting from use of the trisaccharide or pentasaccharide appears to be greater than that of the disaccharide (Cairns et al., 1995; Galili and Matta, 1996; Neethling et al., 1996; Nagasaka et al., 1997; Simon et al., 1998).

In some individuals, an α-gal disaccharide is effective in removal of all anti-Gal antibody, but in others a different or additional α-gal oligosaccharide (e.g. tri or pentasaccharide) is necessary for complete adsorption. Goldberg et al. (1996) provided evidence to suggest that most sera cannot be rendered anti-Gal antibody-free by using only one of the oligosaccharides, but Xu et al. (1998) demonstrated that the trisaccharide alone generally adsorbs the majority of antibody.

As synthetic α-gal oligosaccharides were initially difficult to obtain, purified oligosaccharides from pig stomach mucin were investigated as potential inhibitors of anti-Gal antibodies. Their potency in this respect was demonstrated both *in vitro* and following *in vivo* infusion in baboons (Li et al., 1995, 1996). Recombinant mucin-containing glycoproteins carrying the α-gal epitope, produced by cotransfection of COS cells with expression plasmids encoding a secreted mucin/immunoglobulin chimera and the porcine α1,3GT, were also shown to reduce binding of anti-Gal antibodies and serum cytotoxicity to pig cells (Liu et al., 1997). Synthetic peptides, which mimic the carbohydrate α-gal epitope, also block antibody binding to pig cells (Kooyman et al., 1996; Vaughan et al., 1996).

*In vitro* preincubation (>40 hours) of porcine cultured endothelial cells with commercially available human immunoglobulin preparations has been demonstrated to induce resistance in the porcine cells to complement-mediated injury. Both IgM (Dalmasso et al., 1996) and IgG (Dorling et al., 1996) have been implicated in bringing about this effect. Whether anti-Gal IgM and/or IgG are involved in this effect is unknown. The development of such resistance has not been demonstrated *in vivo*, and the mechanisms remain uncertain.

### 5.2. *Ex Vivo* and *in Vivo* Studies

Depletion of anti-pig (or anti-Gal) antibodies can be achieved by (i) pig organ hemoperfusion, (ii) plasma exchange, (iii) nonspecific immunoadsorption using immunoaffinity columns or anti-IgM (anti-μ) antibody, (iv) specific adsorption using immunoaffinity columns of synthetic α-gal oligosaccharides, or (v) intravenous infusion of synthetic α-gal or anti-idiotypic antibody.

Initial studies were directed towards blood perfusion of isolated pig organs or to plasmapheresis. Anti-pig antibodies were depleted by perfusion of the recipient's plasma or blood through pig livers or kidneys or by plasma exchange, which extended the survival of transplanted pig hearts and kidneys to 5 days and 22 days, respectively (Lexer et al. 1986; Cooper et al., 1988; Alexandre et al., 1989). These methods have subsequently been employed, often combined with immunosuppressive therapy, splenectomy or irradiation (total body, lymphoid or thymic), by several investigators (Platt et al., 1991; Fischel et al., 1992; Roslin et al., 1992; Latinne et al., 1993; Sablinski et al., 1995, 1997).

Immunoaffinity columns conjugated with polyclonal anti-human IgG and IgM have been used for adsorption of anti-pig antibodies (Leventhal et al., 1995; Kroshus et al., 1996; Pascher et al., 1997). Pig heart grafts have functioned <13 days after transplantation into baboons (Leventhal et al., 1995). Anti-human (and anti-baboon) IgM (anti-μ) monoclonal antibodies were developed and infused *in vivo* in nonhuman primates (Latinne et al., 1994; Sablinski et al., 1995) and lowered total IgM (including anti-pig IgM) levels to <1% of baseline level. However, following pig-to-baboon organ transplantation, in combination with irradiation, plasma exchange, organ perfusion, and splenectomy, no clear prolongation of graft survival was obtained (Latinne et al., 1993; 1994; Sablinski et al., 1997).

There are problems with all of the above methods. All immunoglobulins (or all IgM) are removed by nonspecific immunoadsorption; complement and coagulation factors may also be lowered, particularly by plasmapheresis. *Ex vivo* organ hemoperfusion may result in unwanted biological responses due to the antigen-antibody reaction and/or complement activation. A specific immunoadsorption of anti-Gal antibody would clearly be preferable, particularly from the perspective of clinical application.

Immunoadsorption columns consisting of synthetic α-gal oligosaccharides conjugated to a matrix, such as bovine serum albumin, have been developed and were initially investigated by *in vitro* testing (see above). These columns were demonstrated to significantly deplete anti-Gal antibodies from human sera. Initial *in vivo* studies using a specific α-gal disaccharide column were made by Taniguchi and his colleagues and others (Taniguchi et al., 1996b; Cooper et al., 1996a; Kozlowski et al., 1997, 1998). To completely deplete anti-Gal antibodies from baboons, a course of 3 or 4 extracorporeal immunoadsorptions may be necessary. Following such a course, combined with additional immunosuppressive therapy, pig heart transplantation resulted in graft survival of 5 days; return of anti-Gal antibodies resulted in delayed antibody-mediated rejection (Cooper et al., 1996a). To-date, it has proven difficult to prevent the steady return of antibody, which leads to the delayed rejection of the transplanted organ (Lambrigts et al., 1998).

As an alternative approach, the intravenous infusion of specific synthetic α-gal oligosaccharides has been attempted to "neutralize" the anti-Gal antibodies in the blood. Binding of the antibodies to the α-gal oligosaccharides prevents free antibody from binding to a transplanted pig organ. Such an approach has limitations. Oligosaccharides are quickly cleared from the circulation, the half-life being approximately one hour (Nagasaka et al., 1997; Simon et al., 1998). Continuous infusion is therefore essential to maintain an effective concentration of oligosaccharides. Large quantities of the synthetic oligosaccharides are required, which has provided problems with availability and high cost. Initial attempts at the intravenous infusion of large amounts of melibiose (αGal1-6Glc) and arabinogalactan (an impure polysaccharide of plant origin with an α-gal terminal residue) into baboons were by Ye et al. (1994b). Extremely high doses of these relatively non-specific oligosaccharides were required, and they proved only partially successful in inhibiting serum cytotoxicity, although survival of a transplanted pig heart was extended to between 12 and 18 hours in one case. Death of the baboon occurred from the toxicity associated with the high doses of carbohydrate infused.

When the specific α-gal pentasaccharide was infused into baboons for 4 hours, transplanted pig hearts functioned throughout this 4-hour period, but immunohistological study of the grafts showed that low-grade changes of antibody-mediated rejection were occurring during this time (Simon et al., 1998). Romano et al. (1998) have confirmed this observation in one case where the 4-hour infusion consisted of a combination of α-gal disaccharide and trisaccharide. Minimal signs of antibody-mediated rejection were seen on biopsy of the transplanted pig heart at the end of the infusion. Some hours later, the transplanted pig heart underwent antibody-mediated rejection. The affinity of anti-Gal antibodies to these synthetic oligosaccharides would appear to be insufficient to prevent some antibodies from binding to the transplanted pig organ, resulting in the gradual development of changes suggestive of antibody-mediated complement activation.

The addition of anti-idiotypic monoclonal antibodies, produced in mice against human anti-Gal antibodies, to human serum *in vitro* resulted in a dramatic inhibition of cytotoxicity to pig cells. Their intravenous administration into baboons resulted in a 50–80% decrease of serum cytotoxicity that persisted for 48 hours after cessation of the infusion (Koren et al., 1996). Evidence was also provided that these anti-idiotypic antibodies bound to α-gal-reactive B cells, offering the promise of new therapeutic modalities aimed at suppressing this B cell function (see below).

## 6. HUMORAL RESPONSE AFTER XENOTRANSPLANTATION

There is a well-documented potent humoral response to the α-gal epitopes after pig-to-primate transplantation (Satake et al., 1994; Rydberg et al., 1994; Galili et al., 1995; Cooper et al., 1996a; Davis et al., 1996; Taniguchi et al., 1996b; Kobayashi et al., 1997; Kozlowski et al., 1998) or *ex vivo* organ perfusion (Cotterell et al., 1995; Rydberg et al., 1996). Following pig islet cell transplantation into humans, the increase in anti-Gal antibodies ranged from 8- to 64-fold (Galili et al., 1995). Although an elevation in all three immunoglobulin classes (IgG, IgM, IgA) was observed, the highest activity was found within the IgG class. The placement of pig articular cartilage and meniscus tissue into cynomolgus monkeys caused a 30- to 300-fold increase in anti-Gal IgG within four weeks, whereas the increase in IgM was much less (2- to 16-fold) (Galili et al., 1997). The intravenous infusion of pig bone marrow cells into baboons (Kozlowski et al., 1998) and *ex vivo* perfusion of pig livers in humans (Cotterell et al., 1995) similarly resulted in major increases in anti-Gal IgG (200-fold within 7 to 21 days and 60-fold in 10 days, respectively) and a lesser increase in IgM (40-fold and 4-fold, respectively).

These results suggest that mere removal of anti-Gal antibodies from the recipient's serum prior to transplantation will be insufficient to prevent xenograft rejection (Galili et al., 1995). Suppression of anti-Gal antibody production is therefore a key issue. Although potent pharmacologic immunosuppressive therapy, splenectomy and/or irradiation reduce the rate of return of anti-Gal antibodies after antibody depletion, no therapy has to-date been successful in preventing the return of xenoreactive antibodies for more than a few days (Lambrigts et al., 1998). Attention has been directed towards more specific suppression of the subset of B cells that are involved in anti-Gal antibody production. The possibility of destroying this subgroup by an α-gal oligosaccharide or anti-idiotypic antibody conjugated to a toxin is being explored as are other B cell and plasma cell immunotoxins.

The successful establishment of B cell tolerance would inhibit the return of the anti-Gal antibody after depletion. Successful pig bone marrow transplantation in the recipient may be sufficient to "reset" the humoral immune system to an immature state in which the persistent presence of α-gal epitopes would maintain tolerance. As long as bone marrow chimerism is maintained, the recipient would, in theory, remain tolerant to α-gal epitopes (Cooper and Thall, 1997). This state of B cell tolerance has, in fact, been achieved by Sykes' group in α-gal-knockout mice (which produce anti-Gal antibodies) in whom α-gal-positive wild-type mouse bone marrow has been successfully engrafted (DeGoma et al., 1997). B cell tolerance to the α-gal epitope has also been achieved in this model by Iacomini's group who created a state of "molecular chimerism" in α-gal-knockout mice by introducing the gene for α1,3GT into mouse's bone marrow cells using gene therapy techniques (Iacomini, J.—personal communication). The presence of α-gal in the mouse's own cells led to suppression of anti-Gal production.

## 7. COMPARISON WITH ALLOTRANSPLANTATION ACROSS THE ABO HISTO-BLOOD GROUP BARRIER

The basic mechanism of rejection following discordant xenotransplantation has been compared with that following ABO-incompatible allotransplantation (Cooper et al., 1992b). Both have natural antibodies against carbohydrate epitopes (anti-Gal or anti-A/anti-B), which initiate the hyperacute rejection of a transplanted pig or ABO-incompatible organ, respectively. The only significant difference in the respective epitope targets is in the glycosylation of the common precursor, N-acetyllactosamine (Table I). In pigs, α-gal epitopes are synthesized by α1,3GT, while in humans the A or B epitopes are synthesized initially by α1,2 fucosyltransferase (to create the O(H) epitope) and subsequently by A transferase (α1,3 N-acetylgalactosaminyltransferase) or B transferase (α1,3 galactosyltransferase). This latter enzyme is quite different from the porcine α1,3GT in that N-acetyllactosamine cannot be an acceptor for human B transferase (Oriol et al., 1986, 1993; Sandrin et al., 1993; Cooper et al., 1994). With regard to avidity between antibody and antigen, the physiological properties of xenoreactive IgM are similar to those of the isohemagglutinins, anti-A and anti-B. These include high functional avidity in the range of $10^{-8}$ to $10^{-10}$ M and homogeneous interactions with antigens (Economidou et al., 1967a; Parker et al., 1994).

Under unmodified circumstances, the transplantation of an ABO-incompatible organ allograft is associated with an approximate 60% incidence of hyperacute (or at least antibody-mediated) rejection (Cooper, 1990). Both clinical and experimental studies of ABO-incompatible allotransplantation have demonstrated the phenomenon of "accommodation" in which the temporary removal of natural antibody at the time of the transplant results in long-term graft survival. The necessary period of antibody depletion may be no more than a few days or a week or two. Rejection does not occur despite the return of antibody and the presence of normal levels of complement (Alexandre et al., 1987; Palmer et al., 1989; Bach et al., 1991; Cooper et al., 1993c). This state has not yet been conclusively achieved after discordant xenotransplantation.

For example, ABO-incompatible renal transplantation has been successful when it is preceded by plasmapheresis or immunoadsorption to remove anti-A or anti-B antibodies (Alexandre et al., 1987; Bannett et al., 1987; Raja et al., 1987; Toma et al., 1994). Identical or similar treatment to remove anti-Gal antibodies has not led to successful pig-to-primate xenotransplantation, as outlined above. Similarly, whereas the intravenous infusion of α-gal oligosaccharides into baboons resulted only in a delay of rejection of transplanted pig hearts, the infusion of blood group A or B oligosaccharides allowed avoidance of hyperacute rejection and accommodation of ABO-incompatible cardiac allografts in baboons (Cooper et al., 1993c; 1993e).

This discrepancy in the development of accommodation may be ascribed to one or more of the following points: - (i) The species-specificity of complement function. A transplanted allograft is protected by species-specific complement regulatory proteins, whereas a transplanted pig organ has no such protection against human (or baboon) complement (Okada and Tanaka, 1983; Seya et al., 1990; Dalmasso et al., 1991; Rollins et al., 1991; Takizawa et al., 1992). (ii) The relative quantity of anti-donor antibodies in human sera. The level of anti-Gal antibody may be 4 to 5 times as high as that of anti-A or anti-B antibody (Galili, 1993). (iii) The concentration of antigen expression. In comparison with the $10^7$ α-gal epitopes expressed on each pig aortic endothelial cell (Galili et al., 1988b), there are only 0.8–1.2 × $10^6$ A1 or 0.6–0.8 × $10^6$ B epitopes per cell on red blood cells (Economidou et al., 1967b). (The number on human vascular endothelial cells does not appear to have been determined). (iv) Macrophage and NK cell activation by anti-Gal antibodies but not by anti-A/anti-B antibodies (as described below). (v) Molecular incompatibilities between human and pig (Siegel et al., 1997). For example, porcine von Willebrand factor, vasculature-associated tissue factor pathway inhibitor and thrombomodulin cannot regulate interaction with human platelet GPlb, factor Xa, and thrombin/protein C. Consequently, unregulated platelet aggregation and procoagulation could be initiated after pig-to-human organ transplantation.

## 8. INVOLVEMENT OF ANTI-GAL ANTIBODY IN DELAYED XENOGRAFT REJECTION

Even when hyperacute rejection is avoided by complement depletion or inhibition in the recipient, or by the use of organs from pigs transgenic for human complement regulatory proteins, pig xenografts are still typically rejected within days by a process that has been termed "delayed xenograft rejection" or "acute vascular rejection". This is characterized by mononuclear cell activation and graft infiltration (Brauer et al., 1993; Leventhal et al., 1993; Blakely et al., 1994; Candinas et al., 1996; Itescu et al., 1997; Kobayashi et al., 1997) and appears to be a complex mechanism which includes endothelial cell activation, cytokine production, platelet aggregation and coagulation (Bach et al., 1996; Parker et al., 1996b), but is nevertheless primarily related to antigen-antibody binding (Inverardi et al., 1994; Seebach et al., 1996; Galili et al., 1997).

Macrophages and NK cells, which are capable of efficient binding to the antibody-coated pig cells via their Fc receptors, almost certainly play a major role in graft destruction. Both types of cells can directly recognize α-gal or other carbohydrate epitopes through lectin-like receptors on their surfaces. This could contribute to cell lysis or endothelial activation (Bach et al., 1996; Itescu et al., 1996; Inverardi et al., 1997; Lin et al., 1997; Seebach and Waneck, 1997), suggesting the beneficial effects of strategies aimed at reduction of α-gal expression or inhibition of α-gal recognition by these cells.

Anti-Gal IgG molecules might also be involved in the development of chronic xenograft rejection by (i) ADCC, (ii) continuously activating endothelial cells, and (iii) increasing the immunogenicity of xenograft-specific antigens (which are different from the α-gal carbohydrate epitopes). This would be detrimental to long-term graft survival (Galili et al., 1997).

## 9. DONOR MODIFICATION BY GENETIC ENGINEERING

As techniques of molecular biology have advanced, attempts to overcome the problem of antigen-antibody interaction have been directed towards not only the treatment of the recipient by depleting antibodies (as described above), but also the modification of the donor animal by reducing or eliminating α-gal epitopes. Indeed, donor modification would be preferable as it would reduce or negate the need for recipient-directed therapy to prevent xenograft rejection. Such therapy puts the recipient at a serious risk of complications from infection, malignancy and drug-related problems. Genetic engineering has prompted the development of various strategies aimed at preventing rejection of the transplanted pig organ. These have included (i) expression of human complement regulatory proteins in the pig (Cozzi et al., 1997; Platt and Logan, 1997; Squinto and Fodor, 1997), (ii) modification of antigen expression, and (iii) inhibition or suppression of endothelial cell activation (Bach et al., 1997).

Modification of antigen expression is discussed fully elsewhere in this volume, so only brief mention of it will be made here. Attention has been directed toward removal or replacement of the α-gal epitopes (Cooper et al., 1993d; 1996b). Gene knockout technology, by which the gene for α1,3GT might be deleted (Thall et al., 1995), is not achievable in the pig at present due to the unavailability of porcine embryonic stem cells. However, recent progress in nuclear transfer technology (cloning) may make it possible to produce α-gal knockout pigs in the near future without the need for porcine embryonic stem cells. An alternative approach is "competitive glycosylation", a term coined by Sandrin et al. (1995). In this concept, first suggested by Cooper et al. (1993d), the α-gal molecules are replaced by another oligosaccharide by introduction of the gene for the enzyme for that alternative sugar. For example, the introduction of the gene for α1,2 fucosyltransferase results in expression of the O blood group sugar which competes successfully with α1,3GT for the lactosamine substrate (Koike et al., 1996; Sharma et al., 1996; Cohney et al., 1997). Although the inclusion of this gene alone does not replace 100% of the α-gal, the added insertion of the gene for α-galactosidase would appear to result in an almost complete absence of α-gal (Sandrin et al., 1997). Pigs transgenic for α1,2 fucosyltransferase have been documented, but no functional studies have been reported, suggesting that replacement of α-gal was incomplete (Koike et al., 1996; Sharma et al., 1996). Other genes that it has been suggested might be introduced include those for the enzymes β1,4 N-acetylglucosaminyltransferase (to produce N-acetylglucosamine (GlcNAc)) (Tanemura et al., 1997a) and α2,3 or α2,6 sialyltransferase (to produce N-acetylneuraminic acid (NeuAc)) (Tanemura et al., 1997b; Sharma et al., 1997).

Other approaches being explored include inducing the intracellular expression of single chain Fv (ScFv) antibodies against α1,3GT (Vanhove et al., 1997a).

## 10. NON-GAL ANTIBODIES

The observation that an organ from an α-gal-knockout mouse or from a mouse transgenic for α1,2 fucosyltransferase remains susceptible to antibody-mediated rejection (Tearle et al., 1996; Tange et al., 1997b) suggests the existence of recipient antibodies against epitopes other than α-gal (non-Gal antibodies). This problem may be less important when pigs are used as donors as they have a greater expression of α-gal than do mice. Some 10–30% of human xenoreactive antibodies are said to be directed against non-Gal epitopes in pigs, whereas this figure is 40% in mice (Tearle et al., 1996). However, the determinants against which non-Gal antibodies are directed in both mice and pigs remain largely uncertain.

Non-α-gal antigens against which humans are known to have antibodies have been discussed elsewhere (Cooper, 1998; Tange et al., 1997a) and are listed in Table III. They include sulfatide (Holgersson et al., 1992), Galα1,3 Lewis$^{x}$ (Bouhours et al., 1997), Hanganutziu-Deicher (HD) (Bouhours et al., 1994), Thomsen-Friedenreich (T or TF) (Cairns et al., 1996), Tn (TF precursor), I/i (Roelcke, 1989), and P$^{k}$ (Cairns, 1996). If the glycosylation pattern of the vascular endothelium of the pig organ is modified by replacement of a significant amount of α-gal, it is possible that antibody to any of these antigens could result in hyperacute or delayed xenograft rejection. The possibility that the human histo-blood group A antigen might be generated in the endothelial cells of transgenic pigs following the introduction of α1,2 fucosyltransferase has also been suggested (Tange et al.,

**Table III**
**Known Non-αGAL Carbohydrate Antigens against Which Humans Can Have Naturally Occurring Antibodies**[a]

1. A: GalNAcα1-3(Fucα1-2)Galβ1-4GlcNAcβ-R[b]
2. B: Galα1-3(Fucα1-2)Galβ1-4GlcNAcβ-R
3. Thomsen-Friedenreich (T or TF) Galβ1-3GalNAcα1-R
4. Tn (TF precursor) GalNAcα-R
5. Sialosyl-Tn: NeuAcα2-6GalNAcα1-R
6. p$^{k}$: Galα1-4Galβ1-4Glcβ1-R
7. Other P antigens
8. Sulfatide I: $SO_4$-3Gal-R
9. Forssman: GalNAcα1-3GalNAcβ1-3Galα1-4Galβ1-4Glcβ1-R
10. i: Galβ1-4GlcNAcβ1-3Galβ1-4GlcNAcβ-R[c]
11. I: Galβ1-4GlcNAcβ1-3(Galβ1-4Glcβ1-6)Galβ1-4GlcNAcβ-R[c]
12. αRhamnose-containing oligosaccharides
    L-Rhm-α-Rhm
    L-Rhm-α1-3GlcNAcβ1-2L-Rhm-α-R
13. βGlcNAc-containing oligosaccharides
    GlcNAcβ-R
    GlcNAcβ1-4GlcNAcβ-R

[a]Modified from data collected by Tange et al. (1997a), Shinkel et al. (1997), and Cao et al. (1996).
[b]R are glycolipid or glycoprotein carrier molecules anchored in the cell membrane.
[c]The core structures of the ABH antigen system which are fucosylated by H transferase to generate H substance.

1997a). The extent to which these non-Gal antibodies might be involved in the rejection of pig organs by primates has largely been unexplored.

## 11. COMMENT AND CONCLUSIONS

Despite the possible presence of antibodies in humans to non-α-gal determinants, current evidence strongly indicates that it is anti-Gal antibody that is the single most important barrier to successful xenotransplantation in humans. An effective method of depleting the α-gal-reactive B cells and plasma cells in the human host or, alternatively, of deleting expression of the α-Gal epitopes in the donor pig, is urgently needed—and is equally urgently being sought by a number of laboratories. If this problem can be overcome, those working in this field are hopeful that the remaining immunological barriers can be resolved.

However, although clinical xenotransplantation will provide substantial benefits to thousands of patients with end-stage organ failure, it may also create a risk of cross-species infection, particularly of viral transfer from the donor animal (Weiss, 1998). Several of the immunologic manipulations discussed above, whether of the recipient (e.g. depletion of anti-Gal antibody) or of the donor (e.g. modification of vascular endothelial glycosylation or expression of human complement regulatory proteins) may, in fact, increase this risk of infection. For example, the anti-Gal antibody-α-gal antigen interaction is important in human defense mechanisms against microorganisms. Some microbes actually express α-gal on their surface membranes, and may therefore be sensitive to lysis by human serum. Furthermore, the presence of α-gal on the pig host cell may also render an infective virus sensitive to lysis. The removal of this epitope from the surface of the host cell may result in resistance of the viral envelope to such lysis.

The concern that cross-species infection may not only affect the organ recipient but may also be passed from the patient to other members of the community, and therefore prove a risk to the public health, is receiving intensive investigation. Appropriate and reliable systems to monitor possible virus transmission are therefore essential, and need to be in place before xenotransplantation is undertaken in patients. Such monitoring is particularly important in view of the manipulation of the anti-Gal antibody-α-gal antigen response that will almost certainly be necessary if xenotransplantation is to be successful.

## REFERENCES

Alexandre, G. P., Squifflet, J. P., De, B. M., Latinne, D., Reding, R., Gianello, P., Carlier, M., and Pirson, Y., 1987, Present experiences in a series of 26 ABO-incompatible living donor renal allografts. *Transplant. Proc.* **19:** 4538–42.

Alexandre, G. P. J., Gianello, P., Latinne, D., Carlier, M., Dewaele, C. A., Van Obbergh, L., Moriau, M., Marbaix, E., Lambotte, J. L., and Squifflet, L. P., 1989, Plasmapheresis and splenectomy in experimental renal xenotransplantation. In: *Xenograft 25*, (M.A. Hardy, ed), Elsevier, New York, pp. 259–266.

Alvarado, C. G., Cotterell, A. H., McCurry, K. R., Collins, B. H., Magee, J. C., Berthold, J., Logan, J. S., and Platt, J. L., 1995, Variation in the level of xenoantigen expression in porcine organs. *Transplantation.* **59:** 1589–96.

Aspeslet, L. J., Chackowsky, P., Sekhon, H., Malcolm, A. J., Mosleh, Z., Koshal, A., and Yatscoff, R. W., 1996, Identification of porcine membrane antigens involved in the cytotoxic response mediated by human xenoreactive antibodies. *Xenotransplantation.* **3:** 1–10.

Auchincloss, H. J., 1988, Xenogeneic transplantation. A review. *Transplantation.* **46:** 1–20

Bach, F. H., Turman, M. A., Vercellotti, G. M., Platt, J. L., and Dalmasso, A. P., 1991, Accommodation: a working paradigm for progressing toward clinical discordant xenografting. *Transplant. Proc.* **23:** 205–7.

Bach, F. H., Winkler, H., Ferran, C., Hancock, W. W., and Robson, S. C., 1996, Delayed xenograft rejection. *Immunol. Today.* **17:** 379–84.

Bach, F. H., Ferran, C., Soares, M., Wrighton, C. J., Anrather, J., Winkler, H., Robson, S. C., and Hancock, W. W., 1997, Modification of vascular responses in xenotransplantation: inflammation and apoptosis. *Nature Med.* **3:** 944–948.

Bannett, A. D., McAlack, R. F., Raja, R., Baquero, A., and Morris, M., 1987, Experiences with known ABO-mismatched renal transplants. *Transplant. Proc.* **19:** 4543–6.

Blakely, M. L., Van, der, Werf, Wj, Berndt, M. C., Dalmasso, A. P., Bach, F. H., and Hancock, W. W., 1994, Activation of intragraft endothelial and mononuclear cells during discordant xenograft rejection. *Transplantation.* **58:** 1059–66.

Borche, L., Thibaudeau, K., Navenot, J. M., Soulillou, J. P., and Blanchard, D., 1994, Cytolytic effect of human anti-Gal IgM and complement on porcine endothelial cells: a kinetic analysis. *Xenotransplantation.* **1:** 125–131.

Bouhours, D. and Bouhours, J.-F., 1994, Hanganutziu-Deicher xenoantigens. *Xeno.* **2:** 33–34.

Bouhours, D., Liaigre, J., Naulet, J., Maume, D., and Bouhours, J. F., 1997, A novel glycosphingolipid expressed in pig kidney: Galα1-3Lewis(x) hexaglycosylceramide. *Glycoconjugate J.* **14:** 29–38.

Brauer, R. B., Baldwin, W. D., Daha, M. R., Pruitt, S. K., and Sanfilippo, F., 1993, Use of C6-deficient rats to evaluate the mechanism of hyperacute rejection of discordant cardiac xenografts. *J. Immunol.* **151:** 7240–8.

Bruggemann, M., Williams, G. T., Bindon, C. I., Clark, M. R., Walker, M. R., Jefferis, R., Waldmann, H., and Neuberger, M. S., 1987, Comparison of the effector functions of human immunoglobulins using a matched set of chimeric antibodies. *J. Exp. Med.* **166:** 1351–61.

Cairns, T., Hammelmann, W., Gray, D., Welsh, K., and Larson, G., 1994, Enzymatic removal from various tissues of the galactose α1,3-galactose target antigens of human antispecies antibodies. *Transplant. Proc.* **26:** 1279–80.

Cairns, T., Lee, J., Goldberg, L., Cook, T., Simpson, P., Spackman, D., Palmer, A., and Taube, D., 1995, Inhibition of the pig to human xenograft reaction, using soluble Galα1-3Gal and Galα1-3Galβ1-4GlcNAc. *Transplantation.* **60:** 1202–7.

Cairns, T., Lee, J., Goldberg, L. C., Hakim, N., Cook, T., Rydberg, L., Samuelsson, B., and Taube, D., 1996, Thomsen-Friedenreich and PK antigens in pig-to-human xenotransplantation. *Transplant. Proc.* **28:** 795–6.

Candinas, D., Belliveau, S., Koyamada, N., Miyatake, T., Hechenleitner, P., Mark, W., Bach, F. H., and Hancock, W. W., 1996, T cell independence of macrophage and natural killer cell infiltration, cytokine production, and endothelial activation during delayed xenograft rejection. *Transplantation.* **62:** 1920–7.

Cao, Y., Stosiek, P., Springer, G.F., and Karsten, U., 1996, Thomsen-Friedenreich-related carbohydrate antigens in normal adult human tissues: a systematic and comparative study. *Histochem. Cell Biol.* **106:** 197–207.

Chae, S., Kramer, A.D., Zhao, Y., Arn, J.S., Cooper, D.K.C., and Sachs, D.H. Lack of variation in αGal expression on lymphocytes in miniature swine of different genotypes. (Submitted for publication).

Cohney, S., McKenzie, I. F. C., Patton, K., Prenzoska, J., Ostenreid, K., Fodor, W. L., and Sandrin, M. S., 1997, Down-regulation of Galα(1,3)Gal expression by α1,2-fucosyltransferase: further characterization of α1,2-fucosyltransferase transgenic mice. *Transplantation.* **64:** 495–500.

Collins, B. H., Parker, W., and Platt, J. L., 1994, Characterization of porcine endothelial cell determinants recognized by human natural antibodies. *Xenotransplantation.* **1:** 36–46.

Cooke, S. P., Hederer, R. A., Pearson, J. D., and Savage, C. O., 1995, Characterization of human IgG-binding xenoantigens expressed by porcine aortic endothelial cells. *Transplantation.* **60:** 1274–84.

Cooper, D. K. C., Human, P. A., Lexer, G., Rose, A. G., Rees, J., Keraan, M., and Du Toit, E., 1988, Effects of cyclosporine and antibody adsorption on pig cardiac xenograft survival in the baboon. *J. Heart Transplant.* **7:** 238–246.

Cooper, D.K.C., 1990, A clinical survey of cardiac transplantation between ABO-blood group incompatible recipients and donors. *J. Heart Transplant.* **9:** 376–381.

Cooper, D. K. C., Ye, Y., Rolf, L. L. J., and Zuhdi, N., 1991, The pig as potential organ donor for man. In: *Xenotransplantation.* (D.K.C.Cooper, E. Kemp, K. Reemtsma, and D.J.G. White, eds), Springer-Verlag, Heidelberg, pp. 481–500.

Cooper, D. K. C., 1992a, Depletion of natural antibodies in non-human primates-a step towards successful discordant xenografting in humans. *Clin. Transplant.* **6:** 178–183.

Cooper, D. K. C., Ye, Y., Kehoe, M., Niekrasz, M., Rolf, L. J., Martin, M., Baker, J., Kosanke, S., Zuhdi, N., and Worsley, G., et al., 1992b, A novel approach to "neutralization" of preformed antibodies: cardiac allotransplantation across the ABO blood group barrier as a paradigm of discordant transplantation. *Transplant. Proc.* **24:** 566–71.

Cooper, D. K. C., 1993a, Xenografting: how great is the clinical need? *Xeno.* **1:** 25–26.

Cooper, D. K. C., Good, A. H., Koren, E., Oriol, R., Malcolm, A. J., Ippolito, R. M., Neethling, F. A., Ye, Y., Romano, E., and Zuhdi, N., 1993b, Identification of α-galactosyl and other carbohydrate epitopes that are bound by human anti-pig antibodies: relevance to discordant xenografting in man. *Transplant. Immunol.* **1:** 198–205.

Cooper, D. K. C., Good, A. H., Ye, Y., Koren, E., Oriol, R., Ippolito, R. M., Malcolm, A. J., Neethling, F. A., Romano, E., and Zuhdi, N., 1993c, Specific intravenous carbohydrate therapy: a new approach to the inhibition of antibody-mediated rejection following ABO-incompatible allografting and discordant xenografting. *Transplant. Proc.* **25:** 377–378.

Cooper, D. K. C., Koren, E., and Oriol, R., 1993d, Genetically engineered pigs. *Lancet.* **342:** 682–3.

Cooper, D. K. C., Ye, Y., Niekrasz, M., Kehoe, M., Martin, M., Neethling, F. A., Kosanke, S., DeBault, L. E., Worsley, G., and Zuhdi, N., et al., 1993e, Specific intravenous carbohydrate therapy. A new concept in inhibiting antibody-mediated rejection—experience with ABO-incompatible cardiac allografting in the baboon. *Transplantation.* **56:** 769–77.

Cooper, D. K. C., Koren, E., and Oriol, R., 1994, Oligosaccharides and discordant xenotransplantation. *Immunol. Rev.* **141:** 31–58.

Cooper, D. K. C., Cairns, T. D. H., and Taube, D. H., 1996a, Extracorporeal immunoadsorption of αGal antibodies. *Xeno.* **4:** 27–29.

Cooper, D. K. C., Koren, E., and Oriol, R., 1996b, Manipulation of the anti-αGal antibody-αGal epitope system in experimental discordant xenotransplantation. *Xenotransplantation.* **3:** 102–111.

Cooper, D. K. C. and Thall, A. D., 1997, Xenoantigens and xenoantibodies: their modification. *World J. Surg.* **21:** 901–906.

Cooper, D. K. C., 1998, Xenoantigens and xenoantibodies. *Xenotransplantation.* **5:** 6–17.

Cotterell, A. H., Collins, B. H., Parker, W., Harland, R. C., and Platt, J. L., 1995, The humoral immune response in humans following cross-perfusion of porcine organs. *Transplantation.* **60:** 861–8.

Cozzi, E., Yannoutsos, N., Langford, G. A., Pino-Chavez, G., Wallwork, J., and White, D. J. G., 1997, Effect of transgenic expression of human dacay-accelerating factor on the inhibition of hyperacute rejection of pig organs. In: *Xenotransplantation,* 2nd ed., (D.K.C. Cooper, E. Kemp, J.L. Platt, and D.J.G. White, eds), Springer, Heidelberg, pp. 665–682.

Dalmasso, A. P., Vercellotti, G. M., Platt, J. L., and Bach, F. H., 1991, Inhibition of complement-mediated endothelial cell cytotoxicity by decay-accelerating factor. Potential for prevention of xenograft hyperacute rejection. *Transplantation.* **52:** 530–3.

Dalmasso, A. P., Vercellotti, G. M., Fischel, R. J., Bolman, R. M., Bach, F. H., and Platt, J. L., 1992, Mechanism of complement activation in the hyperacute rejection of porcine organs transplanted into primate recipients. *Am. J. Pathol.* **140:** 1157–66.

Dalmasso, A. P., He, T., and Benson, B. A., 1996, Human IgM xenoreactive natural antibodies can induce resistance of porcine endothelial cells to complement-mediated injury. *Xenotransplantation.* **3:** 54–62.

Davis, E. A., Pruitt, S. K., Greene, P. S., Ibrahim, S., Lam, T. T., Levin, J. L., Baldwin, W. R., and Sanfilippo, F., 1996, Inhibition of complement, evoked antibody, and cellular response prevents rejection of pig-to-primate cardiac xenografts. *Transplantation.* **62:** 1018–23.

DeGoma, E., Sykes, M., Thall, A., and Yang, Y-G., 1997, Tolerization of α1,3Gal-specific B cells in lethally irradiated α1,3Gal$^{-/-}$ mice reconstituted with mixed α1,3Gal$^{+}$ and α1,3Gal$^{-}$ BMC. *Fourth International Congress for Xenotransplantation.* Nantes, France. (Abstract P107).

Dorling, A., Stocker, C., Tsao, T., Haskard, D. O., and Lechler, R. I., 1996. In vitro accommodation of immortalized porcine endothelial cells: resistance to complement mediated lysis and down-regulation of VCAM expression induced by low concentrations of polyclonal human IgG antipig antibodies. *Transplantation.* **62:** 1127–36.

Economidou, J., Hughes, J. N., and Gardner, B., 1967a, The functional activities of IgG and IgM anti-A and anti-B. *Immunol.* **13:** 227–34.

Economidou, J., Hughes, J. N., and Gardner, B., 1967b, Quantitative measurements concerning A and B antigen sites. *Vox. Sang.* **12:** 321–8.

Fischel, R. J., Matas, A. J., Platt, J. L., Perry, E., Noreen, H., Shumway, S. J., and Bolman, R. D., 1992, Cardiac xenografting in the pig-to-rhesus monkey model: manipulation of antiendothelial antibody prolongs survival. *J. Heart and Lung Transplant.* **11:** 965–73.

Fournier, A. M., Birchall, I. E., Kyriazis, A. G., Pearse, M. J., and d'Apice, A. J. F., 1993, A human naturally occurring antibody, anti-Gal, recognizes epitopes in pig kidney, heart, and liver and is cytotoxic to endothelial cells in the presence of rabbit complement. *Second International Congress for Xenotransplantation.* Cambridge, UK. (Abstract 113).

Galili, U., Rachmilewitz, E. A., Peleg, A., and Flechner, I., 1984, A unique natural human IgG antibody with anti-α-galactosyl specificity. *J. Exp. Med.* **160:** 1519–31.

Galili, U., Macher, B. A., Buehler, J., and Shohet, S. B., 1985, Human natural anti-α-galactosyl IgG. II. The specific recognition of α(1-3)-linked galactose residues. *J. Exp. Med.* **162:** 573–82.

Galili, U., Buehler, J., Shohet, S. B., and Macher, B. A., 1987a, The human natural anti-Gal IgG. III. The subtlety of immune tolerance in man as demonstrated by crossreactivity between natural anti-Gal and anti-B antibodies. *J. Exp. Med.* **165:** 693–704.

Galili, U., Clark, M. R., Shohet, S. B., Buehler, J., and Macher, B. A., 1987b, Evolutionary relationship between the natural anti-Gal antibody and the Galα1–3Gal epitope in primates. *Proc. Natl. Acad. Sci. USA.* **84:** 1369–73.

Galili, U., Mandrell, R. E., Hamadeh, R. M., Shohet, S. B., and Griffiss, J. M., 1988a, Interaction between human natural anti-α-galactosyl immunoglobulin G and bacteria of the human flora. *Infect. Immun.* **56:** 1730–7.

Galili, U., Shohet, S. B., Kobrin, E., Stults, C. L., and Macher, B. A., 1988b, Man, apes, and Old World monkeys differ from other mammals in the expression of α-galactosyl epitopes on nucleated cells. *J. Biol. Chem.* **263:** 17755–62.

Galili, U., 1991, The natural anti-Gal antibody: evolution and autoimmunity in man. *Immunology Series.* **55:** 355–73.

Galili, U., Anaraki, F., Thall, A., Hill, B. C., and Radic, M., 1993, One percent of human circulating B lymphocytes are capable of producing the natural anti-Gal antibody. *Blood.* **82:** 2485–93.

Galili, U., Tibell, A., Samuelsson, B., Rydberg, L., and Groth, C. G., 1995, Increased anti-Gal activity in diabetic patients transplanted with fetal porcine islet cell clusters. *Transplantation.* **59:** 1549–56.

Galili, U. and Matta, K. L., 1996, Inhibition of anti-Gal IgG binding to porcine endothelial cells by synthetic oligosaccharides. *Transplantation.* **62:** 256–62.

Galili, U., Minanov, O. P., Michler, R. E., and Stone, K. R., 1997, High-affinity anti-Gal immunoglobulin G in chronic rejection of xenografts. *Xenotransplantation.* **4:** 127–131.

Geller, R. L., Bach, F. H., Turman, M. A., Casali, P., and Platt, J. L., 1993, Evidence that polyreactive antibodies are deposited in rejected discordant xenografts. *Transplantation.* **55:** 168–72.

Geller, R. L., Rubinstein, P., and Platt, J. L., 1994, Variation in expression of porcine xenogeneic antigens. *Transplantation.* **58:** 272–7.

Goldberg, L. C., Lee, J., Cairns, T., Weymouth, W. A., Simpson, P., Lawson, C., Hacking, A., Nilsson, R., Hakim, N., and Taube, D., 1996, Polymorphism within the human anti-pig repertoire. *Transplant. Proc.* **28:** 549–50.

Good, A. H., Cooper, D. K. C., Malcolm, A. J., Ippolito, R. M., Koren, E., Neethling, F. A., Ye, Y., Zuhdi, N., and Lamontagne, L. R., 1992, Identification of carbohydrate structures that bind human antiporcine antibodies: implications for discordant xenografting in humans. *Transplant. Proc.* **24:** 559–62.

Hendricks, S. P., He, P., Stults, C. L., and Macher, B. A., 1990, Regulation of the expression of Galα1–3Galβ1–4GlcNAc glycosphingolipids in kidney. *J. Biol. Chem.* **265:** 17621–6.

Holgersson, J., Jovall, P. A., Samuelsson, B. E., and Breimer, M. E., 1990, Structural characterization of non-acid glycosphingolipids in kidneys of single blood group O and A pigs. *J. Biochem.* **108:** 766–77.

Holgersson, J., Cairns, T. D., Karlsson, E. C., Backer, A. E., Breimer, M. E., Taube, D. H., Welsh, K. I., and Samuelsson, B. E., 1992, Carbohydrate specificity of human immunoglobulin-M antibodies with pig lymphocytotoxic activity. *Transplant. Proc.* **24:** 605–8.

Holzknecht, Z. E. and Platt, J. L., 1995, Identification of porcine endothelial cell membrane antigens recognized by human xenoreactive natural antibodies. *J. Immunol.* **154:** 4565–75.

Inverardi, L. and Pardi, R., 1994, Early events in cell-mediated recognition of vascularized xenografts: cooperative interactions between selected lymphocyte subsets and natural antibodies. [Review] [56 refs]. *Immunol. Rev.* **141:** 71–93.

Inverardi, L., Clissi, B., Stolzer, A. L., Bender, J. R., Sandrin, M. S., and Pardi, R., 1997, Human natural killer lymphocytes directly recognize evolutionarily conserved oligosaccharide ligands expressed by xenogeneic tissues. *Transplantation.* **63:** 1318–30.

Itescu, S., Kwiatkowski, P., Wang, S. F., Blood, T., Minanov, O. P., Rose, S., and Michler, R. E., 1996, Circulating human mononuclear cells exhibit augmented lysis of pig endothelium after activation with interleukin 2. *Transplantation.* **62:** 1927–33.

Itescu, S., Minanov, O. P., and Michler, R. E., 1997, Newborn pig-to-baboon cardiac xenotransplantation: a model of delayed xenograft rejection. In: *Xenotransplantation,* 2nd ed.,

(D.K.C. Cooper, E. Kemp, J.L. Platt, and D.J.G. White, eds), Springer, Heidelberg, pp. 478–487.

Joziasse, D. H., Shaper, J. H., Van, den, Eijnden, Dh, Van, T. A., and Shaper, N. L., 1989, Bovine α1–3-galactosyltransferase: isolation and characterization of a cDNA clone. Identification of homologous sequences in human genomic DNA. *J. Biol. Chem.* **264:** 14290–7.

Joziasse, D. H., Shaper, J. H., Jabs, E. W., and Shaper, N. L., 1991, Characterization of an α1–3-galactosyltransferase homologue on human chromosome 12 that is organized as a processed pseudogene. *J. Biol. Chem.* **266:** 6991–8.

Kaplon, R. J., Michler, R. E., Xu, H., Kwiatkowski, P. A., Edwards, N. M., and Platt, J. L., 1995, Absence of hyperacute rejection in newborn pig-to-baboon cardiac xenografts. *Transplantation.* **59:** 1–6.

Karlsson, E., Cairns, T., Holgersson, J., Welsh, K., and Samuelsson, B., 1993, Pig to human xenotransplantation. Confirmation of the major target epitope of pre-formed human natural IgG and IgM anti-pig-teminal galactose 1,3 galactose. *Glycoconjugate J.* **10:** 297.

Katayama, A., Ogawa, H., Kadomatsu, K., Kurosawa, N., Kobayashi, T., Kaneda, N., Uchimura, K., Yokoyama, I., Muramatsu, T., and Takagi, H., 1998, Porcine α1,3-galactosyltransferase: full length cDNA cloning, genomic organization, and analysis of splicing variants. *Glycoconjugate J.* (In press).

Kobayashi, T., Neethling, F. A., Taniguchi, S., Ye, Y., Niekrasz, M., Koren, E., Hancock, W. W., Takagi, H., and Cooper, D. K. C., 1996, Investigation of the anti-complement agents, FUT-175 and K76COOH, in discordant xenotransplantation. *Xenotransplantation.* **3:** 237–245.

Kobayashi, T., Taniguchi, S., Neethling, F. A., Rose, A. G., Hancock, W. W., Ye, Y., Niekrasz, M., Kosanke, S., Wright, L. J., White, D. J. G., and Cooper, D. K. C., 1997, Delayed xenograft rejection of pig-to-baboon cardiac transplants after cobra venom factor therapy. *Transplantation.* **64:** 1255–1261.

Koike, C., Kannagi, R., Takuma, Y., Akutsu, F., Hayashi, S., Hiraiwa, N., Kadomatsu, K., Yamakawa, H., Nagai, T., Kobayashi, S., Okada, H., Nakashima, I., Uchida, K., Yokoyama, I., and Takagi, H., 1996, Introduction of (1,2)-fucosyltransferase and its effect on α-Gal epitopes in transgenic pig. *Xenotransplantation.* **3:** 81–86.

Kooyman, D. L., McClellan, S. B., Parker, W., Avissar, P. L., Velardo, M. A., Platt, J. L., and Logan, J. S., 1996, Identification and characterization of a galactosyl peptide mimetic. Implications for use in removing xenoreactive anti-A Gal antibodies. *Transplantation.* **61:** 851–5.

Koren, E., Neethling, F. A., Ye, Y., Niekrasz, M., Baker, J., Martin, M., Zuhdi, N., and Cooper, D. K. C., 1992, Heterogeneity of preformed human antipig xenogeneic antibodies. *Transplant. Proc.* **24:** 598–601.

Koren, E., Neethling, F. A., Richards, S., Koscec, M., Ye, Y., Zuhdi, N., and Cooper, D. K. C., 1993, Binding and specificity of major immunoglobulin classes of preformed human anti-pig heart antibodies. *Transplant. Int.* **6:** 351–3.

Koren, E., Kujundzic, M., Koscec, M., Neethling, F. A., Richards, S. V., Ye, Y., Zuhdi, N., and Cooper, D. K. C., 1994, Cytotoxic effects of human preformed anti-Gal IgG and complement on cultured pig cells. *Transplant. Proc.* **26:** 1336–9.

Koren, E., Milotic, F., Neethling, F. A., Koscec, M., Fei, D., Kobayashi, T., Taniguchi, S., and Cooper, D. K. C., 1996, Monoclonal antiidiotypic antibodies neutralize cytotoxic effects of anti-αGal antibodies. *Transplantation.* **62:** 837–43.

Kozlowski, T., Fuchimoto, Y., Monroy, R., Bailin, M., Martinez, R. R., Foley, A., Xu, Y., Awwad, M., Fishman, J., Andrews, D., Ritzenthaler, J., Sablinski, T., Ierino, F. L., and Sachs, D. H., 1997, Apheresis and column absorption for specific removal of Gal-α-1,3 Gal natural antibodies in a pig-to-baboon model. *Transplant. Proc.* **29:** 961–962.

Kozlowski, T., Monroy, R., Xu, Y., Glaser, R., Awwad, M., Cooper, D.K.C., and Sachs, D.H., 1998, Anti-αGal antibody response to porcine bone marrow in unmodified baboons and baboons conditioned for tolerance induction. *Transplantation.* (In press).

Kroshus, T. J., Bolman, R. R., and Dalmasso, A. P., 1996, Selective IgM depletion prolongs organ survival in an ex vivo model of pig-to-human xenotransplantation. *Transplantation.* **62:** 5–12.

Kujundzic, M., Koren, E., Neethling, F. A., Milotic, F., Koscec, M., Kujundzic, T., Martin, M., and Cooper, D. K. C., 1994, Variability of anti-αGal antibodies in human serum and their relation to serum cytotoxicity against pig cells. *Xenotransplantation.* **1:** 58–65.

Lambrigts, D., Van Calster, P., Xu, X., Awwad, M., Neethling, F.A., Kozlowski, T., Foley, A., Watts, A., Chae, S., Thall, A., White-Scharf, M., Sachs, D.H., and Cooper, D.K.C., 1998, Pharmacologic immunosuppressive therapy and extracorporeal immunoadsorption in the suppression of anti-αGal antibody in the baboon. (Submitted for publication).

Larsen, R. D., Rajan, V. P., Ruff, M. M., Kukowska, L. J., Cummings, R. D., and Lowe, J. B., 1989, Isolation of a cDNA encoding a murine UDP galactose:β-D-galactosyl-α-1,4-N-acetyl-D-glucosaminide α-1,3-galactosyltransferase: expression cloning by gene transfer. *Proc. Natl. Acad. Sci. USA.* **86:** 8227–31.

Larsen, R. D., Rivera, M. C., Ernst, L. K., Cummings, R. D., and Lowe, J. B., 1990, Frameshift and nonsense mutations in a human genomic sequence homologous to a murine UDP-Gal:β-D-Gal(1,4)-D-GlcNAcα(1,3)-galactosyltransferase cDNA. *J. Biol. Chem.* **265:** 7055–61.

Latinne, D., Gianello, P., Smith, C. V., Nickeleit, V., Kawai, T., Beadle, M., Haug, C., Sykes, M., Lebowitz, E., and Bazin, H., et al., 1993, Xenotransplantation from pig to cynomolgus monkey: approach toward tolerance induction. *Transplant. Proc.* **25:** 336–8.

Latinne, D., Soares, M., Havaux, X., Cormont, F., Lesnikoski, B., Bach, F. H., and Bazin, H., 1994, Depletion of IgM xenoreactive natural antibodies by injection of anti-mu monoclonal antibodies. *Immunol. Rev.* **141:** 95–125.

LaVecchio, J. A., Dunne, A. D., and Edge, A. S., 1995, Enzymatic removal of α-galactosyl epitopes from porcine endothelial cells diminishes the cytotoxic effect of natural antibodies. *Transplantation.* **60:** 841–7.

Leventhal, J. R., Dalmasso, A. P., Cromwell, J. W., Platt, J. L., Manivel, C. J., Bolman, R. D., and Matas, A. J., 1993, Prolongation of cardiac xenograft survival by depletion of complement. *Transplantation.* **55:** 857–65.

Leventhal, J. R., John, R., Fryer, J. P., Witson, J. C., Derlich, J. M., Remiszewski, J., Dalmasso, A. P., Matas, A. J., and Bolman, R. R., 1995, Removal of baboon and human antiporcine IgG and IgM natural antibodies by immunoadsorption. Results of in vitro and in vivo studies. *Transplantation.* **59:** 294–300.

Lexer, G., Cooper, D.K.C., Rose, A.G., Wicomb, W.N., Rees, J., Keraan, M., and du Toit, E., 1986, Hyperacute rejection in a discordant (pig-to-baboon) cardiac xenograft model. *J. Heart Transplant.* **5:** 411–8.

Li, S. F., Neethling, F. A., Taniguchi, S., Yeh, J. C., Kobayashi, T., Ye, Y., Koren, E., Cummings, R. D., and Cooper, D. K. C., 1996, Glycans derived from porcine stomach mucin are effective inhibitors of natural anti-α-galactosyl antibodies in vitro and after intravenous infusion in baboons. *Transplantation.* **62:** 1324–31.

Li, S., Yeh, J-C., Cooper, D.K.C., and Cummings, R.D., 1995, Inhibition of human anti-αGal IgG by oligosaccharides derived from porcine stomach mucin. *Xenotransplantation.* **2:** 279–278.

Lin, Y., Vandeputte, M., and Waer, M., 1997, Contribution of activated macrophages to the process of delayed xenograft rejection. *Transplantation.* **64:** 1677–1683.

Liu, J., Qian, Y., and Holgersson, J., 1997, Removal of xenoreactive human anti-pig antibodies by absorption on recombinant mucin-containing glycoproteins carrying the Gal α1,3Gal epitope. *Transplantation.* **63:** 1673–82.

Lucisano Valim, Y. M. and Lachmann, P. J., 1991, The effect of antibody isotype and antigenic epitope density on the complement-fixing activity of immune complexes: a systematic study using chemaeric anti-NIP antibodies with human Fc regions. *Clin. Exp. Immunol.* **84:** 1–8.

Luo, Y., Cummings, R., Cooper, D. K. C., Katz, E., and Zuhdi, N., 1997, Green coffee bean α-galactosidase: its working condition and potential application in xenotransplantation. *Fourth International Congress for Xenotransplantation.* Nantes, France. (Abstract P106).

McKenzie, I. F. C., Xing, P. X., Vaughan, H. A., Prenzoska, J., Dabkowski, P. L., and Sandrin, M. S., 1994, Distribution of the major xenoantigen (galα1-3gal) for pig to human xenografts. *Transplant. Immunol.* **2:** 81–6.

McMorrow, I. M., Comrack, C. A., Nazarey, P. P., Sachs, D. H., and DerSimonian, H., 1997a, Relationship between ABO blood group and levels of αGal,3galactose-reactive human immunoglobulin G. *Transplantation.* **64:** 546–549.

McMorrow, I. M., Comrack, C. A., Sachs, D. H., and DerSimonian, H., 1997b, Heterogeneity of human anti-pig natural antibodies cross-reactive with the αGal(1,3)Galactose epitope. *Transplantation.* **64:** 501–510.

Minanov, O. P., Itescu, S., Neethling, F. A., Morgenthau, A. S., Kwiatkowski, P., Cooper, D. K. C., and Michler, R. E., 1997, Anti-GaL IgG antibodies in sera of newborn humans and baboons and its significance in pig xenotransplantation. *Transplantation.* **63:** 182–6.

Nagasaka, T., Kobayashi, T., Muramatsu, H., Fujimoto, H., Matsuo, I., Ajisaka, K., Kadomatsu, K., Hayashi, S., Yokoyama, I., Hayakawa, A., Muramatsu, T., and Takagi, H., 1997, α-Galactosyl oligosaccharides conjugated with polyethylene glycol as potential inhibitors of hyperacute rejection upon xenotransplantation. *Biochem. Biophys. Res. Comm.* **232;** 731–6.

Neethling, F.A. and Cooper, D.K.C., 1998, A comparison of the variability and specificity of anti-αGal antibodies in baboons and humans. (Submitted for publication).

Neethling, F. A., Koren, E., Ye, Y., Richards, S. V., Kujundzic, M., Oriol, R., and Cooper, D. K. C., 1994, Protection of pig kidney (PK15) cells from the cytotoxic effect of anti-pig antibodies by α-galactosyl oligosaccharides. *Transplantation.* **57:** 959–63.

Neethling, F. A., Joziasse, D., Bovin, N., Cooper, D. K. C., and Oriol, R., 1996, The reducing end of αGal oligosaccharides contributes to their efficiency in blocking natural antibodies of human and baboon sera. *Transplant. Int.* **9:** 98–101.

Okada, H. and Tanaka, H., 1983 Species-specific inhibition by glycophorins of complement activation via the alternative pathway. *Molecular Immunol.* **20:** 1233–6.

Oriol, R., Le, P. J., and Mollicone, R., 1986, Genetics of ABO, H, Lewis, X and related antigens. *Vox. Sang.* **51:** 161–71.

Oriol, R., Ye, Y., Koren, E., and Cooper, D. K. C., 1993, Carbohydrate antigens of pig tissues reacting with human natural antibodies as potential targets for hyperacute vascular rejection in pig-to-man organ xenotransplantation. *Transplantation.* **56:** 1433–42.

Oriol, R., Barthod, F., Bergemer, A. M., Ye, Y., Koren, E., and Cooper, D. K. C., 1994, Monomorphic and polymorphic carbohydrate antigens on pig tissues: implications for organ xenotransplantation in the pig-to-human model. *Transplant. Int.* **7:** 405–13.

Palmer, A., Taube, D., Welsh, K., Bewick, M., Gjorstrup, P., and Thick, M., 1989, Removal of anti-HLA antibodies by extracorporeal immunoadsorption to enable renal transplantation. *Lancet.* **1:** 10–2.

Parker, W., Bruno, D., Holzknecht, Z. E., and Platt, J. L., 1994, Characterization and affinity isolation of xenoreactive human natural antibodies. *J. Immunol.* **153:** 3791–803.

Parker, W., Lateef, J., Everett, M. L., and Platt, J. L., 1996a, Specificity of xenoreactive anti-Galα1–3Gal IgM for α-galactosyl ligands. *Glycobiol.* **6:** 499–506.

Parker, W., Saadi, S., Lin, S. S., Holzknecht, Z. E., Bustos, M., and Platt, J. L., 1996b, Transplantation of discordant xenografts: a challenge revisited. [Review] [53 refs]. *Immunol. Today.* **17:** 373–8.

Pascher, A., Poehlein, C., Stangl, M., Hoebel, G., Thiery, J., Mueller, D. J., and Hammer, C., 1997, Application of immunoapheresis for delaying hyperacute rejection during isolated xenogeneic pig liver perfusion. *Transplantation.* **63:** 867–75.

Perper, R. J. and Najarian, J. S., 1966a, Experimetal renal heterotransplantaion I. in widely divergent species. *Transplantation.* **4:** 377–388.

Perper, R. J. and Najarian, J. S., 1966b, Experimental renal heterotransplantation II. closely related species. *Transplantation.* **4:** 700–712.

Platt, J. L., Lindman, B. J., Chen, H., Spitalnik, S. L., and Bach, F. H., 1990, Endothelial cell antigens recognized by xenoreactive human natural antibodies. *Transplantation.* **50:** 817–22.

Platt, J. L., Fischel, R. J., Matas, A. J., Reif, S. A., Bolman, R. M., and Bach, F. H., 1991, Immunopathology of hyperacute xenograft rejection in a swine-to-primate model. *Transplantation.* **52:** 214–20.

Platt, J. L. and Holzknecht, Z. E., 1994, Porcine platelet antigens recognized by human xenoreactive natural antibodies. *Transplantation.* **57:** 327–35.

Platt, J. L. and Logan, J. S., 1997, Use of transgenic animals as xenotransplant donors. In: *Xenotransplantation.* 2nd ed., (D.K.C. Cooper, E. Kemp, J.L. Platt, and D.J.G. White, eds), Springer, Heidelberg, pp. 650–658.

Raja, R., McAlack, R., Mendez, M., and Bannett, A., 1987, Technical aspects of antibody immunoadsorption prior to ABO-incompatible renal transplant. *Transplant. Proc.* **19:** 4525–7.

Ratner, A. J., Canhui, H., Pepino, P., Sanchez, J. A., Edwards, N., Watkins, J. F., Xu, H., Reemtsma, K., Rose, E. A., and Berger, C., 1992, Lymphocyte xenoantigens recognized by preformed antibodies. *Transplant. Proc.* **24:** 583–5.

Rieben, R., von Allmen, E., Korchagina, E. Y., Nydegger, U. E., Neethling, F. A., Kujundzic, M., Koren, E., Bovin, N. V., and Cooper, D. K. C., 1995, Detection, immunoabsorption, and inhibition of cytotoxic activity of anti-αGal antibodies using newly developed substances with synthetic Galα1-3Gal disaccharide epitopes. *Xenotransplantation.* **2:** 98–106.

Roelcke, D., 1989, Cold agglutination. *Transfusion Med. Rev.* **3:** 140–66.

Rollins, S. A., Zhao, J., Ninomiya, H., and Sims, P. J., 1991, Inhibition of homologous complement by CD59 is mediated by a species-selective recognition conferred through binding to C8 within C5b-8 or C9 within C5b-9. *J. Immunol.* **146:** 2345–51.

Romano, E., Neethling, F.A., Nilsson, K., Magnusson, S., Svensson, L., Samuelsson, B., and Cooper, D.K.C., 1998, Intravenous synthetic αGal saccharides delay hyperacute rejection of a pig-to-baboon heart transplant. (Submitted for publication).

Roslin, M. S., Zisbrod, Z., Burack, J. H., Tranbaugh, R. F., Strashun, A., Jacobowitz, I. J., Brewer, R. J., Kim, Y., Cunningham, J. N., and Norin, A. J., 1992, 15-day survival in pig-to-baboon heterotopic cardiac xenotransplantation. *Transplant. Proc.* **24:** 572–3.

Ross, J. R., Kirk, A. D., Ibrahim, S. E., Howell, D. N., Baldwin, W. D., and Sanfilippo, F. P., 1993, Characterization of human anti-porcine "natural antibodies" recovered from ex vivo perfused hearts—predominance of IgM and IgG2. *Transplantation.* **55:** 1144–50.

Rydberg, L., Cairns, T. D. H., Groth, C. G., Gustavsson, M. L., Karlsson, E. C., Moller, E., Satake, M., Tibell, A., and Samuelsson, B. E., 1994, Specificities of human IgM and IgG anticarbohydrate xenoantibodies found in the sera of diabetic patients grafted with fetal pig islets. *Xenotransplantation.* **1:** 69–79.

Rydberg, L., Hallberg, E., Bjorck, S., Magnusson, S., Strokan, V., Samuelsson, B. E., and Breimer, M. E., 1995, Studies on the removal of anti-pig xenoantibodies in the human by plasmapheresis/immunoadsorption. *Xenotransplantation.* **2:** 253–263.

Rydberg, L., Bjorck, S., Hallberg, E., Magnusson, S., Sumitran, S., Samuelsson, B. E., Strokan, V., Svalander, C. T., and Breimer, M. E., 1996, Extracorpreal ("ex vivo") connection of pig kidneys to humans. II. The anti-pig antibody response. *Xenotransplantation.* **3:** 340–353.

Sablinski, T., Latinne, D., Gianello, P., Bailin, M., Bergen, K., Colvin, R. B., Foley, A., Hong, H.-Z., Lorf, T., Meehan, S., Monroy, R., Powelson, J. A., Sykes, M., Tanaka, M., Cosimi, A. B., and Sachs, D. H., 1995, Xenotransplantation of pig kidneys to nonhuman primates: I. Development of the model. *Xenotransplantation.* **2:** 264–270.

Sablinski, T., Gianello, P. R., Bailin, M., Bergen, K. S., Emery, D. W., Fishman, J. A., Foley, A., Hatch, T., Hawley, R. J., Kozlowski, T., Lorf, T., Meehan, S., Monroy, R., Powelson, J. A., Colvin, R. B., Cosimi, A. B., and Sachs, D. H., 1997, Pig to monkey bone marrow and kidney xenotransplantation. *Surgery.* **121:** 381–91.

Samuelsson, B. E., Rydberg, L., Breimer, M. E., Backer, A., Gustavsson, M., Holgersson, J., Karlsson, E., Uyterwaal, A. C., Cairns, T., and Welsh, K., 1994, Natural antibodies and human xenotransplantation. *Immunol. Rev.* **141:** 151–68.

Sandrin, M. S., Vaughan, H. A., Dabkowski, P. L., and McKenzie, I. F. C., 1993, Anti-pig IgM antibodies in human serum react predominantly with Gal(α1-3)Gal epitopes. *Proc. Natl. Acad. Sci. USA.* **90:** 11391–5.

Sandrin, M. S., Dabkowski, P. L., Henning, M. M., Mouhtouris, E., and McKenzie, I. F. C., 1994, Characterization of cDNA clones for porcine (1,3)galactosyltransferase: The enzyme generating the Gal(1,3)Gal epitope. *Xenotransplantation.* **1:** 81–88.

Sandrin, M. S., Fodor, W. L., Mouhtouris, E., Osman, N., Cohney, S., Rollins, S. A., Guilmette, E. R., Setter, E., Squinto, S. P., and McKenzie, I. F. C., 1995, Enzymatic remodelling of the carbohydrate surface of a xenogenic cell substantially reduces human antibody binding and complement-mediated cytolysis. *Nature Med.* **1:** 1261–7.

Sandrin, M. S., Cohney, S., Osman, N., and McKenzie, I. F. C., 1997, Overcoming the anti-Gal(1–3)Gal reaction to avoid hyperacute rejection: molecular genetic approaches. In: *Xenotransplantation,* 2nd ed., (D.K.C. Cooper, E. Kemp, J.L. Platt, and D.J.G. White, eds), Springer, Heidelberg, pp. 683–700.

Satake, M., Kawagishi, N., Rydberg, L., Samuelsson, B. E., Tibell, A., Groth, C. G., and Moller, E., 1994, Limited specificity of xenoantibodies in diabetic patients transplanted with fetal porcine islet cell -linked galactose-containing epitopes. *Xenotransplantation.* **1:** 89–101.

Schaapherder, A. F., Daha, M. R., te, B. M., van, der, Woude, Fj, and Gooszen, H. G., 1994, Antibody-dependent cell-mediated cytotoxicity against porcine endothelium induced by a majority of human sera. *Transplantation.* **57:** 1376–82.

Schaapherder, A. F., Gooszen, H. G., te, B. M., and Daha, M. R., 1995, Human complement activation via the alternative pathway on porcine endothelium initiated by IgA antibodies. *Transplantation.* **60:** 287–91.

Seebach, J. D., Yamada, K., McMorrow, I. M., Sachs, D. H., and DerSimonian, H. D., 1996, Xenogeneic human anti-pig cytotoxicity mediated by activated natural killer cells. *Xenotransplantation.* **3:** 188–197.

Seebach, J. D. and Waneck, G. L., 1997, Natural killer cells in xenotransplantation. *Xenotransplantation.* **4:** 201–211.

Seya, T., Okada, M., Hazeki, K., and Nagasawa, S., 1990, Regulatory system of guinea-pig complement C3b: two factor I-cofactor proteins on guinea-pig peritoneal granulocytes. *Biochem. Biophys. Res. Com.* **170:** 504–12.

Sharma, A., Okabe, J., Birch, P., McClellan, S. B., Martin, M. J., Platt, J. L., and Logan, J. S., 1996, Reduction in the level of Gal(α1,3)Gal in transgenic mice and pigs by the expression of an α(1,2)fucosyltransferase. *Proc. Natl. Acad. Sci. USA.* **93:** 7190–5.

Sharma, A., Okabe, J., Birch, P., McClellan, S., Martin, M., Platt, J. L., and Logan, J. S., 1997, Downregulation of the Gal epitope in transgenic mice and pigs by expression of 1,2 fucosyltransferase or 2,6 sialyltransferase. *Fourth International Congress for Xenotransplantation.* Nantes, France. (Abstract O29).

Shinkel, T.A., Chen, C-J., Salvaris, E., Henion, T.R., Barlow, H., Galili, U., Pearse, M.J., and d'Apice, A.J.F., 1997, Changes in cell surface glycosylation in α1,3-galactosyltransferase knock-out and α1,2-fucosyltransferase transgenic mice. *Transplantation.* **64:** 197–204.

Siegel, J. B., Grey, S. T., Lesnikoski, B.-A., Kopp, C. W., Soares, M., am Esch, J. S., Bach, F. H., and Robson, S. C., 1997, Xenogeneic endothelial cells activate human prothrombin. *Transplantation.* **64:** 888–896.

Simon, P. M., Neethling, F. A., Taniguchi, S., Goode, P. L., Zopf, D., Hancock, W. W., and Cooper, D. K. C., 1998, Intravenous infusion of αGal oligosaccharides in baboons delays hyperacute rejeciton of porcine heart xenografts. *Transplantation.* **63:** 346–353.

Squinto, S. P. and Fodor, W. L., 1997, Engineering of xenografts to provide organs for human transplantation. In: *Xenotransplantation*, 2nd ed., (D.K.C. Cooper, E. Kemp, J.L. Platt, and D.J.G. White, eds), Springer, Heidelberg, pp. 659–664.

Strahan, K. M., Gu, F., Preece, A. F., Gustavsson, I., Andersson, L., and Gustafsson, K., 1995, cDNA sequence and chromosome localization of pig α1,3 galactosyltransferase. *Immunogenet.* **41:** 101–5.

Tachi, Y., Kobayashi, T., Yokoyama, I., Hayashi, S., Negita, M., Namii, Y., Katayama, A., Nagasaka, T., Mei, G.-L., Hayakawa, A., Matsuda, H., Muramatsu, T., and Takagi, H., 1998, Variability fo cytotoxicity to pig cultured cells and its determinant factors in human sera. *Transplant. Proc.* **30:** 71–3.

Takizawa, H., Takahashi, K., Murakami, T., Okada, N., and Okada, H., 1992, Species-specific restriction of complement by HRF20 (CD59) generated by cDNA transfection. *Eur. J. Immunol.* **22:** 1943–6.

Tanemura, M., Miyagawa, S., Ihara, Y., Matsuda, H., Shirakura, R., and Taniguchi, N., 1997a, Significant downregulation of the major swine xenoantigen by N-acetylglucosaminyltransferase III gene transfection. *Biochem. Biophys. Res. Com.* **235:** 359–64.

Tanemura, M., Miyagawa, S., Ihara, Y., Matsuda, H., Tsuji, S., Shirakura, R., and Taniguchi, N., 1997b, Coexpression of N-acetylglucosaminyltransferase III (GnT-III) and 2,3 sialyltransferase (2,3 ST) gene for reduciton of xenoantigens. *Fourth International Congress for Xenotransplantation.* Nantes, France. (Abstract O69).

Tange, M. J., Pearse, M. J., and d'Apice, A. J. F., 1997a, Galα1–3Gal xenoepitope: donor-targeted genetic strategies. In: *Xenotransplantation*, 2nd ed., (D.K.C. Cooper, E. Kemp, J.L. Platt, and D.J.G. White, eds), Springer, Heidelberg, pp. 701–713.

Tange, M. J., Tearle, R. G., Aminian, A., Romanella, M., Adam, W. R., Pearse, M. J., and d'Apice, A. J. F., 1997b, Demonstration of the functional importance of the Gal epitope in an ex vivo model of xenotransplantation. *Xenotransplantation.* **4:** 20–24.

Taniguchi, S. and Cooper, D.K.C., 1997, Clinical xenotransplantation—past, present and future. *Ann. R. Coll. Surg. Engl.* **79:** 13–19.

Taniguchi, S., Kobayashi, T., Neethling, F. A., Ye, Y., Niekrasz, M., White, D. J., and Cooper, D. K. C., 1996a, Cobra venom factor stimulates anti-α-galactose antibody production in baboons. Implications for pig-to-human xenotransplantation. *Transplantation.* **62:** 678–81.

Taniguchi, S., Neethling, F. A., Korchagina, E. Y., Bovin, N., Ye, Y., Kobayashi, T., Niekrasz, M., Li, S., Koren, E., Oriol, R., and Cooper, D. K. C., 1996b, In vivo immunoadsorption of antipig antibodies in baboons using a specific Galα1–3Gal column. *Transplantation.* **62:** 1379–84.

Tearle, R. G., Tange, M. J., Zannettino, Z. L., Katerelos, M., Shinkel, T. A., Van, D. B., Lonie, A. J., Lyons, I., Nottle, M. B., Cox, T., Becker, C., Peura, A. M., Wigley, P. L., Crawford, R. J., Robins, A. J., Pearse, M. J., and d'Apice, A. J., 1996, The α-1,3-galactosyltransferase knockout mouse. Implications for xenotransplantation. *Transplantation.* **61:** 13–9.

Thall, A. D., Maly, P., and Lowe, J. B., 1995, Oocyte Gal α1,3Gal epitopes implicated in sperm adhesion to the zona pellucida glycoprotein ZP3 are not required for fertilization in the mouse. *J. Biol. Chem.* **270:** 21437–40.

Thibaudeau, K., Anegon, I., Lemauff, B., Soulillou, J. P., and Blanchard, D., 1994, Human natural antibodies to porcine platelets. *Transplantation.* **57:** 1110–5.

Toma, H., 1994, ABO-incompatible renal transplantation. *Urologic Clinics of North America.* **21:** 299–310.

Turman, M. A., Casali, P., Notkins, A. L., Bach, F. H., and Platt, J. L., 1991, Polyreactivity and antigen specificity of human xenoreactive monoclonal and serum natural antibodies. *Transplantation.* **52:** 710–7.

Tuso, P. J., Cramer, D. V., Middleton, Y. D., Kearns, J. M., Yasunaga, C., Cosenza, C. A., Davis, W. C., Wu, G. D., and Makowka, L., 1993, Pig aortic endothelial cell antigens recognized by human IgM natural antibodies. *Transplantation.* **56:** 651–5.

Vanhove, B. and Bach, F. H., 1993, Human xenoreactive natural antibodies—avidity and targets on porcine endothelial cells. *Transplantation.* **56:** 1251–3.

Vanhove, B., Charreau, B., Cassard, A., Soulillou, J. P., and Pourcel, C., 1997a, α1,3 galactosyltransferase inhibition in the pig: a new approach. *Fourth International Congress for Xenotransplantation.* Nantes, France. (Abstract 040).

Vanhove, B., Goret, F., Soulillou, J. P., and Pourcel, C., 1997b, Porcine α1,3-galactosyltransferase: tissue-specific and regulated expression of splicing isoforms. *Biochem. Biophys. Acta.* **1356:** 1–11.

Vaughan, H. A., Loveland, B. E., and Sandrin, M. S., 1994, Gal α(1,3)Gal is the major xenoepitope expressed on pig endothelial cells recognized by naturally occurring cytotoxic human antibodies. *Transplantation.* **58:** 879–82.

Vaughan, H. A., McKenzie, I. F. C., and Sandrin, M. S., 1995, Biochemical studies of pig xenoantigens detected by naturally occurring human antibodies and the galactose α(1–3)galactose reactive lectin. *Transplantation.* **59:** 102–9.

Vaughan, H. A., Oldenburg, K. R., Gallop, M. A., Atkin, J. D., McKenzie, I. F. C., and Sandrin, M. S., 1996, Recognition of an octapeptide sequence by multiple Galα(1,3)Gal-binding proteins. *Xenotransplantation.* **3:** 18–23.

Watier, H., Guillaumin, J. M., Piller, F., Lacord, M., Thibault, G., Lebranchu, Y., Monsigny, M., and Bardos, P., 1996, Removal of terminal alpha-galactosyl residues from xenogeneic porcine endothelial cells. Decrease in complement-mediated cytotoxicity but persistence of IgG1-mediated antibody-dependent cell-mediated cytotoxicity. *Transplantation.* **62:** 105–13.

Wieslander, J., Mansson, O., Kallin, E., Gabrielli, A., Nowack, H., and Timpl, R., 1990, Specificity of human antibodies against Galα1–3Gal carbohydrate epitope and distinction from natural antibodies reacting with Galα1–2Gal or Galα1–4Gal. *Glycoconjugate J.* **7:** 85–100.

Weiss, R. A., 1998, Transgenic pigs and virus adaptation. *Nature.* **391:** 327–328.

Xu, Y., Lorf, T., Sablinski, T., Gianello, P., Bailin, M., Monroy, R., Kozlowski, T., Cooper, D.K.C., and Sachs, D.H., 1998, Removal of anti-porcine natural antibodies from human and nonhuman primate plasma in vitro and in vivo by a Galα1-3Galβ1-4βGlc-X immunoaffinity column. *Transplantation.* **65:** 172–179.

Xu, H., Kwiatkowski, P., Chen, J. M., Kaplon, R. J., Edwards, N. M., Dong, X., Berger, C., and Michler, R. E., 1994, Abrogation of baboon natural xenoantibody to pig splenocytes by DL-penicillamine. *Transplantation.* **58:** 1299–303.

Ye, Y., Niekrasz, M., Kosanke, S., Welsh, R., Jordan, H. E., Fox, J. C., Edwards, W. C., Maxwell, C., and Cooper, D. K. C., 1994a, The pig as a potential organ donor for man. A study of potentially transferable disease from donor pig to recipient man. *Transplantation.* **57:** 694–703.

Ye, Y., Neethling, F. A., Niekrasz, M., Koren, E., Richards, S. V., Martin, M., Kosanke, S., Oriol, R., and Cooper, D. K. C., 1994b, Evidence that intravenously administered α-galactosyl carbohydrates reduce baboon serum cytotoxicity to pig kidney cells (PK15) and transplanted pig hearts. *Transplantation.* **58:** 330–7.

Yu, P. B., Holzknecht, Z. E., Bruno, D., Parker, W., and Platt, J. L., 1996, Modulation of natural IgM binding and complement activation by natural IgG antibodies: a role for IgG anti-Gal α1–3Gal antibodies. *J. Immunol.* **157:** 5163–8.

*Chapter 11*

# Generation of α1,3Galactosyltransferase Deficient Mice

Aron D. Thall

## 1. INTRODUCTION

All mammals, with the exception of apes, Old World monkeys, and humans possess a functional α1,3GT gene (Galili *et al.*, 1990). Microsomal preparations from tissues of mammalian species as divergent as pigs, cows, New World monkey and mice all contain α1,3GT activity, capable of transferring Gal onto terminal N-acetyllactosamine containing glycolipids and glycoproteins (Thall *et al.*, 1991a). As no other mammalian α1,3GT genes have been cloned, it was assumed but not proven that this α1,3GT gene is responsible for the synthesis of all α-gal epitopes in mammals other than apes, Old World Monkeys and humans. This enzymatic activity is distinct from the blood group B α1,3GT, which requires a terminal α1,2-Fuc linked to N-acetyllactosamine as the acceptor structure and is active only in primates (Elices *et al.*, 1986).

The lack of a functional α1,3GT gene in apes, Old World Monkeys and humans is thought to be the result of mutations in the catalytic domain of enzyme, which has been substantiated through experiments involving mutational analysis

**Aron D. Thall** BioTransplant, Inc., Charlestown Navy Yard, Charlestown, Massachusetts 02129. email: aron.thall@biotransplant.com

*Subcellular Biochemistry, Volume 32: α-Gal and Anti-Gal*, edited by Galili and Avila. Kluwer Academic / Plenum Publishers, New York, 1999.

of a cloned functional α1,3GT (Henion *et al.*, 1994). In the case of humans, the α1,3GT gene has incurred three different mutations, two frame shift mutations and two stop signal mutations (Larsen *et al.*, 1990; Joziasse *et al.*, 1991). Therefore, it is unlikely that the non-functional human α1,3GT gene could revert to functional gene through mutation. This is interesting in view of the fact that some autoimmune diseases, such as Graves' Disease, have been correlated with an increase in serum anti-Gal levels (Etienne-Decerf *et al.*, 1987). Indeed, our past studies suggested that normal human tissues might contain small amounts of epitopes capable of interacting with affinity purified anti-Gal. We speculated that these epitopes might normally be cryptic structures which are exposed during the autoimmune destruction of the thyroid tissue (Thall *et al.*, 1991b). However, we were not able to determine the chemical structures of those epitopes. Recent studies have suggested that such cryptic epitopes capable of interacting with anti-Gal may not be carbohydrate structures but may actually be cross-reactive peptides (Sandrin *et al.*, 1997). Thus, there was no evidence in mammals to support the hypothesis that more than one α1,3GT gene might be responsible for the synthesis of α-gal epitopes.

Adhesion of the mammalian sperm to the zona pellucida of the oocyte is the seminal event in embryogenesis. Evolutionary theory would predict that the interactions responsible for sperm adhesion and penetration of the zona pellucida would be highly conserved. However, *in vitro* experiments in the mouse suggested that adhesion of sperm to the zona pellucida was, at least in part, dependent upon the presence of α-gal epitopes on the zona pellucida (Bliel and Wasserman 1988; Niemela *et al.*, 1995). These experiments involved either the pre-treatment of mouse oocytes with α-galactosidase, or the addition of competing carbohydrates containing α-gal epitopes. A weak point in these experiments was the removal of cumulus cells from the oocyte before addition of sperm, which resulted in a less complex adhesion system than would occur in vivo. Nonetheless, these studies suggested that mammals with a functional α1,3GT gene might use α-gal epitopes as a component of the sperm-oocyte adhesion mechanism, possibly involving a lectin on the sperm interacting with α-gal epitopes on the zona pellucida. Since humans lack α-gal epitopes, this implied that there might be a basic difference in a component of the fertilization process between mice and humans.

Other adhesion mechanisms may also include interactions of α-gal epitopes on the cell surface with mammalian lectins capable of recognizing these epitopes. Studies have suggested that the adhesion of mouse tumor cell lines, such as carcinomas and fibrosarcomas, to collagen is partially dependent on the presence of α-gal epitopes on these cells (Grimstad *et al.*, 1984; Grimstad and Bosnes, 1987; ). F9 teratocarcinoma cell differentiation has been associated with an up-regulation of α1,3GT gene expression and α-gal epitope density (Cummings and Mattox, 1988; Smith *et al.*, 1990). Macrophage differentiation has also been associated with increased numbers of α-gal epitopes on the cell surface (Maddox *et al.*,

1982). If the number of α-gal epitopes on macrophage associated laminin were also increased, this might interfere with the binding to laminin through macrophage Galectin-3, a member of a family of lactosamine binding lectins (Barondes *et al.*, 1994), which binds polylactosamine structures found on laminin (Knibbs *et al.*, 1993), thus, altering the interaction of macrophages with the extracellular matrix. Conversely, animals lacking a functional α1,3GT might express higher amounts of terminal lactosamine structures, resulting in an increase in Galectin interactions. It is likely, however, that other enzymes, such as sialic acid transferases and fucosyltransferases, which recognize terminal lactosamine structures would also modify these adhesive interactions.

An understanding of interactions of human cells and components of the humoral immune system with α-gal epitopes on pig cells will be important for the success of pig to human xenotransplantation of vascularized organs. It has been suggested that natural killer cells (NK cells) in humans might have receptors for α-gal epitopes (Iverardi *et al.*, 1997). T cells have been shown to recognize glycolipids with specific carbohydrate structures through interaction with CD1 molecules presenting these structures (Moody *et al.*, 1997). If one uses Selectin-oligosaccharide facilitated adhesion during leukocyte trafficking as paradigm (McEver *et al.*, 1995) then it is possible that analogous adhesive interactions could occur between T cells and NK cells and the xenogenic endothelium. Therefore, the direct interaction of NK cells , T cells, and perhaps other leukocytes, might explain some of the destructive changes which occur during pig to primate vascularized xenotransplantation (Cooper and Thall, 1997). However, a putative receptor for α-gal epitopes on non-B lineage cells has yet to be characterized.

The interaction of anti-Gal with the porcine organ xenograft endothelium in the pig to primate model has been well characterized (Simon *et al.*, 1998; Platt, 1994; Vaughan *et al.*, 1994; Cooper *et al.*, 1993). This interaction leads to a complement mediated hyperacute rejection of the organ, due to the presence of large amounts of circulating anti-Gal. A subject which is explained in detail elsewhere in this volume.

Another possible consequence of the interaction of anti-Gal with the endothelium is the cross-linking of molecules expressing α-gal epitopes. In theory, this might lead to a direct activation of the xenograft endothelium through an augmented association of signal transduction molecules, since a majority of these molecules have been characterized as glycoproteins. While the direct activation of porcine endothelial cells by anti-Gal has not been well elucidated, we have found that cross-linking molecules on cells expressing α-gal epitopes with the α-gal epitope specific Banderaea simpilicifolia isolectin-$IB_4$ (BS-$IB_4$) can lead to the death of cells in culture. BS-$IB_4$ induced apoptosis of tumor cells has also been observed (Kim *et al.*, 1993).

Since α-gal epitope production has been conserved throughout mammalian evolution in a majority of the mammalian species (Galili *et al.*, 1988), this suggests that there is a strong evolutionary advantage in the maintenance of one or more α1,3GT enzymes. Since humans and Old World primates are a natural "knock-out" for α1,3GT activity (Galili and Swanson, 1991 ; Larsen *et al.*, 1990) it is obviously possible to compensate for the loss of α-gal epitopes. As is described elsewhere in this volume, a consequence of this loss of α1,3GT activity is the production of natural antibodies directed toward α-gal epitopes. Beyond that phenotype, it is uncertain what other consequences would have resulted from the loss of this enzymatic activity in the evolutionary ancestors of the Old World primates.

Thus at least several questions have arisen with respect to the biology of α-gal epitopes in mammals. These include the consequences of the loss of α1,3GT activity, the determination of how many genes are responsible for this activity, and how human immunological interactions with these epitopes can be modified.

In order to address these questions, we needed to develop a model system using a mammal which expresses α-gal epitopes. For this purpose we chose the mouse as a model. We felt that the mouse would be the logical choice for studying α1,3GT activity, since the genetics of the mouse are the best characterized of the mammals. In addition, since mice have a good fecundity, it is possible to rapidly breed them to produce large numbers of experimental animals. With such large numbers of experimental animals it is possible to perform a statistical analysis of the outcome of experiments which would be very difficult to assess using large animal models. An additional important reason to develop a mouse model to study the biology of α-gal epitopes is that there is large body of knowledge of mouse immunology and also a large number of available immunological reagents which would facilitate the study of immunological mechanisms of graft rejection and antibody production in response to these epitopes.

## 2. GENERATION OF A KNOCK-OUT MOUSE MODEL FOR α1,3GT DEFICIENCY

We decided that, in order to determine the biological significance of the mammalian α1,3GT, it would be best to functionally inactivate the endogenous α1,3GT alleles in mice. We felt that this goal would be feasible based upon the previous cloning of the mouse α1,3GT gene in our laboratory (Larsen et al, 1989). In addition, the mouse was and still is the only mammal in which pluripotential embryonic stem cells have routinely been isolated which can give rise to germ line transmission of targeted mutations of alleles. Analysis of the mouse α1,3GT genomic locus indicates that the gene is composed of at least nine exons spread out over at least 35kb (Joziasse et al, 1992) (Figure 1). The amino acid sequences of the mouse and pig α1,3GT exhibit a 76% sequence identity (Strahan *et al.*, 1995). Exons 4 and 5 may be subject to alternative splicing in different tissues giving rise to four different isoforms of α1,3GT in mice (Joziasse et al, 1992) and pigs

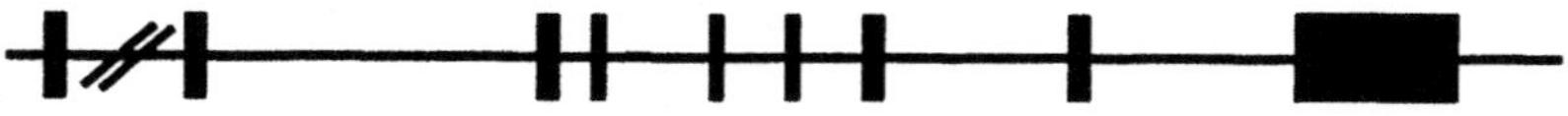


**Figure 1.** Structure of the mouse α1,3GT genomic locus. Black rectangles denote α1,3GT locus exons separated by introns. Cross hatches indicate that the distance between exon 1 and 2 is not drawn to scale.

(Vanhove, *et al.*, 1997). These exons are found within the putative stem domain of the α1,3GT. It has been speculated upon by others that this differential length of the stem domain alters the distance of the α1,3GT catalytic domain from the Golgi membrane and alters the length of the oligosaccharide acceptor structures utilized by the enzyme (Joziasse *et al.*, 1992). However, the catalytic domain of the mouse and pig α1,3GT, found within exon 9, does not appear to be subject to alternative splicing (Joziasse et al, 1992; Vanhove, *et al.*, 1997) (Figure 1). Thus, the genomic organization and processing of the mouse and pig α1,3GT genes are similar. This provides further support for the development of a mouse model for the study of the biological functions of α1,3GT gene.

The catalytic domain of the α1,3GT gene has homology with the human blood group A and B glycosyltransferases (Yamamoto, et al. 1990) and may be a member of the ancestral family of the ABO glycosyltransferases. Based in part by comparison with the ABO enzymes and by analogy with other glycosyltransferases, we reasoned that it was likely that exon 9 was critical for the catalytic function of α1,3GT. Therefore, we decided that it would be appropriate to target a disruption of coding sequence within that exon. Indeed it was subsequently shown by Galili and colleagues that even a DNA point mutation in this catalytic domain has the capacity to abolish α1,3GT enzymatic activity (Henion *et al.*, 1994). Therefore, for the construction of a knock-out vector, we reasoned that the disruption the catalytic domain by the insertion of a foreign DNA sequence would be sufficient to destroy the catalytic activity of the α1,3GT.

A convenient SalI restriction endonuclease cleavage site is found within the 5′ end of exon 9 in a sequence found in the functional α1,3GT cDNA (Figure 1). This site was also utilized by the D'Apice group for the construction of an α1,3GT knock-out targeting vector (Tange *et al.*, 1996).

## 2.1. Design of Targeting Construct

We chose the method of homologous recombination in embryonic stem cells for the generation of mice with a targeted mutation in the α1,3GT gene. In order to accomplish this, a genomic clone of the α1,3GT locus was isolated from a phage lambda 129SV mouse strain genomic library by screening with a mouse α1,3GT

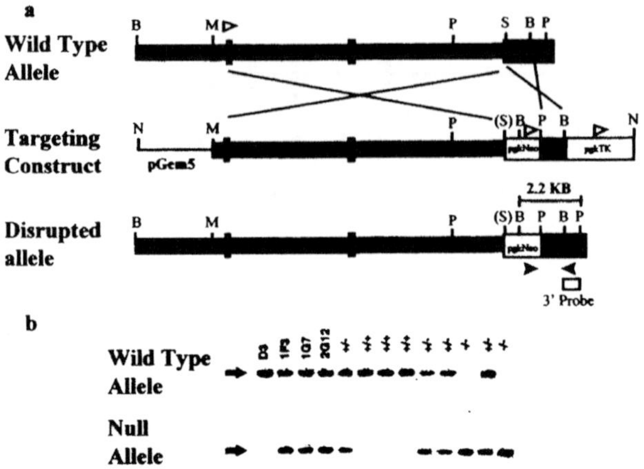


**Figure 2.** Targeted disruption of the murine α1,3GT gene. a, restriction map of the α1,3GT allele, the targeting vector, and the disrupted allele. Black rectanles denote α1,3GT locus exons (thickest portions) and introns (thinnest portions). A neomycin resistance cassette (pgkNeo) was used to disrupt the α1,3GT catalytic domain (found in the largest α1,3GT exon), resulting in a frameshift and premature stop codons relative to both the Neo and 1,3GT translational reading frames. The targeting vector contains 11 kilobases of 5′ genomic DNA and 0.9 kilobases of 3′ sequence flanking the neomycin resistance cassette used to disrupt the α1,3GT catalytic domain exon (largest exon). Restriction sites are indicated by abbreviations (B, BstEII; N, NotI; M, MluI; P, PstI; S; SalI). Restriction sites destroyed during vector construction procedures are in parentheses. Open arrows denote the transcriptional orientation of the α1,3GT locus, the pgkNeo segment, and the pgkTK segment. Solid arrows indicate positions corresponding to polymerase chain reaction primers used to screen targeted ES lines. The position of the BstEII-PstI segment used in Southern blot analyses described in b is indicated below the schematic of the targeted allele (3′ Probe). KB, kilobase(s). b, disruption of the α1,3GT locus. Homologous recombination-mediated replacement of the wild type α1,3GT alleles by the disrupted allele was confirmed by Southern blot analysis. Genomic DNA isolated from the parental ES line, and from three ES clones that gave rise to germ line transmission, is compared to tail genomic DNA from the progeny of F1 heterozygous crosses of littermates. The DNA was digested with PstI and probed with a DNA fragment flanking the 3′ end of the genomic sequence of the targeting vector (3′ probe; see panel a) (Reproduced, with permission, from Thall et al., 1995).

cDNA probe. The clone was first characterized by restriction endonuclease analysis to confirm the identity of the cloned genomic locus. The genomic clone was excised from the phage arms by restriction with NotI and MluI, and the resulting 12kb fragment was cloned into a standard vector (Thall, *et al.*, 1995). This genomic clone was similar to that utilized by the D'Apice group, with the exception that their genomic clone was a 9.5kb fragment and contained approximately 5kb of additional sequence 3′ to the exon 9 catalytic domain (Tange *et al.*, 1996). In order to disrupt the catalytic domain of α1,3GT, a convenient SalI restriction endonuclease cleavage was utilized by both groups, which is located within the 5′ end of exon 9 in a sequence found in the catalytic domain of the functional α1,3GT cDNA (Figure 1 and Figure 2).

The disruption of the functional domains of exons with a drug resistance cassette has historically been an efficient means of selecting embryonic stem cells with a stable introduced genetic disruption. In the case our targeting vector, the neomycin resistance cassette, pgkNeo, was inserted into the SalI site within the catalytic domain, and a 500bp BstEII-NotI fragment was subsequently removed from a position ~900bp 3′ of the pgkNeo insertion. The Neo insertion disrupts the largest coding exon in the α1,3GT gene. Fusion of pgkNeo sequences to this exon yields premature termination codons in all three potential reading frames. The D'Apice group chose to insert a Neo cassette which was flanked by FLP recombinase sites into the SalI site and excised a genomic fragment between 3′ ECORI and BglII sites (Tange *et al.,* 1996). The activity of the yeast FLP recombinase has been shown by others to remove stable integrated sequences from the genome, although in mammalian cells it may not have as high a level of efficacy as the bacteriophage CRE recombinase (Marth, 1996; Buchholz, 1996). Tissue specific recombinase transgenic mice have been used with such constructs to achieve a conditional disruption of genes which might be detrimental if globally disrupted (Betz *et al.,* 1996). Presumably, the investigators were concerned that the inactivation of the α1,3GT gene in mice would be a lethal mutation. Due to the fact that mammals exist with a "natural" knock-out of the α1,3GT we did not expect embryonic lethality to be a likely outcome of the disruption of this gene.

A thymidine kinase cassette (pgkTk) (McBurney, *et al.,* 1991; Tybulewicz, *et al.,* 1991) was then inserted into the targeting vector at a position corresponding to the BstEII site. When the vector is linearized with XbaI the TK gene is on the 3′ terminus of the targeting construct (Figure 2). As homologous integration of the α1,3GT targeting vector would result in the excision of this pgkTK cassette, homologous recombinants can, in many cases, be distinguished from random integrations by negative selection against ES cell lines with stable random integration of the targeting construct by culture in the presence of Gancyclovir (Tybulewicz, *et al.,* 1991). The D'Apice group did not incorporate a similar strategy in the design of their targeting vector. However, at the University of Michigan, our collective experience was that one could achieve an average of a two-fold enrichment for homologous recombinants above that of Neo selection by the incorporation of a pgkTK cassette onto the end of the targeting vector. There are arguments for and against the use of negative selection. However, further discussion of this topic is beyond the scope of this chapter.

## 2.2. Production and Analysis of a Simulated α1,3GT Knock-Out Construct

Before going through the laborious process of producing a knock-out mouse line, we felt that it would be useful to test our strategy for the inactivation of the mouse α1,3GT by constructing simulated α1,3GT genomic and cDNA knock-out vectors. One appropriate vector for this purpose was produced during the process

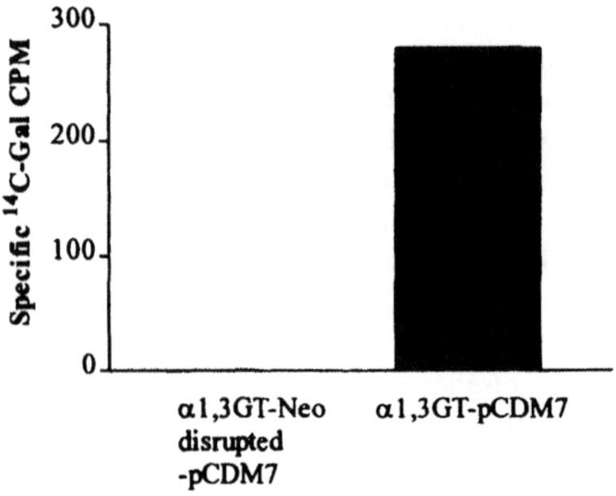


**Figure 3.** Disruption of α1,3GT cDNA results in loss of α1,3GT activity. Cell extracts were prepared from transfected COS-7 cells. The bar indicates the specific incorporation of $^{14}$C-Gal onto an N-Acetyllactosamine acceptor by these cell extracts.

of construction of the α1,3GT knock-out vector for homologous recombination. This vector retained the 3′ sequences of the Neo disrupted genomic clone prior to restriction with BstEII and NotI (Figure 2). This vector was used for the optimization of PCR primers for screening for homologous recombination of the Neo disrupted α1,3GT exon 9. While many groups screen for homologous recombination in ES cells by Southern analyses, we felt that we could streamline the screening process by using a rapid PCR screen to identify candidate ES colonies. PCR primers were chosen for efficiency of amplification by diluting the α1,3GT genomic disruption test construct with ES cell line genomic DNA. The PCR analysis was considered sufficiently sensitive when 5 copies of the genomic test construct could be detected in a background of 10,000 ES cell genomes, while yielding no more than a 5% false positive rate. In order to achieve this level of detection, it was necessary to utilize two amplification steps using a nested series of primers. The PCR primer sets chosen included sequences on the 3′ end of Neo and genomic sequences immediately 3′ to the BstEII restriction site (Figure 2).

With the method of primary screening for α1,3GT knock-out homologous recombinants developed, the focus was shifted to the determination of the function of the disrupted α1,3GT gene. In order to ask this question, a cDNA in a pCDM-7 (Larsen *et al.*, 1989) was utilized as the source of the mouse α1,3GT gene. This gene was disrupted by the insertion of a Neo cassette into the SalI site, as in the planned targeting construct. COS cells, which are of Old World monkey origin, were transfected with the α1,3GT cDNA and the disrupted α1,3GT cDNA. At 72hrs post transfection, the cells were sonicated and extracts were analyzed for α1,3GT activity by the incorporation of $^{14}$C-Gal onto a N-Acetyllactosamine acceptor as described previously (Smith *et al.*, 1990). The results indicated that that the insertion of Neo into the SalI site of exon 9 abolishes the activity of the mouse α1,3GT (Figure 3).

## 2.3. Production of α1,3GT(+/–) ES Cells

As a preliminary step toward the creation of a heritable mutation of the α1,3GT in mice, it was necessary to go through the process of screening different mouse embryonic stem cell lines (ES cells) for their ability to reconstitute the germline lineage of mice. ES cell lines are derived from the inner cell mass of the mouse blastocyst. The majority of ES cell lines are cultured on a feeder layer, which produces factors inhibiting the differentiation of ES cells. In our case, we cultured ES cell lines on monolayers of Neo transgenic fetal fibroblasts, with the addition of mouse Leukemia Inhibitory Factor (Williams *et al.*, 1988). Undifferentiated ES colonies grow as domes on top of the feeder cell monolayer. The process of generating a mouse from ES cells involves injecting the cells into the blastocoel cavity of a day 3.5 mouse blastocyst. The optimal number of ES cells injected per blastocyst is between 10 and 20. The general agreement remains today that ES cells derived from mice of the 129SV background are the most efficient in their ability to incorporate their DNA into the germline of a recipient mouse embryo. Although, recent experiments on the requirements for the establishment of pluripotential ES cell lines have expanded the number of mouse strains which can be utilized in the development of such cell lines (McWhir *et al.*, 1996). We examined several cell lines and found that the D3 ES cell line (Doetschman *et al.*, 1985), one of the first established ES cell lines, gave us an acceptable germline transmission rate of one in three ES colonies selected after maintenance in culture for several weeks.

## 2.4. Transfection of ES Cells with the α1,3GT Targeting Vector

D3 ES cells were cultured on fetal fibroblast monolayers in such a way as to maintain log phase growth in medium consisting of: 15% fetal calf serum, Dulbecco's Minimal Essential Medium, $10^{-4}$M β-mercaptoethanol and 1000U/ml LIF (ES medium). ES colonies were trypsinized extensively to produce a single cell suspension, which was added back to ES culture medium. 30μg of targeting construct was added to $10^7$ ES cells in an electroporation cuvette. Cells were immediately electroporated at 320V/250μF at room temperature. ES cells were added at $3\times10^6$ per 100mm dish and cultured on G418 resistant fetal fibroblast feeder monolayers in ES medium. From 24 hrs until 8 days post electroporation, cells were cultured in the presence of 400μg/ml G418, followed by manual removal of resistant colonies. The criterion for selecting ES colonies at this stage was a dome shaped refractive appearance.

## 2.5. Screening of Transfected ES Cells for Homologous Recombination

Selected colonies were placed in 96 well plates, trypsinized, and split between PCR reaction tubes and replicate culture wells. Nested PCR reactions were conducted and colonies which were PCR positive for the disrupted α1,3GT allele were expanded for Southern analysis of the disrupted allele. ES cells were initially screening with a $^{32}P$ labeled probe containing a genomic DNA sequence within Exon 9, 3′ to the genomic DNA of the targeting construct (Figure 2). This allowed for the discrimination between homologous and random integration of the targeting construct by restricting genomic DNA with PSTI. Homologous recombination between the targeting vector and the ES genome resulted in the appearance of a 2.2kb band by this Southern analysis (Figure 2b). Additional restriction digests were also performed on selected ES cell colonies to confirm the structure of the 5′ region of the disrupted α1,3GT allele. A karyotype analysis was then conducted on ES colonies passing the Southern analysis screening. This was accomplished by the addition of Colcimid for 60–90 min. to replicate cultures in log phase growth. ES colonies showing a grossly normal karyotype by Giemsa staining were expanded for the purpose of mouse blastocyst injection. Colonies which maintained a dome shaped morphology were used to further limit the number suitable ES colonies for injection.

### 2.6. Production of ES Cell Chimeric Mice

Seven ES colonies were selected for injection into blastocysts based on the above criteria. ES colonies possessing a disrupted α1,3GT allele were extensively trypsinized to produce a single cell suspension. ES cell colonies were injected into C57Bl6/J blastocysts at 10 to 20 cells per expanded blastocyst. Blastocysts were implanted into pseudopregnant recipients via the uterine horns at approximately 7 blastocysts per horn.

The D3 ES cells were derived from mice of the black agouti coat color phenotype. Since C57Bl6/J mice have a black coat color, it is simple to score agouti coat color chimerism resulting from the injected ES cells. Upon birth, offspring were scored by two weeks of age based upon the approximate percentage of agouti coat color. A further criterion was that there should be a skewing toward males, since the ES D3 line has the XY genotype. Cell lines gave rise to coat color chimeras ranging from less than 5% to over 90%. Males were chosen from several cell lines which yielded mice with at least 50% coat color chimerism. We observed that the cell lines which gave rise to the high level coat color chimeras did not always yield fertile males, for reasons which can only be speculated upon. However, we have observed that if cell lines result in coat color chimerism of 90% or higher and the males derived from these cells are fertile, then a high percentage of pups born contain the ES cell genome. Thus, the extent of coat color chimerism can be used to make decisions on which ES cell chimeras should be vigorously bred.

### 2.7. Germ Line Transmission of the α1,3GT(+/–) ES Cell Genome

We used two criteria for examining germ line transmission of the α1,3GT(+/–) (heterozygous) ES cell genome. The first and most simple method of analysis was the appearance of black agouti coat color phenotype, which is evident on pups as early as day 8. Three out of the seven ES cell lines injected gave rise to germ line transmission of the agouti coat color when mated with (C57Bl6/JXDBA2/J)$F_1$ females. It was suggest by Tak Mak, one of the early pioneers in the field of homologous recombination in mice, that $F_1$ females should be used for the first matings. The logic being that, if the disruption of the α1,3GT, compromised the viability of the mice, we would have a greater chance of obtaining homozygous knock-out mice if they were bred into a robust hybrid background. Therefore, the preliminary analyses of the ability of chimeric males to transmit the ES cell genome was by breeding to these black coat color $F_1$ hybrid female. However, since the ES cell were heterozygous for the α1,3GT disruption, the presence of the ES cell genome did not guarantee that the progeny of a cross between a wild type oocyte and ES cell derived sperm would contain this mutation. Therefore, all agouti coat color F1 progeny were subjected to examination for the α1,3GT(+/–) genotype by Southern analysis from tail DNA. The restriction digests and probes used comprised the same reagents used for detection of the α1,3GT disruption in ES lines (Figure 2). Heterozygote male and female mice were then bred to yield F2 generation progeny. These α1,3GT(+/–) mice were of a mixed genetic background, since they contained the genomes of the ES cells (129SV) and the oocytes (C57Bl6/JXDBA2/J)$F_1$. Figure 2b shows the results of a genomic Southern analysis of pup tail DNA from the progeny of a representative heterozygote cross, compared with the bands obtained from the three parental ES lines which gave rise to germ line chimeras. F1 α1,3GT (+/–) littermates were intercrossed to yield viable progeny with genotype frequencies of 22% (–/–), 50% (+/–), and 28% (+/+) out of 300 pups subjected to Southern analysis. This corresponds to a Mendelian inheritance pattern, indicating that homozygosity for the null α1,3GT allele is compatible with essentially normal intrauterine development. The D'Apice group found a distribution of 33%(+/+), 43%(+/–), and 24% (–/–) with a P-value of less than 0.03 (Tearle *et al.*, 1996), suggesting a skewing toward the wild type allele. After examining several hundred additional mice in different genetic backgrounds, we have yet to find such a skewing. Therefore, the data presented by the D'Apice group is either a statistical anomaly, a phenotype exhibited in the Balb/c background, or a problem associated with their founder ES cell lines.

## 2.8. Analysis of the Phenotype of α1,3GT (–/–) Mice

Mice that are homozygous for the null allele do not differ in size or appearance from their wild type litter mates. The life span of the α1,3GT (–/–) mice does not appear to be compromised compared with wild type mice. We have maintained male breeders of the hybrid background for over one year without a loss of fecundity. In fact, as mentioned in the introduction, the issue of the role of α-gal structures in sperm-egg adhesion in the mouse was one of the controversies which we wished to address using α1,3GT (–/–) mice. *In vitro* studies suggested that terminal α-Gal epitopes displayed by O-linked glycans on the mouse zona pellucida

**Table I**
**Comparison of the Fecundity of α1,3GT Null Mice with Wild Type Mice**

| Mating | Average litter size |
|---|---|
| Maternal transmission (–/–) × (+/+) | 7±2 (24) |
| Paternal Transmission (–/–) × (+/+) | 5±2 (48) |
| Heterozygous cross (+/–) × (+/–) | 7±1 (160) |
| Null cross (–/–) × (–/–) | 5±2 (47) |

Chimeras which transmitted the inactive allele were derived from ES lines 1G7 and 1F3. Male chimeras were mated with female F1(C57Bl/6J × DBA/2J) mice to yield F1(129SV × C57Bl/6J × DBA/2J) offspring. The percentages of different genotypes in heterozygous F1 crosses were 28%, 50%, and 22%, wild type, heterozygous and null, respectively. Null crosses were performed with male and female littermates. Wild type crosses with null mice were performed with proven fertile wild type animals of the same F2(129SV × C57Bl/6J x DBA/2J) genetic background. Parentheses indicate the number of pups born.

glycoprotein ZP3 are required for the binding of sperm to the oocyte (Wassarman, 1988). These glycoconjugates are easily demonstrated on the zona pellucida of wild type oocytes (Thall *et al.*, 1995), using a lectin (*Bandeiraea simplicifolia* isolectin B4; BSIB4) that specifically recognizes these molecules ( McCoy, *et al.*, 1983). By contrast, oocytes obtained from α1,3GT (–/–) females did not stain with this lectin. The same result was also observed by staining oocytes with human anti-Gal. The loss of the ability to detect oocyte α-galactosides was is not due to a blocking effect of maternal anti-Gal immunoglobulins bound to the oocyte, since anti-mouse immunoglobulins did not interact with these oocytes (Thall *et al.*, 1995). These observations directly demonstrate that the α1,3GT locus determines oocyte expression of terminal α-galactosides.

Table I summarizes breeding studies completed to determine if fertility is affected by absence of zona pellucida terminal α-galactosides consequent to nullizygosity at the α1,3GT locus. In matings between α1,3GT (–/–) females and fertile wild type males of the same genetic background, we observed fertility rates and litter sizes equivalent to those observed in control matings involving α1,3GT (-/+) and α1,3GT (+/+) females. These observations demonstrate that absence of zona pellucida terminal α-galactosides is compatible with normal fecundity, and indicate that terminal α-galactosides do not represent an essential component of the mouse oocyte sperm receptor(s). This conclusion leaves open the possibility that Galβ1,4GlcNAc-terminated blood group I-related oligosaccharides capable of blocking sperm-egg binding (Litscher *et al.*, 1995) are instead responsible for sperm-egg adhesion during fertilization. Absence of an essential role for α-gal epitopes in fertilization is also consistent with an alternative hypothesis that murine sperm-egg adhesion during fertilization is accomplished through an interaction between terminal N-Acetylglucosamine moieties on the oocyte and surface-localized β(1,4)galactosyltransferase on murine spermatids (Miller *et al.*, 1992). Since the *in vitro* fertilization studies involved the removal of cumulus cells from the oocyte prior to sperm adhesion, this in vitro model lacked the full complexity of interactions which occur *in vivo* (Bleil & Wassarman,1988). Therefore, α-gal epitopes may not be necessary for mammalian fertilization due to a cumulus cell mediated increased avidity toward additional molecular interactions, such as terminal β-galactosides on the oocyte. In mammals without a functional α1,3GT, the high affinity interactions with α-gal epitopes may be replaced by a sperm Selectin-like molecule interacting with sialyl-Lewis$^x$ or Lewis$^x$ (Johnston *et al.*, 1988).

In addition to oocytes, we used human anti-Gal antibodies to confirm that the α1,3GT(–/–) mice are deficient in α-gal epitope expression on other tissues. As explained in greater detail elsewhere in this volume, substantial amounts of circulating IgG and IgM class antibodies, are directed against α-gal epitopes (Galili *et al.*, 1984; Sandrin *et al.*, 1993) and human tissues are essentially devoid of these epitopes, since the α1,3GT gene in humans is a non-functional. In humans, anti-Gal antibodies are produced, in part, as a consequence of continuous immunization by gastrointestinal flora containing glycoconjugates with terminal α-gal epitopes (Galili *et al.*, 1988b) and perhaps, also as a result of dietary exposure to α-gal epitopes on mammalian cells. We have found that α-gal epitopes are present on a variety of cell types in the wild type mouse and are present on most leukocytes, with the exception of T cells. The D'Apice group showed an $IB_4$ lectin binding profile for α1,3GT (+/+) splenocytes in which roughly half of the cells did not have sufficient numbers of α-gal epitopes to interact with the lectin, possibly representing the T cells (Tearle *et al.*, 1996). The α-gal epitopes are also absent within the parenchyma of the liver. As noted previously, human anti-Gal antibody detects α-galactosidase-susceptible α-gal epitopes on a variety of wild type mouse cells (Galili *et al.*, 1988a), including vascular endothelium (Thall *et al.*, 1995). By contrast, we have found that the vascular endothelium of α1,3GT (–/–) mice, as well as all tissues examined, are devoid of detectable α-gal epitopes. We have also observed that α1,3GT enzymatic activity, normally present in murine spleen cells, is not detectable in α1,3GT (–/–) splenocytes. In addition, there has been a focus for several years on enzymatic competition between different terminal glycosyltransferases for common acceptors. Two enzymes which may compete for LacNAc terminal oligosaccharide acceptors with the mammalian α1,3GT are α2,3sialyltransferase and α1,2fucosyltransferase. When we examined α1,3GT (–/–) mouse blood leukocytes, using a periodic acid/thiobarbituric acid assay and a fluorometric assay, for changes in the amount of sialic acid on these cells only a minor increase was observed. Given the fact that in the mouse the majority of blood leukocytes are T cells, this observation is not unexpected. The reciprocal relationship between the absence of α-gal epitopes and the presence of anti-Gal antibody observed in humans and some other primates (Galili *et al.*, 1988a) is recapitulated in the α1,3GT (–/–) mice. Sera from α1,3GT (–/–) mice directly agglutinate α-gal epitope-positive (Galili *et al.*, 1984; Suzuki and Naiki, 1984) rabbit erythrocytes, whereas sera from α1,3GT (+/+) mice, as expected, do not, and thus are devoid of anti-Gal antibody activity. The rabbit erythrocyte hemagglutinating activity present in α1,3GT (–/–) sera can be removed by pre-incubation of the sera with immobilized synthetic Galα1,3Galβ1,4GlcNAc structures, whereas removal of hemagglutinating activity does not occur when the sera are pre-absorbed with Fucα1,2Galβ1,4GlcNAc structures. Sera prepared from α1,3GT(–/–) mice contains antibodies that also bind to α-gal-BSA, a neoglycoprotein containing at least 17 α-gal epitopes per molecule. By contrast, sera from α1,3GT (+/+) mice do not bind to this neoglycoprotein. These observations indicate that the α1,3GT (–/–) mice maintain naturally occurring anti-Gal antibodies, and indicates that these mice are therefore essentially deficient in the expression of α-gal epitopes.

Other than the absence of α-Gal epitopes in α1,3GT (–/–) mice, the only other phenotype which our group as well as the D'Apice group have observed as a consequence of this mutation in mice is cataract development (Tearle *et al.*, 1996). We have observed that cataract development begins histologically at around 3 weeks of age. The development of the cataract is manifested histologically as progressive disorganization of the cortical lens fibers which is especially marked in the bow region of the lens where proliferating cells are located. In the mice in our colony, the cataract formation is grossly visible by 5 to 6 weeks of age (Figure 4). It has been reported that mammals which possess a functional α1,3GT have α-Gal epitope containing neotetraosylceramide in the inner cortical and nuclear lens fibers (Ogiso *et al.*, 1988). Our staining of the mouse lens with BSI-B4-FITC revealed a similar pattern of α-gal epitope expression in the wild type mouse lens, which was absent in α1,3GT (–/–) mice. This is a region of the lens in which Lewis$^x$ and sialyl-Lewis$^x$ are expressed on the surface of the lens fibers (Ogiso *et al.*, 1998). It is interesting to note that the area where we first noticed a disorganization of lens fibers in α1,3GT (–/–) mice is in the region where the highest intensity of staining of α-gal epitopes occurs in wild type mice. Thus, these data suggest that α-gal epitopes are important for the structural integrity of lens fibers in animals with a functional α1,3GT. Since α-gal epitopes may be localized on the surface of the lens fibers, it is plausible that these structures are involved in cell-cell or cell matrix interactions. This involvement might be at the level of a direct association of α-gal epitopes with cell surface lectins or in the regulation of charge density by competition with sialic acid transferases. One ramification of this hypothesis is

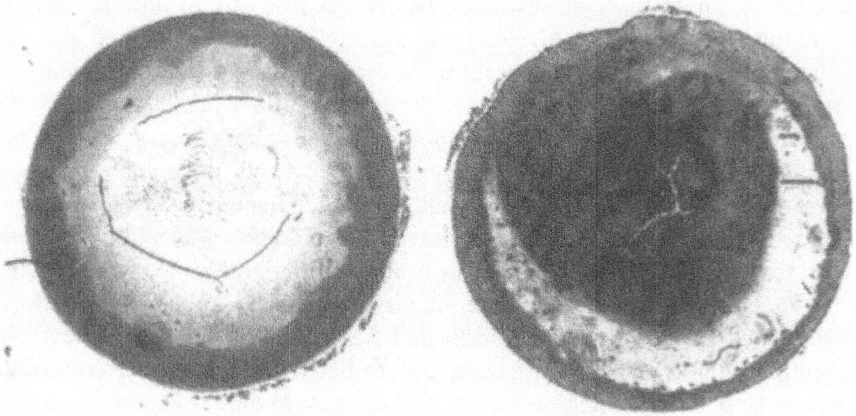

**Figure 4.** Cataract development in α1,3GT(–/–) mice. Lenses were removed from α1,3GT(+/+)and α1,3GT(–/–)littermates at 12 weeks of age, left and right, respectively. Lenses were, fixed, embedded in paraffin, sectioned and stained with H&E.

that, if pigs are generated with a non-functional α1,3GT, then these pigs would also be predicted to generate cataracts in an accelerated fashion. Other cellular adhesive interactions involving α-gal epitopes might be a components of redundant systems, thus masking the role that these epitopes might play in the structural integrity of other tissues.

Human anti-Gal antibodies have been found to be cytotoxic for endothelial cells from mammals, including the pig, which express cell surface α-gal epitopes. These antibodies are thought to be largely responsible for the hyperacute response against xenoantigens on porcine endothelial cells. To determine the similarity in the activity of the murine natural anti-Gal antibodies to the human natural anti-Gal antibodies, cytotoxicity studies were performed on endothelial cell monolayers using sera from α1,3GT (–/–) mice or wild type control mice in the presence or absence of complement. All sera from null mice contained antibodies which were capable of mediating complement dependent lysis of wild type mouse aortic endothelial cells (Thall *et al.*, 1996). The cytotoxicity was found to be complete at the highest serum concentration and decreased in a concentration dependent fashion. These sera were completely inactive toward α1,3GT (–/–) aortic endothelial cells. Wild type control sera were unable to mediate complement dependent lysis of wild type or α1,3GT (–/–) endothelial cells. It was also observed that α1,3GT (–/–) mouse sera could mediate the complete complement dependent lysis of porcine aortic endothelial cells. However, wild type control sera were unable to mediate such complement dependent lysis of porcine endothelial cells. These observations suggest that the mouse natural anti-Gal antibodies are functionally similar to human natural anti-Gal antibodies in their ability to mediate a complement dependent response directed against cells expressing cell surface α-gal epitopes.

Since anti-Gal antibodies are thought to be largely responsible for the hyperacute rejection seen in vascularized xenotransplants into humans and these mouse anti-Gal antibodies are functionally similar to human anti-Gal antibodies, we wondered whether the α1,3GT (–/–) mouse could be used as a small animal model for studying approaches that can diminish anti-Gal-dependent organ transplant rejection. The goal of our preliminary studies was to begin to develop a small animal model which would be useful for evaluating different therapeutic interventions aimed at ameliorating anti-Gal mediated graft rejection. Heart transplants have been used in mouse models in the past to model various aspects of allograft rejection (Russell *et al.*, 1997; Wong *et al.*, 1996; Colson *et al.*, 1995). We had previously determined that the endothelium of the mouse heart vessels expressed α-gal epitopes at high levels. Therefore, we were interested in observing the pathology associated with the transplantation of wild type mouse hearts to an α1,3GT (–/–) mice as an extension of earlier established heart allograft models. It was of particular interest to determine whether α1,3GT (–/–) mice producing measurable anti-Gal could hyperacutely reject wild type mouse hearts, as is seen in the pig heart to primate xenotransplant model. This mouse model is analogous to the transplantation of an ABO mismatched human organ to a pre-sensitized recipient.

The results of our preliminary experiments suggest that wild type hearts transplanted to α1,3GT (–/–) mice do not hyperacutely reject, unless the α1,3GT (–/–) recipient is augmented by the addition of a serum containing a level of complement before the wild type mouse heart transplant. Older mice, with higher natural anti-Gal levels than younger mice will acutely reject wild type mouse hearts with a pathology which is consistent with a diagnosis of acute vascular rejection. Young mice, less than 4 months old, which are congenic for the α1,3GT (–/–) allele do not appear to acutely reject wild type mouse hearts. Studies to examine the pathology associated with this transplant combination remain in progress and this subject will be discussed in greater detail elsewhere in this volume. In collaboration with Megan Sykes' group (Yong *et al.*, 1998) we have shown that wild type mouse bone marrow chimerism can lead to a tolerant state in α1,3GT (–/–) mice toward α-gal epitopes. Our collaborators have recently used this model as an endpoint for tolerance induction via α1,3GT (+/+) bone marrow chimerism. It is our opinion that, since we can achieve an anti-Gal based acute rejection of wild type mouse hearts, this transplant model is a useful means to determine the efficacy of therapeutic interventions aimed at lowering anti-Gal levels.

# 3. CONCLUSIONS

The ability to generate a mouse model of α1,3GT deficiency has been shown by two separate groups. In the process of breeding these mice it was demonstrated that, contrary to then existent theories, the presence of α-gal epitopes on the zona pellucida is not required for the fertilization process. Therefore, these studies indicate that disruption of the α1,3GT could be a viable strategy for the generation of pigs deficient in the activity of this gene. In the mouse, a disruption in the activity of α1,3GT leads to a progressive cataractogenesis. Based on the distribution of α-gal epitopes in the pig lens, one would predict that α1,3GT (–/–) would also generate cataracts. In spite of this abnormality, it is known that mice have naturally poor eyesight and rely mainly upon their hearing and olfactory senses. Since we have had actively breeding α1,3GT (–/–) mice in our colony of greater than one year of age, it would appear that eyesight in laboratory mice is not an important factor in fecundity. The same might be said for the pig, which interacts with its environment largely through its sense of hearing and olfaction. One of the consistent findings among the various groups now working with α1,3GT (–/–) mice is the fact that they produce levels of natural anti-Gal IgM which increase with the age of the mouse, as in the case of human anti-Gal following birth. This renders the α1,3GT (–/–) mouse a potential animal model for the study of anti-Gal production.

Several laboratories have used the α1,3GT (–/–) mouse model for studies of anti-Gal mediated vascularized grafts. The natural antibody titers, in conjunction with a relatively low classical complement activity in these mice are not sufficient hyperacutely reject α1,3GT (+/+) mouse hearts. However, as recently reported by several groups, including our own, at the International Congress on Xenotransplantion in Nantes, France, α1,3GT (–/–) mice can acutely reject α1,3GT (+/+) mouse hearts if anti-Gal levels are augmented by immunization with cells expressing α-gal epitopes. The addition of xenogenic complement to this system results in a hyperacute rejection. Thus, this model appears to resemble anti-Gal mediated rejection, as seen in pig to primate vascular transplants but has the advantage of being amenable to the dissection of the events which lead to rejection of the organ.

As has been mentioned elsewhere, the production of anti-Gal antibodies, which has to date been refractory to a variety of interventions in primates, is a major impediment to pig-to-primate xenotransplantation. In collaboration with our colleagues at the Transplantation Biology Research Center of Massachusetts General Hospital, we have been investigating the phenotype of the cells responsible for these antibodies using the α1,3GT (–/–) mouse model. Recently we published a study in which we demonstrated that α1,3GT (+/+) mouse bone marrow chimerism in α1,3GT (–/–) mice leads to tolerance toward anti-Gal production in response to xenogenic cells expressing α-gal epitopes. We are continuing to refine techniques aimed at tolerance induction toward α-gal epitopes and our near term plans are to expand these endeavors to primate studies.

Thus, the derivation of α1,3GT (–/–) mice has proven to be useful for the determination of the role that α-gal epitopes play in the development of mammals, as well as for the role of anti-Gal antibodies in resistance to engraftment of cells expressing these epitopes.

## REFERENCES

Barondes, S.H., Cooper, D.N., Gitt, M.A., and Leffler, H., 1994, Galectins. Structure and function of a large family of animal lectins, *J. Biol. Chem.*, **269**: 20807–20810.

Betz, U.A., Vosshenrich, C.A., Rajewsky, K., and Muller, W., 1996, Bypass of lethality with mosaic mice generated by Cre-loxP-mediated recombination, *Curr. Biol.*, **6**:1307–1316.

Bleil, J.D. and Wassarman, P.M., 1988, Galactose at the nonreducing terminus of O-linked oligosaccharides of mouse egg zona pellucida glycoprotein ZP3 is essential for the glycoprotein's sperm receptor activity, *Proc. Natl. Acad. Sci.*, **85**: 6778–6782.

Buchholz, F., Ringrose, L., Angrand, P.O., Rossi, F., and Stewart, A.F., 1996,

Different thermostabilities of FLP and Cre recombinases: implications for applied site-specific recombination, *Nucleic Acids Res.*, **24**: 4256–4262.

Colson, Y.L., Wren, S.M., Schuchert, M.J., Patrene, K.D., Johnson, P.C., Boggs, S.S., and Ildstad, S.T., 1995, A nonlethal conditioning approach to achieve durable multilineage mixed chimerism and tolerance across major, minor, and hematopoietic histocompatibility barriers, *J. Immunol.*, **155**: 4179–4188.

Cooper, D.K., Good, A.H., Koren, E., Oriol, R., Malcolm, A.J., Ippolito, R.M., Neethling, F.A., Ye, Y., Romano, E., and Zuhdi, N., 1993, Identification of alpha-galactosyl and other carbohydrate epitopes that are bound by human anti-pig antibodies: relevance to discordant xenografting in man, *Transpl. Immunol.*, **1**: 198–205.

Cooper, D.K. and Thall, A.D., 1997, Xenoantigens and xenoantibodies: their modification, *World J. Surg.*, **21**: 901–906.

Cummings, R.D. and Mattox, S.A., 1988, Retinoic acid-induced differentiation of the mouse teratocarcinoma cell line F9 is accompanied by an increase in the activity of UDP-galactose: beta-D-galactosyl-alpha 1,3-galactosyltransferase, *J. Biol. Chem.*, **263**: 511–519.

Doetschman, T.C., Eistetter, H., Katz, M., Schmidt, W., and Kemler, R. J.,1985, The in vitro development of blastocyst-derived embryonic stem cell lines: formation of visceral yolk sac, blood islands and myocardium, *Embryol. Exp. Morph.*, **87**: 27–45.

Elices, M.J., Blake, D.A., and Goldstein, I.J., 1986, Purification and characterization of a UDP-Gal:beta-D-Gal(1,4)-D-GlcNAc α(1,3)-galactosyltransferase from Ehrlich ascites tumor cells, *J. Biol. Chem.*, **261**: 6064–6072.

Etienne-Decerf, J., Malaise, M., Mahieu, P., and Winand, R., 1987, Elevated anti-alpha-galactosyl antibody titres. A marker of progression in autoimmune thyroid disorders and in endocrine ophthalmopathy?, *Acta Endocrinol. (Copenh)*, **115**: 67–74.

Galili, U., Mandrell, R.E., Hamdeh, R.M., Shohet, S.B. and Griffis, J.M., 1988b, Interaction between human natural anti-alpha-galactosyl immunoglobulin G and bacteria of the human flora, *Infect. Immun.*, **57**: 1730–1737.

Galili, U., Rachmilewitz, E.A., Peleg, A. and Flechner, I., 1984, A unique natural human IgG antibody with anti-alpha-galactosyl specificity, *J. Exp. Med.*, **160**: 1519–1531.

Galili, U. and Swanson, K., 1991, Gene sequences suggest inactivation of α-1,3-galactosyltransferase in catarrhines after the divergence of apes from monkeys, *Proc. Natl. Acad. Sci.*, **88**: 7401–7404.

Galili, U., Thall, A., and Macher, B.A., 1990, Evolution of the Galα1®3Galβ1®4GlcNAc epitope in mammals, *Trends in Glycoscience and Glycotechnology*, **2**: 308–318.

Galili,U., Shohet, S.B., Kobrin, E., Stults, C.L. and Macher, B.A., 1988a, Man, apes, and Old World monkeys differ from other mammals in the expression of alpha-galactosyl epitopes on nucleated cells,*J. Biol. Chem.*, **263**: 17755–17762.

Grimstad, I.A. and Bosnes, V. , 1987, Cell-surface laminin-like molecules and alpha-D-galactopyranosyl end-groups of cloned strongly and weakly metastatic murine fibrosarcoma cells, *Int. J. Cancer*, **40**: 505–510.

Grimstad, I.A., Varani J., and McCoy, J.P. Jr, 1984, Contribution of α-D-galactopyranosyl end groups to attachment of highly and low metastatic murine fibrosarcoma cells to various substrates, *Exp. Cell Res.*, **155**: 345–358.

Henion, T.R., Macher, B.A., Anaraki, F., and Galili U., 1994, Defining the minimal size of catalytically active primate alpha 1,3galactosyltransferase: structure-function studies on the recombinant truncated enzyme, *Glycobiology*, **4**: 193–201.

Inverardi L, Clissi B, Stolzer AL, Bender JR, Sandrin MS, and Pardi R., 1997, Human natural killer lymphocytes directly recognize evolutionarily conserved oligosaccharide ligands expressed by xenogeneic tissues, *Transplantation*, **63**: 1318–1330.

Johnston, D.S., Wright, W.W., Shaper, J.H., Hokke, C.H., Van den Eijnden, D.H., and Joziasse, D.H., 1998, Murine sperm-zona binding, a fucosyl residue is required for a high affinity sperm-binding ligand. A second site on sperm binds a nonfucosylated, beta-galactosyl-capped oligosaccharide, *J. Biol. Chem.*, **273**: 1888–1895.

Joziasse, D.H., Shaper, J.H., Jabs, E.W., and Shaper, N.L, 1991, Characterization of an α1,3-galactosyltransferase homologue on human chromosome 12 that is organized as a processed psuedogene, *J. Biol. Chem.*, **266**: 6991–6998.

Joziasse, D.H., Shaper, N.L., Shaper, J.H., and Kozak, C.A., 1991, Gene for murine α1,3-galactosyltrasferase is located in the centromeric region of chromosome 2. *Soma.t Cell Mol. Genet*, **17**: 201–205.

Joziasse, D.H., Shaper, N.L., Kim, D., Van den Eijnden, D.H., and Shaper, J.H., 1992, Murine α1,3-galactosyltransferase. A single gene locus specifies four isoforms of the enzyme by alternative splicing, *J. Biol. Chem.*, **267**: 5534–5541.

Kim, M., Rao, M.V., Tweardy, D.J., Prakash, M., Galili, U., and Gorelik, E., 1993, Lectin-induced apoptosis of tumour cells, *Glycobiology*, **3**:447–453.

Knibbs, R.N., Agrwal, N., Wang, J.L., and Goldstein, I.J., 1993, Carbohydrate-binding protein 35. II. Analysis of the interaction of the recombinant polypeptide with saccharides, *J. Biol. Chem.*, **268**: 14940–14947.

Larsen R.D., Rivera-Marrero C.A., Ernst, L.K., Cummings R.D., and Lowe, J.B., 1990, Frameshift and nonsense mutations in a human genomic sequence homologous to a murine UDP-Gal:β-D-Gal(1,4)-D-GlcNac α(1,3)-galactosyltransferase cDNA, *J. Biol. Chem.*, **265**:7055–7061.

Larsen, R.D., Rajan, V.P., Ruff, M.M.., Kukowska-Latallo, J., Cummings, R.D., and Lowe, J.B., 1989 Isolation of a cDNA encoding a murine UDP-galactose:β- D -galactosyl-1,4-N-acetyl-D-glucosaminide α-1,3-galactosyltransferase: Expression cloning by gene transfer, *Proc. Natl. Acad. Sci. USA*, **86**: 8227–8231.

Litscher, E. S., Juntunen, K., Seppo, A., Penttila, L. Niemela R., Renkonen, O., and Wassarman, P. M., 1995, Oligosaccharide constructs with defined structures that inhibit binding of mouse sperm to unfertilized eggs in vitro, *Biochemistry*, **34**: 4662–4669.

Maddox, D.E., Shibata, S., Goldstein, I.J., 1982, Stimulated macrophages express a new glycoprotein receptor reactive with Griffonia simplicifolia I-B4 isolectin, *Proc. Natl. Acad. Sc., USA*, **79**:166–170.

Marth, J.D., 1996, Recent advances in gene mutagenesis by site-directed recombination, *J. Clin. Invest.*, **97**: 1999–2002.

McBurney, M.W., Sutherland, L.C., Adra, B., Leclair, B., Rudnicki, M.A., and Jardine, K.X., 1991, The mouse Pgk-1 gene promoter contains an upstream activator sequence. *Nucleic Acids Res.*, **19**: 5755–5761.

McCoy, J.P. Jr., Varani, J., and Goldstein, I.J., 1983, Enzyme-linked lectin assay (ELLA): use of alkaline phosphatase-conjugated Griffonia simplicifolia B4 isolectin for the detection of alpha-D-galactopyranosyl end groups, Anal. Biochem., **130**: 437–444.

McEver, R.P., Moore, K.L., and Cummings, R.D., 1995, Leukocyte trafficking mediated by selectin-carbohydrate interactions, *J. Biol. Chem.*, **270**: 11025–11028.

McWhir, J., Schnieke, A.E., Ansell, R., Wallace, H., Colman, A., Scott, A.R., and Kind, A.J., 1996, Selective ablation of differentiated cells permits isolation of embryonic stem cell lines from murine embryos with a non-permissive genetic background, *Nat. Genet.*, **14**: 223–226.

Miller, D.J. Macek, M.B., and Shur, B.D.,1992, Complementarity between sperm surface beta-1,4-galactosyltransferase and egg-coat ZP3 mediates sperm-egg binding, *Nature*, **357**: 589–593.

Moody, D.B., Reinhold, B.B., Guy, M.R., Beckman, E.M., Frederique, D.E., Furlong, S.T., Ye, S., Reinhold, V.N., Sieling, P.A., Modlin, R.L., Besra, G.S., and Porcelli, S.A., 1997, Structural requirements for glycolipid antigen recognition by CD1b-restricted T cells, *Science*, **278**: 283–286.

Niemela, R., Penttila, L., Seppo, A., Helin, J., Leppanen, A., Rabina, J., Uusitalo, L., Maaheimo, H., Taskinen, J., Costello C.E., and Renkonen, O., 1995, Enzyme-assisted synthesis of a bivalent high-affinity dodecasaccharide inhibitor of mouse gamete adhesion. The length of the chains carrying distal α1,3-bonded galactose residues is critical, *FEBS Lett*, **367**: 67–72.

Ogiso, M., Hidehiko, S., and Hoshi, M., 1998, Localization of Lewis$^x$, sialyl-Lewis$^x$ and α-galactosyl epitopes on glycosphingolipids in lens tissues, *Glycobiology*, **8**: 95–105.

Platt, J.L., 1994, A perspective on xenograft rejection and accommodation, *Immunol. Rev.*, **141**: 127–149.

Russell, P.S., Chase, C.M., and Colvin, R.B., 1997, Alloantibody- and T cell-mediated immunity in the pathogenesis of transplant arteriosclerosis: lack of progression to sclerotic lesions in B cell-deficient mice, *Transplantation*, **64**: 1531–1536.

Sandrin, M.S., Vaughan, H.A., Dabkowski, P.L., and McKenzie, I.F.C., 1993, Anti-pig IgM antibodies in human serum react predominantly with Gal(α1–3)Gal epitopes, *Proc. Natnl Acad. Sci.*, **90**: 11391–11395.

Sandrin, M.S., Vaughan, H.A., Xing, P.-X., and McKenzie, I.F.C., 1997, Natural anti-Galα1,3Gal antibodies react with human mucin peptides, *Glycoconj. J.*, **14**: 97–105.

Simon, P.M., Neethling, F.A., Taniguchi, S., Goode, P.L., Zopf, D., Hancock, W.W., and Cooper, D.K., 1998, Intravenous infusion of Gal alpha1–3Gal oligosaccharides in baboons delays hyperacute rejection of porcine heart xenografts, *Transplantation*, **65**: 346–353.

Smith, D.F., Larsen, R.D., Mattox, S., Lowe, J.B., and Cummings, R.D., 1990, Transfer and expression of a murine UDP-Gal:β-D-Gal-α-1,3-galactosyltransferase gene in transfected chinese hamster ovary cells, *J. Biol. Chem.*, **265**: 6225–6234.

Strahan, K.M., Gu, F., Preece, A.F., Gustavsson, I., Andersson, L., and Gustafsson, K., 1995, cDNA sequence and chromosome localization of pig alpha 1,3 cDNA sequence and chromosome localization of pig alpha 1,3galactosyltransferase, *Immunogenetics*, **41**: 101–105.

Suzuki, E. and Naiki, M., 1984, Heterophile antibodies to rabbit erythrocytes in human sera and identification of the antigen as a glycolipid, *J. Biochem.*, **95**: 103–108.

Tearle, R.G., Tange, M.J., Zannettino, Z.L., Katerelos, M., Shinkel, T.A., Van Denderen, B.J., Lonie, A.J., Lyons, I., Nottle, M.B., Cox, T., Becker, C., Peura, A.M., Wigley, P.L., Crawford, R.J., Robins, A.J., Pearse, M.J., and d'Apice, A.J., 1996, The α-1,3-galactosyltransferase knockout mouse. Implications for xenotransplantation, *Transplantation*, **61**: 13–19.

Thall, A., Etienne-Decerf, J., Winand, R.J., and Galili, U., 1991a, The α-galactosyl epitope on mammalian thyroid cells, *Acta Endocrinologica*, **124**: 692–699.

Thall, A., Etienne-Decerf, J., Winand, R.J., and Galili, U., 1991b, The α-galactosyl epitope on human normal and autoimmune thyroid cells: possible relationship to autoimmunity, *Autoimmunity*, **10**: 81–87.

Thall, A., Murphy, H., and Lowe, J.B., 1996, α1,3-galactosyltransferase deficient mice produce cytotoxic natural anti-Gal antibodies. *Transplantation Proceedings*, **28**:561–62.

Tybulewicz VL, Crawford CE, Jackson PK, Bronson RT, and Mulligan, R.C., 1991, Neonatal lethality and lymphopenia in mice with a homozygous disruption of the c-abl proto-oncogene, *Cell*, **65**:1153–1163.

Vanhove, B., Goret, F., Soulillou, J.P., and Pourcel, C., 1997, Porcine alpha1,3-galactosyltransferase: tissue-specific and regulated expression of splicing isoforms, *Biochim Biophys Acta*, **1356**: 1–11.

Vaughan, H.A., Loveland, B.E., and Sandrin, M.S., 1994, Gal α(1,3)Gal is the major xenoepitope expressed on pig endothelial cells recognized by naturally occurring cytotoxic human antibodies, *Transplantation*, **58**: 879–882.

Wassarman, P. M., 1988, Zona pellucida glycoproteins, *Ann. Rev. Biochem.* **57**, 415–422.

Williams, R.L., Hilton, D.J., Pease, S., Willson, T.A., Stewart, C.L., Gearing, D.P., Wagner, E.F., Metcalf, D., Nicola, N.A., and Gough, N.M., 1988, Myeloid leukaemia inhibitory factor maintains the developmental potential of embryonic stem cells, *Nature*, **336**: 684–687.

Wong, W., Morris, P.J., and Wood, K.J., 1996, Syngeneic bone marrow expressing a single donor class I MHC molecule permits acceptance of a fully allogeneic cardiac allograft, *Transplantation*, **62**:1462–1468.

Yamamoto, F.-i., Clausen, H., White, T., Marken, J., and Hakamori, S.-i., 1990, Molecular genetic basis of the histo-blood group ABO system, *Nature*, **345**: 229–233.

Yong, Y.-G., de Goma, E., Ohdan, H., Bracy, J.L., Xu, Y., Iacomini, J., Thall, A.D., and Sykes, M., 1998, Tolerization of anti-Galα1–3Gal natural antibody-forming B cells by induction of mixed chimerism, *J. Exp. Med.*, **187**: 1335–1342.

*Chapter 12*

# Anti-Xenograft Immune Responses in α1,3-Galactosyltransferase Knock-Out Mice

Martin J. Pearse, Peter J. Cowan, Trixie A. Shinkel, Choa-Guang Chen, and Anthony J. F. d'Apice

## 1. INTRODUCTION

There is now an abundant of evidence (briefly reviewed below), derived from a variety of *in vitro, ex vivo* and *in vivo* studies, to demonstrate that Galactose α1, 3 galactose (α-Gal) is the major xenoantigen on murine and porcine cells and tissues recognised by naturally occurring xenoantibody in human plasma (Sandrin et al., 1993; Cooper et al., 1993a; Galili et al., 1988b). The α1,3 galactosyltransferase (α1,3GT) enzyme that forms this linkage is present in all mammals with the notable exceptions of human and Old World monkeys, in which the α1,3GT gene has been inactivated as the result of frame shift and nonsense mutations (Galili and Swanson, 1991). As a consequence of not expressing the α-Gal epitope, humans develop high titre anti-Gal antibodies (IgG, IgM and IgA) due to exposure to α-

**Martin J. Pearse, Peter J. Cowan, Trixie A. Shinkel, Choa-Guang Chen, and Anthony J. F. d'Apice** Immunology Research Centre, St Vincent's Hospital Melbourne, Fitzroy, Australia, 3065.

*Subcellular Biochemistry, Volume 32: α-Gal and Anti-Gal,* edited by Galili and Avila.
Kluwer Academic / Plenum Publishers, New York, 1999.

Gal on the surface of enteric bacteria and other pathogens (Galili et al., 1988a). In order to establish a small animal model to study the role of anti-Gal antibodies in xenograft rejection we generated a line of mice lacking the α-Gal epitope by inactivating the α1,3GT gene (Tearle et al., 1996). These mice will subsequently be referred to as GalT KO. A similar line of mice were generated at approximately the same time by Thall et al (1995), primarily to investigate the proposed role of the α1,3galactosyltransferase enzyme in fertilization. The main focus of this chapter will be to review the biology of these mice, with particular emphasis on their application as a model to study anti-Gal antibody mediated anti-xenograft immune responses.

### 1.1. Evidence that α-Gal Epitope Is the Major Xenoantigen in Mice and Pigs

The importance of the α-Gal epitope to xenograft rejection was initially demonstrated by carbohydrate competition studies in which it was shown that only carbohydrates containing terminal α-Galactose residues were capable of blocking human xenoantibody reactivity with porcine endothelial cells (Neethling et al., 1994; Sandrin et al., 1993; Cooper et al., 1993a). These studies were confirmed by Vaughan et al (1994) who demonstrated that transfection of COS cells (α-Gal epitope negative cell line of Old World monkey origin) with α1,3GT cDNA rendered them sensitive to human serum-mediated cytotoxicity. Furthermore, absorption of human serum against the α1,3GT transfected COS cells stripped the serum of xenoantibodies capable of reacting with porcine endothelial cells (EC). An indication of the functional importance of the α-Gal epitope was provided by Tange et al (1997) who demonstrated using a Langendorff *ex vivo* perfusion model that hearts from α-Gal epitope positive wild-type mice perfused with human serum depleted of anti-Gal antibodies survived for 44.5 min compared with 24 min hearts perfused with control untreated serum (Figure 1). Consistent with this, hearts from wild-type mice perfused with human serum containing an α-Gal-containing sugar such as melibiose survived for significantly longer than hearts perfused with human serum alone or with serum containing a non-α-Gal sugar such as lactose (Tange:unpublished). In both of these studies the increase in heart survival was accompanied by a decrease in the deposition of human immunoglobulin, in particular IgM and of the activated complement component C3c. These *ex vivo* perfusion results indicate a role for anti-Gal antibodies in mediating humoral xenogeneic immune responses. However, it was not possible to determine whether the partial nature of the protective effect observed was due to incomplete depletion of the anti-Gal antibodies, inefficiency of meliobiose at blocking the interaction of the anti-Gal antibodies (Neethling et al., 1994) with the α-Gal epitopes on the endothelium of mouse heart, or the presence of non-Gal xenoantibodies in human serum. As discussed in section 2.1.3, the availability of GalT KO mice has enabled these issues to be examined directly.

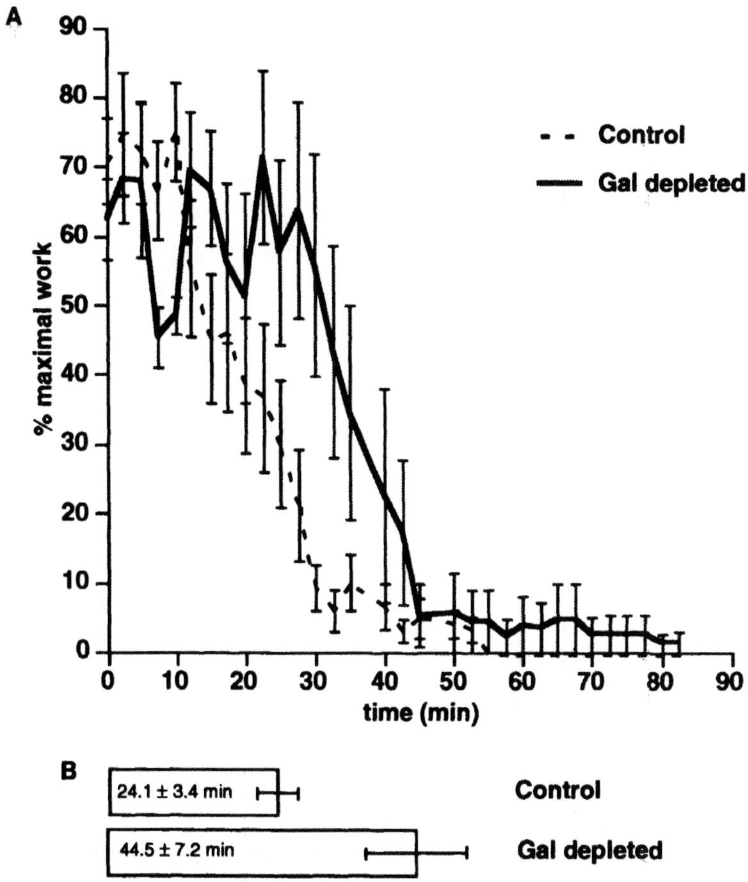


**Figure 1.** A: % maximal work achieved after the addition of normal human plasma against time. The solid line represents hearts perfused with normal human plasma (n=12) and the broken line hearts perfused with anti-Gal depleted plasma (n=6). B: The means ± sem of the time to 90% reduction in work (P=0.04). Reprinted from Tange et al (1997) with permission.

## 1.2. Biosynthetic Pathway for Synthesis of the α-Gal Epitope

The biosynthetic pathway for the synthesis of the α-Gal epitope (Gal α1,3Galβ1,4GlcNAc-R) is represented in Figure 2. The α-Gal epitope has significant homology to human blood group B substance and is generated by the addition of a galactose residue to terminal N-acetyllactosamine residues on glycolipids and glycoproteins. This reaction is catalysed by the enzyme α1,3GT, which is present in all mammals, but is absent in Old World monkeys and primates, including humans (Galili and Swanson, 1991). In humans N-acetyllactosamine is also the sub-

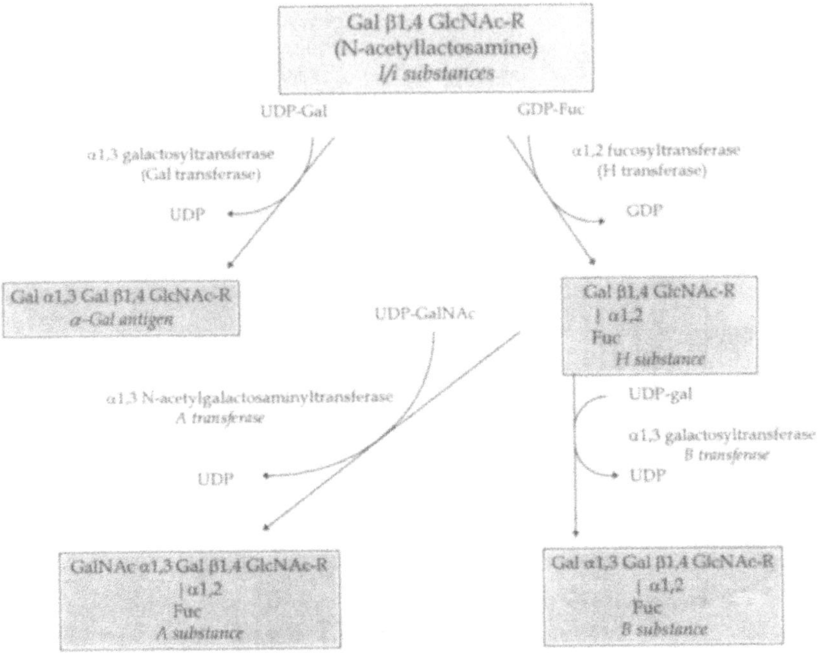


**Figure 2.** Biosynthesis of the α-Gal antigen.

strate for α1,2fucosyltransferase (H-transferase, HTF).Once N-acetyllactosamine has been fucosylated it is unable to serve as a substrate for α1,3GT. As shown in Figure 2, the further addition of N-acetyllactosamine or Gal to H-substance results in the generation of A and B blood group substances, respectively.

In an analogous manner to the development of isoagglutinins in humans, species which do not express the α-Gal epitope, including humans, develop antibodies against it after birth (Galili et al., 1988a; Galili et al., 1988b). In humans, apes and Old World Monkeys it has been estimated that approximately 1% of serum immunoglobulin (Meri et al., 1993; Galili et al., 1987) is anti-Gal specific. This is predominantly IgG, but both IgM and IgA isotypes are also represented (Galili et al., 1995; Parker et al., 1994). As far as hyperacute xenograft rejection (HAR) is concerned anti-Gal antibodies of the IgM isotype (major complement fixing antibody isotype) have been shown to be the key mediators (David et al., 1996; Sandrin and McKenzie, 1994). In vivo, rapid complement-mediated damage to the vascular endothelium leads either to EC death or activation, resulting in the loss of the natural anticoagulant properties of the endothelium and leading to intravascular coagulation, necrosis and ultimately loss of the organ. IgM xenoantibodies, however, are not alone in mediating xenograft damage. In settings where HAR has been overcome, the binding of IgG anti-Gal antibodies (the most abundant anti-Gal antibody isotype) to the vascular endothelium, as discussed below, has the capacity to initiate antibody-dependent cell-mediated cytotoxicity (ADCC) via the Fc receptor-mediated binding of cytotoxic leukocytes, including macrophages, NK-cells and granulocytes. In discordant xenograft models in which HAR is prevented by the removal of xenoantibody (Lesnikoski et al., 1995), decomplementation with cobra venom factor (Azimzadeh et al., 1996; Leventhal et al., 1993) or a combination of both (Hancock et al., 1993; Rooney et al., 1993), organs are still rejected by a process referred to as delayed xenograft rejection (DXR) (Bach et al., 1994) or acute vascular rejection (Platt, 1996). One of the key features of this form of rejection, which will subsequently be referred to as DXR, is the infiltration of the graft with host macrophages and NK-cells (Blakely et al., 1994). However, in pre-clinical trials in which organs from transgenic pigs expresing human complement regulators were transplanted into non-human primates under the cover of "very heavy" immunosuppression, the organs survived for 5–60 days depending on the organ transplanted and on whether it was transplanted orthotopically or heterotopically (unpublished: results presented at the 4$^{th}$ International Xenotransplantation Congress, Nantes, Sept. 1997). The mechanism of rejection was not clear from the results presented, but in the majority of cases there was little or no evidence of a cellular infiltrate. Unfortunately, however, none of the studies reported the results of transplants performed in the absence of immunosuppression, which invariably involved combination therapy, including clinically unacceptably high doses of cyclophosphamide. Therefore it is not possible at present to determine whether the lack of cellular infiltrate observed in these transplant studies is due to the action of the transgene, immunosuppression or a combination of both. Alternatively it may indicate that the cellular features associated with DXR that have been observed in other species combinations, including guinea pig-to-rat (Candinas et al., 1996) and rat-to-primate (Azimzadeh et al., 1996), are not relevant in the pig-to-primate setting. However this is not consistent with the demonstration by Lesniskoski et al (1995) that pig organs transplanted into antibody-depleted baboons were heavily infiltrated by activated mononuclear (non-T) cells at the time of rejection.

## 2. GENERATION AND CHARACTERISATION OF THE GalT KO MICE

As mentioned above GalT KO mice have been generated by two groups (Tearle et al., 1996; Thall et al., 1995). The principles of generating gene knock-out animals by homologous recombination are discussed in detail elsewhere in this volume. Therefore the following discussion is limited to a brief overview of the

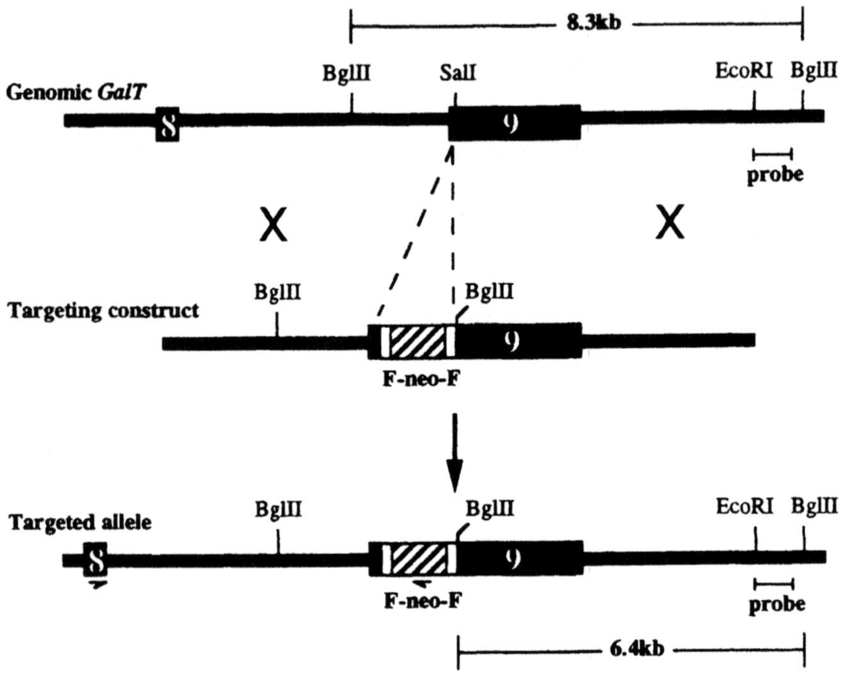


**Figure 3.** Disruption of the mouse α1,3GalT gene by gene targeting. A targeting construct was generated from fragments of genomic DNA from exons 8 and 9 region of the mouse α1,3GalT gene. The diagram shows the wild-type GalT gene region, the targeting construct and the disrupted allele generated by homologous recombination with the targeting construct. The neomycin resistance gene (neo) used to disrupt exon 9 of the α1,3GalT gene is indicated. Also shown are the positions of the probe used for Southern analysis, and primers used for PCR analysis. Reprinted from Tearle et al (1996) with permission.

generation of the GalT KO mice produced by Tearle et al (1996), which have been used in many of the studies described below.

The α1,3GT targeting construct was prepared by interrupting exon 9 of the mouse α1,3GT gene (coding for the catalytic domain of the enzyme) with the neomycin resistance gene (neo$^R$) (Figure 3). The construct was then electroporated into murine ES cells (129SV) and recombinants selected for resistance to G418. The disrupted α1,3GT allele was identified by Southern analysis on genomic DNA by the presence of a diagnostic 6.4kb *Bgl*II fragment. Cells from 16 independently targeted clones were microinjected into Balb/c blastocysts to generate chimeric mice, 3 of which transmitted the interrupted α1,3GT allele through the germline. Chimeras from two of these cell lines were used to generate mice homozygous for the inactive α1,3GT gene (α1,3GT$^{-/-}$).

## 2.1. Characterisation of GalT KO Mice

### 2.1.1. Phenotypic Effects of Deleting the α1,3GT Gene

The GalT KO mice are born healthy and remain healthy throughout life, with the notable exception that they invariably develop cortical cataracts by 4–6 weeks of age. Cataract formation is associated with significant membrane leakage, suggesting that the mouse lens is very sensitive to biochemical disturbance (unpublished observation). Consistent with this, cataract formation has been used in a number studies as a readout for mutagenesis (West and Fisher, 1985). Whether or not the pig lens is similarly sensitive to biochemical/physicochemical changes induced by α1,3GT inactivation remains to be determined. The life span and health status of GalT KO mice, which have now been observed for up to 2 years, are otherwise unaffected. Furthermore elimination of α1,3GT has no adverse effect on fertility, which is consistent with the observations of Thall et al (1995). Together these studies demonstrate that the interaction of α1,3GT on sperm with N-acetyllactosamine residues on the zona-pelucida is not, at least in mice, a critical interaction for fertilisation, as proposed by (Shur and Hall, 1982).

### 2.1.2. Effect of GalT KO on the α-Gal Epitope and Reactivity with Human Serum

RNA for the α1,3GT enzyme was not detected in cells or tissues from GalT KO mice by either Northern analysis or PCR amplification of purified RNA. Consistent with this the α-Gal epitope was not detected on either cells (peripheral blood

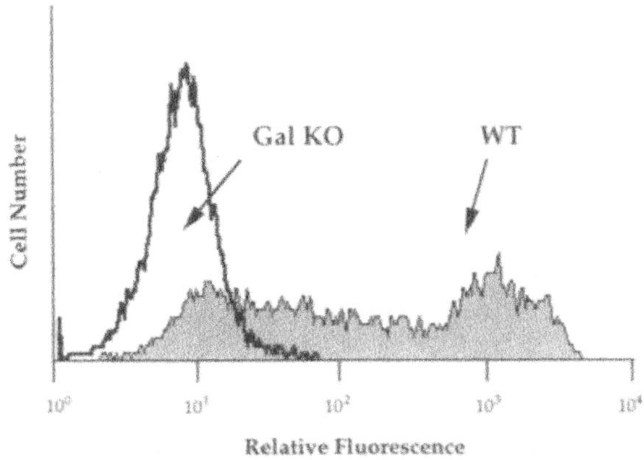


**Figure 4.** Flow cytometric profile of α-Gal epitope expression on splenocytes from wild-type (WT) and GalT KO mice as indicated. α-Gal expression was detected using the α-Gal specific lectin GS-IB4.

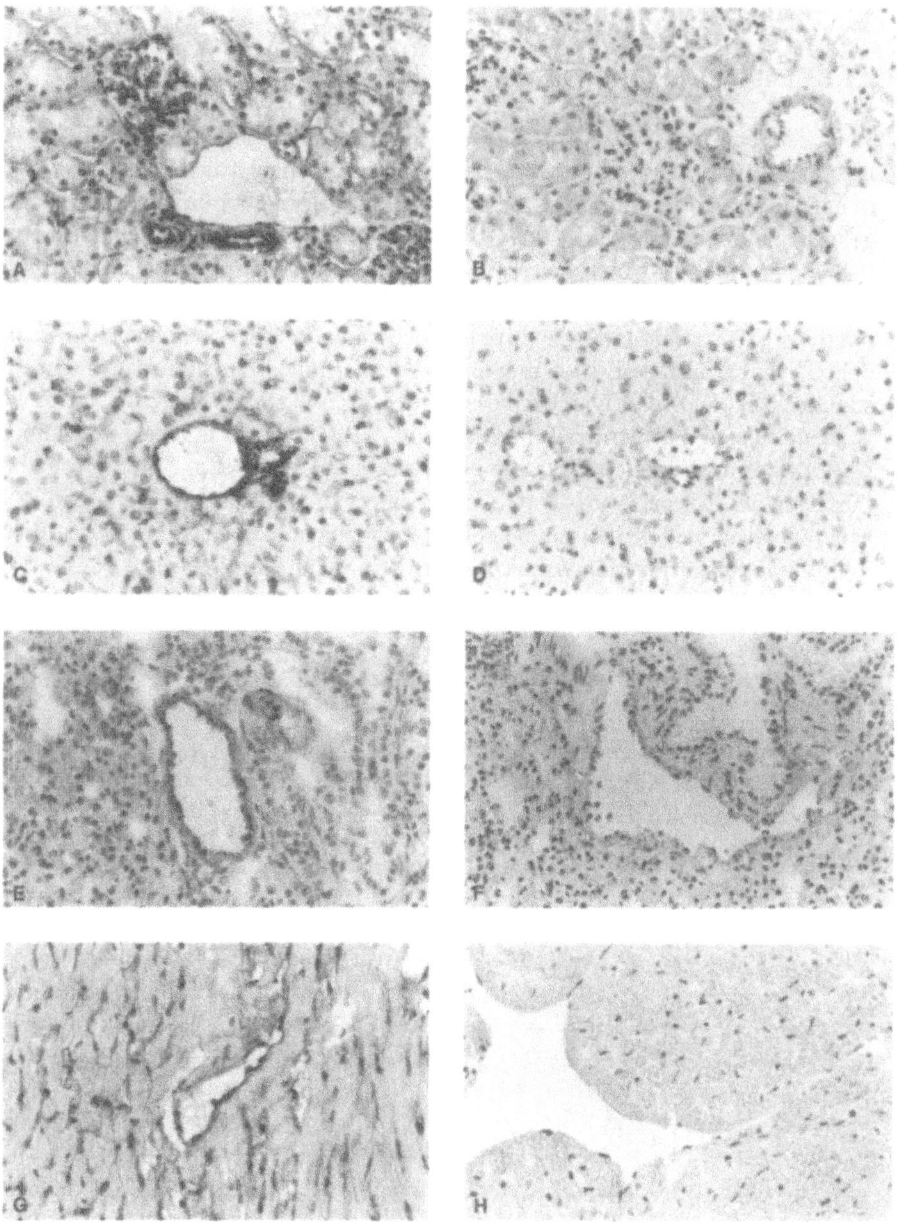


**Figure 5.** Histological sections of kidney (A and B), liver (C and D), lung (E and F), and heart (G and H) from wild-type (left-hand column) and $GalT^{-/-}$ mice (right-hand column) stained with $GS\text{-}IB_4$ lectin by the avidin-biotin-peroxidase technique. Reprinted from Tearle et al (1996) with permission.

leukocytes and splenocytes) (Figure 4) or tissues (heart, lung, kidney, liver, pancreas) (Figure 5) of GalT KO mice, using either the α-Gal-specific lectin, *Griffonia simplicifolia* isolectin B4 (GSI-B4) or affinity purified anti-Gal antibodies.

Elimination of the α-Gal epitope resulted in a significant reduction in the binding of both human IgG and IgM xenoantibodies to cells and tissue from GalT KO mice compared with wild-type controls. Using flow cytometric analysis (FACS) in which cells from GalT KO and wild-type mice were incubated with various concentrations of human serum and bound IgG and IgM were detected with fluorescein-labelled anti-human Ig reagents, it has been estimated that both IgG and IgM binding to GalT KO cells is reduced by approximately 60% compared with wild-type cells (Tearle et al., 1996). In contrast, in pigs it has been estimated that 90% or greater of human naturally occurring xenoantibody recognises the α-Gal epitope (Sandrin et al 1994). This difference should be kept in mind when interpreting the results of studies using the GalT KO mouse to investigate human anti-xenograft immune responses. Consistent with the demonstrated reduction in antibody binding, hearts from GalT KO mice also demonstrated reduced (30–50%) activation of complement (detected using an antibody to the complement activation component C3c), following incubation with human serum. The specificities of the mouse-reactive, non-Gal xenoantibodies are currently unknown. Similarly, it is not yet known whether non-Gal xenoantigens in mice are the same as those in the pig. A comparison of the reactivity of anti-Gal antibody depleted human serum with cell surface proteins isolated from GalT KO mice and an α1,3GT KO porcine cell line, such as PK-1, which is currently being generated, would provide a useful model to address this issue. It is very unlikely in the pig-to-human combination that non-Gal antibodies would play a major role in mediating hyperacute rejection, however the possibility that they may contribute to post-hyperacute rejection mechanisms such as ADCC should not be ignored.

### 2.1.3. Functional Effect of GalT KO

In order to determine the effect of eliminating α-Gal epitope expression on human xenoantibody and complement-mediated injury, hearts from GalT KO mice were perfused *ex vivo* with human plasma using a Langendorff isolated apparatus, as described by Romanella et al. (1996). Hearts were perfused with buffer alone for 20 min to enable the heart to "stabilize." Human plasma was then added to the perfusate pool to achieve a final concentration of 6%, and the heart rate and force of contraction were monitored continuously using MacLab. As shown in Figure 6, hearts from GalT KO mice survived for significantly longer than hearts from wild-type control mice, with the time to reduction in work output to 10% being $10.2 \pm 0.6$ and $33.8 \pm 6.6$ respectively (Tange et al., 1997). Consistent with the prolongation in survival, GalT KO hearts examined histologically at the end of the perfusion period (60 mins) demonstrated a reduction in the level of xenoantibody binding, in particular IgM, and complement (C3c) activation. The effectiveness of

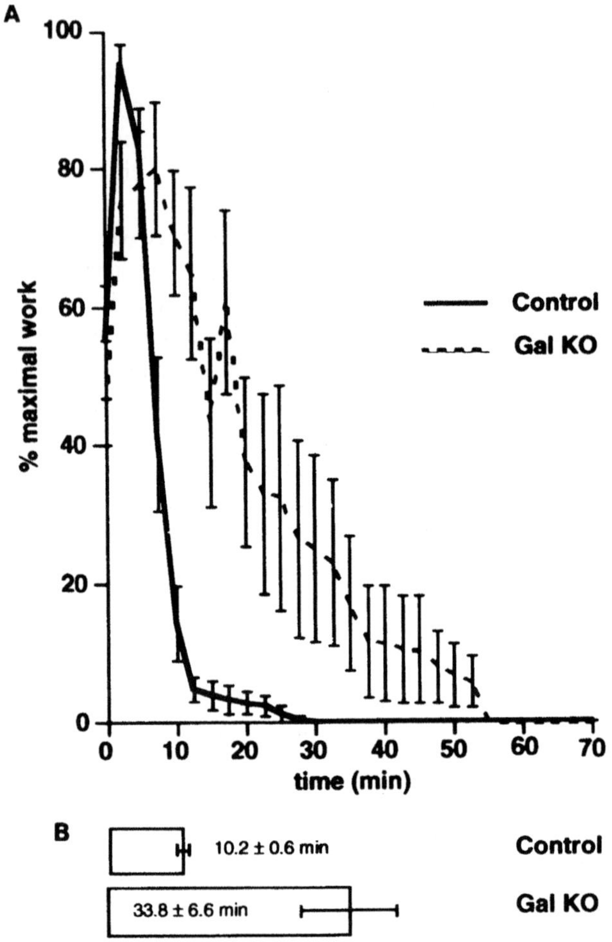


**Figure 6.** A: % maximal work achieved after the addition of normal human plasma against time. The solid line represents wild type control hearts and the broken line hearts from Gal T knockout mice. N=6 in both groups. B: The means ± SEM of the time to 90% reduction in work (P=0.02). Reprinted from Tange et al (1997) with permission.

the GalT KO strategy compared with other transgenic strategies aimed at ameliorating HAR, such as the expression of the human cell surface complement regulators (CD55 and CD59) or H-transferase, either alone or in combination is discussed below.

### 2.2. Effect of GalT KO on Residual Cell Surface Glycosylation: Comparison with α1,2-Fucosyltransferase (HTF) Transgenic Mice

Before comparing the effects of GalT KO and HTF expression on residual cell surface glycosylation, it is worthwhile examining the reasons why the HTF-strategy has been proposed and the effect of HTF transgene expression on α-Gal epitope levels.

### 2.2.1. H-Transferase Strategy

Transgenic expression of H-Transferase (HTF) has been proposed as an alternative strategy to GalT KO as a means to reduce α-Gal epitope expression, because porcine embryonic cells, which are critical for gene knock-out technology (homologous recombination) are not yet available. As shown if Figure 2, HTF and α1,3GT utilise a common acceptor substrate (N-acelyllactosamine), adding terminal fucose (αFuc1-2βGal1-4 GlcNAc-R) and Galactose (αGal1-3βGal1-4GlcNAc-R) respectively. It was originally proposed by Cooper et al (1993b) and later by Sandrin et al (1995), that over expression of HTF would reduce the level of α-Gal epitope expression by "competing" with α1,3GT for the common acceptor substrate. Subsequently it was elegantly shown by Sandrin (1996) using *in vitro* transient transfection of COS cells (African Green Monkey fibroblast cell line) that HTF does not compete with α1,3GT, but rather has preferential access to the acceptor substrate, presumably as a result of being positioned in an "earlier" golgi compartment. To confirm this, Sandrin et al (1996) demonstrated that switching the amino-terminal cytoplasmic domains (which contains the golgi localisation signal) of HTF and α1,3GT reversed the relative dominance of the two enzymes, such that α1,3GalT was dominant over HTF. However, analysis of the ability of transgenically expressed HTF to suppress α-Gal epitope levels in mice would suggest that this compartmentalisation of the two enzymes is not absolute, and furthermore that it is variable depending on cell type. For example, α-Gal epitope expression in splenocytes isolated from transgenic mice expressing HTF under the control of the H-2Kb promoter, was reduced by greater than 90% (Chen et al., 1996). However with endothelial cells isolated from the same mice, α-Gal epitope expression levels were only reduced by 70% (Figure 7). Consistent with this, tissues from HTF-transgenic mice demonstrated considerable reactivity with the α-Gal epitope-specific lectin GSI-B4, compared with tissues from GalT KO mice which were completely negative. The reduction in α-Gal epitope expression in endothelial cells from HTF transgenic mice was found to be variable depending on the endogenous level of α-Gal expression. In small blood vessel endothelial cells, which express low-to-moderate levels of the α-Gal epitope, HTF-expression virtually eliminated GSI-B4 reactivity. However, in larger vessels which in general express high levels of the α-Gal epitope, expression was reduced but clearly not eliminated (Chen et al., 1996).

The explanation for the partial reduction of α-Gal epitope expression in EC compared with splenocytes is not obvious. One possibility is that there is a difference in the golgi localisation of α1,3GT and HTF in splenocytes and EC, such that HTF is not "as dominant" over α1,3GT in the latter. However this seems unlikely because the level of H-substance expression on splenocytes and EC was shown to

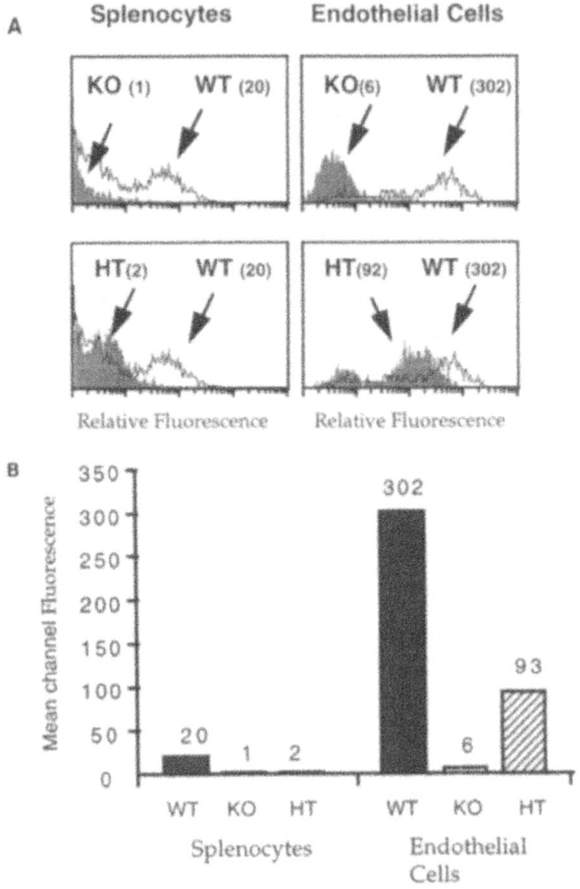


**Figure 7.** *a*) FACS profiles of splenocytes and cardiac endothelial cells from HT-transgenic (HT), GalT KO (KO) and wild type (WT) mice stained with FITC-$IB_4$. The mean channel fluorescence values are in brackets. *b*) representation of FITC-$IB_4$ mean channel fluorescence values. Reprinted from Chen et al (1998) with permission.

be the same in both HTF transgenic and HTF transgenic mice which had been crossed with GalT KO mice (Figure 7). In these mice, the absence of the GalT enzyme presumably results in HTF having "free" access to the substrate. Alternatively it is possible that compared with splenocytes, EC contain more substrate (N-acetyllactosamine) which cannot be fully utilised at the level of HTF expression achieved in the transgenic mice used in this study. The obvious solution to this problem would be to express HTF at a higher level. However, despite considerable effort, we have been unable to generate homozygous HTF transgenic mice from this line, or produce transgenic mice expressing high levels of HTF using the strong endothelial cell-specific ICAM-2 promotor (Cowan et al., 1996). Together these observations suggest that there may be "toxicity" problems associated with high level HTF expression during embryogenesis. This may, as discussed in greater detail below, be related to the marked distortion of cell surface carbohydrates that has been observed on cells from HTF-transgenic mice (Shinkel et al., 1997). Problems of this kind may provide an explanation for the paucity of data, either published or anecdotal, reporting the functional effect of HTF expression in pigs, despite the fact that the generation of HTF transgenic pigs has been reported by at least two groups (Koike et al., 1996; Sharma et al., 1996). This combined with the fact that we have been unable to generate transgenic pigs expressing high levels of HTF, despite considerable effort, suggests, at least to the authors, that there may be serious problems with the application of the HTF-transgenic strategy as a means of reducing α-Gal epitope levels in pigs. Clearly however, this will require further investigation before a definitive answer is available.

### 2.2.2. HTF Expression Affords Partial Protection from Human Xenoantibody and Complement-Mediated Damage

The ability of HTF expression to protect against human xenoantibody mediated damage was examined using the Langendorff *ex vivo* perfusion model described in 2.1.3. As shown in Figure 8, hearts from HTF transgenic mice demonstrated a similar prolongation in survival as hearts from GalT KO mice, when perfused with human plasma. Furthermore, there was no added benefit when the HTF and GalT KO modifications were combined (HTF/GalT KO in Figure 8), providing additional evidence that HTF has preferential access to, and does not

**Figure 8.** *Ex vivo* perfusion. Hearts from HT-transgenic, GalT KO, HT-transgenic/Gal KO and wild type mice were perfused with human plasma. Work performed by the hearts were calculated as the product of the rate and force of contraction. Data is expressed as the percentage of work at the point shown, compared to maximal work at t=0, (time of the first plasma addition). Results are means ± SEM, n=6 in each group. The time points at which human plasma was added are indicated (arrow head). Reprinted from Chen et al (1998) with permission.

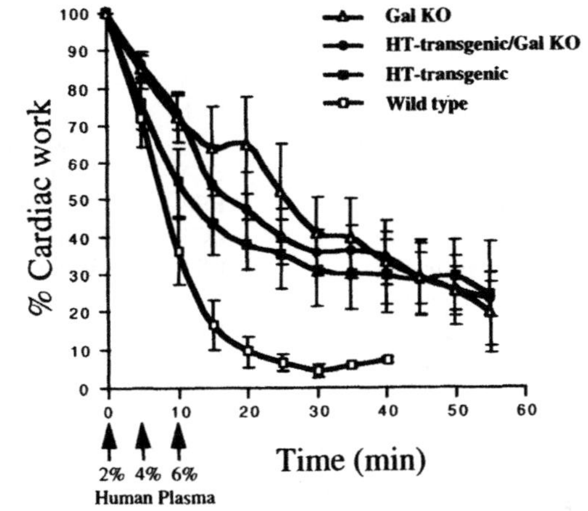

compete with α1,3GalT for the common acceptor substrate (N-acetyl-lactosamine). Consistent with the similar prolongation in survival times, HTF and GalT KO hearts examined post-perfusion with 6% human plasma demonstrated a similar degree of deposition of human immunoglobulin (IgG and IgM) and activated complement (C3c)

### 2.2.3. Analysis of Cell Surface Glycosylation in GalT KO and HTF Transgenic Mice

In order to characterise the cell surface glycosylation changes in GalT KO and HTF-transgenic mice compared with non-transgenic littermates, splenocytes from the 3 groups of mice were examined by flow cytometry for reactivity with a panel of 19 lectins. The choice of lectins, the specificities of which are summarised in Table I, was based on the studies of Oriol et al (1994). The pattern of lectin reactivity relative to non-transgenic littermates is summarised in Figure 9. As expected, GalT KO mice demonstrated complete elimination of the α-Gal epitope, as indicated by loss of reactivity of cells and tissues from GalT KO mice with the α-Gal-specific lectins GSI-B4 and VRA. Surprisingly however, this was associated with only a modest increase in N-acetyllactosamine residues as detected by a

**Table I**
**Lectin Specificity**[a]

| Lectin | Abbreviation | Inhibitor oligosaccharides |
|---|---|---|
| *Griffonia simplicifolia 1B$_4$* | GSIB$_4$ | αGal1-3βGal-R |
| *Artocarpus intergrifolia* | Jacalin | αGal-R |
| *Vigna radiata* | VRA | αGal-R |
| *Ulex europaeus* | UEA1 | αFuc1-2βGal1-4βGlcNAc-R |
| *Anguilla anguilla* | AAA | αFuc-R |
| *Maackia amurensis* | MAA | αNeuAc2-3βGal1-4βGlcNAc/βGlc-R |
| *Sambucus nigra* | SNA | αNeuAc2-6βGal-R |
| *Limulus polyphemus* | LPA | αNeuAc2-6GalNAc>αNeuAc2-3GalNAc>NeuAc |
| *Limax flavus* | LFA | αNeuAc>αNeuGc |
| *Lycopersicon esculentum* | LEL | (βGal1-4βGlcNAc)n>(βGlcNAc1-4βGlcNAc)n |
| *Solanum tuberisum* | STA | (βGal1-4βGlcNAc)n>(βGlcNAc1-4βGlcNAc)n |
| *Euonymus europaeus* | EEA | βGal1-4βGlcNAc<αFuc1-2βGal1-R |
| *Erythrina cristagalli* | ECA | βGal1-4βGlcNAc>βGal1-4Glc>GalNAc>Gal |
| *Cytisus scoparius* | CSA | GalNAc>βGal1-4Glc>αGal1-6Glc>Gal |
| *Sophora japonica* | SJA | GalNAc>Gal |
| *Helix pomatia* | HPA | GalNAcα1-3GalNAc>αGalNAc-R>GlcNAc>αGal |
| *Arachis hypogaea* | PNA | βGal1-4Glc>βGal1-R and αGal |
| *Ricinus communis* | RCA | βGal1-4Glc |
| *Triticum vulgare* | WGA | (βGlcNAc)3>(βGlcNAc)2>βGlcNAc>NeuAc |

[a]Modified from Oriol et al.1993 with permission.

slight increase in the reactivity with the lectins LEL, STA and EEA. In contrast the pattern of lectin binding to cells from HTF-transgenic mice, as indicated by the major shifts in the shaded bars in Figure 9, was more profoundly disturbed, and indicated that in addition to the expected expression of HTF and the reduction in α-Gal, that there was a marked reduction in sialylation, as detected with the lectin MAA. Consistent with this, neuraminidase treatment of cells from HTF mice resulted in only a small increase in recombinant α1,3GT-mediated transfer of UDP-[$^3$H] Gal, compared with similarly treated cells from either control wild-type or GalT KO mice. This suggests that the HTF-mediated fucosylation of N-acetlyllactosamine prevents its subsequent sialylation by α2,3sialytransferase. Whether this is the result of direct competition, that is both enzymes are in the same golgi compartment and therefore compete for the common acceptor substrate, or that HTF is localised to an earlier golgi compartment than α2,3sialy-

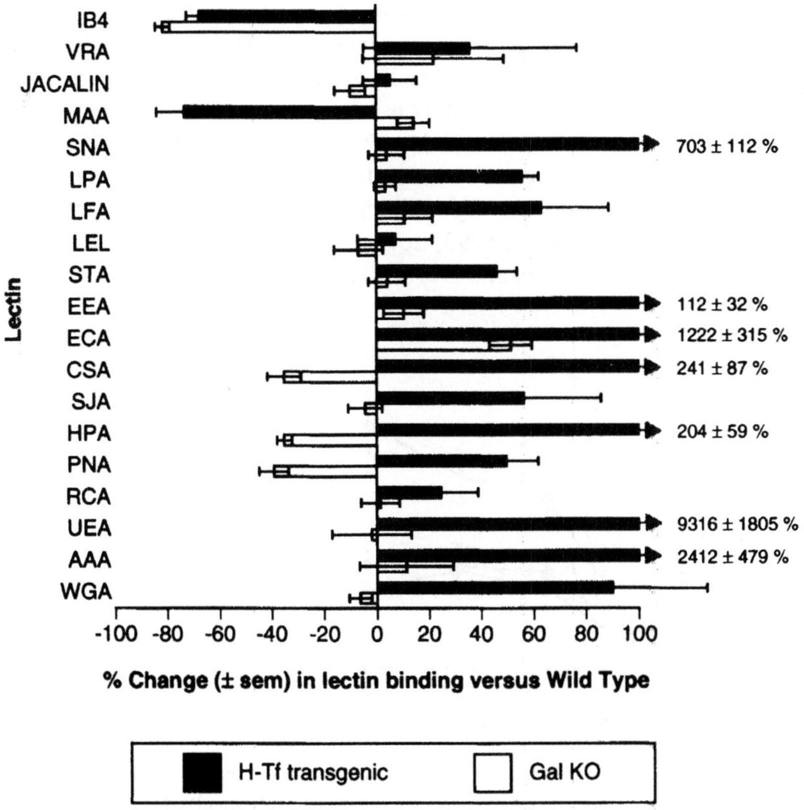


**Figure 9.** Percentage change in binding of lectins to splenocytes of Gal KO and H-transferase transgenic mice compared to non-transgenic wild type mice.

transferase and therefore has preferential access to N-acetlyllactosamine, is yet to be determined.

Cells from HTF-transgenic mice also demonstrated increased reactivity with the lectins STA and ECA, both of which are specific for N-acetyllactosamine (Galβ1-4GlcNAc-R), the acceptor substrate for HTF, α1,3GT and α2,3sialyl-transferase. Large increases in binding (200–700%) were also observed with SNA, which has affinity for the sialyated form of Tn (αNeuAc2-6βGalNAc-R); CSA, which recognises terminal GalNAC residues and so detects both Tn and Forssman antigens; and HPA, which reacts strongly with Forssman antigen (3GalNAcα1-3GalNAcβ1-3Gal-R). Consistent with the cell staining data, there was also an increase in binding of these lectins to heart sections from HTF-transgenic mice compared with GalT KO and wild-type controls (Shinkel et al., 1998). Our working hypothesis is that sialylated Tn and Forssman, which are normally crypt antigens are exposed on the surface of cells from HTF transgenic mice because of a reduction in the zeta potential (electronegativity) as the result of reduced sialylation. In accordance with this, an increase in binding of the lectins specific for Tn and Forssman to pig cells was significantly increased following treatment with neuraminidase to remove the terminal sialic acid residues (Shinkel et al., 1997). Gorelik et al (1995) have made a similar observation with melanoma cell lines following transfection with α1,3GT cDNA. The exposure of these crypt antigens may have significance for long term xenograft survival, because it has been shown that humans have naturally occurring antibodies to some of these epitopes, including sialylated Tn (Thurner et al., 1993; Levene et al., 1988) and Forssman antigen (Cooper et al., 1993a; Young et al., 1979; Hakamori et al., 1977). Consistent with

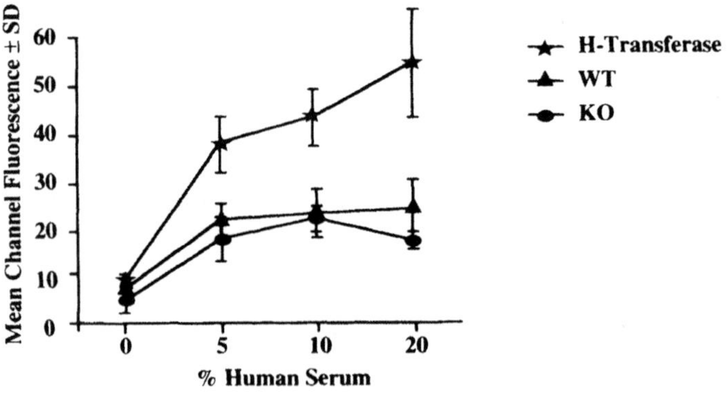


**Figure 10.** Binding of anti-Gal antibody depleted pooled human serum with splenocytes from Wild type, GalT KO and H-transferase transgenic mice. There is significantly more binding to HTF splenocytes than Wild type or control splenocytes. Reprinted from Shinkel et al (1997) with permission.

this Shinkel et al (1997) have shown that cells from HTF but not GalT KO or wild-type (control) mice demonstrate increased reactivity with human serum which has been depleted of anti-Gal antibodies, indicating that the expression of HTF results in the exposure of antigens, most likely crypt antigens, which are recognised by naturally occurring antibodies in human serum (Figure 10). It is unlikely that the titre of these antibodies would be sufficient to have a major impact as far as HAR is concerned, however the possibility that they will contribute to graft damage by mediating antibody-dependent cellular cytotoxicity (ADCC) cannot be excluded.

### 2.2.4. Production of Anti-Gal Antibodies by GalT KO Mice

The reciprocal relationship between the absence of the α-Gal epitope and the presence of α-Gal antibodies observed in Old world monkeys and primates including humans, was first demonstrated in GalT KO mice by Thall et al (1996), who showed that serum from GalT KO but not wild-type littermate mice was capable of causing agglutination of α-Gal epitope-positive rabbit erythrocytes. Furthermore, they demonstrated that the haemagglutinating activity of serum from GalT KO mice was removed by immunoabsorption against immobilised Galα1,3Galβ1,4GlcNAc but not an irrelevant sugar. They further demonstrated that serum from GalT KO but not wild-type mice bound to mouse laminin, a glycoprotein with terminal α-Gal epitopes, and that this reactivity could be eliminated by pre-treatment of the laminin with α-galactosidase. However, using the line of GalT KO mice generated by Tearle et al (1996), we have found that GalT KO mice housed in a "clean" but not SPF mouse facility, have anti-Gal antibody titres of 1:320 [as determined by rabbit red blood cell (RRBC) agglutination], compared with 1:160 in wild-type littermates. This small increase in anti-Gal antibody titre was eliminated when the assay was performed in the presence of melibiose but not lactose, consistent with the increase being due to anti-Gal antibodies (Figure 11). However, immunisation of GalT KO mice with a single dose ($5\times10^6$) of heat-killed *leishmania major* promastigotes (Pearse et al. 1998) or purified rabbit red blood cell membranes (RRBC) (Galili: personnel communication), both of which are heavily glycosylated, results in a significant increase in anti-Gal antibody titre. With the Leishmania-primed (single dose) GalT KO, mice anti-Gal antibody titres ranged from 1:640 to 1:2,560. However, with repeated immunisation titres of up to 1:20,480 can be achieved (unpublished observation). This compares with 1:1,280–2,560 for pooled human serum. The induced anti-Gal antibodies contain all immunoglobulin isotypes with the largest increases in titre observed with IgM and IgG3 (Pearse et al., 1998). A similar induction in anti-Gal antibody titre is also seen in Leishmania-infected humans (Avila et al., 1989).

Therefore, as discussed in greater detail below, the GalT KO mice represent a unique small animal model with which to study the role of anti-Gal antibodies in xenograft rejection and other physiological processes in which anti-Gal antibodies

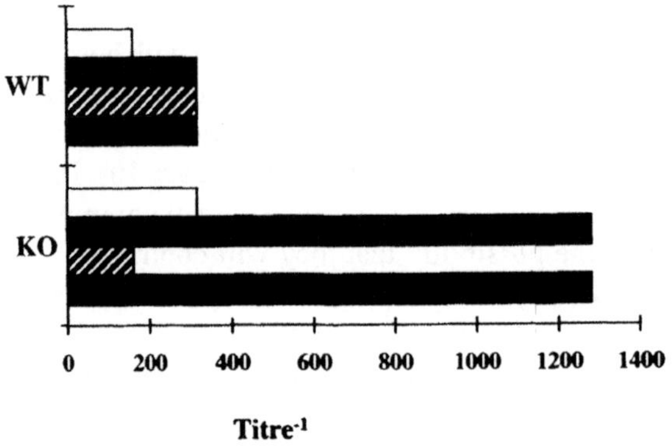


**Figure 11.** RRBC haemagglutination titre of serum from control (non-immunised) wild-type (WT) and Gal KO (KO) mice (open bars), Leishmania-primed mice (black bars) in the presence of 10mM melibiose (hatched bars) or 10 mM lactose (stippled bars). Results are presented as the highest dilution at which haemagglutination was observed. Reprinted from Pearse et al (1998) with permission.

are thought to play a role, such as tumor metastasis. The latter facet of anti-Gal antibody biology is currently being examined by Galili et al (personal communication).

## 3. GalT KO MICE AS MODEL OF XENOGRAFT REJECTION

In an analogous manner to humans, GalT KO mice develop antibodies to the α-Gal epitope, most probably due to exposure to normal gut flora bearing Galα1,3Gal or related α-Gal sugars. Therefore, GalT KO mice represent a unique, inexpensive, small animal model with which to investigate the role of anti-Gal antibodies in xenograft rejection. The alternative is to use primates, however the expense, combined with the considerable logistical and ethical problems associated with primate experimentation, severely limits their utility as a research tool.

### 3.1. GalT KO Mice as Model of HAR

Hearts from α-Gal epitope bearing wild-type mice transplanted into untreated GalT KO mice, with low titre anti-Gal antibodies (1:320) survived long term (>90 days). One possible explanation for the prolonged survival of these grafts is that the GalT KO mice have low complement activity. However, we have shown that the serum from GalT KO mice contains a similar level of classical complement pathway activity as human serum (Pearse et al. 1998). Furthermore, we have shown (unpublished) that hearts from wild-type mice transplanted into GalT KO mice which have been hyper-immunised to achieve anti-Gal antibody titres of 1:10,400–

1:20,800 are rejected in less than an hour, with the hallmark features of HAR, including intravascular coagulation and neutrophil accumulation. One possible explanation for these observations is that when hearts are transplanted to low anti-Gal antibody titre GalT KO mice, the activity of the endogenous mouse complement regulatory proteins is sufficient to regulate the extent of complement activation that is initiated by this level of antibody binding. However, in GalT KO recipients with high anti-Gal titres, complement activation overwhelms the endogenous regulators, and as a consequence the hearts are hyperacutely rejected. Consistent with this McKenzie et al (1998) have shown that the administration of rabbit complement alone to the GalT KO recipients of wild-type (Gal+/+) mouse hearts did not lead to HAR. However, HAR was observed when both normal human serum (NHS: 0.5–1ml) and rabbit complement (0.5ml) were administered to the recipient at the time of receiving the wild-type heart transplant. In the same study it was also shown that rat hearts (which express higher amounts of α-Gal epitope than wild-type mice), transplanted into GalT KO mice treated with NHS were consistently hyperacutely rejected, in the absence of additional complement. On the basis of these findings the authors suggested that rat hearts are superior to mouse hearts in the GalT KO mouse model of α-Gal antibody-mediated HAR. The features of HAR in this model include progressive blackening of the heart and haemorrage. The changes were focal, with haemorrhage being predominantly pericapilliar and perivascular, with surrounding oedema. Histological examination of grafts demonstrated deposition of human immunoglobulin (IgG and IgM) and complement. Heating the serum to 56°C to inactivate complement or removing the anti-Gal antibody by immunoabsorption with Galα1,3Gal-coated beads prevented HAR rejection, demonstrating the dependence on both complement and antibody. IgG-depleted serum was able to cause HAR, however, when IgM antibodies were destroyed by pretreatment of the serum with the reducing agent 2-mercaptoethanol HAR rejection did not occur. Furthermore, hearts from transgenic mice expressing H-transferase survived for greater than 100 min without evidence of HAR. This is consistent with the prolongation in survival of hearts from HTF transgenic mice perfused with human plasma (Chen et al., 1998). Together these results suggest that the HTF and GalT KO strategies to reduce or eliminate expression of the α-Gal epitope, respectively, may be similarly effective in preventing HAR. However, as discussed above and in more detail by Galili et al (1997), for long term xenograft survival it may be critical to generate pigs in which α-Gal epitope expression is completely eliminated in order to prevent the residual α-Gal epitope from inducing an increase in α-Gal antibody titre. The capacity of the α-Gal epitope to induce an increase in anti-Gal antibody titre was investigated by Galili et al (1997), who demonstrated that Cynomolgus monkeys exposed to the α-Gal epitope (in the form of bovine or porcine cartilage), even at low levels, showed a massive increase in anti-Gal antibody titre (10–100 fold pre-sensitisation levels). A similar increase in anti-Gal antibody titre was also observed in diabetic human recipients of fetal pig pancreas (Galili et al., 1995; Groth et al., 1994; Satake et al., 1994). The capacity of cells and tissues from HTF-transgenic mice to induce anti-Gal antibody production in GalT KO mice is currently being examined.

### 3.2. GalT KO Mice as Model of Delayed Xenograft Rejection (DXR)

There is now a significant body of evidence, demonstrating that anti-Gal xenoantibody is a key mediator of hyperacute xenograft rejection. However, the role(s) of anti-Gal xenoantibodies in the post-HAR phases of xenograft rejection are poorly understood, primarily due to the lack of a suitable small animal model which isolates the rejection pathology caused by the anti-Gal antibodies from other mediators of rejection, such as complement. Studies in which guinea pig hearts were transplanted into complement (C6) deficient or cobra venom factor (CVF)-treated rats have shown that in the absence of complement-mediated HAR, xenografts are rejected in hours to days by a rejection process characterised by endothelial cell activation and infiltration of the graft with large numbers of host macrophages and NK-cells (Blakely et al., 1994). However, the major limitation of this species combination is that it does not involve anti-Gal antibodies, which is the dominant xenoantibody specificity of relevance in the porcine-to-human combination. To determine the role of anti-Gal antibody in post-hyperacute rejection, hearts from WT (α-Gal epitope positive) mice were transplanted heterotopically into age- and sex-matched Leishmania-immunised GalT KO mice. Cardiac function was monitored daily by palpation, the results of which are summarised in Table II. Graft survival was found to be dependent on the anti-Gal antibody titre of the recipient at the time of transplant. WT hearts transplanted into GalT KO mice with an antibody titre of 1:1,280 or greater ("high titre") were rejected between 8–12 days of engraftment, with a mean survival time of 10.2 ± 1.2 days. Rejected grafts demonstrated areas of oedema, tissue disruption, myonecrosis and a marked cellular infiltrate, primarily composed of macrophages (90%) and NK-cells (10%), with few T (1%) and B-cells (1%). However, there was no evidence of interstitial

**Table II**
**Graft Survival**

| Donor | Recipient | Anti-rabbit RBC(anti-Gal) titre | Graft survival (days) |
|---|---|---|---|
| WT | Gal KO | 1:320 (4), 1:320 (8) | All> 90 |
| | | 1:1280 (6), 2560 (3) | 8,9,10,10,10,11,11,11,12 |
| WT | WT | 1:160 (5), 320 (3) | All> 90 |
| Gal KO | Gal KO | 1:320 (1), 1:1280 (4), 1:2560 (1) | All> 90 |

Legend: Graft survival in days for donor/recipient combinations indicated. All recipients were immunised with Leishmania promastigotes. Anti-Gal antibody titres at the time of transplantation are indicated.

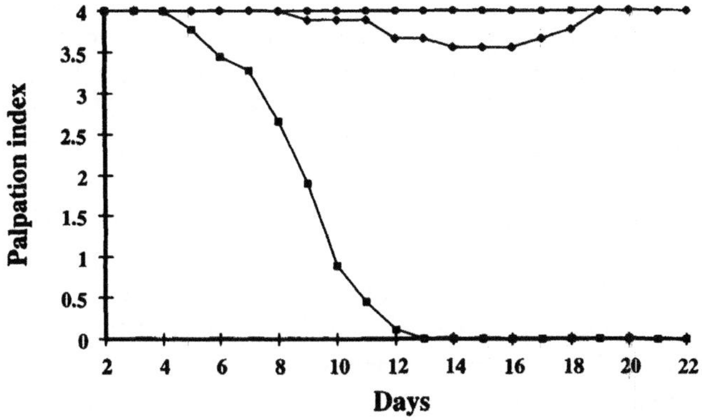


**Figure 12.** Heterotopic cardiac transplantation. Survival (days) of hearts transplanted from WT mice to Gal KO recipients with high (squares) and low (diamonds) titre anti-Gal antibody. Hearts transplanted from WT to Leishmania-primed WT mice (circles) were included as a control. Hearts were monitored daily by abdominal palpation and the rate and strength of contraction graded using a scale from 0–4, with 4 being strong regular contraction and 0 complete cessation of function. Reprinted from Pearse et al (1998) with permission.

haemorrhage. This composition of leukocytes is very similar to that observed in disordant xenograft models in which HAR has been prevented (Azimzadeh et al., 1996; Leventhal et al., 1993; Hancock et al., 1993;), including pig-to-primate (Lesnikoski et al., 1995). In contrast, WT hearts transplanted into recipients with an antibody titre of 1:640 or less ("low titre"), all demonstrated long term survival (>90 days) (Figure 12). However, a proportion of these hearts (30–40%) experienced a rejection "crisis," commencing 10–13 days post transplantation, during which the heart rate was reduced by approximately 50%. Histological examination of hearts harvested during the rejection crisis revealed a modest macrophage and NK-cell infiltrate, with little or no interstitial oedema or myonecrosis. These hearts subsequently recovered, suggesting some form of graft "accommodation" (although this remains to be formally demonstrated), and were still beating strongly at 90 days. As controls, hearts from GalT KO and WT mice were transplanted into Leishmania-primed GalT KO and WT recipients respectively, all of which survived for greater than 90 days, irrespective of the anti-Gal antibody titre.

### 3.3. GalT KO Mice as Model to Study the Role of Anti-Gal Antibodies in Xenogeneic Islet Rejection

It is generally considered that because pancreatic islets are i) non-vascularised, and ii) express low levels of the α-Gal epitope that anti-Gal antibodies will be irrelevant to xenogeneic islet rejection. However, without access to a suitable small animal model it has been difficult to examine this directly. Before discussing the use of GalT KO mice to address this issue, it is worthwhile addressing some of the data on which this assumption is based. Firstly, the expression of the α-Gal epitope in the pancreas has been examined by several groups, including McKenzie et al. (1994) and Oriol et al. (1993) with essentially the same findings, namely that little α-Gal expression was detected in the endocrine /exocrine tissues using either the α-Gal-specific lectin GSI-B4 (McKenzie et al., 1994)) or purified anti-Gal antibodies (Oriol et al., 1993). Strong staining was only detected on the endothelium, including intra-islet capillary endothelial cells (Mirenda et al., 1997), with weak staining on the cuboidal lining cells of the small pancreatic ducts. On the basis of this pattern of expression it is unlikely that anti-Gal antibodies will be relevant to HAR. However, to address this issue directly islets from wild-type mice were transplanted under the kidney capsule of streptozotocin-treated GalT KO mice. Islets transplanted into "non-primed" GalT KO recipients were not subjected to accelerated rejection compared with similar grafts transplanted into wild-type littermates (O'Connell: Personal Communication). Furthermore, there was no evidence of antibody or complement deposition in the long term surviving islet grafts in the GalT KO recipients (similar results were presented by Korsgren and Thall at the 4$^{th}$ International Congress for Xenotransplantation: Nantes, France, 1997). However, as previously mentioned GalT KO mice housed in a "clean" facility have relatively low anti-Gal antibody titres (1:160–1:320) compared with humans (1:1,280–1:2,560). Therefore to investigate the effect of more physiologically relevant anti-Gal antibody titres on islet xenograft rejection, islets from wild-type mice were transplanted into Leishmania-primed GalT KO mice with anti-Gal antibody titres up to 1:20,640. Preliminary results from this study have shown that in recipients with anti-Gal antibody titres of 1:10,320 or greater, the wild-type islet grafts are rejected at approximately 10–12 days post-transplant, which is similar to the time to rejection when hearts from WT mice are transplanted into GalT KO mice with anti-Gal antibody titres between 1:1,240–1:2,480 (Pearse et al. 1998). Histological examination of the grafts revealed an intense leukocyte infiltrate, the composition of which is currently being investigated. These results would suggest that although not relevant to HAR, anti-Gal antibodies at a sufficiently high titre, could contribute to accelerated rejection by mediating antibody-dependent cellular cytoxicity (ADCC). It could be argued with porcine-to-human islet transplants that at the time of transplant the anti-Gal antibody titre in the serum of most human recipients would not be sufficiently high to induce ADCC. However, it has been shown in a number of studies that the anti-Gal antibody titre is massively increased in human recipients of fetal porcine islet grafts (Groth et al., 1994; Satake et al., 1994; Andersson et al., 1992). Furthermore the elicited anti-Gal antibodies, which are evident within 7 weeks of engraftment, possess increased affinity for the α-Gal epitope (Andersson et al., 1992), and include all the major immunoglobulin isotypes. The consequences of the elicited anti-Gal antibodies to xenograft survival are not clear. One possibility, as discussed above, is that they would mediate cell mediated graft rejection, as observed in the high anti-Gal antibody titre GalT KO recipients of islet grafts from WT mice. Alternatively, the delay in the elevation of the anti-Gal antibody titre, such that the grafts exist for a prolonged time in the presence of low titre anti-Gal antibodies, may cause a state of accommodation, similar to that observed in anti-ABH antibody depleted recipients of ABO-mismatched human allografts (Tanabe et al., 1998) or xenoantibody- or complement-depleted recipients of xenografts (Bach et al., 1997). Whatever the explanation, there is sufficient data to indicate that ignoring the potential of anti-Gal antibodies to mediate anti-islet xenograft immune responses, until this issue has been addressed more thoroughly is unwise. The GalT KO mouse provides a useful small animal model in which these issues can be addressed.

### 3.4. Identification of the Optimal Strategy to Overcome HAR

Clearly the pig is the relevant species with respect to a potential source of tissues and organs for transplantation into humans, however the cost involved in making transgenic pigs, and their long gestation times preclude them from being a useful "test-bed" for either therapeutic or transgenic-based strategies. In contrast the short gestation time of mice, combined with the relative ease in performing germ line modifications (both insertion and deletion of genes), makes them an ideal small animal to identify the optimal strategy to overcome HAR and post-hyperacute rejection mechanisms. As discussed above, we have shown using an *ex vivo* Langendorff heart perfusion model that hearts from both GalT KO and HTF transgenic mice are partially protected from human plasma- mediated damage (Figures 1 and 6). In addition we have generated a large number of transgenic mouse lines expressing the human cell surface complement regulatory molecules (CRFs) CD46, CD55 and CD59, either alone or in combination to identify the optimal combination of human CRFs and the level of CRF expression required to afford maximal protection from HAR. Using the Langendorff *ex vivo* heart perfusion model we have shown that expression of very high levels (25–30 fold human), but not low levels of CD59 (1 x human) afforded significant protection when hearts were perfused with 20% human plasma. The high level EC-specific expression described in this study was achieved using the ICAM-2 promoter (Cowan et al., 1996). Furthermore, it has been shown that the co-expression of high levels of CD55 and CD59 provides greater, but still not complete, protection from human complement-mediated injury, than CD59 alone (Cowan et al., 1998b). In order to determine if the combination of an α-Gal epitope reduction strategy with the expression of human CRFs provides greater protection from human xenoantibody and complement that CRF expression alone, transgenic mice expressing high lev-

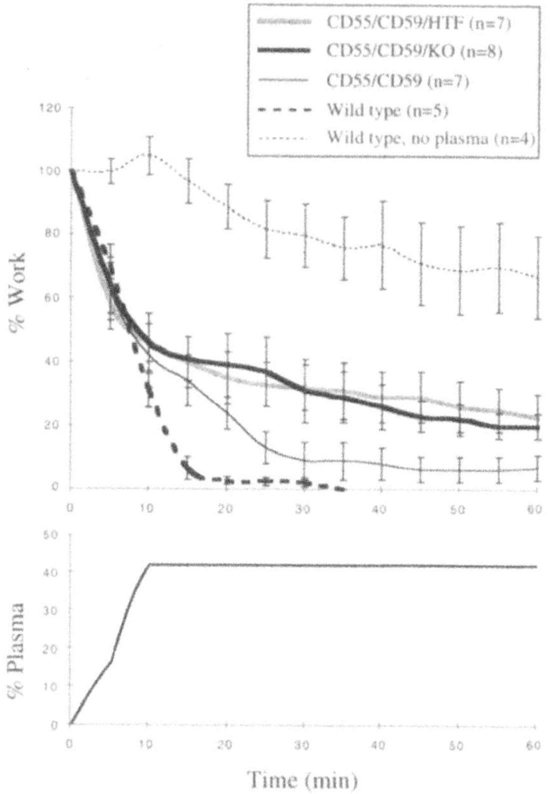


**Figure 13.** (Upper Plot) Function of mouse hearts perfused ex vivo with human plasma. 100% work is defined as heart work measured immediately before time=0. Points are shown as mean ±SEM. (Lower plot) Approximate concentration of plasma in the perfusate. Reprinted from Cowan et al (1998) with permission.

els of both CD55 and CD59 were bred with either the GalT KO or HTF-transgenic mice. Hearts from both CD55/CD59/α1,3GT KO and CD55/CD59/HTF mice demonstrated a similar prolongation in function when perfused with 40% human plasma, compared to hearts from mice expressing the CRFs (CD55/CD59) alone (Figure 13) (Cowan et al., 1998a). It would be interesting to investigate the effect of perfusing the hearts with undiluted human plasma. However, it is difficult for practical reasons to perform this type of analysis using the Langendorff *ex vivo* perfusion apparatus. Although technically demanding, an alternative which is currently being explored is to test the genetically-modified mouse hearts by transplanting them to primates, similar to the studies performed by Azimadeh et al. (1996) using rat hearts.

Together these results suggest that the optimal genetic strategy to overcome HAR rejection would be to express high levels of the human complement regulators CD55 and CD59 in combination with either HTF expression or α1,3GT KO. However, as discussed above, the HTF-transgenic approach has several limitations. Firstly it is highly likely that due to the reduction but not elimination of α-Gal epitope expression that the residual α-Gal epitope will induce an increase in anti-Gal titre. Secondly, as shown by Shinkel et al (1998), H-transferase expression results in the exposure of crypt antigens, against which there are antibodies in human. Thirdly, the inability to generate pigs expressing high levels of HTF suggests that this may be embronically lethal. Accordingly, it is our view that the generation of GalT KO pigs, or the ability to induce stable, long term tolerance of anti-Gal antibody producing B-cells in the recipient (as discussed below), remains one of the essential goals which must be achieved if long term xenograft survival is to be achieved.

### 3.5. Tolerisation of Anti-Gal Producing B-Cells in GalT KO by Induction of Mixed Chimaerism

Yang and deGoma et al (Yang et al., 1998) have recently shown that lasting mixed chimaerism was achieved following the co-transfer of bone marrow from GalT KO and wild-type (GalT-positive) mice to lethally irradiated GalT KO mice. The proportion of WT donor cells in these mice increased between 2 and 8 weeks after BM transplantation, and was subsequently maintained at a steady state. Mixed chimaerism was observed among B, T-cells and myeloid lineages in the periphery, bone marow and spleen. The level of anti-Gal antibodies in the serum of the mixed chimaeras was reduced significantly by 2 weeks, and was virtually undetectable 4 weeks after receiving the bone marrow transplant. In contrast, the level of anti-Gal antibody in GalT KO mice receiving only GalT KO bone marrow was similar to that in untreated GalT KO mice. It was subsequently demonstrated by ELISPOT assays that the reduction in anti-Gal antibody titre was due to a reduction in the number of B-cells producing anti-Gal antibodies. However, the mice retained the capacity to raise antibodies against non-Gal antigens, including porcine antigens, indicating that the mixed chimaeras are functionally and specifically tolerant to the α-Gal epitope. As indicated by the authors, mixed chimaerism represents the first approach to achieve efficient and permanent inhibition of antidonor natural antibody production. Similar studies are currently being pursued using the pig-to-primate species combination to determine the potential of this strategy to induce T- and B-cell tolerance in primates. If successful, the key to this strategy finding clinical applicability will be the ability to induce a state of mixed chimaerism without the need for myeloablation of the recipient. The GalT KO mouse provides an ideal test bed to develop and test non-myeloablative strategies for the generation of mixed bone-marrow chimaerism, prior to evaluation in the pig-to-primate combination.

## 4. SUMMARY

Although originally generated to test the effect of eliminating the α-Gal epitope on HAR, it is becoming increasingly clear that GalT KO mice offer a convenient and inexpensive model to investigate many aspects of the anti-xenorgraft immune response. Clearly, not all aspects of anti-xenograft rejection responses are identical in mice and primates, which should be kept in mind when interpreting results of GalT KO mouse studies. However, with this and other mouse models it is possible to test a large number of variables, which is impractical for both logistical and financial reasons with primates. Furthermore the short gestation time and large litter size of mice means that genetic strategies targeting different aspects of the anti-xenograft immune response can be combined and subsequently tested to identify the optimal combination of genetic and therapeutic approaches to achieve long term xenograft survival. In this regard the GalT KO mouse has been and will continue to be a valuable small animal model for the study of all facets of xenograft rejection involving anti-Gal antibodies.

## REFERENCES

Andersson, A., Groth, C. G., Korsgren, O., Tibell, A., Tollemar, J., Kumagai, M., Moller, E., Bolinder, J., Ostman, J., Bjoersdorff, A., and Hellerstrom, C., 1992, Transplantation of Porcine Fetal Islet-Like Cell Clusters to Three Diabetic Patients. *Transplant.Proc.* **24:**677–678.

Avila, J. L., Rojas, M., and Galili, U., 1989, Immunogenic Galα1–3Gal Carbohydrate Epitopes are present on Pathogenic American Trypanosoma and Leishmania. *J.Immunol.* **142:**2828–2834.

Azimzadeh, A., Wolf, P., Dalmasso, A. P., Odeh, M., Beller, J. P., Fabre, M., Charreau, B., Thibaudeau, K., Cinqualbre, J., Soulillou, J. P., and Anegon, I., 1996, Assessment of hyperacute rejection in a rat-to-primate cardiac xenograft model, *Transplantation* **61:**1305–1313.

Bach, F. H., Ferran, C., Hechenleitner, P., Mark, W., Koyamada, N., Miyatake, T., Winkler, H., Badrichani, A., Candinas, D., and Hancock, W. W., 1997, Accommodation of vascularized xenografts: Expression of "protective genes" by donor endothelial cells in a host Th2 cytokine environment, *Nature Med.* **3:**196–204.

Bach, F. H., Robson, S. C., Ferran, C., Winkler, H., Millan, M. T., Stuhlmeier, K. M., Vanhove, B., Blakely, M. L., Vanderwerf, W. J., Hofer, E., Demartin, R., and Hancock, W. W., 1994, Endothelial cell activation and thromboregulation during xenograft rejection, *Immunol Rev.* **141:**5–30.

Blakely, M. L., Vanderwerf, W. J., Berndt, M. C., Dalmasso, A. P., Bach, F. H., and Hancock, W. W., 1994, Activation of intragraft endothelial and mononuclear cells during discordant xenograft rejection, *Transplantation* **58:**1059–1066.

Candinas, D., Lesnikoski, B. A., Robson, S. C., Miyatake, T., Scesney, S. M., Marsh, H. C., Ryan, U. S., Dalmasso, A. P., Hancock, W. W., and Bach, F. H., 1996, Effect of repetitive high-dose treatment with soluble complement receptor type 1 and cobra venom factor on discordant xenograft survival, *Transplantation* **62:**336–342.

Chen, C. G., Fisicaro, N., Shinkel, T. A., Aitken, V., Katerelos, M., Vandenderen, B. J. W., Tange, M. J., Crawford, R. J., Robins, A. J., Pearse, M. J., and d'Apice, A. J. F., 1996, Reduction in Gal-alpha 1,3-Gal epitope expression in transgenic mice expressing human H-transferase, *Xenotransplantation* **3:**69–75.

Chen, C. G., Salvaris, E., Romanella, M., Aminian, A., Katerelos, M., Fisicaro, N., d'Apice, A. J., and Pearse, M. J., 1998, Transgenic expression of human α1,2fucosyltransferase prolongs mouse heart survival in an ex vivo model of xenograft rejection. Transplantation **65**:832–837..
Cooper, D. K. C., Good, A. H., Koren, E., Oriel, R., Malcom, A. J., Ippolito, R. M., Neethling, F. A., Romano, E., and Zuhdi, N., 1993, Identification of α-galactosyl and other carbohydrate epitopes that are bound by human anti-pig antibodies: relevance to discordant xenografting in man., *Transplant Immunology* **1:**198–205.

Cooper, D. K. C., Koren, E., and Oriol, R., 1993, Genetically Engineered Pigs, *Lancet* **342:**682–683.

Cowan, P. J., Chen, C. G., Shinkel, T. A., Fisicaro, N., Salvaris, E., Aminian, A., Romanella, M., Pearse, M. J., and d'Apice, A. J., 1998, Knock-out of α1,3galactosyltransferase or expression of α1,2 fucosyltransferase further protects CD55- and CD59-expressing mouse hearts in an ex vivo model of xenograft rejection, *Transplantation:In Press.*

Cowan, P. J., Shinkel, T. A., Aminian, A., Romanella, M., Wigley, P. L., Lonie, A. J., Nottle, M. B., Pearse, M. J., and d'Apice, A. J., 1998, High-level co-expression of complement regulators on vascular endothelium in transgenic mice: CD55 and CD59 provide greater protection from human complement-mediated injury than CD59 alone, *Xenotransplantation:In Press*

Cowan, P. J., Shinkel, T. A., Witort, E. J., Barlow, H., Pearse, M. J., and Dapice, A. J. F., 1996, Targeting gene expression to endothelial cells in transgenic mice using the human intercellular adhesion molecule 2 promoter, *Transplantation* **62:**155–160.

David, K., Ollert, M. W., Juhl, H., Vollmert, C., Erttmann, R., Vogel, C. W., and Bredehorst, R., 1996, Growth arrest of solid human neuroblastoma xenografts in nude rats by natural IgM from healthy humans, *Nature Medicine* **2:**686–689.

Galili, U., 1998, Anti-α-galactosyl (anti-Gal) antibody damage beyond hyperacute rejection, in Xenotransplantation, Second Edition (D.K. Cooper, E. Kemp, J.L. Platt and D.J. White, eds), Springer, Berlin, pp. 95–103.

Galili, U., Clark, M. R., Shohet, S. B., Buehler, J., and Macher, B. A., 1987, Evolutionary relationship between natural anti-Gal antibody and the Galα1,3Gal epitope in primates, *Proc. Natl.Acad.Sci.USA.* **84:**1369–1373.

Galili, U., La Temple, DC, Walgenbach, A. W., and Stone, K. R., 1997, Porcine and bovine cartilage in cynomolgus monkeys: ii. Changes in anti-Gal response during chronic rejection, *Transplantation* **63:**46–51.

Galili, U., Mandrell, R. E., Hamedeh, R. M., Shohet, S. B., and Griffiths, J. McL, 1988, Interaction between natural anti-α-galactosyl immunoglobulin G and bacterial of the human flora, *Immunity and Infection* **56:**1730–1737.

Galili, U., Shohet, S. B., Korbin, E., Stults, C. L. M., and Macher, B. A., 1988, Man, apes and old world monkeys differ from other mammals in the expression of α-galactosyl epitopes on nucleated cells, *J.Biol.Chem.* **263:**17755–17762.

Galili, U. and Swanson, K., 1991, Gene Sequences Suggest Inactivation of alpha-1,3-Galactosyltransferase in Catarrhines After the Divergence of Apes from Monkeys, *Proc.Natl.Acad.Sci.USA.* **88:**7401–7404.

Galili, U., Tibell, A., Samuelsson, B., Rydberg, L., and Groth, C. G., 1995, Increased anti-Gal activity in diabetic patients transplanted with fetal porcine islet cell clusters, *Transplantation* **59:**1549–1556.

Gorelik, E., Duty, L., Anaraki, F., and Galili, U., 1995, Alterations of cell surface carbohydrates and inhibition of metastatic property of murine melanomas by alpha 1,3 galactosyltransferase gene transfection, *Cancer Res.* **55:**4168–4173.

Groth, C. G., Korsgren, O., Tibell, A., Tollemar, J., Moller, E., Bolinder, J., Ostman, J., Reinholt, F. P., Hellerstrom, C., and Andersson, A., 1994, Transplantation of porcine fetal pancreas to diabetic patients, *The Lancet* **344:**1402–1404.

Hakamori, S., Wang, S.-N., and Young Jr., W. W., 1977, Isoantigenic expression of Forssman glycolipid in human gastric and colonic mucosa: its possible identity with "A-like antigen" in human cancer., *Proc.Natl.Acad.Sci.USA.* **74:**3023.

Hancock, W. W., Blakely, M. L., Vanderwerf, W., and Bach, F. H., 1993, Rejection of Guinea Pig Cardiac Xenografts Post-Cobra Venom Factor Therapy Is Associated with Infiltration by Mononuclear Cells Secreting Interferon-Gamma and Diffuse Endothelial Activation, *Transplant.Proc.* **25:**2932–2932.

Koike, C., Kannagi, R., Takuma, Y., Akutsu, F., Hayashi, S., Hiraiwa, N., Kadomatsu, K., Muramatsu, T., Yamakawa, H., Nagai, T., Kobayashi, S., Okada, H., Nakashima, I., Uchida, K., Yokoyama, I., and Takagi, H., 1996, Introduction of alpha(1,2)-fucosyltransferase and its effect on alphα-Gal epitopes in transgenic pig, *Xenotransplantation* **3:**81–86.

Lesnikoski, B. A., Shaffer, D. A., Vanderwerf, W. J., Dalmasso, A. P., Soares, M. P., Latinne, D., Bazin, H., Hancock, W. W., and Bach, F. H., 1995, Endothelial and host mononuclear cell activation and cytokine expression during rejection of pig-to-baboon discordant xenografts, *Transplant Proc.* **27:**290–291.

Levene, C., Levene, N. A., Buskila, D., and Manny, N., 1988, Red Cell Polyagglutination, *Transfusion Medicine Reviews* **2:**176–185.

Leventhal, J. R., Dalmasso, A. P., Cromwell, J. W., Platt, J. L., Manivel, C. J., Bolman, R. M. III., and Matas, A. J., 1993, Prolongation of cardiac xenograft survival by depletion of complement., *Transplantation* **55:**857–866.

McKenzie, I. F. C. Li, Y. Q., Patton, K., Thall, A. D., and Sandrin, M. S. A murine model for antibody mediated hyperacute rejection by Galα(1–3)Gal antibodies in Gal o/o mice, 1998, *J. Exp. Med:In Press.*

Mckenzie, I. F. C., Xing, P. X., Vaughan, H. A., Prenzoska, J., Dabkowski, P. L., and Sandrin, M. S., 1994, Distribution of the major xenoantigen (gal(α1–3)gal) for pig to human xenografts, *Transplant Immunology* **2:**81–86.

Meri, S., Mattila, P., and Renoken, R., 1993, Regulation of CD59 expression on the human endothelial cell line EA.hy 926., *Eur.J Immunol.* **23:**2511–2516.

Mirenda, V., Lemauff, B., Cassard, A., Huvelin, J. M., Boeffard, F., Faivre, A., Soulillou, J. P., and Anegon, I., 1997, Intact pig pancreatic islet function in the presence of human xenoreactive natural antibody binding and complement activation, *Transplantation* **63:**1452–1462.

Neethling, F. A., Koren, E., Ye, Y., Richards, S. V., Kujundzic, M., Oriol, R., and Cooper, D. K. C., 1994, Protection of Pig Kidney (PK15) Cells from the Cytotoxic Effect of Anti-Pig Antibodies by alphα-Galactosyl Oligosaccharides, *Transplantation* **57:**959–963.

Oriol, R., Barthod, F., Bergemer, A. M., Ye, Y., Koren, E., and Cooper, D. K. C., 1994, Monomorphic and polymorphic carbohydrate antigens on pig tissues: implications for organ xenotransplantation in the pig-to-human model, *Transplant International* **7:**405–413.

Oriol, R., Ye, Y., Koren, E., and Cooper, D. K. C., 1993, Carbohydrate Antigens of Pig Tissues Reacting with Human Natural Antibodies as Potential Targets for Hyperacute Vascular Rejection in Pig-to-Man Organ Xenotransplantation, *Transplantation* **56:**1433–1442.

Parker, W., Bruno, D., Holzknecht, Z. E., and Platt, J. L., 1994, Characterization and affinity isolation of xenoreactive human natural antibodies, *J Immunol.* **153**:3791–3803.

Pearse, M. J., Witort, E., Mottram, P., Han, W., Murray-Segal, L., Romanella, R., Salvaris, E., Handman, E., Goodman, D. J., and d'Apice, A. J. F., 1997, Anti-Gal antibody mediated allograft rejection in Gal-KO mice: A model of delayed xenograft rejection and accommodation, *Transplantation:In press*

Platt, J. L., 1996, The immunological barriers to xenotransplantation, *Crit Rev Immunol.* **16**:331–358.

Romanella, M., Aminian, A., Adam, W. R., Pearse, M. J., and d'Apice, A. J. F., 1996, Involvement of both the classical and alternate pathways of complement in an ex vivo model of xenograft rejection, *Transplantation,* **63**:1021–1025.

Rooney, I. A., Oglesby, T. J., and Atkinson, J. P., 1993, Complement in Human Reproduction: Activation and Control, *Immunology Research* **12**:276–294.

Sandrin, M. S., Fodor, W. L., Cohney, S., Mouhtouris, E., Osman, N., Rollins, S. A., Squinto, S. P., and Mckenzie, I. F. C., 1996, Reduction of the major porcine xenoantigen Gal alpha(1,3)Gal by expression of alpha(1,2)fucosyltransferase, *Xenotransplantation* **3**:134–140.

Sandrin, M. S. and Mckenzie, I. F. C., 1994, Gal alpha(1,3)Gal, the major xenoantigen(s) recognised in pigs by human natural antibodies, *Immunol Rev.* **141**:169–190.

Sandrin, M. S., Mouhtouris, E., Osman, N., Fodor, W. L., Squinto, S. P., and Mckenzie, I. F. C., 1995, Elimination of the major porcine xenoantigen, Galα(1,3) Gal, by expression of α(1,2) fucosyltransferase., *Glycoconjugate J.* **12**:468–468.

Sandrin, M. S., Vaughan, H. A., Dabkowski, P. L., and Mckenzie, I. F. C., 1993, Anti-pig IgM antibodies in human serum react predominantly with Gal(α1–3)Gal epitopes., *Proc.Natl.Acad. Sci.USA.* **90**:11391–11395.

Satake, M., Kawagishi, N., Rydberg, L., Samuelsson, B. E., Tibell, A., Groth, C. G., and Moller, E., 1994, Limited specificity of xenoantibodies in diabetic patients transplanted with fetal procine islet cell clusters. Main antibody reactivity against α-linked galactose-containing epitopes, *Xenotransplantation* **1**:89–101.

Sharma, A., Okabe, J., Birch, P., Mcclellan, S. B., Martin, M. J., Platt, J. L., and Logan, J. S., 1996, Reduction in the level of Gal(alpha 1,3)Gal in transgenic mice and pigs by the expression of an alpha(1,2)fucosyltransferase, *Proc.Natl.Acad.Sci.USA.* **93**:7190–7195.

Shinkel, T. A., Chen, C. G., Salvaris, E., Henion, T. R., Barlow, H., Galili, U., Pearse, M. J., and d'Apice, A. J., 1997, Changes in cell surface glycosylation in alpha 1,3-galactosyltransferase knock out and alpha 1,2-fucosyltransferase transgenic mice, *Transplantation* **64**:197–204.

Shur, B. D. and Hall, N. G., 1982, A Role for Mouse Sperm Surface Galactosyltransferase in Sperm Binding to the Egg Zona Pellucida, *The Journal of Cell Biology* **95**:574–579.

Tanabe K, Takahashi K, and Sonda K, 1998, ABO-incompatible liver kidney donor transplantation:results and immunological aspects, *Transplantation Proceedings* **27**:1020–1023.

Tange, M. J., Tearle, R. G., Aminian, A., Romanella, M., Adam, W. R., Pearse, M. J., and d'Apice, A. J. F., 1997, Demonstration of the functional importance of the Gal epitope in an ex vivo model of xenotransplantation, *Xenotransplantation.* **4**:20–24.

Tearle, R. G., Tange, M. J., Zannettino, Z. L., Katerelos, M., Shinkel, T. A., Vandenderen, B. J. W., Lonie, A. J., Lyons, I., Nottle, M. B., Cox, T., Becker, C., Peura, A. M., Wigley, P. L., Crawford, R. J., Robins, A. J., Pearse, M. J., and d'Apice, A. J. F., 1996, The α-1,3-galactosyltransferase knockout mouse - Implications for xenotransplantation, *Transplantation* **61**:13–19.

Thall, A. D., Maly, P., and Lowe, J. B., 1995, Oocyte Gal alpha 1,3Gal epitopes implicated in sperm adhesion to the zona pellucida glycoprotein ZP3 are not required for fertilization in the mouse, *Joornal of Biological Chemistry* **270**:21437–21440.

Thall, A. D., Murphy, H. S., and Lowe, J. B., 1996, alpha 1,3-Galactosyltransferase-deficient mice produce naturally occurring cytotoxic anti-Gal antibodies, *Transplant Proc.* **28**:556–557.

Thurner, M., Rusconi, S., and Berger, E. G., 1993, Persistent repression of a functional allele can be responsible for galactosyltransferase deficiency in Tn syndrome., *J.Clin.Invest.* **91**:2103–2110.

Vaughan, H. A., Loveland, B. E., and Sandrin, M. S., 1994, Gal alpha(1,3)Gal is the major xenoepitope expressed on pig endothelial, cells recognized by naturally occurring cytotoxic human antibodies, *Transplantation* **58**:879–882.

West, J. D. and Fisher, G., 1985, Inherited cataracts in inbred mice., *Genetic Research*. **46**:45–56.

Yang, Y-C, deGoma, E., Ohdan, H., Bracy, J., Xu, Y., Iacomini, J., Thall, A. D., and Sykes, M., 1998, Tolerization of anti-Galα1,3Gal natural antibody-forming B cells by induction of mixed chimaerism, *Journal of Experimental Medicine* **187**:1335–1342.

Young Jr., W. W., Hakamori, S., and Levine, P., 1979, Characterization of anti-Forssman (anti-Fs) antibodies in human sera: their specificity and possible changes in patients with cancer., *J.Immunol.* **123**:92.

*Chapter 13*

# Modulation of αGal Epitope Expression on Porcine Cells

Mauro S. Sandrin and Ian F. C. McKenzie

## 1. INTRODUCTION

Xenotransplantation, the transplanting of organs from species other than humans, is now seen as a viable solution to the world wide problem of lack of supply of suitable human donors (Auchincloss, 1988; Auchincloss, 1990; Auchincloss and Sachs, 1998; Sachs, 1994). The major barrier to successful clinical xenotransplantation is the lack of an effective way of eliminating antibody and complement, which mediate hyperacute rejection involving natural human antibodies to Galα(1,3)Gal (αGal). Numerous studies have shown that the major, antigen to which all humans have naturally occurring antibodies, and therefore of importance when pig tissue is transplanted to humans, is the αGal carbohydrate epitope (Good *et al.*, 1992; Sandrin *et al.*, 1993; Cooper *et al.*, 1994; Sandrin and McKenzie, 1994; McKenzie *et al.*, 1994b; Sandrin *et al.*, 1994b). The phenomenon of hyperacute graft rejection and the importance of αGal in the pig-to-primate xenograft hyperacute rejection has been reviewed elsewhere (Cooper *et al.*; 1994; Sandrin and McKenzie; 1994; Sandrin *et al.*, 1994b). Here we examine the importance of the αGal epitope in xenograft rejection; describe the α1,3galactosyltransferase enzyme (αGT) responsible for generating αGal, and describe transgenic strategies designed to eliminate or reduce expression of αGal, such that

**Mauro S. Sandrin and Ian F. C. McKenzie** Molecular Immunogenetics Laboratory, Austin Research Institute, Austin and Repatriation Medical Centre, Heidelberg, Victoria, 3084, Australia.

*Subcellular Biochemistry, Volume 32: α-Gal and Anti-Gal,* edited by Galili and Avila.
Kluwer Academic / Plenum Publishers, New York, 1999.

the epitope is no longer recognised by natural human antibodies or indeed by human NK cells, which have been recently reported to recognise αGal (Artrip *et al.*, 1998; Inverardi *et al.*, 1997).

## 2. αGAL EPITOPE EXPRESSION ON PORCINE CELLS

The distribution of αGal on different tissues is of relevance to xenotransplantation, and has been examined both histologically and biochemically.

### 2.1. Histological Distribution

The distribution of αGal in pig tissues has been examined by immunohistological techniques using biotinylated Griffonia Simplicifolia lectin (IB4) which binds only to αGal (McKenzie *et al.*, 1994a; McKenzie *et al.*, 1995a) (Table I) and also by using anti-Gal containing human serum or purified human anti-αGal antibodies (McKenzie *et al.*, 1994a; Oriol *et al.*, 1993). The most striking findings were: a) endothelial cells in all small vessels (capillaries, arterioles and venules) had a uniform and dense expression of αGal; b) the highest concentrations of αGal were found in the kidney and the liver parenchyma—particularly in the brush border of the proximal convoluted tubules of the kidney; c) heart muscle was non-reactive; d) in contrast to these tissues was the distribution in the pancreas, which, apart from vessels and the lining of ducts, was non-reactive ie., adult islet cells were essentially lacking αGal. The studies present a visualisation of the problems that will occur on xenografting vascularized organs from pig to humans: hyperacute rejection will occur due to the reaction of naturally occurring antibodies with vascular endothelium and parenchyma. By contrast, the studies predict that the transplantation of isolated pancreatic islets (as opposed to a vascularized graft) would not lead to αGal antibody mediated rejection, a prediction that has been confirmed in a small animal model of αGal dependent hyperacute rejection (see below).

The distribution of αGal in pig islet cells are of particular interest (Table 1), and extensive studies (Koulmanda *et al.*, 1995; McKenzie *et al.*, 1995b; McKenzie *et al.*, 1995c) showed that:- a) adult pig islet cells do not express αGal (other than in the capillaries of the islet—which are not likely to be relevant if isolated non-vascularised islets are used); b) fetal islet cells also do not express αGal although there appears to be more αGal in the islet as the capillary staining is stronger than the adult; c) after tissue culture αGal is strongly expressed in the cells in the islet, but not on differentiated endocrine cells—in particular is not found on the insulin producing β cells. The expression is endogenous and likely to be due to up regulation of the αGT (Ariga *et al.*, 1993; Cummings and Mattox, 1988); d) staining for endocrine hormone producing cells (insulin, glucagon stated in c above or

**Table I**
**Tissue Distribution of αGal**

| Tissue | Intensity of staining |
|---|---|
| Endothelium | |
| capillaries | +++ |
| arterioles | +++ |
| arteries | +++ |
| aorta | +/– |
| Liver | |
| parencyma | ++ |
| Kidney | |
| proximal convoluted tubules | ++ |
| glomerulii | ++ |
| collecting ducts | ++ |
| Lung | |
| alveoli | + |
| Heart | |
| myocardium | – |
| Pancreas | |
| ducts | + |
| acini | – |
| islets | – |
| Islets | |
| adult | – |
| foetal | – |
| cultured | +++ |
| cultured and transplanted to nude mice | – |

Staining recorded as – = absent, + = weak, to +++ = strongly staining. Adapted from McKenzie *et al.* 1994, 1995a,b..

somatostatin) indicate that hormone secreting cells do not express αGal; e) after transplantation to mice, the islet matures and αGal expression of the cultered islet is lost. A major question to answer is to determine the relevance of αGal expression on islet graft survival; this is discussed below.

## 2.2. Molecular Distribution

The tissue distribution studies of αGal describe which cells express αGal and several studies have now identified the molecules bearing αGal epitopes or pig cells $\alpha Gal^+$. Glycoproteins of molecular weights 115 kDa, 125 kDa and 135 kDa have been identified as the major targets on the surface of pig endothelial cells to which natural human antibodies react (Holzknecht and Platt, 1995; Platt *et al.*, 1990). The cell surface molecules bearing αGal epitopes were also examined after cell surface iodination and performing 2D SDS-PAGE after immuno-precipitation with either Gal antibodies in normal human serum or using the IB4 lectin (Vaughan *et al.*, 1995a). The studies demonstrated that pig endothelial cells and peripheral blood lymphocytes have at least 30 αGal⁺ glycoproteins recognised by Gal antibodies and the IB4 lectin, of which the major glycoproteins are of molecular weights: 200kD; 100–140kD; 50–90kD and Mr = 45kD. Some of these proteins have been identified; e.g. von Willebrand's factor; a 95-kDa glycoprotein homologous with chick DM-GRASP (an immunoglobulin superfamily cell adhesion molecule that is expressed in the nervous and immune systems); and the integrins:- α1, αv, α3/α5, β1 and β3 (Holzknecht and Platt, 1995; Platt and Holznecht, 1994). It is likely that αGal is present on both N-linked and O-linked carbohydrate side chains on these molecules.

Glycoproteins are not the only molecules that express αGal as glycolipids also carry αGal. Galili originally demonstrated that human αGal IgG antibodies react with a major glycolipid found on the surface of rabbit erythrocytes (Galili *et al.*, 1985; Galili *et al.*, 1984). Similar studies from Samuelsson demonstrated that naturally occurring human IgM reacts with glycolipids extracted from pig cells, and showed that αGal is the major epitope of these glycolipids using a combination of NMR and mass spectroscopy (Samuelsson *et al.*, 1994). In addition, αGal occurs on the carbohydrate side chains of secreted proteins including mouse IgG (Thall and Galili, 1990) and laminin (Arumugham *et al.*, 1986)—indeed many mouse monoclonal antibodies are αGal⁺, which may contribute to their short 1/2 life in humans (Borrebaeck *et al.*, 1993).

As αGal is expressed on many different molecules and in many different cell types the elimination of any one of these molecules will not remove αGal—the gene encoding the transferase which α-galactosylates *all* molecules needs to be targeted as discussed below.

## 3. αGAL EPITOPE EXPRESSION ON HUMAN CELLS IS DUE TO A CROSS-REACTION

Humans do not express αGal as there are multiple major alterations in the human αGT gene (see below) to render it non-functional. However it has recently been proposed that the αGT gene can become functional in humans in cancer and auto-immune diseases, as anti-αGal antibodies have been found to react with breast cancer cell lines (Castronovo *et al.*, 1989; Galili, 1989; Petryniak *et al.*, 1991) and with proteins which are targets of antibodies in autoimmune disease (Galili, 1989).

### 3.1. Expression of αGal on Human Tumors

Castronovo used cytofluorographic analysis with several lectin and human anti-αGal antibodies to demonstrate the presence of αGal on human carcinoma cell lines and on 50% of malignant breast carcinoma cells, but not on normal breast epithelial cells (Castronovo *et al.*, 1989). In the same study, the presence of αGal antibodies inhibited the attachment of malignant cells to human umbilical vein endothelium and to laminin, which is thought to be involved in metastases, and suggested a possible role for anti-αGal antibodies in the natural immunity to cancer (Castronovo *et al.*, 1989). In accord with these observations, Galili demonstrated that the human breast cancer cell line MCF-7 (which lacks αGT mRNA) was lysed by purified anti-αGal in an antibody-dependent cell-mediated cytolysis assay (Galili, 1989). The explanation for these findings was that in the cancer cells, the human αGT genes were 'derepressed' allowing for expression of the enzyme and synthesis of αGal. However, the lack of expression of the human αGT genes is not due to gene repression, but rather, is the result of multiple mutations within the gene (see below). It is difficult to see how several mutations could occur so regularly to allow the αGT to be expressed by a majority of cancer cells, particularly as no αGT mRNA was present (Galili, 1989). We have another explanation for the appearance of αGal on tumors—the cross reaction of Gal antibodies with cellular mucins—shown by the description of a αGal peptide mimic.

### 3.2. Peptide Mimics for αGal

There is now convincing data that peptides can mimic carbohydrates and interact specifically with the carbohydrate binding regions of both lectins and anti-carbohydrate antibodies (Hoess *et al.*, 1993; Oldenburg *et al.*, 1992; Scott *et al.*, 1992). One approach to block the interaction of αGal antibodies with the αGal on endothelial cells would be to use a peptide which mimics αGal; we have demonstrated that a synthetic octapeptide, (DAHWESWL), isolated from a peptide library using the α-galactosyl specific lectin IB4, "mimics" the carbohydrate epitope αGal (Vaughan *et al.*, 1995b); a second peptide has a similar effect (Kooyman et al., 1996). *In vitro* studies demonstrated that the binding of the IB4 lectin or human natural αGal antibodies to pig cells could be blocked by the mimic peptide, in several serological assays:- a) the hemagglutination of pig erythrocytes; b) cytofluorographic analysis of pig lymphocytes and endothelial cells; c) cytotoxicity of human serum on pig endothelial cells; d) in an ELISA assay. The relative affinity of the peptide for the IB4 lectin was similar to that of α-galactosyl sugars, although it was bound at a lower affinity by human antibodies (Vaughan *et al.*, 1995b). In testing the specificity of these reactions it was apparent that human Mucin 1 peptides could also react with Gal antibodies, and providing an explanation for the cross reactivity (Sandrin *et al.*, 1997b) as the breast cancer tumors are known to express MUC1. These Gal antibodies cross react with MUC1 present in the cancer cells.

### 3.3. Cross Reaction with Human Mucin 1

Thus, not only do anti-αGal antibodies and IB4 lectin react with the DAHWESWL peptide, they also react with peptides encoded by mucin genes (particularly Mucin 1) (Sandrin *et al.*, 1997b)—sequences known to be rich in serine, threonine and proline. This activity was demonstrated by blocking the reaction of antibody and lectin with a αGal⁺ pig endothelial cell line by mucin derived peptides; the reactions were specific and did not occur with a random peptide sequence containing the same amino acids. In addition, Mucin 1 antibodies could react with the αGal expressed after transfection of COS cells with αGT cDNA. Furthermore, the IB4 lectin and anti-αGal antibodies could react with Mucin 1 found on the surface of tumours cells. The findings are of interest and serve to explain the previously reported findings that human cells can, at times, express αGal, in that the anti-αGal antibodies directly cross react with deglycosylated mucins expressed on the cancer cells.

## 4. THE α1,3GALACTOSYLTRANSFERASE ENZYME (αGT)

The αGal sugar is the terminal sugar at the non-reducing end of N-glycans, O-glycans and glycolipids on many different molecules, and is used as an alternative to sialic acid as an uncharged terminal carbohydrate residue. The α1,3galactosyltransferase (αGT) is the enzyme which catalyses the synthesis of αGal, and as the knowledge of this enzyme/gene is crucial to the studies on alteration of the expression of αGal, some background is given here but see also Chapters of Joziasse *et al.*, and Henion and Galili.

### 4.1. Biochemical Reaction

The αGal residues are synthesised by αGT, which catalyses the addition of a terminal α-linked galactose (Gal) to a β-linked Gal, such as N-acetyl lactosamine (Fig. 1). The enzyme requires $Mn^{2+}$ ions for catalytic activity (Blanken and Van den Eijnden, 1985), and uses UDP-Gal as the sugar donor. The substrate specificity of the αGT is any unbranched galactose in β-linkage with either N-acetyl glucosamine or N-acetyl galactosamine (Blanken and Van den Eijnden, 1985). The αGT of calf thymus can utilise all 4 types of acceptor substrates: Type I (Galβ(1,3)GlcNAc), Type II (Galβ(1,4)GlcNAc), Type III (Galβ(1,3)GalNAc) and Type IV (Galβ(1,4)GalNAc), but not branched acceptors, such as H substance, which is of importance when designing transgenic strategies for the reduction of αGal (see below). We have examined the substrate specificity of the pig αGT and shown that it can add αGal to both Type 1 and Type 2 acceptor substrates (unpublished data).

### 4.2. Molecular Cloning

cDNA clones encoding the αGT have been isolated from the mouse (Joziasse *et al.*, 1992; Larsen *et al.*, 1989), ox (Joziasse *et al.*, 1989), pig (Sandrin

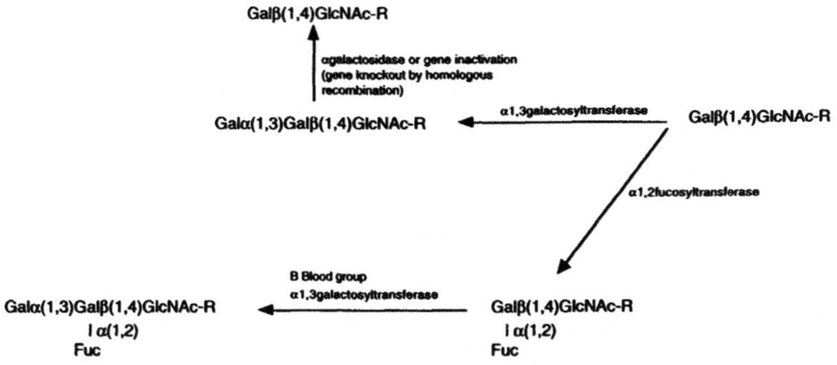


**Figure 1.** Biosynthetic pathway for synthesis of αGal. Pathway begins with N-acetyl-lactosamine (Galβ(1,4)GlcNAc), and the αGT adds galactose to generate αGal. Both gene inactivation by homologous recombination and transgenic approaches using αGaldase would eliminated αGal. An alternative transgenic approach to utilise the substrate of αGT by αFT, produces H substance, the precursor of human A and B blood groups.

*et al.*, 1994a; Strahan *et al.*, 1995) and New World monkeys (Henion *et al.*, 1994), and all have sequence homology (Fig. 2) and have similar requirements for acceptors (ie non-branched β-linked Gal). The amino acid sequence of αGT suggests it to be a type II membrane protein (amino terminal orientated towards the cytoplasm) comprising of a short cytoplasmic tail, a transmembrane region (16 amino acids), a stalk region and a carboxyl-terminal catalytic domain (that resides within the lumen of the Golgi). This predicted structure is typical of all glycosyltransferases isolated thus far (Joziasse, 1992). The murine αGal transferase gene (which is the best characterised) is a single gene composed of 9 exons spanning ~ 35 kb of chromosome 2 (Joziasse *et al.*, 1992): the first three exons encode the 5′ untranslated sequence of mRNA, with the next six exons encoding the protein. The catalytic domain of the αGT is encoded within the last exon (exon 9) (Joziasse *et al.*, 1992), which makes this an ideal exon to target for gene knockout strategies (see below). Studies in the mouse have demonstrated 4 isoforms of this enzyme (Joziasse *et al.*, 1992), and 2 isoforms have been identified in the pig (Sandrin *et al.*, 1994a; Strahan *et al.*, 1995), although the significance of these isoforms is unknown.

## 4.3. A Family of α1,3Galactosyltransferases

Comparison of the αGT amino acid sequence shows homology with the human ABO blood group transferases (A transferase transfers an α-linked GalNAc and the B transferase (Yamamoto *et al.*, 1990) transfers an α-linked Gal to H sub-

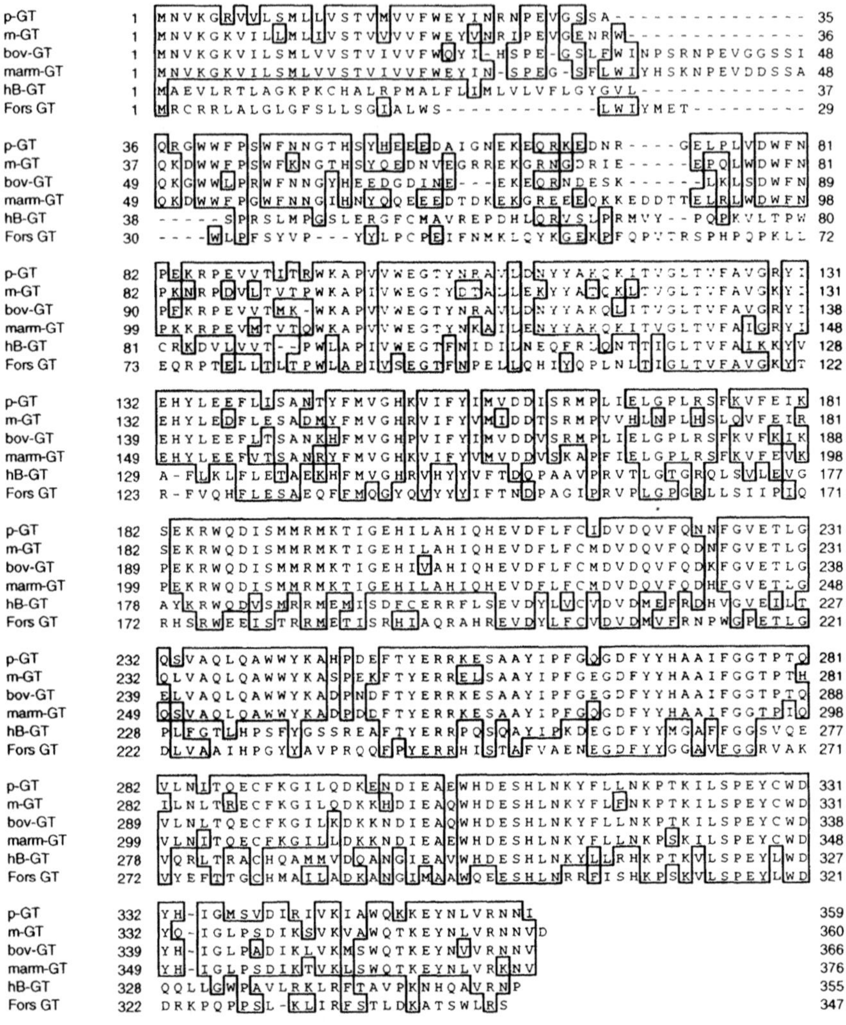

**Figure 2.** Alignment of the amino-acid sequences comparison of αGTs. Single letter amino acid code used for the pig (p-GT), mouse (m-GT), bovine (b-GT), marmoset (marm-GT), human B transferase (UDP-Gal: Fuc(α1,2)Gal(α1,3)Gal transferase) (hB-GT) and dog Forssman transferases (Fors-GT). Identical amino acid boxed,—represents break introduced into the amino acid sequence for maximal alignment.

stance) and the Forssman transferase (Haslam and Baenziger, 1996) (which transfers an α-linked GalNAc) (Fig. 2) which suggests a common evolutionary origin of the molecules. Comparison of the amino acid sequences of pig, mouse, ox and marmoset show that the highest sequence identity between all the cloned αGT is in the carboxyl portion which contains the catalytic domain, with an identity of 78%. Using a series of truncated cDNA clones (Henion *et al.*, 1994) it was shown that the active site (catalytic domain) of αGT commences 67 amino acids from the putative transmembrane domain, the 277 carboxyl terminal amino acids being required for activity. Although the ABO blood group transferases share amino acid homology with αGT, they differ in their substrate specificity in one very important aspect; both the human transferases have an absolute requirement for the subterminal Gal to be branched and fucosylated; the opposite is true for the murine, bovine and pig αGT as these transferases will only transfer a Gal to a non-branched Gal. The precursor for the A and B transferases is H substance which is produced by the *H* gene which is an α1,2fucosyltransferase (αFT) (see below).

## 5. GENE INACTIVATION OF THE α1,3GALACTOSYLTRANSFERASE

As anti-αGal antibody removal or blocking can inhibit hyperacute rejection (Simon *et al.*, 1998; Xu *et al.*, 1998), it follows that antigen removal would also prevent the occurrence of hyperacute rejection. The most obvious way of doing this would be to destroy the αGT gene. Apart from the technical difficulty of performing gene inactivation in pigs, what would the consequences of inactivating the αGT gene in pigs? Would this be lethal as is the inactivation of some glycosyltransferases (Asano *et al.*, 1997; Ioffe and Stanley, 1994; Lu *et al.*, 1997; Metzler *et al.*, 1994)? Natural occurring mutations in humans and knockout mice have given insights into these questions.

### 5.1. Gene Inactivation of Human αGT

The αGT gene in humans contains multiple frame shift and non-sense mutations and is therefore a non-functional pseudogene (Joziasse *et al.*, 1991; Larsen *et al.*, 1990b); it has been localised to human chromosome 9 (Shaper *et al.*, 1992), which is syntenic with mouse chromosome 2 where the murine αGT gene is located. The genes for the human A and B transferases have also been mapped to human chromosome 9q (Ferguson-Smith *et al.*, 1976), linked to the human αGT pseudogene. It is likely that a tandem duplication occurred in the gene encoding the αGT in an ancestor of the higher primates, and through random mutation, one of the αGT became dependant on the presence of a fucosylated galactose (H substance) on the acceptor molecule for functional activity ie the B transferase

(Sandrin *et al.*, 1994b). This ancestor probably produced both B blood group and the αGal epitope. Another mutation occurred in the αGT gene which introduced an in-phase termination codon, thus producing a pseudogene and an ancestor which no longer expressed the αGal epitope and made anti-αGal antibodies (Sandrin *et al.*, 1994b). A second copy of a human αGT pseudogene maps to human chromosome 12 (Shaper *et al.*, 1992), although the relationship between the two pseudogenes is unclear. Similar events leading to αGT gene inactivation by frame shift mutation have been shown in Old World monkeys (Galili and Swanson, 1991).

### 5.2. Consequences of Gene Inactivation

The inactivated αGT gene, arising by naturally occurring mutations in humans and Old World monkeys apparently has no deleterious effects. In addition, humans with blood group O have several nucleotide substitutions which introduce in-phase termination codons into the B transferase gene, giving rise to a non-functional enzyme or gene (Yamamoto *et al.*, 1990) and to a null blood group phenotype. Furthermore, humans with the "Bombay" phenotype lack a functional αFT (Kelly *et al.*, 1994) and do not have a branched terminal Gal with a Fuc residue. Thus humans and Old World monkeys are viable without an αGT and there are Bombay phenotype humans without an αFT with which to "cap" terminal Gal residues. As humans and Old World monkeys no longer express αGal, they produced anti-αGal antibodies. Thus one would predict that other species deficient in αGT would be viable and produce αGal antibodies, as shown in chickens and Gal o/o mice.

### 5.3. Gal o/o Mice

Mice deficient in αGT (Gal o/o mice) after gene inactivation by homologous recombination have been made in two centres (Tearle *et al.*, 1996; Thall *et al.*, 1995), see also Chapters of Thall and of Pearce et al. targeting exon 9 (which encode the catalytic domain) for disruption. Several important observations arise from the studies of these mice: as predicted, inactivation of the αGT gene is not lethal (ie like humans), and also like humans the mice completely lack αGal in all tissues and produce natural anti-αGal antibodies. Whether the antibodies are induced by bacterial infection, or without antigenic stimulus is not clear, but the occurrence of antibodies in the Gal o/o mice parallels the observations in humans and Old World monkeys. Although the tissues of Gal o/o mice do not bind purified human anti-αGal antibodies, when examined with human serum some antibody binding was still observed (Tearle *et al.*, 1996), due probably to the binding of natural human anti-N-acetyl-lactosamine antibodies which are present in substantial amounts in all humans (McKenzie *et al.*, 1996). The Gal o/o mice have proven to be useful for the development of a small animal model to study αGal dependent hyperacute rejection (see below).

## 6. MODULATION OF αGAL EPITOPE EXPRESSION BY αGALACTOSIDASE

Another approach from the removal of the αGal carbohydrate is using αgalactosidase (αGaldase), a lysosomal exoglycosidase, to cleave the terminal α-D-galactosyl residues from αGal.

### 6.1. *In Vitro* and *ex Vivo* Studies

Several groups have shown that αGaldase (of coffee bean or bacterial original) treatment of erythrocytes, lymphocytes or endothelial cells removes the αGal and eradicates the reaction with human serum (Holzknecht and Platt, 1995; Oriol *et al.*, 1993). Other studies have shown that perfusion of tissue with bacterial αGaldase prior to transplantation delayed the onset of hyperacute rejection (Cairns *et al.*, 1994). However, an *ex vivo* approach does not solve the problem of continual resynthesis and replacement of αGal by αGaldase treated cells and it would not be possible to ensure total and permanent eradication of αGal by such a method.

### 6.2. Transgenic Approaches

Our studies have examined the efficacy of using αGaldase cDNA as a transgene to remove αGal in situ (Osman *et al.*, 1997; Sandrin *et al.*, 1997a). Initially we tested whether co-expression of human αGaldase and pig αGT in mammalian cells could alter the level of cell surface αGal (Table II). COS cells transiently-transfected with αGT cDNA alone and stained with IB4 lectin showed~ 60% of cells expressing αGal, while cells expressing αGaldase alone showed no surface staining with IB4. The co-expression of αGaldase and αGT cDNAs resulted in a 75% reduction in IB4 staining (Osman *et al.*, 1997) and the level of αGaldase activity in transiently-transfected COS cells was at ~ three times higher than background level ie. expression of αGaldase after transfection of cDNA can significantly reduce levels of αGal. These studies were extended to the $\alpha Gal^+$ pig endothelial cell line (PIEC) by generating stable cell lines which express human αGaldase (Osman *et al.*, 1997) which had a 10-fold decrease in antibody binding when compared to control PIEC cells. Thus, expression of the αGaldase enzyme after transfection by constitutive $\alpha Gal^+$ cells leads to a reduced level of cell surface αGal.

Transgenic mice expressing the human αGaldase gene were produced and analysis of their splenocytes showed a 25% reduction in binding of Gal antibodies

**Table II**
**Summary of Serology of Transfected COS Cells**

| Enzyme expressed | Surface expression of αGal | Surface expression of H substance | Lysis by human antibodies |
|---|---|---|---|
| — | – | – | – |
| αGT | +++ | – | +++ |
| αGaldase | – | – | – |
| αGT+αGaldase | ++ | – | ++ |
| αFT | – | +++ | – |
| αGT+αFT | + | +++ | + |
| αGT+αFT+αGaldase | + – | +++ | – |

Staining recorded as – = absent, + = weak, to +++ = stronlgy staining. Adapted from Osman *et al.* 1997, Sandrin *et al.* 1995a,b..

(Osman *et al.*, 1997) and a major decrease in complement-mediated cytolysis of αGaldase transgenic splenocytes compared with that obtained with normal mice. However using αGaldase in this manner did not completely eliminate αGal, and therefore greater expression of the αGaldase or combination with other approaches (see next section) is required for the total removal of αGal.

## 7. MODULATION OF αGAL EPITOPE EXPRESSION BY α1,2FUCOSYLTRANSFERASE

Another strategy to decrease αGal expression would be to use another transferase which could "deviate" the glycosylation pathway away from αGal expression and lead to the production of another carbohydrate not recognised by natural antibodies. It is known that different glycosyltransferases can compete for the same substrate and this information was used to develop a strategy to produce less αGal. We had showed that the simultaneous expression of two glycosyltransferases [αFT (H transferase) and αGT] does not lead to equal synthesis of each monosaccharide, but the activity of the αFT is given preference over that of the αGT, so that the expression of αGal is almost entirely suppressed (Sandrin *et al.*, 1995). This was shown both *in vitro* and *in vivo* in transgenic mice.

### 7.1. Enzyme

The αFT (H transferase) was chosen as an appropriate enzyme to decrease the expression of αGal, as both the αFT and the αGT use N-acetyl lactosamine (NAcLac) as an acceptor substrate, transferring fucose or galactose to generate fucosylated N-acetyl lactosamine (H substance) or αGal, respectively (Fig 1). Furthermore, the αGT of most animals cannot use the H substance as an acceptor to transfer the terminal galactose, but will only transfer to NAcLac residues. In contrast the human B blood group transferase, which is an αGT, has an absolute requirement for the acceptor to be H substance and will not transfer a Gal to NAcLac. Our hypothesis was that the overexpression of αFT may lead to a reduction in the formation of αGal epitopes (Sandrin *et al.*, 1995) and data showed that there is indeed a hierarchy of glycosyltransferases that are expressed within the same cell, and that the αFT takes precedence over the αGT (Sandrin *et al.*, 1996; Sandrin *et al.*, 1995).

### 7.2. Transfection

COS cells simultaneously transfected with cDNA clones encoding αFT and the αGT show preferential expression of the H substance rather than αGal , even though αGT mRNA and functional enzyme was present (Sandrin *et al.*, 1996; Sandrin *et al.*, 1995) (Table II). Furthermore, human antibody binding by the COS cells transfected with both αFT and αGT was decreased. In a pig kidney cell line, which expresses both the αGal and H substance, the increased expression of H substance induced by the transfection and stable expression of αFT cDNA, resulted in decreased expression of αGal, decreased human antibody binding and decreased complement mediated cell lysis. We obtained similar results using the pig homologue of the human H type αFT, PFUT1 (Cohney *et al.*, 1996), and pig endothelial cell lines expressing the pig αFT show a decreased binding of human anti-αGal antibodies, and of human NK cells (Artrip *et al.*, 1998).

### 7.3. Transgenesis

Transgenic mice have been produced which ubiquitously express the αFT enzyme (Cohney *et al.*, 1997; Sandrin *et al.*, 1995). The spleen cells of such mice show a major decrease in αGal expression and decreased anti-αGal natural antibody binding. This was also observed in other tissues, including endothelial cells (Cohney *et al.*, 1997). The results have been confirmed by others (Chen *et al.*, 1996; Koike *et al.*, 1996; Sharma *et al.*, 1996) and more recently transgenic pigs expressing the αFT have been reported (Koike *et al.*, 1996; Sharma *et al.*, 1996) which show similar results to transgenic mice, ie. reduction in the cell surface expression of αGal. It is not clear whether the remaining amount of αGal present in an αFT transgenic animal would be sufficient to react with human antibody to trigger hyperacute rejection, and it remains to be seen if organs from H transgenic pigs do not undergo hyperacute rejection when transplanted into primates. However based on our findings in the Gal o/o mice (see below) we would predict that they may not.

### 7.4. Combination with αGalactosidase

A combined approach using α-galactosidase and α1,2fucosyltransferase was tested (Osman *et al.*, 1997). COS cells expressing αGT, αFT and α-galactosidase cDNAs showed negligible cell surface staining with the IB4 lectin specific for αGal (Table II) and also lysis by human serum containing antibody and complement. The results clearly demonstrate that αFT and α-galactosidase have an additive effect in reducing the expression of αGal on the cell surface, and have the potential to be used together in transgenic pigs.

### 7.5. Other Fucosyltransferases

The αFT strategy outlined in the preceding sections led to significant but not an absolute reduction of αGal expression. Can other glycosyltransferases be used to give improved αGal reduction? It is known that the αGT enzyme can galactosylate two types of precursor chains: Type 1: Galβ(1,3)GlcNAc and Type 2: Galβ(1,4)GlcNAc (Blanken and Van den Eijnden, 1985). Similarly αFTs can transform both Type 1 and Type 2 precursor chains into H substance. Two human αFTs have been described, H-transferase or FUT1 (Larsen *et al.*, 1990a) and secretor (Se) transferase or FUT2 (Kelly *et al.*, 1995), and while both enzymes can use both types of precursors, FUT1 preferentially utilises Type 2 precursor chains and FUT2 utilises both Type 1 and Type 2. We have isolated a second αFT cDNA from the pig (Cohney *et al.*, 1998), PFUT2, the homologue of the Secretor gene, and which shows a high level of identity with all the previously cloned fucosyltransferases, but highest with the human "Sec" αFT and Rabbit RFT-III. COS cells transfected with PFUT2 showed cell surface staining for H substance with UEA-I lectin and kinetic studies with a number of substrates are consistent with PFUT2 having a preference for type 1 and 3 acceptors (Cohney *et al.*, 1998) as found with the human and rabbit counterparts. It was found that the coexpression of PFUT2 and αGT in COS cells leads to preferential expression of H substance and dramatic reduction in αGal and further, using a combination of FUT1 and FUT2 had an additive effect in suppressing expression of the αGal of >99% (Cohney *et al.*, 1998). Thus the dominance originally described with the human αFT is not unique to that enzyme, but is a property of all αFTs.

## 8. MOLECULAR MECHANISM OF TRANSFERASE DOMINANCE

A possible mechanism to account for the apparent dominance of one transferase over another is the location of the enzyme within the Golgi apparatus (Sandrin *et al.*, 1995). Perhaps if αFT was proximal to αGT, an fucosylation would occur before galactosylation and result in high H and low αGal expression. We examined the effect of Golgi localisation by exchanging the cytoplasmic tails of αFT

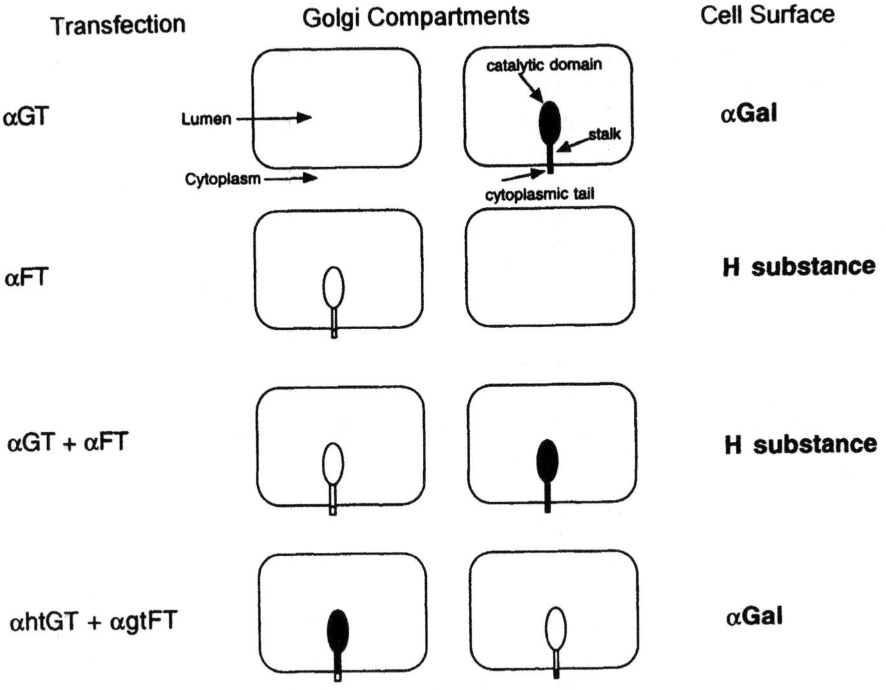


**Figure 3.** Schematic diagram of wild-type and chimeric glycosyltransferases. Diagram depicting interpretation of Golgi localization of wild-type and chimeric transferases. COS cells were transfected with cDNA clones encoding: αGT, αFT, αhtGT (in which the cytoplasmic domain of αGT has been completely replaced by the cytoplasmic domain of αFT), αgtFT (in which the cytoplasmic domain of αFT has been entirely replaced by the cytoplasmic domain of αGT). The clear pattern represents αFT, filled pattern represents αGT represents carbohydrate epitope expressed on the cell surface. Based on the data of Osman *et al.* 1996.

and αGT; if Golgi targeting signals are contained within the cytoplasmic tail sequences of the enzymes then a "tail switch" would permit αGT first access to the substrate and thereby reverse the observed dominance of αFT (Osman *et al.*, 1996). Two chimeric transferase proteins were constructed which consisted of the $NH_2$ terminal cytoplasmic tail of one transferase coupled to the transmembrane and catalytic domain of the other (Fig. 3). On transfection a complete reversal in the staining pattern of transfected COS cells was seen with the chimeric transferase compared with the wild-type transferases, demonstrating that the cytoplasmic domains of αFT and αGT are sufficient for the retention, localisation and activity of these two enzymes. This finding can be used to target another αFT to the same compartment as the αGT, and in conjunction with an αFT transgene would lead to greater reduction in αGal expression. The chimeric enzymes were fully functional showed Km values and cell surface carbohydrate expression comparable to normal glycosyltransferases and co-expression of the two chimeric glycosyltransferases in COS cells resulted in cell surface expression of αGal and virtually no H substance (Osman *et al.*, 1996). Furthermore, coexpression in COS cells of glycosyltransferases with the same cytoplasmic tail resulted in equal expression of both αGal and H substance. Further studies which localized enzymes using confocal microscopy, and mutational analysis of the cytoplasmic tails (in preparation) demonstrated that the cytoplasmic tail of the glycosyltransferases determines their site of location, the temporal order of action, and therefore "dominance," of the two enzymes.

## 9. MODULATION OF αGAL EXPRESSION BY OTHER GLYCOSYLTRANSFERASES

Can other glycosyltransferases decrease the level of αGal expression? Possible glycosyltransferases are those which also utilise NAcLac as a substrate (Sandrin and McKenzie, 1994; Sandrin *et al.*, 1994b), and sialyltransferases (αST). The results using STs differ, one study using α2,3ST showed some reduction in the levels of αGal, but not as dramatic as with the αFT (Tanemura *et al.*, 1997a; Tanemura *et al.*, 1997b). Another study found that neither the α2,3ST nor the α2,6ST resulted in any significant reduction of αGal (Sharma *et al.*, 1996). It is interesting to note that in earlier studies, Smith demonstrated that the α2,3ST and the αGT compete for the same substrate within a cell (Smith *et al.*, 1990)—as the α2,3ST, like the αGT, is involved in transferring a terminal sugar, and is located in the same Golgi compartment. The efficiency of αST may be improved by the redirection of this enzyme to an earlier Golgi compartment using a chimeric construct with the cytoplasmic tail of αFT, as discussed above.

Another approach is to over express N-acetlyglucosaminyl transferase III (GNTIII) which effectively reduces αGal expression both *in vitro* (Tanemura *et al.*, 1997a) and *in vivo* in transgenic mice (Tanemura *et al.*, 1997b). While the mechanism of this suppression of αGal expression is not clear, it is possibly due to the branched GlcNAc residue inhibit oligosaccharide processing by other glycosyltransferases (Tanemura *et al.*, 1997a; Tanemura *et al.*, 1997b), as the GNTIII is involved in the production of bisecting GlcNAc residues in branched N-linked oligosaccharides.

## 10. SMALL ANIMAL MODELS OF αGAL EPITOPE DEPENDENT HYPERACUTE REJECTION

It is evident that to overcome the hyperacute rejection of vascularised organs in pig to primate/human transplants, αGal expression in the donor has to be decreased or the αGal antibodies removed from the recipient. What is not clear is what is the minimal level of expression to cause hyperacute rejection, and what is the optimal combination of transferases, galactosidases or complement regulatory proteins required to avoid HAR. *In vitro* experiments can only suggest what might happen *in vivo*. To overcome the problems of preclinical studies in primates, we have developed models using the Gal o/o mice to examine the αGal/anti-αGal dependent hyperacute rejection. Gal o/o mice are similar to humans in that they have naturally occurring αGal antibodies and thus are suitable recipients of αGal$^+$ tissues to provide models to study both the early and later phases of αGal antibody mediated rejection which is of direct relevance to pig xenografts to humans.

### 10.1. Skin Grafts

Our first studies of anti-αGal antibody mediated rejection consisted of transplanting skin from Gal$^+$ mice and rats to Gal o/o mice (McKenzie *et al.*, 1998b). In all cases αGal$^+$ skin was not rejected hyperacutely by Gal o/o mice, but rejected at the same rate as skin grafts from normal αGal$^+$ mice, and by cellular, and not humoral mechanisms. This is not unexpected as the studies of Winn and colleagues (Baldmus *et al.*, 1973; Winn *et al.*, 1973) demonstrated antibody alone will not usually cause allograft rejection in mice without additional complement being given (as mouse complement is very poor at lysing cells), or when high titered xeno-antibody such as mouse anti-rat serum is given, then additional complement may not be necessary. Furthermore, rejection is dependant on the density of antigen to fix a sufficient amount of antibody and complement to causes damage. To address these issues, the Gal o/o mice were immunosuppressed (to eradicate cellular rejection), additional anti-αGal antibody and complement given (or the titer and affinity of the endogenous αGal antibody increased by immunization of the Gal o/o mice) and the mice grafted with skin from rats or mice: rat skin grafts rejected slowly whereas mouse skin grafts did not. Thus, skin grafts are relatively resistant to anti-αGal antibody mediated hyperacute rejection and are not a suitable model to examine HAR.

### 10.2. Heart Grafts

We next examined a more relevant model, of heart transplants, wherein Gal$^+$ hearts were heterotopically transplanted into the abdomen of Gal o/o mice and observed (McKenzie *et al.*, 1998b). Heart transplants from Gal$^+$ C57BL/6 mice were not rejected by 5–7 week old Gal o/o mice, unless additional anti-αGal antibody and complement were given; although older recipients rejected αGal$^+$ mouse hearts if additional αGal antibody was given. Similar results were obtained in 5–7 week old Gal o/o mice using rat hearts, which have a higher level of αGal, ie. additional anti-αGal antibody and complement were required. By contrast, rat hearts were rejected by older Gal o/o recipients, without additional anti-αGal antibody

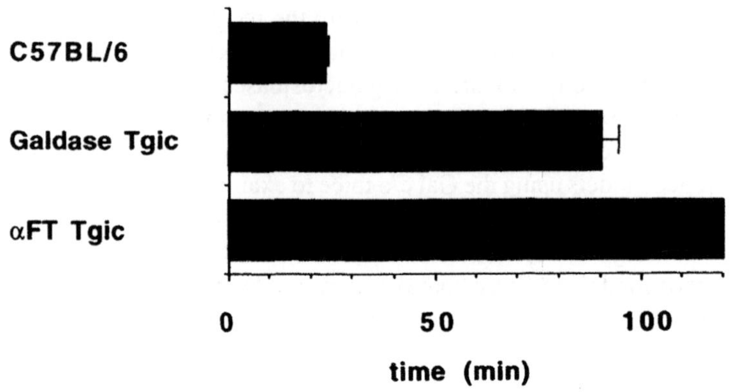


**Figure 4.** Survival times of heart grafts in vivo. Hearts from C57BL/6, the Galdase transgenic and αFT transgenic were transplanted to Gal o/o mice, human serum was injected and survival times (minutes) monitored. Adapted from McKenzie *et al.* 1998a.

and complement. The rejection process occured within 10–20 minutes, although the histological features were not those of classical hyperacute rejection with intravascular thrombosis and polymorph accumulation, but rather, consisted mostly of intravascular hemorrhage, with the deposition of antibody and complement. While both IgG and IgM were deposited in the tissue when human serum was used as a source of anti-αGal antibody, the major class of antibody which mediated rejection was IgM anti-αGal antibodies; IgG anti-αGal antibodies had no effect.

Transgenic mice with αGal antigen modification by expression of the αFT and αGaldase were examined in this model: donor hearts from αGaldase mice were hyperacutely rejected, although the survival time was extended from 30 minutes to 90 minutes (Fig. 4), perhaps not a clinically exciting observation, but encouraging in that the decrease in cell surface αGal significantly prolonged graft survival *in vivo*. By contrast, αFT transgenic mouse hearts showed "indefinite" survival (>120mins, Fig. 4) with no evidence hyperacute rejection (McKenzie *et al.*, 1998b). Furthermore αFT transgenic mouse hearts were not rejected by antibody mediated mechanisms when transplants were examined up to 100 days, at which time they were macroscopically and microscopically normal, even though the levels of anti-αGal antibodies increased dramatically within the first 21 days (unpublished results).

Several conclusions can be drawn from these studies: firstly the heart transplants to Gal o/o mice are a suitable model to study αGal antibody mediated rejection phenomena of relevance to pig-to-human xenografts; secondly the studies demonstrate that in experimental systems, hyperacute rejection can be overcome, and suggests that the use of αFT transgenic pigs for heart transplants to humans may be feasible.

### 10.3. Islet Grafts

As discussed above, pig pancreatic islets, which do not express αGal, show a marked increase in αGal expression after culture. Fetal pig islets have been transplanted to humans (Groth *et al.*, 1996), and there was a suggestion of function (production of C-reactive protein and residual tissue found) though the level and affinity of the αGal antibodies greatly increased (Galili *et al.*, 1995). Several questions remain to be answered: Are islets intrinsically resistant after transplantation to hyperacute rejection, or specifically resistant to anti-αGal antibody destruction? To address this question we transplanted cultured foetal pig islets to Gal o/o mice, and showed that pig islets are not rejected by Gal o/o mice (McKenzie *et al.*, 1998a)—not even in the presence of large amounts of αGal antibodies and complement; some damage was sustained in anti-αGal hyperimmune mice which may be involved in subsequent rejection. However the islets were found not to be intrinsically resistant to hyperacute rejection, as they could be destroyed by hyperimmune mouse anti-pig serum. The Gal system does not appear to be involved in the rejection of pig islets by humans.

### 10.4. Bone Marrow Grafts

Bone marrow xenotransplantation will be an important part of tolerance induction to xenografts, however, destruction by αGal antibodies will eradicate the transplanted marrow and must be overcome for the transplant to survive. We have examined the effects of αGal antibodies on the ability of $\alpha Gal^+$ bone marrow to engraft Gal o/o mice (Sandrin *et al.*, 1998). Bone marrow from $\alpha Gal^+$ mice and αFT transgenic mice is immediately rejected by natural Gal antibodies found in Gal o/o mice; lymphocytes from αFT transgenic mice can recirculate to lymphoid organs in the short term—unless the Gal o/o mice are made hyperimmune and have excess anti-αGal antibodies. Thus, while αFT transgenic hearts may not be rejected by existing αGal antibodies, bone marrow is more susceptible to the Gal antibodies and is "rejected." The effect of the Abs could be overcome by depleting the anti-αGal antibodies, and we note that giving high doses of immunosuppression and high numbers of bone marrow cells ($5 \times 10^8$ cells/kg) has led to a transient low level of pig cell chimera in a cynomologous monkey (Sachs *et al.*, 1995).

The results of our studies suggest that Gal o/o mice are a valid model to use for potential pig-to-human transplants and can be designed to study the early and later phases of antibody mediated rejection. It is clear that there is a hierarchy of tissues to hyperacute rejection—islets being totally resistant, skin partially so, whereas heart is susceptible and bone marrow entirely susceptible to pre existing antibody; hyperimmune animals (made by immunising with either a transplant or with αGal$^+$ antigen) makes islet cells partially susceptible to rejection.

## 11. ROLE OF αGAL IN NATURAL IMMUNITY

Whether the anti-αGal immune response plays a protective role in natural immunity is not unclear. Not only has αGal been found on cell surface glycolipids, cell surface glycoproteins and secreted glycoproteins of animals, but also on infectious agents such as retroviruses (Friend murine leukaemia virus), bacteria (e.g., *E.coli*, *Klebsiella* and *Salmonella*) and protozoa (e.g., *Plasmodium falciparum*, *Trypanosoma cruzi* and *Leishmania mexicana*) (Galili, 1993), indicating that the encounter of these pathogens with the anti-αGal antibodies could lead to inactivation by complement mediated lysis or by Fc receptor mediated phagocytosis. αGal antibodies can neutralize viruses *in vitro*, and their passage through human cell lines abrogates the ability of human serum to neutralize and kill these viruses, although the sensitivity to αGal antibodies is virus dependant (Welsh *et al.*, 1998). Is the anti-αGal response a barrier for cross species infection of certain viruses? It is perhaps fortuitous that a number of attenuated viruses produced for human vaccines have been propagated in chicken embryos, a species which does not express αGal and makes anti-αGal antibodies (Galili *et al.*, 1988; McKenzie *et al.*, 1998c).

What is the stimulus for humans and animals which do not express αGal for antibody production. It is considered that the stimulus is gut bacteria, or complex plant polysaccharides which contain α-linked Gal. Furthermore the nature of the response is not clear, although both IgM and IgG antibodies are present—the role of T cells, if any, in generating the IgG response is unknown.

Another aspect of natural immunity, particularly in the xenogeneic response, is to consider the role of NK cells. NK cells have been implicated in the delayed xenograft response (Goodman *et al.*, 1996; Malyguine *et al.*, 1996; Schaapherder *et al.*, 1994; Waiter *et al.*, 1996), either through direct recognition of antigen on the xenogeneic cell or by NK Fc receptor recognition of cell surface bound IgG. Naturally occurring and induced anti-αGal IgG antibodies, as shown in pig-to-human islet xenografts (Galili *et al.*, 1995), may play a role in such a response. With regards to direct recognition of xenogeneic cells, human NK cells have recently been shown to recognise the αGal epitope (Artrip *et al.*, 1998; Inverardi *et al.*, 1997), although the nature of the NK receptor which recognises αGal is not known. The demonstration that peptides can mimic αGal (Vaughan *et al.*, 1995b), and in particular Mucin peptides (Sandrin *et al.*, 1997b), leads to the intriguing possibility that NK cell αGal receptor may also recognise these peptide epitopes, and thus play an important role in the immune surveillance for tumor cells in which these epitopes are expressed.

## 12. CONCLUSION

It is clear that the αGal is the major barrier to pig-to-human xenotransplantation. To reduce αGal to a level which does not react in deleterious ways with antibodies, it will be necessary to use several strategies in combination; such as gene knockout [which may become feasible with advances in animal cloning (Schnieke *et al.*, 1997; Wilmut *et al.*, 1997)], or αGaldase transgene (both of which may unmask other potentially immunogenic epitopes), together with the αFT transgene which would result in the conversion of any free N-acetyl lactosamine groups to H substance. The role of human complement regulators (Cozzi and White, 1995; Ryan, 1995) in a composite transgenic animal is likely to be of importance. Once an animal is engineered that has no target for antibody, no complement activation and therefore no hyperacute rejection, then the subsequent phases of xenograft rejection (DXR and cell mediated rejection) can be studied.

ACKNOWLEDGMENTS. This work was supported by funds obtained from the National Health and Medical Research Council of Australia and Juvenile Diabetes Foundation.

## REFERENCES

Ariga, T., Yoshino, H., Ren, S., Pal, S., Katoh-Semba, R., and Yu, R. K., 1993, Activation of UDP-galactose:globotriaosylceramide α1–3-galactosyltransferase during PC12D cell differentiation induced by galactosylceramide. *Biochemistry* **31**: 7904–7908.

Artrip, J. H., Kwiatkowski, P., Wang, S.-F., Tugulea, S., Ankersmit, J., Chisholm, L., Michler, R. E., McKenzie, I. F. C., Sandrin, M. S., and Itescu, S., 1998, Target cell susceptibility to lysis by human natural killer cells is augmented by α(1,3)-galactosyltransferase and reduced by α(1,2)-fucosyltransferase. : (submitted).

Arumugham, R. G., Hsieh, T. C., Tanzer, M. L., and Laine, R. A., 1986, Structures of the asparagine-linked sugar chains of laminin. *Biochim. Biophys. Acta* **883**: 112–126.

Asano, M., Furukawa, K., Kido, M., Matsumoto, S., Umesaki, Y., Kochibe, N., and Iwakura, Y., 1997, Growth retardation and early death of β-1,4-galactosyltransferase knockout mice with augmented proliferation and abnormal differentiation of epithelial cells. *EMBO J.* **16**: 1850–1857.

Auchincloss, H., 1988, Xenogeneic transplantation. A review. *Transplantation* **46**: 1–20.

Auchincloss, H., 1990, Xenografting: a review. *Transplantation Reviews* **4**: 14–27.

Auchincloss, H. and Sachs, D. H., 1998, Xenogeneic transplantation. *Annu. Rev. Immunol.* **16**: 433–470.

Baldmus, C. A., McKenzie, I. F. C., Winn, H. J., and Russell, P. S., 1973, Acute destruction by humoral antibody of rat skin grafted to mice. *J. Immunol.* **110**: 1532–1541.

Blanken, W. M. and Van den Eijnden, D. H., 1985, Biosynthesis of terminal Galα1–3Galβ1–4GlcNAc oligosaccharide sequences on glycoconjugates. Purification and acceptor specificity of a UDP-Gal:N-acetyllactosaminide α1–3-galactosyltransferase from calf thymus. *J. Biol. Chem.* **260**: 12927–12934.

Borrebaeck, C. A. K., Malmborg, C., and Ohlin, M., 1993, Does endogenous glycosylation prevent the use of mouse monoclonal antibodies as cancer therapeutics? *Immunology Today* **14**: 477–479.

Cairns, T., Hammelmann, D., Gray, K., Welsh, K., and Larson, G., 1994, Enzymatic removal from various tissues of galactoseα1,3-galactose target antigens of human antispecies antibodies. *Transplant. Proc.* **26**: 1279–1280.

Castronovo, V., Colin, C., Parent, B., Foidart, J., Lambotte, R., and Mahieu, P., 1989, Possible role of human natural anti-Gal antibodies in the natural antitumor defense system. *J. Natl. Canc. Inst.* **81**: 212–216.

Chen, C.-G., Fisicaro, N., Shinkel, T. A., Aitken, V., Katerelos, M., van Dendren, B. J. W., Tange, M. J., Crawford, R. J., Robbins, A. J., Pearce, M. J., and d'Apice, A. J. F., 1996, Reduction in Galα(1,3)Gal epitope expression in transgenic mice expressing human H-transferase. *Xenotransplantation* **3**: 69–75.

Cohney, S., McKenzie, I. F. C., Patton, K., Prenzoska, J., Ostenreid, K., Fodor, W. L., and Sandrin, M. S., 1997, Down regulation of Galα(1,3)Gal expression in mice by the α1,2fucosyltransferase transgene. *Transplantation* **64**: 495–500.

Cohney, S., Mouhtouris, E., McKenzie, I. F. C., and Sandrin, M. S., 1996, Molecular cloning of a pig α1,2fucosyltransferase. *Immunogenetics* **44**: 76–79.

Cohney, S., Mouhtouris, E., McKenzie, I. F. C., and Sandrin, M. S., 1998, The cloning and characterisation of the pig secretor α1,2fucosyltransferase (FUT2). *Biochem J.* : (submitted).

Cooper, D. K., Koren, E., and Oriol, R., 1994, Oligosaccharides and discordant xenotransplantation. *Immunol. Rev.* **141**: 31–58.

Cozzi, E. and White, D. J. G., 1995, The generation of transgenic pigs as potential organ donors for humans. *Nature Medicine* **1**: 964–966.

Cummings, R. D. and Mattox, S. A., 1988, Retinoic acid-induced differentiation of the mouse teratocarcinoma cell line F9 is accompanied by an increase in the activity of UDP-galactose: β-D-galactosyl-α1,3-galactosyltransferase. *J. Biol. Chem.* **263**: 511–519.

Ferguson-Smith, M. A., Aitken, D. A., Turleau, C., and de Grouchy, J., 1976, Localisation of the human ABO: Np-1: AK-1 linkage group by regional assignment of AK-1 to 9q34. *Hum. Genet.* **34**: 35–43.

Galili, U., 1989, Abnormal expression of alpha-galactosyl epitopes in man. A trigger for autoimmune processes? *Lancet* **2**: 358–361.

Galili, U., 1993, Evolution and pathophysiology of the human natural anti-α-galactosyl IgG (anti-Gal) antibody. *Springer Semin. Immunopathol.* **15**: 155–171.

Galili, U., Macher, B. E., Buehler, J., and Shohet, S. B., 1985, Human natural anti-α-galactosyl IgG II. the specific recognition of α(1,3)-linked galactose residues. *J. Exp. Med.* **162**: 573–582.

Galili, U., Rachmilewitz, E. A., Peleg, A., and Flechner, I., 1984, A unique natural human IgG antibody with anti-α-galactosyl specificity. *J. Exp. Med.* **160**: 1519–1531.

Galili, U., Shohet, S. B., Korbin, E., Stults, C. L. M., and Macher, B. A., 1988, Man, apes and Old World monkeys differ from other mammals in the expression of the α-galactosyl epitopes on nucleated cells. *J. Biol. Chem.* **263**: 17755–17762.

Galili, U. and Swanson, K., 1991, Gene sequences suggest inactivation of α-1,3-galactosyltransferase in catarrhines after divergence of apes from monkeys. *Proc. Natl. Acad. Sci. USA* **88**: 7401–7404.

Galili, U., Tibell, A., Samuelsson, B., Rydberg, L., and Groth, C. G., 1995, Increased anti-Gal activity in diabetic patients transplanted with fetal porcine islet cell clusters. *Transplantation* **59**: 1549–1556.

Good, A. H., Cooper, D. K. C., Malcolm, A. J., Ippolito, R. M., Koren, E., Neethling, F. A., Ye, Y., Zuhdi, N., and Lamontagne, L. R., 1992, Identification of carbohydrate structures that bind human antiporcine antibodies: implications for discordant xenografting in humans. *Transplant. Proc.* **24**: 559–562.

Goodman, D. J., von Albertini, M., Willson, A., Millan, M. T., and Bach, F. H., 1996, Direct activation of porcine endothelial cells by human natural killer cells. *Transplantation* **61**: 763–771.

Groth, C. G., Korsgren, O., Wennberg, L., Tibell, A., Zhu, S., Sundberg, B., Soderlund, J., Biberfeld, P., Satake, M., Moller, E., Wallgren, A. C., and Karlsson-Parra, A., 1996, Xenoislet rejection following pig-to rat, pig-to-primate, and pig-to-man transplantation. *Transplant. Proc.* **28**: 538–539.

Haslam, D. B. and Baenziger, J. U., 1996, Expression cloning of Forssman glycolipid synthetase: a novel member of the histo-blood group ABO gene family. *Proc. Natl. Acad. Sci. USA* **93**: 10697–10702.

Henion, T. R., Macher, B. A., Anaraki, F., and Galili, U., 1994, Defining the minimal size of catalytically active primate α1,3galactosyltransferase: structure-function studies on the recombinant truncated enzyme. *Glycobiology* **4**: 193–201.

Hoess, R., Brinkman, U., Handel, T., and Pastan, I., 1993, Identification of a peptide which binds to the carbohydrate-specific monoclonal antibody B3. *Gene* **128**: 43–49.

Holzknecht, Z. E. and Platt, J. L., 1995, Identification of porcine endothelial cell membrane antigens recognized by human xenoreactive natural antibodies. *J. Immunol.* **154**: 4565–4575.

Inverardi, L., Clissi, B., Stolzer, A. L., Bender, J. R., Sandrin, M. S., and Pardi, R., 1997, Human natural killer lymphocytes directly recognize evolutionarily conserved oligosaccharide ligands expressed by xenogeneic tissues. *Transplantation* **63**: 1318–1330.

Ioffe, E. and Stanley, P., 1994, Mice lacking N-acetylglucosaminyltransferase I activity die at midgestation, revealing an essential role for complex or hybrid N-linked carbohydrates. *Proc. Natl. Acad. Sci. USA* **91**: 728–732.

Joziasse, D. H., 1992, Mammalian glycosyltransferases: genomic organization and protein structure. *Glycobiology* **2**: 271–277.

Joziasse, D. H., Shaper, J. H., Jabs, E. W., and Sharper, N. L., 1991, Characterisation of an α1,3-galactosyltransferase homologue on human chromosome 12 that is organised as a processed pseudogene. *J. Biol. Chem.* **266**: 6991–6998.

Joziasse, D. H., Shaper, J. H., Van den Eijnden, D. H., Van den Tunen, A. J., and Sharper, N. L., 1989, Bovine α1,3 galactosyltransferase: isolation and characterisation of a cDNA clone. Identification of homologous sequences in human genomic DNA. *J. Biol. Chem.* **264**: 14290–14297.

Joziasse, D. H., Shaper, N. L., Kim, D., Van den Eijnden, D. H., and Shaper, J. H., 1992, Murine α1,3 galactosyltransferase: A single gene locus specifies four isoforms of the enzyme by alternative spicing. *J. Biol. Chem.* **267**: 5534–5541.

Kelly, R. J., Ernst, L. K., Larsen, R. D., Bryant, J. G., Robinson, J. S., and and Lowe, J. B., 1994, Molecular basis for H blood group deficiency in Bombay ($O_h$) and para-Bombay individuals. *Proc Natl Acad Sci USA* **91**: 5843–5847.

Kelly, R. J., Rouquier, S., Giorgi, D., Lennon, G. G., and Lowe, J. B., 1995, Sequence and expression of a candidate for the human Secretor blood group α(1,2)fucosyltransferase gene (FUT2). Homozygosity for an enzyme-inactivating nonsense mutation commonly correlates with the non- secretor phenotype. *J Biol Chem* **270**: 4640–4649.

Koike, C., Kannagi, R., Takamura, Y., Akutsu, F., Hayashi, S., Hiraiwa, N., Kadomatsu, K., Muramatsu, T., Yamakawa, H., Nagui, T., Kobayshi, S., Okada, H., Nakashima, I., Uchida, K., Yokoyama, I., and Takagi, H., 1996, Introduction of α(,2)-fucosyltransferase and its effect on α-Gal epitopes in transgenic pig. *Xenotransplantation* **3**: 81–86.

Kooyman, D. L., McClellan, S. B., Parker, W., Avissar, P. L., Velardo, M. A., Platt, J. L., and Logan, J. S., 1996, Identification and characterization of a galactosyl peptide mimetic. Implications for the use in removing xenoreactive anit-αGal antibodies. *Transplantation* **61**: 851–855.

Koulmanda, M., McKenzie, I. F. C., Sandrin, M. S., and Mandel, T. E., 1995, Fetal pig islet xenografts in NOD/Lt mice: The effects of peritransplant anti-CD4 monoclonal antibody and graft immunomodulation on graft survival, and lack of expression of Galα(1–3)Gal on endocrine cells. *Xenotranplantation* **2**: 295–305.

Larsen, R. D., Ernst, L. K., Nair, R. P., and Lowe, J. B., 1990a, Molecular cloning, sequence, and expression of a human GDP-L- fucose:beta-D-galactoside 2-alpha-L-fucosyltransferase cDNA that can form the H blood group antigen. *Proc Natl Acad Sci U S A* **87**: 6674–6678.

Larsen, R. D., Rajan, V. P., Ruff, M., Kukowska-Latallo, J., Cummings, R. D., and Lowe, J. B., 1989, Isolation of a cDNA encoding a murine UDPgalactose:β-D-galctosyl-1,4-N-acetylglucosaminide -α-1,3-galactosyltransferase: Expression cloning by gene transfer. *Proc. Natl. Acad. Sci. USA* **86**: 8227–8231.

Larsen, R. D., Riverra-Marrero, C. A., Ernst, L. K., Cummings, R. D., and Lowe, J. B., 1990b, Frameshift and non sense mutations in a human genomic sequence homologous to a murine UDP-Gal:β-D-Gal 1,4-D- GlcNAcα1,3-galactosyl-transferase cDNA. *J. Biol. Chem.* **265**: 7055–7061.

Lu, Q., Hasty, P., and Shur, B. D., 1997, Targeted mutation in β1,4-galactosyltransferase leads to pituitary insufficiency and neonatal lethality. *Dev. Biol.* **181**: 257–267.

Malyguine, A. M., Saadi, S., Platt, J. L., and Dawson, J. R., 1996, Human natural killer cells induce morphologic changes in porcine endothelial cell monolayers. *Transplantation* **61**: 161–164.

McKenzie, I. F., Koulmanda, M., Mandel, T., Xing, P. X., and Sandrin, M. S., 1995a, Comparative studies of the major xenoantigen gal alpha(1,3)gal in pigs and mice. *Transplant Proc* **27**: 247–248.

McKenzie, I. F., Xing, P. X., Vaughan, H. A., Prenzoska, J., Dabkowski, P. L., and Sandrin, M. S., 1994a, Distribution of the major xenoantigen (gal (α1–3)gal) for pig to human xenografts. *Transpl. Immunol.* **2**: 81–86.

McKenzie, I. F. C., Koulmanda, M., Mandel, T. E., and Sandrin, M. S., 1995b, The expression of Galα(1–3)Gal by porcine islet cells and its relevance to xenotransplantation. *Xenotransplantation* **2**: 139–142.

McKenzie, I. F. C., Koulmanda, M., Mandel, T. E., Xing, P.-X., and Sandrin, M. S., 1995c, Pig to human xenotransplantation: the expression of Galα(1–3)Gal epitopes on pig islet cells. *Xenotransplantation* **2**: 1–7.

McKenzie, I. F. C., Koulmanda, M., Mandell, T. E., and Sandrin, M. S., 1998a, Pig islet xenografts are susceptible to "anti-pig" but not to anti Galα(1,3)Gal antibody and complement in Gal o/o mice. *J. Immunol.*: (submitted).

McKenzie, I. F. C., Li, Y. Q., Patton, K., Thall, A., and Sandrin, M. S., 1998b, A murine model for antibody mediated hyperacute rejection due to anti-Galα(1,3)Gal antibodies in Gal o/o mice. *Transplantation* : (in press).

McKenzie, I. F. C., Osman, N., Cohney, S., Vaughan, H. A., Patton, K., Mouhtouris, E., Atkin, J., Elliot, E., Fodor, W. L., Squinto, S. P., Burton, D., Gallop, M. A., Oldenburg, K. R., and Sandrin, M. S., 1996, Strategies to overcome the anti-Galα(1,3)Gal reaction in xenotransplantation. *Transplant. Proc.* **28**: 567.

McKenzie, I. F. C., Patton, K., Smit, J. A., Mouhtouris, E., Xing, P.-X., Myburgh, J. A., and Sandrin, M. S., 1998c, Definition and characterization of chicken Galα(1,3)Gal antibodies. *Transplantation* : (in press).

McKenzie, I. F. C., Vaughan, H. A., and Sandrin, M. S., 1994b, How important are anti-Galα(1–3)Gal antibodies in pig to human xenotransplants? *Xeno* **2**: 107–110.

Metzler, M., Gertz, A., Sarkar, M., Schachter, H., Schrader, J. W., and Marth, J. D., 1994, Complex asparagine-linked oligosaccharides are required for morphogenic events during post-implantation development. *EMBO J.* **13**: 2056–2065.

Oldenburg, K. R., Loganathan, D., Goldstein, I. J., Schultz, P. G., and Gallop, M. A., 1992, Peptide ligands for a sugar-binding protein isolated from a random peptide library. *Proc. Natl. Acad. Sci. USA* **89**: 5393–5397.

Oriol, R., Ye, Y., Koren, E., and Cooper, D. K., 1993, Carbohydrate antigens of pig tissues reacting with human natural antibodies as potential targets for hyperacute vascular rejection in pig-to-man organ xenotransplantation. *Transplantation* **56**: 1433–1442.

Osman, N., McKenzie, I. F. C., Mouhtouris, E., and Sandrin, M. S., 1996, Switching amino terminal cytoplasmic domains of α1,2fucosyltransferase and α1,3galactosyltransferase alters the expression of H substance and Galα(1,3)Gal. *J. Biol. Chem.* **271**: 33105–33109.

Osman, N., McKenzie, I. F. C., Ostenreid, K., Ioannou, Y. A., Desnick, R. J., and Sandrin, M. S., 1997, Combined transgenic expression of α-galactosidase and α1,2fucosyltransferase leads to optimal reduction in the major xenoepitope Galα(1,3)Gal. *Proc. Natl. Acad. Sci. USA* **94**: 14677–14682.

Petryniak, J., Varani, J., Ervin, P. R., and Goldstein, I. J., 1991, Differential expression of glycoproteins containing α-D-galactosyl groups on normal human breast epithelial cells and MCF-7 human breast carcinoma cells. *Cancer Lett.* **60**: 59–65.

Platt, J. L. and Holznecht, Z. E., 1994, Porcine platelet antigens recognised by human xenoreactive natural antibodies. *Transplantation* **57**: 327–335.

Platt, J. L., Lindman, B. J., Chen, H., Spitalnik, S. L., and Bach, F. H., 1990, Endothelial cell antigens recognized by xenoreactive human natural antibodies. *Transplantation* **50**: 817–822.

Ryan, U. S., 1995, Complement inhibitory therapeutics and xenotransplantation. *Nature Medicine* **1**: 967–968.

Sachs, D. H., 1994, The pig as a potential xenograft donor. *Vetinary Immunology and Immunopathology* **43**: 185–191.

Sachs, D. H., Sykes, M., Greenstein, J. L., and Cosimi, A. B., 1995, Tolerance and xenograft survival. *Nature Med.* **1**: 969.

Samuelsson, B. E., Rydberg, L., Breimer, M. E., Backer, A., Gustavsson, M., Holgersson, J., Karlsson, E., Uyterwaal, A. C., Cairns, T., and Welsh, K., 1994, Natural antibodies and human xenotransplantation. *Immunol. Rev.* **141**: 151–168.

Sandrin, M. S. and McKenzie, I. F. C., 1994, Galα(1,3)Gal, the major xenoantigen(s) recognised in pigs by human natural antibodies. *Immunol. Rev.* **141**: 169–190.

Sandrin, M. S., Cohney, S., Osman, N., and McKenzie, I. F. C.: Overcoming the anti-Galα(1,3)Gal rejection to avoid hyperacute reaction: molecular genetic approaches. *In* D. K. C. Cooper, E. Kemp, J. L. Platt, and D. J. G. White (eds.): *Xenotransplantation: The transplantation of organs and tissues between species.*, pp. 683–700, Springer-Verlag, Berlin, 1997a

Sandrin, M. S., Dabkowski, P. L., Henning, M. M., Mouhtouris, E., and McKenzie, I. F. C., 1994a, Characterization of cDNA clones for porcine α1,3 galactosyltransferase: the enzyme generating the Galα(1,3)Gal epitope. *Xenotransplantation* **1**: 81–88.

Sandrin, M. S., Fodor, W. L., Cohney, S., Mouhtouris, E., Osman, N., Rollins, S. A., Squinto, S. P., and McKenzie, I. F. C., 1996, Reduction of the major porcine xenoantigen Galα(1,3)Gal by expression of α(1,2)fucosyltransferase. *Xenotransplantation* **3**: 134–140.

Sandrin, M. S., Fodor, W. L., Mouhtouris, E., Osman, N., Cohney, S., Rollins, S. A., Guilmette, E. R., Setter, E., Squinto, S. P., and McKenzie, I. F. C., 1995, Enzymatic remodeling of the carbohydrate surface of a xenogenic cell substantially reduces human antibody binding and complement-mediated cytolysis. *Nature Medicine* **1**: 1261–1267.

Sandrin, M. S., Patton, K., and McKenzie, I. F. C., 1998, Inability to alter the rejection of Galα(1,3)Gal$^+$ bone marrow by α1,2fucosyltransferase transgene. Transplantation (submitted).

Sandrin, M. S., Vaughan, H. A., and McKenzie, I. F. C., 1994b, Identification of Galα(1,3)Gal as the major epitope for pig-to-human vascularised xenografts. *Transplant. Rev.* **8**: 134–149.

Sandrin, M. S., Vaughan, H. A., Dabkowski, P. L., and McKenzie, I. F. C., 1993, Anti-pig IgM antibodies in human serum reacts predominantly with Galα(1,3)Gal epitopes. *Proc. Natl. Acad. Sci. USA* **90**: 11391–11395.

Sandrin, M. S., Vaughan, H. A., Xing, P.-X., and McKenzie, I. F. C., 1997b, Natural human anti-Galα(1,3)Gal antibodies react with human mucin peptides. *Glycoconjugate J.* **14**: 97–105.

Schaapherder, A. F. M., Daha, M. R., te Bulte, M.-J., van der Woude, F. J., and Gooszen, H. G., 1994. Anti-body-dependant cell-mediated cytotoxicity against porcine endothelium induced by a majority of human sera. *Transplantation* **57**: 1376–1382.

Schnieke, A. E., Kind, A. J., Ritchie, W. A., Mycock, K., Scott, A. R., Ritchie, M., Wilmut, I., Colman, A., and Campbell, K. H., 1997, Human factor IX transgenic sheep produced by transfer of nuclei from transfected fetal fibroblasts. *Science* **278**: 2130–2133.

Scott, J. K., Loganathan, D., Easley, R. B., Gong, X., and Goldstein, I. J., 1992, A family of concanavalin A-binding peptides from a hexapeptide epitope library. *Proc. Natl. Acad. Sci. USA* **89**: 5398–5402.

Shaper, N. L., Lin, S. P., Joziasse, D. H., Kim, D. Y., and Yang-Feng, T. L., 1992, Assignment of two human alpha-1,3-galactosyltransferase gene sequences (GGTA1 and GGTA1P) to chromosomes 9q33-q34 and 12q14-q15. *Genomics* **12**: 613–615.

Sharma, A., Okabe, J., Birch, P., McClellan, S. B., Martin, M. J., Platt, J. L., and Logan, J. S., 1996, Reduction in the level of Gal(α1,3)Gal in transgenic mice and pigs by the expression of an α(1,2)fucosyltransferase. *Proc. Natl. Acad. Sci. USA* **93**: 7190–7195.

Simon, P. M., Neethling, F. A., Taniguchi, S., Goode, P. L., Zopf, D., Hancock, W. W., and Cooper, D. K., 1998, Intravenous infusion of Galα1–3Gal oligosaccharides in baboons delays hyperacute rejection of porcine heart xenografts. *Transplantation* **65**: 346–353.

Smith, D. F., Larsen, R. D., Mattox, S., Lowe, J. B., and Cummings, R. D., 1990, Transfer and expression of a murine UDP-Gal:β-D-Gal-α1,3-galactosyltransferase gene in transfected Chinese hamster ovary cells. Competition reactions between the α1,3-galactosyltransferase and the endogenous α2,3-sialyltransferase. *J. Biol. Chem.* **265**: 6225–6234.

Strahan, K. M., Gu, F., Preece, A. F., Gustavsson, I., Andersson, L., and Gustafsson, K., 1995, cDNA sequence and chromosome localization of pig alpha 1,3 galactosyltransferase. *Immunogenetics* **41**: 101–105.

Tanemura, M., Miyagawa, S., Ihara, Y., Mikata, S., Matsuda, H., Shirakura, R., and Taniguchi, N., 1997a, Reduction of the major swine xenoantigen Galα(1,3)Gal by transfection of N-acetylglucosaminyltransferase III (GnT-III) gene. *Transplant. Proc.* **29**: 891–892.

Tanemura, M., Miyagawa, S., Ihara, Y., Nishikawa, A., Suzuki, M., Yamamura, K., Matsuda, H., Shirakura, R., and Taniguchi, N., 1997b, Suppression of the xenoantigen Galα(1,3)Gal by N-acetylglucosaminyltransferase III (GnT-III) in transgenic mice. *Transplant. Proc.* **29**: 895–896.

Tanemura, M., Miyagawa, S., Ihara, Y., Matsuda, H., Tsuji, S., Shirakura, R., and Taniguchi, N., 1997a, Coexpression of N-acetylglucosaminyltransferase III (GnTIII) and α2,3sialyltransferase (α2,3ST) gene for reduction of xenoantigens. : (in press).

Tanemura, M., Miyagawa, S., Ihara, Y., Matsuda, H., Tsuji, S., Shirakura, R., and Taniguchi, N., 1997b, Effects of Galβ1,4GlcNAc3-α-D-sialyltransferase on reduction of the swine xenoantigen. (in press).

Tearle, R. G., Tange, M. J., Zannettino, Z. L., Katerelos, M., Shinkel, T. A., Van Denderen, B. J. W., Lonie, A. J., Lyons, I., Nottle, M. B., Cox, T., Becker, C., Peura, A. M., Wigley, P. L., Crawford, R. J., Robins, A. J., Pearse, M. J., and d'Apice, A. J. F., 1996, The a-1,3-galactosyltransferase knockout mouse: implications for xenotransplantation. *Transplantation* **61**: 13–19.

Thall, A. and Galili, U., 1990, Distribution of Galα1→3Galβ1→4GlcNAc residues on secreted mammalian glycoproteins (thyroglobulin, fibrinogen, and immunoglobulin G) as measured by a sensitive solid-phase radioimmunoassay. *Biochemistry* **29**: 3959–3965.

Thall, A. D., Maly, P., and Lowe, J. B., 1995, Oocyte Galα1,3Gal Epitopes implicated in sperm adhesion to the zona pellucida glycoprotein ZP3 are not required for fertilization in the mouse. *J. Biol. Chem.* **270**: 21437–21440.

Vaughan, H. A., McKenzie, I. F. C., and Sandrin, M. S., 1995a, Biochemical studies of pig xenoantigens detected by naturally occurring human antibodies and the galactoseα(1–3)galactose reactive lectin. *Transplantation* **59**: 102–109.

Vaughan, H. A., Oldenburg, K. R., Gallop, M. A., Atkin, J. D., McKenzie, I. F. C., and MS., S., 1995b, Recognition of an octapeptide sequence by multiple Galα(1,3)Gal-binding proteins. *Xenotransplantation* **3**: 18–23.

Waiter, H., Gullaumin, J. M., Vallee, I., Thibault, G., Gruel, Y., Leraunchu, Y., and Bardos, P., 1996, Human NK cell-mediated direct and IgG-dependent cytotoxicity against xenogeneic porcine endothelial cells. *Transplant. Immunol.* **4**: 293–299.

Welsh, R. M., O'Donnell, C. L., Reed, D. J., and Rother, R. P., 1998, Evaluation of the Galα1–3Gal epitope as a host modification factor eliciting natural humoral immunity to enveloped viruses. *J. Virol.* **72**: 4650–4656.

Wilmut, I., Schnieke, A. E., McWhir, J., Kind, A. J., and Campbell, K. H. S., 1997, Viable offspring derived from fetal and adult mammalian cells. *Nature* **385**: 810–813.

Winn, H. J., Baldmus, C. A., Jooste, S. V., and Russell, P. S., 1973, Acute destruction by humoral antibody of rat skin grafted to mice: The role of complement and polymorphonuclear granulocytes. *J. Exp. Med.* **137**: 893–910.

Xu, Y., Lorf, T., Sablinski, T., Gianello, P., Bailin, M., Monroy, R., Kozlowski, T., Awwad, M., Cooper, D. K., and Sachs, D. H., 1998, Removal of anti-porcine natural antibodies from human and nonhuman primate plasma in vitro and in vivo by a Galα1–3Galβ1–4βGlc-X immunoaffinity column. *Transplantation* **65**: 172–179.

Yamamoto, F., Clausen, H., White, T., Marken, J., and Hakmori, S., 1990, Molecular genetic basis of the histo-blood group ABO system. Nature 345: 229–233.

*Chapter 14*

# Graves' Disease as a Model for Anti-Gal Involvement in Autoimmune Diseases

Uri Galili

## 1. INTRODUCTION

Anti-Gal, which is the most abundant antibody in humans appeared relatively recently in the evolution of Old World monkeys and apes, i.e., approximately 20 million years ago, subsequent to the suppression of α-gal epitope expression (Galili et al., 1988; see the chapter of Galili on the "Evolution of α1,3galactosyltransferase and of the α-gal epitope" in this book). It is well established that humans do not produce α-gal epitopes because they lack the glycosylation enzyme α1,3galactosyltransferase (α1,3GT). This enzyme synthesizes α-gal epitopes on glycolipids and glycoproteins in nonprimate mammals and New World monkeys (Galili et al., 1987a; 1988; Spiro and Bhyroo 1984; Thall et al., 1991). Nevertheless, several studies have described the *in vitro* interaction of anti-Gal with various human cells. These included binding of the antibody to: placental cells (Christiane et al., 1992), the mammary carcinoma cell line MCF7 (Galili 1989; Petriniyak et al., 1991), freshly obtained mammary carcinoma cells (Castronovo et al., 1989), human normal senescent red cells (Galili et al., 1986), and some pathologic red cells including

**Uri Galili** Department of Microbiology and Immunology, MCP Hahnemann School of Medicine, Philadelphia, Pennsylvania 19129.

*Subcellular Biochemistry, Volume 32: α-Gal and Anti-Gal*, edited by Galili and Avila. Kluwer Academic / Plenum Publishers, New York, 1999.

red cells from patients with β-thalassemia (Galili et al., 1983; 1984) and sickle cell anemia (Galili et al., 1986). Other studies demonstrated marked increase in anti-Gal activity in patients with autoimmune diseases such as Graves' Disease (Etienne-Decerf et al., 1987) and scleroderma (Gabrielli et al., 1991). Taken together, all these observations suggest that despite the evolutionary suppression of α-gal epitope expression in ancestral Old World primates, anti-Gal may interact with human cells under certain circumstances, which are poorly understood at present. Such interaction may inevitably result in autoimmunity (Galili, 1989).

This chapter describes studies performed to elucidate the possible role of anti-Gal in the pathogenesis of one autoimmune disorder—Graves' Disease (autoimmune thyrotoxicosis). Although the mechanism underlying the aberrant interaction of anti-Gal with Graves' Disease thyrocytes is far from being understood, the observed effects of anti-Gal on these thyrocytes may serve as a model for possible contribution of this antibody to other, so far undefined autoimmune states. Several hypothetical mechanisms that may result in anti-Gal mediated autoimmunity are discussed at the end of this chapter.

## 2. GRAVES' DISEASE

Graves' Disease is an autoimmune disease in which autoantibodies mimic the stimulatory effect of thyroid stimulating hormone (TSH) (also termed thyrotropin) on thyroid cells (termed here thyrocytes). TSH is produced in the pituitary under control of the hypothalamic secretion of thyrotropin releasing hormone (TRH), which in turn is controlled in negative feedback by the blood level of the thyroid hormones $T_3$ and $T_4$. Increased secretion of $T_3$ and $T_4$ by the thyroid gland results in decrease in TRH secretion by the hypothalamus and therefore decreased secretion of TSH by the pituitary. Binding of TSH to TSH receptors on thyrocytes generates intracellular signals associated with the G protein system and cAMP production. This signal transduction pathway ultimately activates the genes involved in the unique metabolic activity of thyrocytes including: iodine uptake, iodine organification (i.e. linking iodine to tyrosine and ultimately to thyroglobulin), thyroglobulin production, as well as $T_3$ and $T_4$ production. When the pituitary decreases TSH secretion because of high blood level of $T_3$ and $T_4$, the metabolic activity of thyrocytes also decreases, thus lowering $T_3$ and $T_4$ production.

In Graves' Disease patients the immune system generates autoantibodies that interact with a variety of molecules produced by thyrocytes including thyroglobulin, peroxidase, cell surface TSH receptors and other cell surface molecules. Binding of autoantibodies to TSH receptors induces chronic activation of the thyrocytes for continuous production of thyroglobulin and $T_3$ and $T_4$ and for cell proliferation (Strakosch et al., 1982; Weetman and McGregor 1984; Zakarija and McKenzie 1987; Smith et al., 1988). This results in enlargement of the thyroid, abnormally high level of $T_3$ and $T_4$ which induce systemic increase in metabolic rate that is associated with heat intolerance, weight loss, nervousness, hyperdynamic cardiovascular status, tachycardia, and exophthalmos. In severe cases, Graves' Disease can be life threatening. Treatment modalities for Graves' Disease patients include drugs that inhibit organification of iodine, administration of radioactive iodine for the destruction of the thyroid gland, or surgical removal of the gland.

The causes for appearance of autoantibodies to thyroid associated antigens are not clear. Their production may be induced by cross reactive bacterial antigens (Shenkman and Bottom, 1976; Weiss et al., 1983; Luo et al., 1993), aberrant HLA class II expression on thyrocytes due to inflammation induced by viral infection and local interferon-γ production (Bottazo et al., 1983), or the result of aberrant regulatory function of the immune system (Volpe, 1986). The possible involvement of anti-Gal in Graves' Disease pathogenesis was suggested by Winand and colleagues (Etienne-Decerf et al., 1987) who reported that anti-Gal titers in patients with active Graves' Disease increase by four to eight fold when compared with the titers found in healthy individuals. Furthermore, anti-Gal titers in Graves' Disease patients return to normal when the patients are in remission. These observations raised the possibility that thyrocytes in the active stage of the disease may aberrantly express α-gal epitopes. These epitopes would stimulate the immune system to produce increased amounts of anti-Gal above the physiologic level induced by antigens on bacteria of the normal gastrointestinal flora (see the chapter of Galili et al., on "The natural anti-Gal antibody").

We hypothesized that if such epitopes are aberrantly expressed on TSH receptors, then binding of anti-Gal to these epitopes will mimic the effect of TSH interaction with the receptors, i.e., it would result in stimulation of the Graves' Disease thyrocytes (Winand et al., 1994). Therefore, it was of interest to determine whether interaction of anti-Gal with α-gal epitopes on TSH receptors that express this epitope can indeed mimic the effect of TSH.

## 3. INTERACTION OF ANTI-GAL WITH PORCINE TSH RECEPTORS

The mammalian TSH receptor is a membrane bound glycoprotein molecule with seven transmembrane domains, an extracellular domain comprising of approximately 415 amino acids that include the binding site for TSH, and a cytoplasmic tail containing the peptide motifs that activate the G protein system upon the binding of TSH to the receptor. The G protein activation initiates the signal transduction process. Sequences of TSH receptor in various species indicated that there are five to seven potential N-glycosylation sites (Nagayoma et al., 1989; Libert et al., 1989; Parmantier et al., 1989; Akamizu et al., 1990; Misrahi et al., 1990). In species that have active α1,3GT some of these N-linked carbohydrate chains are likely to have the α-gal epitope (Figure 1).

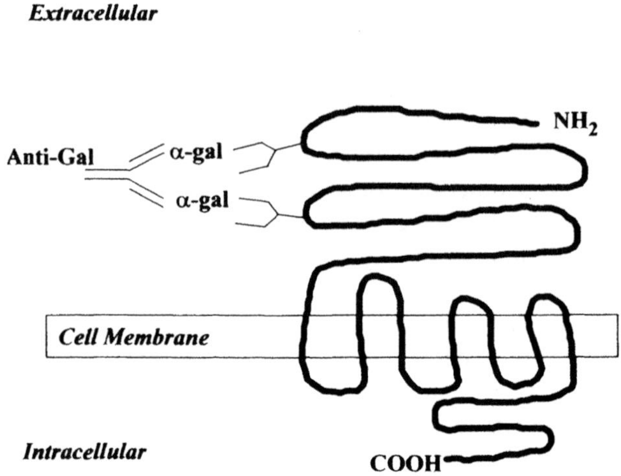


**Figure 1.** Membrane topology of the mammalian TSH receptor with two of the 5 to 7 putative sites of N-linked carbohydrate chains, on which the α-gal epitope may be expressed. It is hypothesized that binding of anti-Gal to these epitopes causes the activation of the TSH receptor (modified after Kosugi et al., 1992).

Thyroid cells of nonprimate mammals and New World monkeys were indeed found to contain α1,3GT and to synthesize α-gal epitopes on various membrane bound glycoproteins (Thall et al., 1991; Edge and Spiro, 1985) and on thyroglobulin molecules (Spiro and Bhoyroo, 1984; Thall and Galili 1990). Furthermore, based on the interaction of α-gal epitopes with the *Bandeiraea (Griffonia) simplicifolia* IB4 lectin, Kress and Spiro (1986) demonstrated presence of α-gal epitopes on TSH receptors of bovine origin. Based on these considerations we hypothesize that α-gal epitopes are also likely to be expressed on porcine TSH receptors. We, therefore, explored the biological outcome of the interaction of anti-Gal with these receptors. For this purpose anti-Gal was isolated from normal human sera by affinity chromatography (Galili et al., 1987b) and primary cultures of porcine thyrocytes were established from freshly obtained pig thyroid glands as previously described (Winand and Kohn, 1975).

## 3.1. Induction of cAMP Synthesis in Porcine Thyrocytes

The interaction of TSH with its receptor on the thyrocyte membrane induces the cytoplasmic activation G proteins and subsequently of adenylate cyclase. This results in the synthesis of cAMP that functions as an intracellular second messenger, inducing phosphorylation of proteins involved in the transduction of the signal into the nucleus (Dumont et al., 1989). This synthesis of cAMP can be readily demonstrated in primary cultures of porcine thyrocytes after the administration of TSH

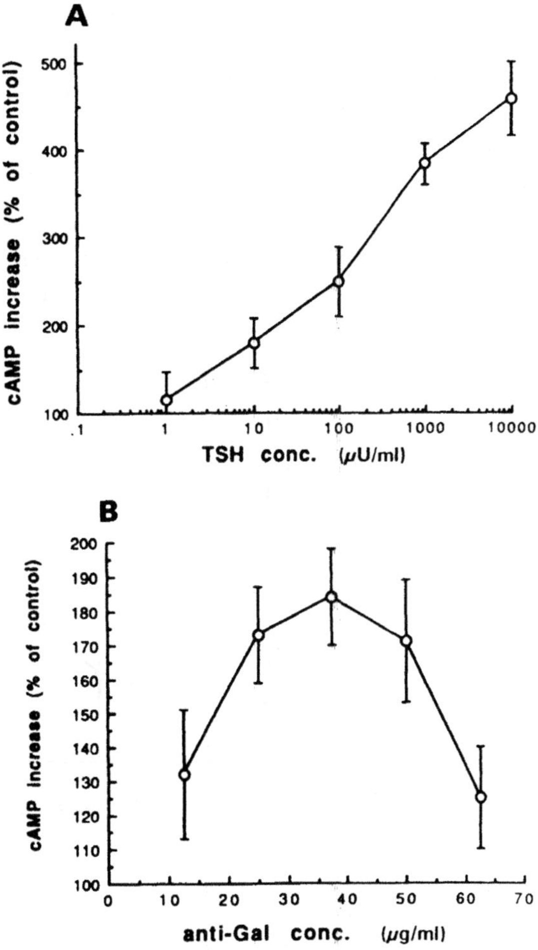


**Figure 2.** Increase in cAMP content of porcine thyrocytes in presence of TSH (A), or of anti-Gal (B) (Mean of six experiments) (from Winand et al. 1993, with permission).

(Figure 2). The produced cAMP was measured by a radioimmunoassay (Etienne-Decerf and Winand, 1981) with homogenates of $5 \times 10^6$ thyrocytes. The thyrocytes were incubated for 48 hr with TSH or with anti-Gal and subjected to measurement of cAMP (Winand et al., 1993). The basal level (i.e., background control) defined as 100% was 2.0 nM of synthesized cAMP and it represented the stimulatory effect of TSH present within the fetal calf serum supplementing the tissue culture medium. When thyrocytes were incubated for 2 hr in presence of 10 μU/ml of TSH, a concentration that corresponds to the upper physiologic level, cAMP synthesis was two fold the basal level, whereas 100 μU/ml of TSH increased cAMP stimulation by 2.5 fold over basal level. Incubation of thyrocytes with 12.5 μg/ml anti-Gal resulted in 30% increase in cAMP synthesis, and anti-Gal at the physiologic concentration of 35 μg/ml induced an almost two-fold increase in cAMP synthesis (Figure 2). For unknown reasons the antibody at higher concentrations was less effective in stimulating porcine thyrocytes for cAMP synthesis.

### 3.2. Induction of $^{125}$I Uptake into Porcine Thyrocytes by Anti-Gal

As indicated above, one of the metabolic outcomes of TSH binding to its receptor on thyrocytes is the stimulation for increased uptake of iodine into these cells. This effect is demonstrated in Figure 3 where $7.5\times10^5$ porcine thyrocytes

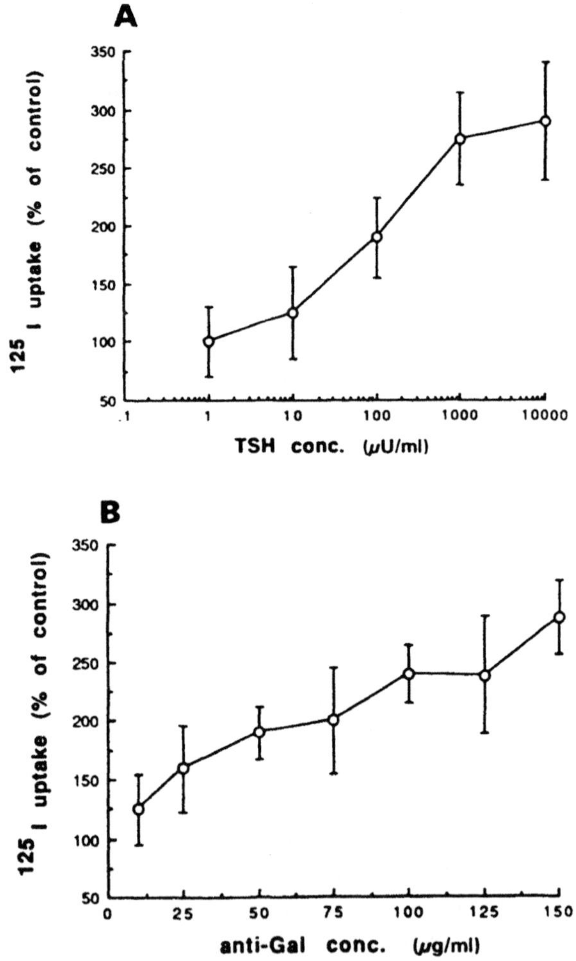


**Figure 3.** Increase in $^{125}$I uptake by porcine thyrocytes in presence of TSH (A), or of anti-Gal (B). (Mean of six experiments) (from Winand et al., 1993, with permission).

were incubated in tissue culture wells with TSH or anti-Gal for 48 hr, then in presence of Na $^{125}I$ for 5 hr. Cells were then washed extensively, detached from the plates and their radioactivity measured (Winand et al., 1993). The basal level of $^{125}I$ uptake was found to be 7025 cpm. Incubation of thyrocytes with 10 μU/ml of TSH induced 75% increase in uptake (Figure 3). Anti-Gal also induced iodine uptake by the porcine thyrocytes. Incubation of the cells with 12.5 μg/ml of anti-Gal resulted in 50% increase in $^{125}I$ uptake whereas incubation with 50 μg/ml of the antibody increased the uptake by 100% (Figure 3).

### 3.3. Anti-Gal Inhibition of TSH Binding to the TSH Receptor

The TSH-mimicking effects of anti-Gal, described above, suggested that porcine TSH receptors express α-gal epitopes as described in Figure 1 and similar to the expression of these epitopes on other porcine glycoproteins (Thall et al., 1991; Spiro and Bhoryoo, 1984; Edge and Spiro, 1985; Thall and Galili, 1990). Thus the binding of anti-Gal to these epitopes may elicit biological effects similar to those of TSH. It was therefore of interest to determine whether α-gal epitopes of the porcine TSH receptor are located in the TSH binding domain. In order to test this possibility, anti-Gal was assayed for its ability to inhibit the binding of radiolabelled TSH to thyrocyte cell membrane in a binding assay previously described by Macchia et al., (1983). As shown in Figure 4, anti-Gal displayed a partial inhibitory effect on the interaction of TSH with its receptor on porcine thyrocytes (Winand et al., 1993). Inhibition of 10% of TSH binding was observed at the physiologic concentration of 50 μg/ml of the antibody. At a concentration of 150 μg/ml, which is 2-4 fold higher than the physiologic concentration, anti-Gal inhibited TSH binding by no more than 25%. This inhibition is highly specific since human IgG depleted of anti-Gal had no inhibitory effect on TSH binding to

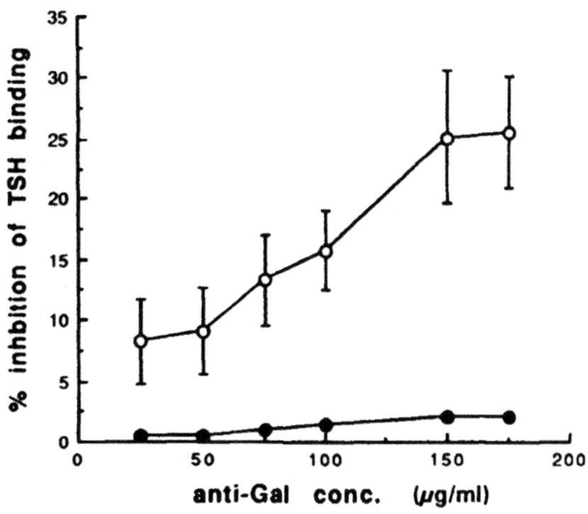


**Figure 4.** Inhibitory activity of anti-Gal (○) and of human IgG depleted of anti-Gal (●) on the interaction of TSH with porcine TSH receptors. (Mean ± SD of six experiments) (from Winand et al., 1993, with permission).

the thyrocyte membranes. Overall, these data implied that the α-gal epitope is not within the binding domain of the TSH receptor. It is probably expressed on the TSH receptor molecule at a site near the TSH binding domain, so that anti-Gal binding to this epitope results in a partial hindrance effect on TSH interaction with the receptor. These findings are in accord with previous observations which indicated that no carbohydrate chains are present in the TSH binding domain of the receptor (Atassi et al., 1991).

## 4. INTERACTION OF ANTI-GAL WITH RECOMBINANT HUMAN TSH-RECEPTORS

The stimulatory effect of anti-Gal on porcine thyrocytes and the partial inhibition of TSH binding to its receptor on porcine cells in presence of anti-Gal, strongly suggested that this antibody binds to α-gal epitopes on the porcine TSH receptor. However, these data did not offer a direct proof for the interaction. Direct demonstration was achieved by measuring the interaction of anti-Gal with human recombinant TSH receptors expressed on cells that produce α-gal epitopes. We hypothesized that anti-Gal will induce cAMP synthesis in cells expressing *de novo* the recombinant TSH receptor, if the cells are capable of synthesizing α-gal epitopes. Conversely, no cAMP production would occur in cells that lack α1,3GT, and thus the *de novo* expressed recombinant TSH receptor would not bind anti-Gal.

The cells studied were mouse 3T3 cells which synthesize an abundance of α-gal epitopes on surface glycoproteins (Santer et al., 1989) and hamster CHO cells, which, unlike other hamster cells, they completely lack this epitope (Smith et al., 1990). Both cells are of nonthyroid origin and do not express TSH receptors. The two cell types were transfected with recombinant human TSH receptor (Nagayama et al., 1989) in a plasmid containing a selectable marker and the stable transfectants were confirmed for expression of human TSH receptors by measuring the binding of radiolabelled TSH (Winand et al., 1993). As shown in Figure 5A, only residual binding of radiolabelled TSH was detected with either parental 3T3 or CHO cells. In contrast, both types of transfected cells readily bound TSH. The transfection of cells did not alter the expression of α-gal epitopes on cell surface glycoconjugates, as assessed by the binding of labeled anti-Gal (Figure 5A). Anti-Gal readily bound to parental or transfected 3T3 cells but not to the corresponding CHO cells.

The activity of the *de novo* expressed recombinant TSH receptor molecules, i.e. their coupling to the G protein system and ability to activate adenylate cyclase, was assessed by measuring the synthesis of cAMP in the transfected cells subsequent to TSH binding to the cells. TSH at the concentrations of 10 μU/ml and 50 μU/ml readily induced cAMP synthesis within 2 hr in both types of transfected

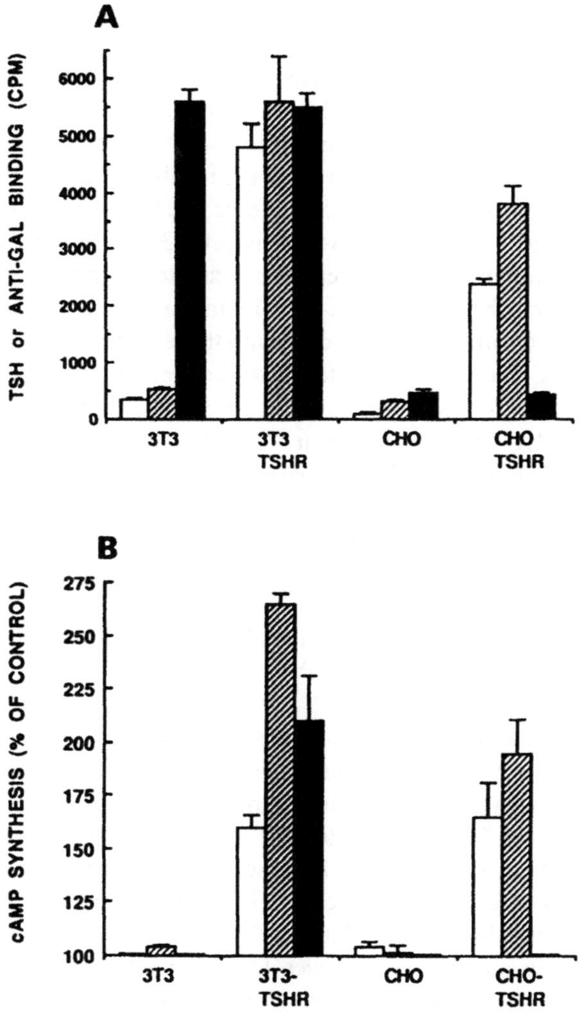


**Figure 5.** (A) Binding of TSH or anti-Gal to nontransfected and transfected (TSHR) 3T3 and CHO cells. Open columns—biotinylated TSH concentration of 10 μU/ml; hatched columns—biotinylated TSH concentration of 50 μU/ml; closed columns—biotinylated anti-Gal at a concentration of 50 μg/ml. Binding was assessed by subsequent incubation with $^{125}$I-streptavidin. (B) Stimulation of nontransfected or transfected cells for cAMP production. Concentrations of TSH and anti-Gal in the various columns as in Figure 5A. (Mean ± SD of three experiments) (from Winand et al., 1993, with permission).

cells, at levels ranging between 50 and 165% above the basal level (which was referred to as 100% synthesis of cAMP) (Figure 5B). As expected no increase in cAMP production was observed in parental 3T3 or CHO cells incubated with TSH.

The parental and transfected cells were also incubated with anti-Gal (50 μg/ml) for 2 hr and assayed for cAMP synthesis. Despite the interaction of anti-Gal with 3T3 cells (Figure 5A), no increase in cAMP synthesis was observed in nontransfected cells (Figure 5B). In contrast, incubation of the transfected 3T3 cells with anti-Gal resulted in a 110% increase in cAMP synthesis. The findings that anti-Gal bound to both parental and transfected 3T3 cells, whereas cAMP synthesis occurred only in the transfected cells incubated with anti-Gal, implied that this stimulation was the result of anti-Gal binding to α-gal epitopes on the recombinant TSH receptor molecule. In accord with these findings, neither the parental CHO cells, nor CHO cells transfected with the recombinant TSH receptor displayed any increase in cAMP synthesis when incubated with anti-Gal. This is to be expected since CHO cells lack the ability to synthesize α-gal epitopes, thus the *de novo* expressed TSH-receptors lack α-gal epitopes and do not bind anti-Gal. Overall these observations implied that anti-Gal binding to α-gal epitopes on TSH receptor molecules can indeed mimic the stimulatory effects of TSH.

## 5. STIMULATION OF PORCINE THYROCYTES BY ANTI-GAL IN SERA OF GRAVES' DISEASE PATIENTS

Part of the diagnostic tests for Graves' Disease has been the assessment of the stimulatory effect of sera from the patients on thyrocytes from various species including pig thyrocytes. It has been assumed that this stimulatory activity is of autoantibodies to the TSH receptor, termed thyroid stimulatory immunoglobulins (TSI). The activity of TSI in Graves' Disease sera has been assessed by measuring cAMP synthesis as previously described (Etienne-Decerf and Winand, 1981). Since anti-Gal mimicked the stimulatory effect of TSH and since anti-Gal activity was found to be elevated in sera of Graves' Disease patients, it was of interest to determine what proportion of the TSI activity in such sera may be attributed to anti-Gal.

For this purpose TSI activity with porcine thyrocytes was assessed in these sera before and after depletion of anti-Gal. The antibody was removed by adsorption on a column with synthetic α-gal epitopes (Galili, 1993). The sera were in parallel adsorbed on a column with N-acetyllactosamine (Galβ1-4GlcNAc-R) that served as adsorption control. The sera at a final dilution of 1:5 were added to the tissue culture medium of the thyrocytes for 48 h. Subsequently, the thyrocytes were detached and assayed for cAMP content. Porcine thyrocytes incubated with 1:5 diluted normal human sera pooled from 100 donors provided the base line

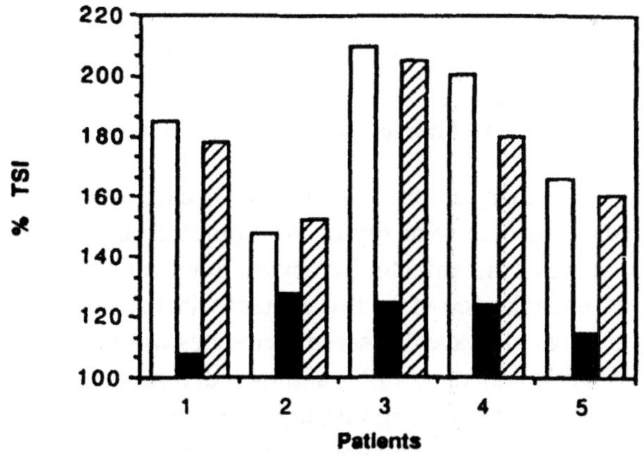


**Figure 6.** Effect of anti-Gal depletion on TSI activity in sera of five patients with Graves' Disease. Open columns—TSI activity in the various serum samples; closed columns—TSI activity in sera depleted of anti-Gal by affinity chromatography on columns with α-gal epitopes; hatched columns—TSI activity in sera passed on control N-acetyllactosamine columns. The value of 100% TSI activity is that observed with pooled normal sera (from Winand et al., 1993, with permission).

value (i.e. 100%) of cAMP synthesis in sera from healthy individuals. This basal level of cAMP synthesis in presence of normal pooled sera was 3.7 nM. The sera of all five patients studied displayed a TSI activity that resulted in 50 to 100% increase in cAMP synthesis above the basal level observed in presence of normal pooled serum (Figure 6). Adsorption on a column with α-gal epitopes resulted in the removal of 50-80% of the TSI activity. In contrast, adsorption of the sera on a N-acetyllactosamine column resulted in no significant changes in TSI activity (Winand et al., 1993). These data are in accordance with the data on anti-Gal ability to mimic TSH stimulatory effect. Much of the TSI activity in the sera of patients with Graves' Disease tested with nonprimate mammalian thyrocytes should be attributed to the elevated anti-Gal activity in these sera.

## 6. STIMULATION OF THYROCYTES FROM GRAVES' DISEASE PATIENTS BY ANTI-GAL

The observations on increased activity of anti-Gal in patients with Graves' Disease and on the ability of this antibody to stimulate porcine thyrocytes, raised the question of whether anti-Gal can also stimulate the thyrocytes from patients with Graves' Disease and thus contribute to the pathogenesis of this autoimmune disease. Since α-gal epitopes are not expressed on normal human thyrocytes (Thall et al., 1991), it could be expected that normal human thyrocytes will not be affected by anti-Gal. It was not clear, however, whether the thyrocytes from Graves' Disease patients are affected by this antibody.

## 6.1. Binding of Anti-Gal to Graves' Disease Thyrocytes

Preparation of primary cultures of Graves' Disease thyrocytes is much more difficult than preparation of porcine thyrocyte cultures, since the patients' sera contain antibodies that are detrimental to thyrocytes. Such antibodies, if present in the culture media, prevent adhesion of the thyrocytes to the tissue culture plates. Establishing primary cultures of Graves' Disease thyrocytes was performed according to the method of Winand and Kohn (1975) and included proteolytic digestion of the gland obtained from patients undergoing thyroidectomy, isolation of single cells followed by ten washes to remove all antibodies prior to culturing. Normal human thyrocytes were obtained from thyroid specimens of patients with thyroid "cold" nodules (i.e. thyroid glands that did not display increased iodine uptake) and from brain-dead individuals with no apparent thyroid disorders (Winand et al., 1994).

To determine whether anti-Gal displays differential binding to Graves' Disease thyrocytes, primary cultures of such thyrocytes or of thyrocytes from normal thyroid tissue were incubated with biotinylated anti-Gal and subsequently with $^{125}$I-labeled streptavidin. The radioactivity on the detach cells was measured after washing of the cells. As expected, binding of anti-Gal to $1 \times 10^6$ normal human thyrocytes was marginal and did not exceed 450 cpm (Figure 7). For comparison,

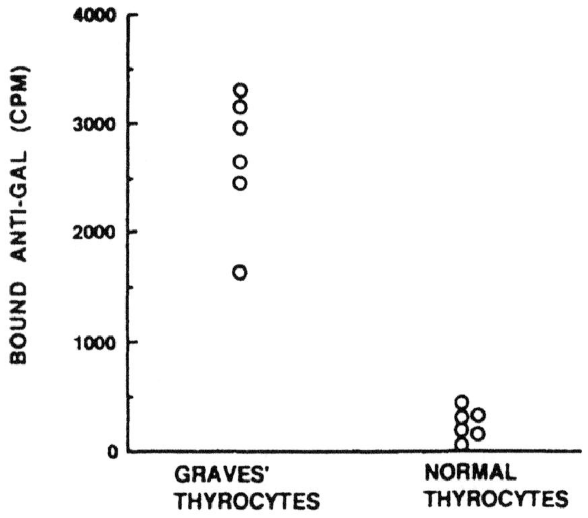


**Figure 7.** Binding of biotinylated anti-Gal (10 μg/ml) (and subsequently of $^{125}$I-labeled strepavidin) to $1 \times 10^6$ thyrocytes in primary cultures. Each circle represents the mean data from two cultures of each individual studied (from Winand et al., 1994, with permission).

anti-Gal binding to porcine thyrocytes under the same conditions was found to correspond to 25,000 cpm (not shown). In contrast to normal human thyrocytes, Graves' Disease thyrocytes bound anti-Gal in amounts that corresponded to 1650 to 3300 cpm (Winand et al., 1994).

### 6.2. Anti-Gal Stimulation of Graves' Disease Thyrocytes

In order to address the question of whether the binding of anti-Gal to Graves' Disease thyrocytes results in the stimulation of these cells, production of cAMP was measured in these thyrocytes following incubation with anti-Gal. As shown in Figure 8A, anti-Gal at a concentration of 50 μg/ml induced an increase of 100 to 350% in cAMP synthesis over the basal level in thyrocytes from patients with Graves' Disease. The basal level defined as 100% synthesis corresponded to 2.7 pmol cAMP per $1 \times 10^6$ cells. The basal level of cAMP synthesis in normal thyrocytes did not differ from that of Graves' Disease thyrocytes. However, no significant increase in cAMP synthesis was observed in the normal thyrocytes incubated with anti-Gal.

Another indication for the stimulation of Graves' Disease thyrocytes by anti-Gal was obtained by measuring uptake of $^{125}I$ into these cells. The uptake of $^{125}I$ by thyrocytes in the absence of anti-Gal did not differ in Graves' Disease and in normal thyrocytes and was found in all thyrocyte cultures to be in the range of 1500 to 2100 cpm per $1 \times 10^6$ cells (not shown). However, the two types of thyrocytes greatly differed in $^{125}I$ uptake after incubation for 72 hr with 50 μg/ml of anti-Gal (Figure 8B). The data in this figure are expressed after subtraction of the spontaneous uptake of $^{125}I$ in the absence of the antibody. Incubation of normal human thyrocytes with anti-Gal did not increase $^{125}I$ uptake whereas increased $^{125}I$ uptake was observed in thyrocytes of all Graves' Disease patients studied.

The internalized $^{125}I$ in the Graves' Disease thyrocytes was followed by effective organification of the iodine, i.e. formation of iodoamino acids and ultimate linkage to thyroglobulin. This could be demonstrated by size fractionation of the cytosolic content of these thyrocytes on Sephadex G-10 column. As shown in Figure 9, such a procedure performed with Graves' Disease thyrocytes yielded two peaks, of which the smaller one corresponded to macromolecule fractions with the average size of thyroglobulin and the larger peak represented free $^{125}I$ or iodoamino acids (Wadeleux et al., 1978). These fractionation data indicate that a significant proportion of the $^{125}I$ internalized by Graves' Disease thyrocytes as a result of stimulation by anti-Gal, was further subjected to organification in a pattern similar to that observed in normal thyrocytes incubated with TSH (Wadeleux et al., 1978).

The cytosolic content and the organification of $^{125}I$ in normal thyrocytes incubated with anti-Gal were found to be much lower than in Graves' Disease thyrocytes. These findings are in accord with the data in Figure 8B which demonstrate no increased $^{125}I$ uptake in normal human thyrocytes incubated with anti-Gal.

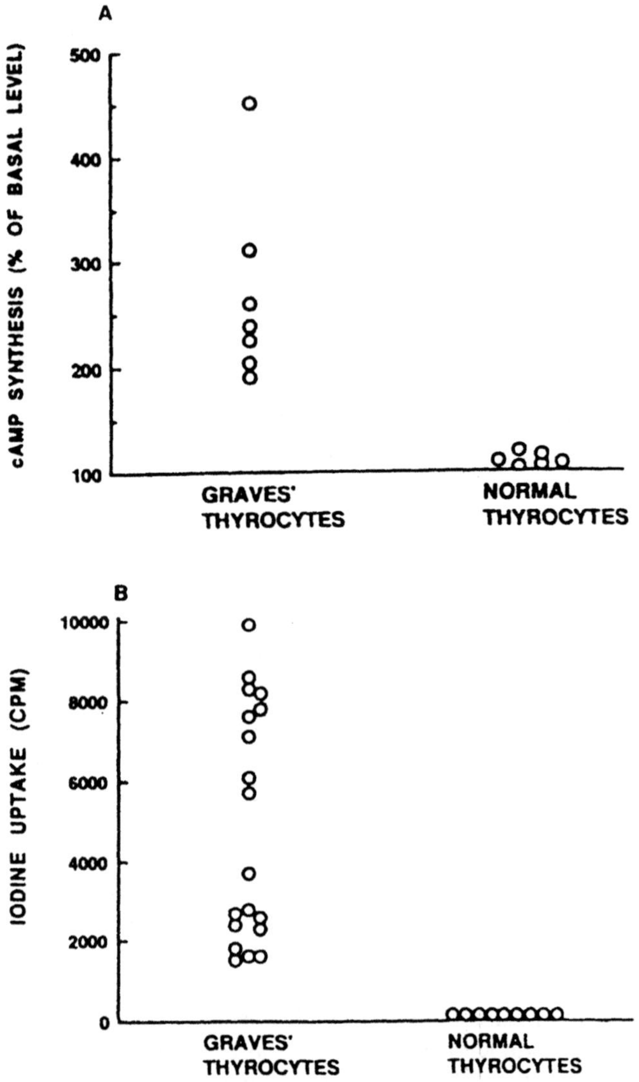


**Figure 8.** Anti-Gal mediated stimulation of cAMP production (A) and of $^{125}I$ uptake (B) by thyrocytes from Graves' Disease patients and normal individuals. Measurements of cAMP synthesis and of $^{125}I$ uptake were performed after incubation of thyrocytes with 50 μg/ml of anti-Gal. Each circle represents data from one donor (from Winand et al., 1994, with permission).

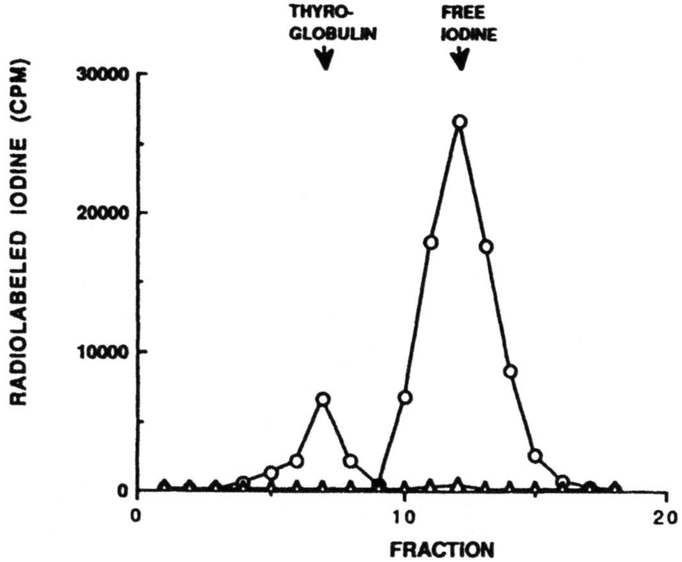


**Figure 9.** Stimulatory effect of anti-Gal (50 μg/ml) on the organification of $^{125}I$ in Graves' Disease thyrocytes (O) and in normal thyrocytes (Δ). Thyrocyte homogenates were chromatographed on Sephadex G-10 to separate free $^{125}I$ taken up by the cells from $^{125}I$ that was incorporated into macromolecules. Arrows represent the peak fractions in which thyroglobulin and Na $^{125}I$ are detected (from Winand et al., 1994, with permission).

## 7. STIMULATION OF GRAVES' DISEASE THYROCYTES BY ANTI-GAL IN AUTOLOGOUS SERA

The observed selective TSH-like effects of anti-Gal on Graves' Disease thyrocytes but not on normal human thyrocytes suggested that this antibody may play a role in the *in vivo* stimulation of thyrocytes in Graves' Disease. Therefore, it was of interest to determine whether anti-Gal within Graves' Disease sera can stimulate cAMP synthesis and iodine uptake in cultured autologous thyrocytes. These thyrocytes were incubated in medium containing 10% autologous serum or autologous serum depleted of anti-Gal and assayed for cAMP synthesis. Anti-Gal depletion was performed by adsorption of the sera on rabbit red cells since α-gal epitopes are abundantly expressed on these cells (Eto et al., 1968). Human blood group O red cells served as adsorption control.

Autologous sera induced increase of 137 to 221% in cAMP synthesis over the basal 100% level (i.e., synthesis observed with 10% fetal calf serum) in cell cultures from six Graves' Disease patients (Figure 10A). Removal of anti-Gal re-

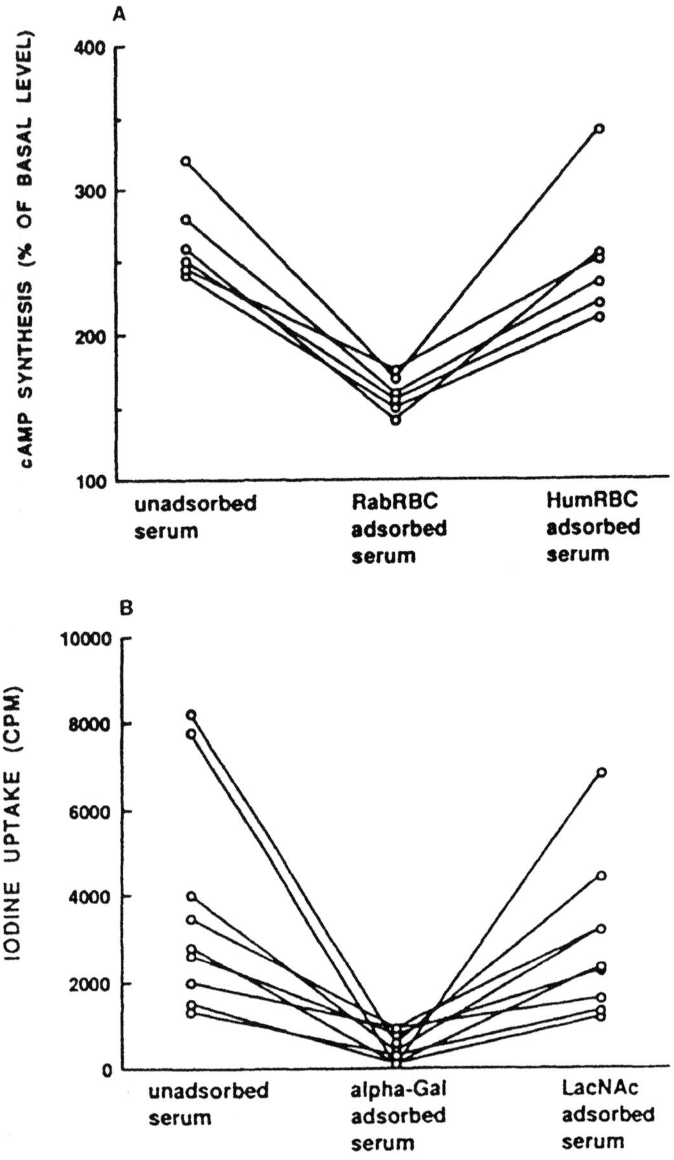


**Figure 10.** Stimulation of Graves' Disease thyrocytes by autologous sera for cAMP production (A) and for $^{125}I$ uptake (B), in the presence or absence of anti-Gal. The stimulatory effects of the sera were measured before and after removal of anti-Gal by adsorption on rabbit red blood cells (Rab RBC) (A), or on columns with α-gal epitopes (B). Human red blood cells (Hum RBC) and N-acetyllactosamine (LacNAc) columns served as adsorption controls, respectively. Data from each serum sample are represented as circles connected for the three conditions, i.e., preadsorption, adsorption on α-gal epitopes and control adsorption (from Winand et al., 1994, with permission).

sulted in a marked decrease in the stimulatory activity of the sera. Adsorption of the sera on human red cells did not alter the stimulatory activity. It should be stressed that incubation of the Graves' Disease thyrocytes with 10% pooled normal human sera had no significant stimulatory effect on these thyrocytes, as cAMP synthesis increased by no more than 24% (not shown).

Similar stimulatory effects were observed when $^{125}$I uptake was measured in the thyrocytes under the same conditions (Figure 10B). Variable uptake of $^{125}$I was observed when Graves' Disease thyrocytes were incubated in culture media containing 10% autologous sera. This uptake, however, was greatly reduced when the sera were depleted of anti-Gal by adsorption on a column with synthetic α-gal epitopes, whereas adsorption on a column with *N*-acetyllactosamine residues had no such effect (Figure 10B).

Overall these data indicate that anti-Gal in Graves' Disease sera plays an important role in the stimulation of autologous thyrocytes. This is inferred from the findings that specific removal of anti-Gal from such sera results in the decrease of 50-80% in the stimulatory effect on autologous thyrocytes. Graves' Disease sera contain therefore, two types of antibodies capable of stimulating thyrocytes. One type is the well documented anti-thyroid autoantibodies that are likely to bind peptide antigens on the TSH receptor and on other cell surface molecules (Weetman and McGregor, 1984; Smith et al., 1988; DeGroot and Quintans, 1989; Volpe, 1986). These antibodies can be observed in Graves' Disease sera depleted of anti-Gal. These antibodies also are capable of stimulating normal human thyrocytes since such cells express the antigens recognized by the anti-thyroid autoantibodies. The second type of thyroid stimulating antibodies is anti-Gal antibodies that display elevated activity in Graves' Disease patients. Anti-Gal in these patients interacts with, so far, undefined epitopes that are expressed on Graves' Disease thyrocytes. Anti-Gal in these patients, however, does not bind to normal human thyrocytes. Activation of Graves' Disease thyrocytes by anti-Gal requires the elevated activity of this antibody beyond that in normal human sera. As previously reported (Etianne-Decerf et al., 1987) sera of Graves' Disease patients contain anti-Gal with high activity, which is probably the result of increased amount, as well as increased affinity of the antibody (see the chapter of Galili et al., on "The anti-Gal antibody"). In contrast, normal human sera, in which anti-Gal activity is significantly lower than in Graves' Disease sera, fail to stimulate the pathologic thyrocytes. Isolated anti-Gal from normal sera is enriched for high affinity anti-Gal clones because of the affinity chromatography process. Because of its high affinity and relatively high concentration, this isolated antibody also can stimulate the pathologic thyrocytes. Although the stimulatory activity of anti-Gal in various Graves' Disease sera is highly variable, it is at least as potent as the stimulatory activity of the anti-thyroid autoantibodies. Therefore, anti-Gal probabely plays a major role in the chronic stimulation of the thyroid gland in patients with Graves' Disease.

## 8. POSSIBLE IDENTITY OF ANTI-GAL BINDING EPITOPES ON GRAVES' DISEASE THYROCYTES

The epitopes on Graves' Disease thyrocytes that interact with anti-Gal have not been characterized as yet. The relative small number of these epitopes makes the task of their identification very difficult. It is unlikely that these epitopes are the α-gal epitope. As discussed in the chapter of Galili on the "Evolution of α-galactosyltransferase and of the α-gal epitope," two types of mutations prevent expression of α1,3GT in humans and subsequent synthesis of α-gal epitopes: regulatory mutations that prevent the activity of the enhancer/promoter domains, and structural mutations that cause the production of a truncated enzyme molecule which lacks catalytic activity. Theoretically, aberrant expression of α-gal epitopes because of rare correction of these two types of mutations would have resulted in expression of the α-gal epitope in several tissues in addition to its expression on thyrocytes. This may result in the occurrence of autoimmune processes throughout the body and not only in the thyroid gland.

Two types of molecules produced in small amounts in humans were shown to interact with anti-Gal and thus may serve as possible ligands for anti-Gal on Graves' Disease thyrocytes:

1. Teneberg et al., (1996) reported on the presence in small amounts of a glycolipid designated $x_2$ with the terminal structure GalNAcβ1-3Gal-R on human red blood cells. This glycolipid is capable of interaction with anti-Gal, albeit at a much lower affinity than the α-gal epitope. The $x_2$ epitopes are normally cryptic on human red blood cells and are thought to be exposed when the red blood cells become senescent (Galili, 1997). It is possible that increased expression of $x_2$ on human thyrocytes could result in excessive binding anti-Gal and stimulation of the cells.
2. Sandrin et al., (1997) have shown that the peptide DAHWESWL, normally present in human mucin can interact with anti-Gal because it somewhat mimics the three dimensional structure of the α-gal epitope (see the chapter of Sandrin and McKenzie on "Modulation of α-gal epitope expression on porcine cells"). The increased expression of such a peptide may explain the observations of Castronovo et al., (1989) on the binding of anti-Gal to human mammary carcinoma tumors, and of Galili (1989) on the binding of anti-Gal to the human mammary carcinoma cell line MCF7. Increased expression of such a peptide on Graves' Disease thyrocytes may also result in the ability of anti-Gal to bind to these cells and stimulate them. Both these possibilities should be explored with thyrocytes obtained from Graves' Disease patients.

## 9. CONCLUSIONS

The natural anti-Gal antibody is ubiquitously present in all humans, whereas its ligand, the α-gal epitope, is not produced by humans because of the lack of α1,3GT activity. Nevertheless, several studies reported on the interaction of anti-Gal with various human cells or with molecules produced in humans, as the $x_2$ glycolipid on red cells and the peptide DAHWESWL in human mucin. Normally, the interaction with the $x_2$ glycolipid or with the mucin peptide is prevented *in vivo* because these ligands are cryptic on the membrane of cells. However, increased expression of such molecules may lead to binding of anti-Gal that would mediate autoimmune processes against tissues expressing these molecules. The most distinct association between anti-Gal and an autoimmune disease is in Graves' Disease. Studies with porcine thyrocytes expressing α-gal epitopes, or with recombinant human TSH receptors engineered to express these epitope have shown that binding of anti-Gal to the TSH receptor activates it in a manner similar to the activation by TSH. Furthermore, binding of anti-Gal to Graves' Disease thyrocytes results in their stimulation. Whereas no such anti-Gal mediated stimulation has been observed with normal human thyrocytes, most of the stimulatory activity of Graves' Disease sera on autologous thyrocytes could be attributed to anti-Gal activity.

Interaction of anti-Gal with aberrantly expressed ligands may theoretically occur in various tissues, and result in autoimmunity. The likelihood of anti-Gal initiating autoimmune processes may be increased in tissues exposed to bacteria such as the gastrointestinal mucosa or kidney cells (e.g., in pyelonephritis). We have previously shown that bacterial cell wall components that interact with anti-Gal may bind to human cells and thus induce anti-Gal mediated destruction of such cells (Galili et al., 1988). The anti-Gal mediated inflammatory reaction generated under such circumstances would be analogous to that induced by the interaction of this antibody with α-gal epitopes on *Trypanosoma cruzi* cell membrane molecules in patients with Chagas' Disease (see the chapter of Avila on "α-galactosyl bearing epitopes as potent immunogens in Chagas' Disease"). Moreover, bacterial α1,3GT may be capable of utilizing human cell surface carbohydrate acceptors for *de novo* synthesis of α-gal epitopes. The activity of bacterial α1,3GT was previously demonstrated in *Salmonella* (Endo and Rothfield, 1969). The result of an inflammation with bacteria that have α1,3GT may thus be the synthesis of α-gal epitopes on human cells adjacent to the inflammatory site, the subsequent binding of anti-Gal to such cells, and the induction of an autoimmune inflammatory reaction.

Overall, the initial observations on the association between anti-Gal and autoimmunity suggest that ligands for anti-Gal may aberrantly appear on human cells by a variety of mechanisms which should be explored. Understanding of these mechanisms would provide insight into the role of this unique natural antibody in the pathogenesis of various autoimmune disorders.

DEDICATION This chapter is dedicated to the memory of Professor Roger J. Winand of the University of Liege Hospital in Belgium. Professor Winand, who was one of the pioneers in research on pathogenesis of Graves' Disease, performed a major part of the studies described here.

## REFERENCES

Akamizu T, Ibuyama S, Saji M, Kosugi S, Kozak C, McBride OW and Kohn LD. 1990. Cloning, chromosomal assignment, and regulation of the rat thyrotropin receptor: Expression of the gene is regulated by thyrotropin, ag ents that increase cAMP levels, and thyroid autoantibodies. *Proc Natl Acad Sci USA 87*:5677–5681.

Atassi MZ, Manshouri T and Sakata S. 1991. Localization and synthesis of the hormone binding regions of the human thyrotropin receptor. *Proc. Natl. Acad. Sci. USA 88*:3613–3617.

Bottazzo GF, Pujol-Borrell R and Hanafusa T. 1983. Role of aberrant HLA-DR expression and antigen presentation in induction of endocrine autoimmunity. *Lancet 2*:1115–1119.

Castronovo V, Parent B, Steeg PS, Colin C, Foidart JM, Lambotte R and Mahieu P. 1989. Human natural anti-Gal antibodies may play a role in the natural anti-tumor defense system. *J. Natl. Cancer Inst. 81*:212–216.

Christiane Y, Aghayan M, Emonard H, Lallemand AL, Mahieu P and Foidart JM. 1992. Galactose alpha 1-3 galactose and anti-alpha galactose antibody in normal and pathologic pregnancies. *Placenta 13*:475–487.

DeGroot LJ and Quintans J. 1989. The causes of autoimmune diseases. *Endocrin. Rev. 10*:537–562.

Dumont JE, Jauniaux J-C and Roger PP. 1989. The cyclic AMP-mediated stimulation of cell proliferation. *TIBS 14*:67–71.

Edge ASB and Spiro RG. 1985. Thyroid cell surface glycoproteins. Nature and disposition of carbohydrate units and evaluation of their blood group I activity. *J. Biol. Chem. 260*:6893–6898.

Endo A and Rothfield L. 1969. Studies on phospholipid requiring bacteria enzymes. I. Purification and properties of uridine disphosphate galactose: lipopolysaccaride α-galactosyltransferase. *Biochemistry 8*:3500–3512.

Etienne-Decerf J, Malaise M, Mahieu P and Winand R. 1987. Elevated anti-α-galactosyl antibody titers. A marker for progression in autoimmune thyroid disorders and in endocrine ophthalmopathy. *Acta. Endocrin. 115*:67–74.

Etienne-Decerf, J., and R. J. Winand. 1981. A sensitive technique for determination of thyroid-stimulating immunoglobulin (TSI) in unfractionated serum. *Clin. Endocrin. 14*:83–91.

Eto T, Iichikawa Y, Nishimura K, Ando S and Yamakawa T. 1968. Chemistry of lipids of the posthemolytic residue or stroma of erythrocytes. IVI. Occurrence of ceramide pentasaccharide in the membrane of erythrocytes and retriculocytes in rabbit. *J. Biochem. (Tokyo) 64*:205–213.

Gabrielli A, Leoni P, Danieli G. Herrmann K, Krieg T and Wieslander J. 1991. Antibodies against galactosyl α(1-3) galactose in connective tissue disease. *Arthrit. Rheumat. 34*:375–376.

Galili U. 1989. Abnormal expression of α-galactosyl epitopes in man. A trigger for autoimmune processes? *Lancet 2*:358–361.

Galili U. 1993. Evolution and pathophysiology of the human natural anti-Gal antibody. *Springer Seminars in Immunopathology 15*:155–171.

Galili, U.1997. The α-galactosyl epitope (Galα1-3Galß1-4GlcNAc-R) and the natural anti-Gal antibody. In: Molecular Biology and Evolution of Blood Groups and MHC Antigens in Primates, Eds. W.W. Socha and A. Blancher, Springer Verlag, pp. 236–253.

Galili U, Buehler J, Shohet SB, and Macher BA. 1987b. The human natural anti-Gal IgG. III. The subtlety of immune tolerance in man as demonstrated by crossreactivity between natural anti-Gal and anti-B antibodies. *J. Exp. Med. 165*:693–704.

Galili U, Clark MR, Shohet SB,Buehler J, and Macher BA. 1987a. Evolutionary relationship between the anti-Gal antibody and the Galα1-3Gal epitope in primates. *Proc. Natl. Acad. Sci (USA) 84*:1369–1373.

Galili U, Flechner I, Kniszinski A,Danon D, and Rachmilewitz ER. 1986. The natural anti-α-galactosyl IgG on human normal senescent red blood cells. *Br. J. Haematol. 62*:317–324.

Galili U, Mandrell RE, Hamahdeh RM, Shohet SB and Griffis JM. 1988a. Interaction between human natural anti-α-galactosyl immunoglobulin G and bacteria of the human flora. *Infect. Immun.* *56*:1730–1737.

Galili, U, E.A. Rachmilewitz, A. Peleg, and I. Flechner. 1984. A unique natural human IgG antibody with anti-α-galactosyl specificity. *J. Exp. Med.* *160*:1519–1531.

Galili U, Shohet SB, Kobrin K, Stults CLM and Macher BA. 1988b. Man, apes, and Old World monkeys differ from other mammals in the expression of α-galactosyl epitopes on nucleated cells. *J. Biol. Chem.* *263*:17755–17762.

Kosugi S, Okajima F, Ban T, Hidaka A, Shenker A and Kohn LD. 1992. Mutation of alanine 623 in the third cytoplasmic loop of the rat-thyrotropin (TSH) receptor results in a loss in the phosphoinositide but not cAMP signal induced by TSH and receptor autoantibodies. *J. Biol. Chem.* *267*:24153–24156.

Kress B and Spiro RG. 1986. Studies on the glycoprotein nature of the thyrotropin receptor: interaction with lectins and purification of bovine protein with the use of *Bandeiraea simplicifolia I* affinity chromatography. *Endocrinology* *118*:974–979.

Libert F, LeFort A, Gerard C, Parmentier M, Perret J, Ludgate M, Dumont JE and Vassart G. 1989. Cloning, sequencing and expression of human thyrotropin (TSH) receptor: Evidence for binding of autoantibodies. *Biochem Biophys Res Commun* *165*:1250–1255.

Luo G, Fan J-Z, Seetharamaiah GS, Desai RK, Dallas JS, Wagle N, Doan R, Niesel DW, Klimpel GR and Prabhakar BS. 1993. Immunization of mice with *Yersinia enterocolitica* leads to the induction of antithropin receptor antibodies. *J. Immunol.* *151*:922–928.

Macchia E, Fenzi GF, Monzami F, Bartalena L, Lippi F, Aloisio V, Cupini C, Baschien L and Pinchera A. 1983. TSH-displacing activity versus TSH binding inhibiting activity of immunoglobulins from patients with Graves' disease. *J Endocrinol. Invest.* *6*:375–378.

Misrahi M, Loosfelt H, Atger M, Sar S, Guichon-Mantel A and Milgrom E. 1990. Cloning sequencing and expression of human TSH receptor. *Biochem Biophys Res Comm* *166*:394–403.

Nagayama Y, Kaufmann KD, Seto P and Rapoport B. 1989. Molecular cloning, sequence and functional expression of the cDNA for the human thyrotropin receptor. *Biochem Biophys Res Commun* *165*:1184–1190.

Parmentier M, Libert F, Maehaut C, Lefort A, Gerard C, Perret J, VanSande J, Dumont JE and Vassart G. 1989. Molecular cloning of the thyrotropin receptor. *Science* *245*:1620–1622.

Petryniak J, Varani J, Ervin PR and Goldstein IJ. 1991. Differential expression of glycolipids containing α-D-galactosyl groups on normal human breast epithelial cells and MCF-7 human breast carcinoma cells. *Cancer Ltrs.* *60*:59–65.

Sandrin MS, Vaughan HA, Xing P-X and McKenzie IFC. 1997. Natural human anti-Galα(1,3)Gal antibodies react with human mucine peptides. *Glycoconjugates J.* *14*:97–105.

Santer UV, DeSantis R, Hård KJ, van Kuik JA, Vligenthart JFG, Won B and Glick MC. 1989. N-linked oligosaccharide changes with oncogenic transformation require sialylation of multiantennae. *Eur J Biochem* *181*:249–260.

Shenkman L and Bottom EJ. 1976. Antibodies to *Yersinia enterocolitica* in thyroid disease. *Ann. Int. Med.* *85*:735–739.

Smith DF, Larsen RD, Mattox S, Lowe J and Cummings RC. 1990. Transfer and expression of a murine UDP-Gal: ß-D-Gal-α1,3galactosyltransferase gene in transfected Chinese hamster ovary cells. Competition reaction between the α1,3galactosyltransferase and endogenous α2,3-sialyltransferase. *J Biol Chem* *265*:6225–6234.

Rees-Smith B, McLachlan SM and Furmaniak J. 1988. Autoantibodies to thyrotropin receptor. *Endocrin. Rev.* *9*:106–121.

Spiro RG and Bhoyoo VD. 1984. Occurrence of α-D-galactosyl residues in the thyroglobulins from several species. Localization in the saccharide chains of the complex carbohydrate units. *J. Biol. Chem.* *259*:9858–9566.

Strakosch CR, Wenzel BE, Row VV and Volpe R. 1982. Immunology of autoimmune thyroid diseases. *New. Engl. J. Med.* *307*:1499–1507.

Teneberg S, Lönnroth I, Torres Lopez JF, Galili U, Olwegard Halvarsson M, Angstrom J and Karlsson K-A. 1996. Molecular mimicry in the recognition of glycosphinolipids by Galα3Galβ4GlcNAc-β-binding *Clostridium difficile* toxin A, human natural anti-α-galactosyl IgG and the monoclonal antibody Gal-13: characterization of a binding-active human glycosphingolipid, non-identical with the animal receptor. *Glycobiology* *6*:599–609.

Thall A, Etienne-Decerf J, Winand RJ and Galili U. 1991. The α-galactosyl epitope on mammalian thyroid cells. *Acta Endocrin.* *124*:692–699.

Thall A and Galili U. 1990. Distribution of Galα1-3Galß1-4GlcNAc residues on secreted mammalian glycoproteins (thyroglobulin, fibrinogen and immunoglobulin G) as measured by a sensitive solid-phase radioimmunoassay. *Biochem.* *29*:3951–3965.

Volpe R. 1986. Autoimmune thyroid disease: A perspective. *Mol. Biol. Med.* *3*:25–51.

Wadeleux PA, Etienne-Decerf J, Winand RJ, and Kohn LD. 1978. Effects of thyrotropin on iodine metabolism of dog thyroid cells in tissue culture. *Endocrinol.* *102*:889–902.

Weetman AP and McGregor AM. 1984. Autoimmune thyroid disease: Development in our understanding. *Endocrin. Rev.* *5*:309–355.

Weiss M, Ingbar SH, Winblad S and Kasper DL. 1983. Demonstration of a saturable binding site for thyrotropin in *Yersinia enterocolitica. Science* *219*:1331–1333.

Winand RJ, Anaraki F, Etienne-Decerf J, and Galili U. 1993. Xenogeneic TSH like activity of the human natural anti-Gal antibody: Interaction of anti-Gal with porcine thyrocytes and with recombinant human TSH receptors expressed on mouse cells. *J. Immunol.* *151*:3923–3934.

Winand RJ and Kohn LD. 1975. Thyrotropin effects on thyroid cells in culture: Effect of trypsin on the thyrotropin receptor and on thyrotropin-mediated cyclic 3'-5' AMP changes. *J. Biol. Chem.* *250*:6534–6540.

Winand RJ, Winand-DevigneJ, Meurisse M. and Galili U .1994. Specific stimulation of Graves' disease thyrocytes by the natural anti-Gal antibody from normal and autologous serum. *J. Immunol.* *153*:1386–1395.

Zakarija M and McKenzie JM. 1987. The spectrum and significance of autoantibodies reacting with the thyrotropin receptor. *Endocrin. Metabol. Clin. North. Am.* *16*:343–363.

*Chapter 15*

# Enhancement of Autologous Tumor Vaccine Immunogenicity by Anti-Gal

Denise C. LaTemple and Uri Galili

## 1. INTRODUCTION

Anti-Gal is considered to be the most prevalent naturally occurring antibody common to all humans. It represents approximately 1% of total IgG in the serum in humans and interacts specifically with the α-gal epitope of cell surface glycoproteins and glycolipids (see the chapter on "The Natural Anti-Gal Antibody"; Galili, 1993; Galili *et al.*, 1985; Galili *et al.*, 1984). Understandably, the α-gal epitope is not expressed on tissues or cells derived from human sources since the enzyme synthesizing this epitope, the α1,3 galactosyltransferase, is not active in humans (see the chapter on "Evolution of α1,3 Galactosyltransferase and the α-Gal Epitope"; Galili *et al.*, 1988). But imagine the consequences of a human cell genetically or enzymatically engineered to express the α-gal epitope on its surface and then introduced into a normal individual. A prediction of the outcome lies in studies in xenotransplantation where non-primate mammalian tissues, which normally express the α-gal epitope, succumb to hyperacute immune rejection when trans-

---

**Denise C. LaTemple** Department of Medicine, Division of Hematology and Oncology, University of Pennsylvania, Philadelphia, Pennsylvania. **Uri Galili** Department of Microbiology and Immunology, MCP Hahnemann School of Medicine, Philadelphia, Pennsylvania 19129.

*Subcellular Biochemistry, Volume 32: α-Gal and Anti-Gal*, edited by Galili and Avila.
Kluwer Academic / Plenum Publishers, New York, 1999.

planted into a human recipient (see the chapter of Kobayashi and Cooper on "Anti-Gal in Xenotransplantation"; Collins *et al.*, 1994; Galili 1993b; Sandrin *et al.*, 1993; Good *et al.*, 1992). Anti-Gal thus forms an immunological barrier preventing xenotransplantation facilitated by binding to α-gal epitopes on the surface of the xenograft. Both IgM and IgG isotypes of anti-Gal bind the α-gal epitopes on the xenograft and destroy it through antibody mediated effector mechanisms such as antibody dependent cell cytotoxicity (ADCC) and complement fixation and lysis (Galili 1993b; Sandrin *et al.*, 1993; Good *et al.*, 1992). Whereas anti-Gal has a detrimental effect on xenografts, the immunologic potential of anti-Gal might be exploited for beneficial clinical use. Because of the ubiquitous occurrence of anti-Gal in large amounts in all humans and because of its restricted specificity, it is possible to manipulate anti-Gal by administration of various glycoconjugates or cells expressing α-gal epitopes. The interaction of such glycoconjugates with anti-Gal would result in the formation of many immune complexes in situ and thus would present a unique variety of therapeutic opportunities, most of which are unexplored as yet. One example for such manipulation is the formation of immune complexes between various vaccines expressing α-gal epitopes and anti-Gal to target these vaccines to antigen presenting cells (APCs). As discussed below, effective uptake of vaccines complexed with anti-Gal would result in increased immunogenicity of the vaccine because of better processing and presentation of the vaccinating antigen(s) by APCs.

We studied the possible use of anti-Gal for targeting autologous tumor vaccines to APCs by engineering such vaccines to express α-gal epitopes. Administration of α-gal-expressing autologous tumor vaccines into cancer patients may result in an effective immune response to autologous tumor-associated antigens (TAAs), ultimately inducing destruction of metastatic tumor cells expressing these antigens. This chapter discusses the rationale for using tumor vaccines expressing α-gal epitopes and the possible strategies for engineering tumor vaccines to express this epitope. In addition, we describe preliminary studies demonstrating anti-Gal-mediated targeting of tumor cells to APCs.

## 2. CANCER IMMUNOTHERAPY

Immunotherapy, along with surgery, radiation, and chemotherapy, is one of four major approaches for the treatment of cancer. Currently, the treatment of cancer patients involves the removal of the primary tumor by surgery followed by regimens of radiation or chemotherapy to reduce the recurrence of the tumor. These treatments are very effective when the tumor is localized and much less effective for disseminated tumors and micrometastases. Cancer immunotherapy, on the other hand, involves the stimulation of a patient's natural defenses to seek out and specifically destroy metastatic tumor cells. Successful activation of an anti-tumor immune response relies on the ability of the immune system to recognize distinct surface TAAs. The notion that the immune system can be instructed to recognize TAAs and mediate tumor rejection is supported by experimental evidence from early studies in mice. Immunization of mice with tumor cells derived from a chemically-induced tumor, developed in a syngeneic strain, protected these mice from tumor growth upon rechallenge with the parental tumor cells (Prehn and Main, 1957; Foley, 1953). The immune system, particularly T-cells, was shown to be the major mediator in tumor destruction which relied upon the recognition of unique tumor antigens (Kast *et al.*, 1989). Alternatively, tumors of spontaneous origin in mice failed to protect mice against subsequent tumor challenge suggesting that these types of tumors do not express antigens recognizable by the immune system (Klein and Klein, 1977). Since the majority of tumors in humans arise spontaneously, it was believed that the immune system would be unable to mount an anti-tumor immune response similar to what was observed in mice. However, studies in humans revealed not only the infiltration of host mononuclear cells and lymphocytes within various tumors, but also the demonstration that T-cells isolated from cancer patients displayed cytotoxic activity against autologous tumor cells in vitro (Old, 1996; Herberman, 1992; Knuth *et al.*, 1989; Herberman, 1977). These studies suggested that T-cells react to unique antigens expressed on the tumor cells which led to the identification of these antigens. In general, TAAs that can be recognized by lymphocytes are unique to the tumor cells (e.g. viral proteins, mutated normal proteins), or they can be present on normal cells, but are overexpressed (e.g. gangliosides), or abnormally exposed (e.g. differentiation antigens, mucins) once normal cells have transformed into neoplastic cells (Sznol and Holmlund, 1997; Roth *et al.*, 1994). Many TAAs have been identified (e.g. MAGE-1, tyrosinase, MART-1/Melan-A, HER-2/neu, MUC-1) and are known to be present on a large proportion of human tumors (Wang and Rosenberg, 1996; Van den Eynde and Brichard, 1995; Boon *et al.*, 1995; Pardoll, 1994), yet the majority of TAAs have yet to be identified or characterized with respect to their ability to elicit an immune response.

### 2.1. T-cells and Antigen Presenting Cells: The Key to a Successful Anti-Tumor Immune Response

T-cells, with help from B cells, natural killer (NK) cells, and macrophages, are the crucial cell population involved in an anti-tumor immune response. The importance of these cells is exemplified in cell subset depletion studies where mice given tumors are also depleted of the various immune system cells by specific antibody ablation. Depending on the tumor model used, tumor growth remains unchecked when either cytotoxic (CD8+) or helper (CD4+) T-cells are absent (Chen *et al.*, 1997; Dranoff *et al.*, 1993; Restifo *et al.*, 1992; Colombo *et al.*, 1992; Hock *et al.*, 1991; Golumbek *et al.*, 1991; Gansbacher *et al.*, 1990a; Fearon *et al.*, 1990). These results have been confirmed also in nude mice (mice lacking T-cells), as well as with adoptive transfer studies in which CD8+ and/or CD4+ T-cells from a mouse who successfully cleared a tumor can protect a naïve mouse from tumor challenge (Greenberg, 1991; Rosenberg *et al.*, 1986; Mitchison, 1955).

Both CD8+ and CD4+ T-cells have distinct effector functions that lead to the eradication of tumor cells. CD8+ T-cells recognize TAA peptides in association with cell surface class I major histocompatibility complex (MHC) molecules and their main effector function is the direct lysis of tumor cells subsequent to the engagement of their T-cell receptor (TCR) with the corresponding ligand on the tumor cells. In contrast, CD4+ T-cells are capable of interaction with TAA peptides only if such peptides are presented on cells in association with class II MHC molecules. Because of these constraints on CD4+ T-cell recognition, these cells cannot directly recognize tumor cells since many tumors are MHC class II deficient. Therefore, CD4+ T-cells specific for TAAs interact with the corresponding TAA peptides when presented on APCs since APCs are the main cells that express class II MHC molecules. As a result of such interactions, CD4+ T-cells mediate the effector functions of other immune system cells through the secretion of cytokines. Cytokines, which are small protein hormones involved in regulating immune and inflammatory responses, such as interleukin 2 (IL-2), are produced by CD4+ T-cells and aid CD8+ T-cells to become activated and to proliferate. Other cytokines produced or induced by CD4+ T-cells, such as tumor necrosis factor alpha (TNFα) and interferon gamma (IFNγ), are capable of directly lysing tumor cells or slowing tumor growth , respectively (Boehm *et al.*, 1997; Asher *et al.*, 1991).

After completing development in the thymus, naïve T-cells are normally located within secondary lymphoid organs, such as lymph nodes and spleen. Only following their specific activation within these organs can TAA-specific T-cells migrate out and circulate in the periphery to encounter tumor cells expressing TAAs and mediate effector mechanisms. In order to carry out their respective effector functions, T-cells must be activated in a specific manner. The activation of both subsets of T-cells requires two signals coordinated by a variety of cell surface proteins on both the T-cell to be activated and APCs such as B-cells, macrophages, and dendritic cells. The TCR mediates the recognition of antigens that have been processed into small peptides and are normally associated with MHC molecules expressed on the surface of APCs (signal 1). Costimulation of the T-cell (signal 2) is provided by the interaction of the CD28 molecule on the T-cell with the B7 molecule (CD80 or CD86) on the APC. These two signals are crucial for naïve T-cell activation since the delivery of signal 1 in the absence of signal 2 results in T-cell anergy caused by a block in IL-2 production (Schwarz, 1992; Schwarz, 1990). Anergized T-cells can no longer respond to their appropriate stimuli (i.e. antigenic peptides) and thereby remain ineffective in an immune response.

## 2.2. Autologous Cancer Vaccines

Immunization of cancer patients with autologous tumor cells as vaccines serves as a strategy to achieve stimulation of TAA-specific T-cells for the induction of an anti-tumor immune response. The rationale for vaccination of patients with autologous tumor cells or tumor-cell lysates obtained from primary tumors is that tumor cells express all potential TAAs necessary to induce an immune response. This procedure is advantageous since it negates the need to identify specific TAAs and uses the primary tumor which is surgically removed from the patient in course of the treatment regimen. The process of autologous tumor cell vaccination involves the surgical removal of a patient's primary tumor followed by ex vivo manipulation to create a single cell suspension. The tumor cells are then irradiated and injected back into the patient. Theoretically, this procedure should stimulate the immune system to seek out and destroy metastatic tumor cells within the patient. Over the past century, this method has been used usually in combination with immune adjuvants, but responses in patients were varied and overall this approach met with limited success (Hollinshead et. al., 1987; Hoover et. al., 1984; McCune *et al.*, 1981; Zuhrie *et al.*, 1980; Vogler *et al.*, 1978; Powles *et al.*, 1977; Gutterman *et al.*, 1974). The reason for the relatively poor success of autologous cancer vaccines is that tumor cells by themselves are not inherently immunogenic enough to induce an immune response. Tumor cells are considered to be antigenic since they express a variety of TAAs potentially recognizable by T-cells, but these TAAs do not elicit an anti-tumor immune response. Tumor cells have evolved ways in which to evade recognition by an immune system normally capable of eradicating them (Herin *et al.*, 1987; Knuth *et al.*, 1984). Evasive mechanisms by tumor cells include lack of costimulatory molecules (Schwarz, 1992), lack or variable expression of MHC molecules (Schrier and Peltenburg, 1993; Moller and Hammerling, 1992; Browning and Bodmer, 1992), production of suppressive cytokines (Sulitzeanu, 1993), and lack of uptake by host APCs (Pardoll, 1994; Lanzavecchia, 1993). These properties contribute to the autologous tumor cell's low immunogenicity and diminish the activation of TAA-specific naïve T-cells since they prevent presentation of their own TAAs as peptides recognizable by T-cells. In the absence of such activation no induction of an anti-tumor immune response is achieved.

Several approaches have been developed for improvement of tumor cell immunogenicity by genetic modification of tumor cells including the introduction of MHC genes (Wallich *et al.*, 1985; Hui *et al.*, 1984), B7 genes to provide signal 2, (Baskar *et al.*, 1993; Townsend and Allison 1993; Chen *et al.*, 1992), and various cytokine genes (Dranoff *et al.*, 1993; Golumbek *et al.*, 1991; Fearon *et al.*, 1990). Results using these vaccines in mouse models are promising. However, genetic modification is hard to perform with human solid tumors since it requires the manipulation of live tumor cells and such cells are usually difficult to isolate from solid tumors (Pardoll, 1993). Overall, the challenge presented to cancer immunotherapists is to enhance the immunogenicity of autologous tumor cells in such a way as to enable vaccinating tumor cells to effectively stimulate T-cells.

### 2.3. The Role of APCs in Vaccination

Tumor cells lack any labels that will identify them for APCs as cells or cell membranes that ought to be ingested for antigenic processing and presentation (Grabbe *et al.*, 1995; Nossal, 1993; Lanzavecchia, 1985). Effective uptake of tumor vaccines by APCs is a prerequisite for the induction of an anti-tumor immune response since the immune response does not occur at the vaccination site. The vaccinating tumor cell or cell membranes have to be internalized by professional APCs such as macrophages, dendritic cells, or Langerhans skin cells and transported to the adjacent lymph node or to the spleen where they present the processed TAA peptides to naïve TAA-specific T-cells for the activation of these cells (Lord and Frelinger, 1998; Zinkernagel *et al.*, 1997).

The process of naïve T-cell activation for the induction of systemic anti-tumor immunity, as a result of tumor cell vaccination, is thought to occur in three steps: (1) the uptake and processing of tumor cells at the vaccination site by host APCs, (2) the migration of APCs to draining lymph nodes where naïve T-cell priming occurs, and (3) the circulation of activated (effector) T-cells to the periphery to seek and destroy TAA-bearing metastatic tumor cells (Lord and Frelinger, 1998; Schweighoffer *et al.*, 1996; Maass *et al.*, 1995). Support of this model came from studies showing that tumor vaccines induced the infiltration of monocytes and granulocytes, but rarely T-cells, to the site of vaccination (Maass *et al.*, 1995). Lack of T-cells at these sites is consistent with the idea that naïve T-cells are only found in the blood and lymph, whereas only activated or memory T-cells are found in tissues (Mackay *et al.*, 1992; Mackay *et al.*, 1990). Therefore, it was unclear as to how T-cells become activated after tumor cell vaccination. However, examination of the draining lymph nodes of vaccinated mice showed various mRNA transcripts of T-cell activation-specific cytokines and phenotypic markers such as IL-2, IL-4, IFN$\gamma$, and IL-2 receptor $\alpha$ (Maass *et al.*, 1995). Another study showed that protective immunity by a tumor vaccine could be abrogated by depleting naïve T-cells in draining lymph nodes implicating the lymph node as an important anatomical site for T-cell activation (Schweighoffer *et al.*, 1996).

The importance of APCs in this role is further supported by other studies in vaccine therapy. It has been shown that the most successful vaccines against tumor growth in mice have been tumor cell vaccines transduced with cytokines that recruit APCs to the site of vaccination such as GM-CSF (Dranoff *et al.*, 1993). However, in addition to the APCs being recruited to the site of vaccination by GM-CSF, they require signals to internalize the vaccinating tumor cells. Therefore, it is important to improve the uptake of vaccinating autologous tumor cells by APCs once they reach the site of vaccination.

### 2.4. Enhancement of Tumor Vaccine Immunogenicity Using Antigen/Antibody Complexes

A potentially effective way to increase the uptake of tumor vaccines by APCs is to complex them with an IgG antibody, a process called opsonization. The resulting immune complexes would be targeted to Fc receptors on APCs that can bind to the Fc portion of the complexed IgG. This interaction leads to phagocytosis of the vaccinating tumor cells or membranes by APCs. The three major types of Fcγ receptors are FcγI, II, and III (CD64, CD32, and CD16 respectively) and are normally found on activated B-cells (CD16), monocytes/macrophages (CD64, CD16), as well as blood dendritic cells and skin Langerhans cells (CD64, CD32) (Fanger *et al.*, 1996; Sallusto and Lanzavecchia, 1994; Schmitt *et al.*, 1990; Unkeless, 1989). Uptake of tumor cells by APCs via these receptors would result in proteolytic degradation of the TAAs on the cells, followed by presentation of the TAA peptides in association with MHC class I and class II molecules. During this process APCs transport TAAs to draining lymph nodes where these TAA peptides, together with costimulatory molecules provided by the APCs, would activate naïve, TAA-specific CD4+ (Lanzavecchia, 1996; Guinan *et al.*, 1994; Pardoll, 1993; Schwarz, 1992; Chen *et al.*, 1992) and CD8+ T-cells (Falo *et al.*, 1995; Huang *et al.*, 1994; Schwarz, 1992). This process ultimately leads to an immune response directed against the TAAs on the vaccinating tumor cells and which prime T-cells to kill metastatic tumor cells expressing the same TAAs.

This hypothesis of enhancing tumor vaccine immunogenicity is further supported by many studies that have shown that the administration of an antigen complexed with a specific opsonizing antibody elicits a much higher immune response than when the antigen is administered alone. These studies have shown that the augmented immune response is the result of efficient processing and presentation of the antigen. Targeting tetanus toxoid (TT) to Fcγ receptors by complexing with anti-TT IgG results in a 10- to 1000-fold increase in processing and presentation of this antigen as measured by helper T-cell activation (Liu *et al.*, 1996; Sallusto and Lanzavecchia, 1994; Gosselin *et al.*, 1992; Manca *et al.*, 1991). This enhancement was also observed in vitro using hepatitis B antigen complexed with its corresponding antibody (Celis *et al.*, 1987; Celis and Chang, 1984). The use of $F(ab)_2$ fragments in these experiments fails to induce a response indicating that the Fc portion of the opsonizing antibody interacts with the Fcγ receptor which is necessary for the antigen to be phagocytosed by APCs (Gosselin *et al.*, 1992; Manca *et al.*, 1991;Celis *et al.*, 1987). In mouse models, vaccination with TT complexed with anti-TT IgG (Stoner and Terres, 1963), as well as with antibody-complexed hepatitis B vaccines (Celis *et al.*, 1987) were found to elicit a much higher antibody response compared to vaccination with TT or hepatitis B alone. Antibody-complexed Venezuelan equine encephalomyelitis virus vaccine in monkeys was also more successful in enhancing the immune response when compared with monkeys receiving the virus vaccine alone (Houston *et al.*, 1977). These studies indicate the importance of immune complexes in enhancing the immune response through effective uptake, processing, and presentation of antigens by APCs. We propose to exploit the natural anti-Gal antibody for the formation of such immune complexes in situ.

## 3. ENHANCING CANCER VACCINE IMMUNOGENICITY BY ANTI-GAL

In order to overcome the low immunogenicity of autologous tumor vaccines, we have developed methods in which anti-Gal can directly target tumor cell vaccines to APCs. Tumor cell vaccines are engineered to express α-gal epitopes, resulting in opsonization of the vaccinating tumor cells by anti-Gal in vivo (Figure

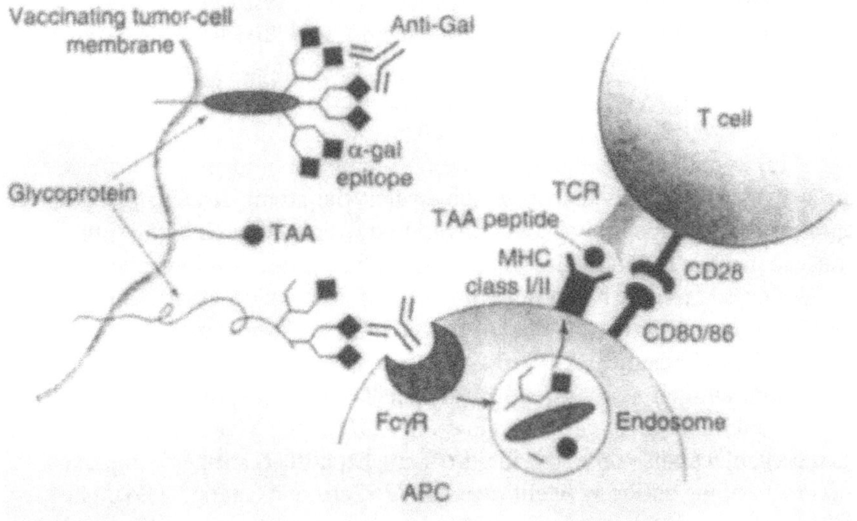


**Figure 1.** Anti-Gal-mediated augmentation of TAA immunogenicity. Vaccinating tumor-cell membranes are engineered to express α-gal epitopes. Host anti-Gal opsonizes these membranes, resulting in Fcγ receptor-mediated phagocytosis by APCs. The effective processing and presentation of the internalized TAA peptides may ultimately lead to TAA-specific T-cell activation (from Galili and LaTemple, 1997 with permission).

1). Subsequently, the interaction between the Fc portion of the antibody and the Fcγ receptor on APCs leads to the uptake of vaccinating tumor cells by APCs.

Normally, human tumor cells do not express α-gal epitopes, but the carbohydrate makeup of human cell membranes makes the synthesis of the epitope possible. Many glycoproteins and glycolipids of human cell membranes contain N-linked carbohydrate chains in which their terminal structure is N-acetyl-lactosamine that is capped or non-capped by sialic acid (SA). N-acetyl-lactosamine is a potential acceptor site for the action of the glycosylation enzyme α1,3 galactosyltransferase. As described in previous chapters in this book, this enzyme catalyzes the transfer of the sugar donor UDP-Gal to N-acetyl-lactosamine molecules in an α1,3 linkage. Normally, this enzyme is active in the Golgi apparatus in cells of New World monkeys and non-primate mammals synthesizing α-gal epitopes, but it is not active in humans. However, the production of a soluble recombinant form of the α1,3 galactosyltransferase (see the chapter on "New World Monkey α1,3 Galactosyltransferase: Structure-Function Studies"; Henion *et al.*, 1994) made it possible to synthesize the α-gal epitope on human tumor cells in vitro by simply incubating tumor cells with the recombinant enzyme as well as adding exogenous UDP-Gal to the reaction mixture. Synthesis of α-gal epitopes on human tumor cells can take place in a two-step reaction: (1) since many of the N-acetyl-lactosamine residues on human cell membranes are capped with SA, the cells must first be incubated with neuraminidase to remove the SA and to expose the underlying N-acetyl-lactosamine molecules, (2) cells are then incubated with recombinant α1,3 galactosyltransferase and UDP-Gal to form α-gal epitopes (Figure 2).

### 3.1. Synthesis of α-Gal Epitopes on Human Tumor Cells and Cell Membranes

We demonstrated that α-gal epitopes could indeed be synthesized on human tumor cells in the two-step reaction process described in Figure 2. Four human tumor cell lines were subjected to the two-step synthesis reaction in vitro. Using radiolabeled UDP-[$^3$H]Gal in the reaction, we were able to measure the number of α-gal epitopes synthesized per cell based on the specific activity of [$^3$H]Gal. This process resulted in the expression of as many as $5 \times 10^5$ α-gal epitopes per desialylated tumor cell (LaTemple *et al.*, 1996; Figure 3).

These newly synthesized α-gal epitopes were tested for their accessibility for interaction with the anti-Gal antibody. Anti-Gal was capable of binding the α-gal expressing tumor cells. This interaction resulted in phagocytosis by human macrophages, which are very effective APCs. In this experiment, the process of phagocytosis was dependent on the presence of α-gal epitopes on the tumor cells and the anti-Gal antibody in the culture medium. Almost no phagocytosis was observed with tumor cells lacking α-gal epitopes that were incubated with anti-Gal

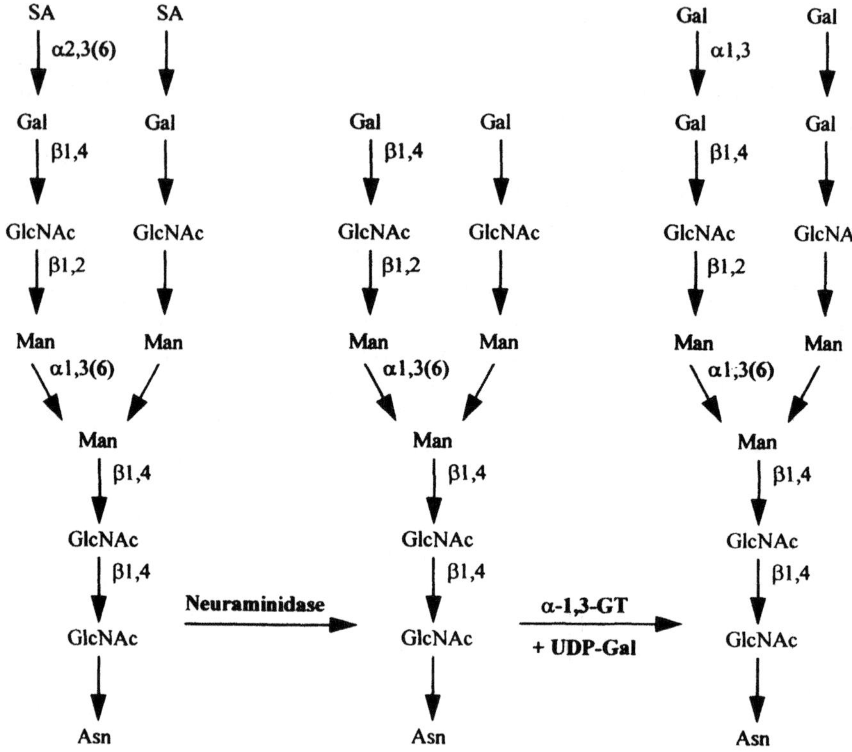


**Figure 2.** Synthesis of α-gal epitopes on human cells by recombinant α1,3 galactosyltransferase (rα1,3GT) in a two-step reaction. Sialic acid (SA) is removed by neuraminidase to expose penultimate N-acetyl-lactosamine structures (Galβ1,4GlcNAc-R). Galactose (Gal) is transferred from uridine diphosphate galactose (UDP-Gal) to N-acetyl-lactosamine by rα1,3GT. This results in the formation of α-gal epitopes (from Galili and LaTemple, 1997 with permission).

(LaTemple *et al.*, 1996; Figure 4). Therefore, anti-Gal mediated uptake of tumor cells is dependent on the expression of α-gal epitopes by the tumor cells.

Since most tumors cannot be isolated in single cell form, with the exception of leukemias and lymphomas, we tested whether cell membranes obtained from homogenized solid tumors could also serve as substrates for α-gal epitope synthesis. Homogenized colon and mammary carcinomas were subjected to the same synthesis process described above. We showed that anti-Gal was capable of binding to the various tumor membranes after α-gal epitopes synthesis (Galili and LaTemple, 1997; Figure 5). Therefore, the tumor removed from each individual patient may be processed by homogenization and isolation of the membranes. Such membranes would be subjected to synthesis of α-gal epitopes by incubation with recombinant α1,3 galactosyltransferase and UDP-Gal. Subsequently, the

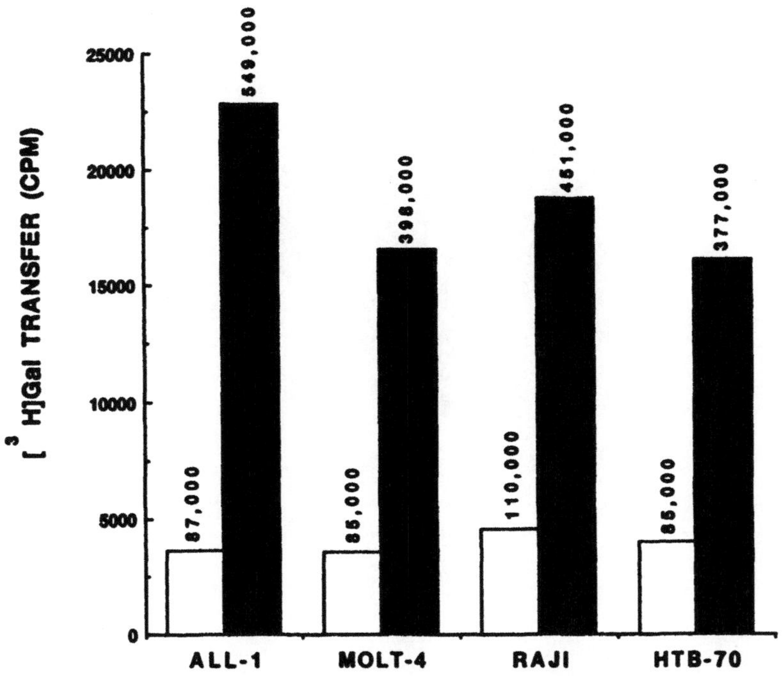


**Figure 3.** Synthesis of α-gal epitopes on human tumor cell lines by 10 μg/ml α1,3 galactosyltransferase, as measured by the transfer of [$^3$H]Gal from 0.4 μCi/ml UDP-[$^3$H]Gal to the cell membrane of untreated cells (open columns), or cells that were pretreated with neuraminidase (closed columns). Data above the columns represent the number of α-gal epitopes synthesized per cell, (from LaTemple *et al.*, 1996 with permission).

membranes would be washed, irradiated to kill any remaining live cells, and administered repeatedly to the patient as an autologous tumor vaccine. In situ binding of the circulating anti-Gal IgG to these membranes will result in the formation of immune complexes that would target autologous vaccine to APCs, as described in Figure 1.

## 3.2. Alternative Method for Expression of α-Gal Epitopes on Tumor Cells

Although α-gal epitopes can be successfully synthesized on human tumor membranes by an enzymatic process, engineering tumor cells in other ways to express this epitope may be benifitial for preparation of autologoud vaccines, in particular when many intact tumor cells are available . Genetic modification of tumor cells has proven to be useful for expression of MHC molecules, cytokines, and B7 molecules, therefore it is possible to transduce tumor cells with the α1,3 galactosyltransferase gene so as to allow endogenous synthesis of the α-gal

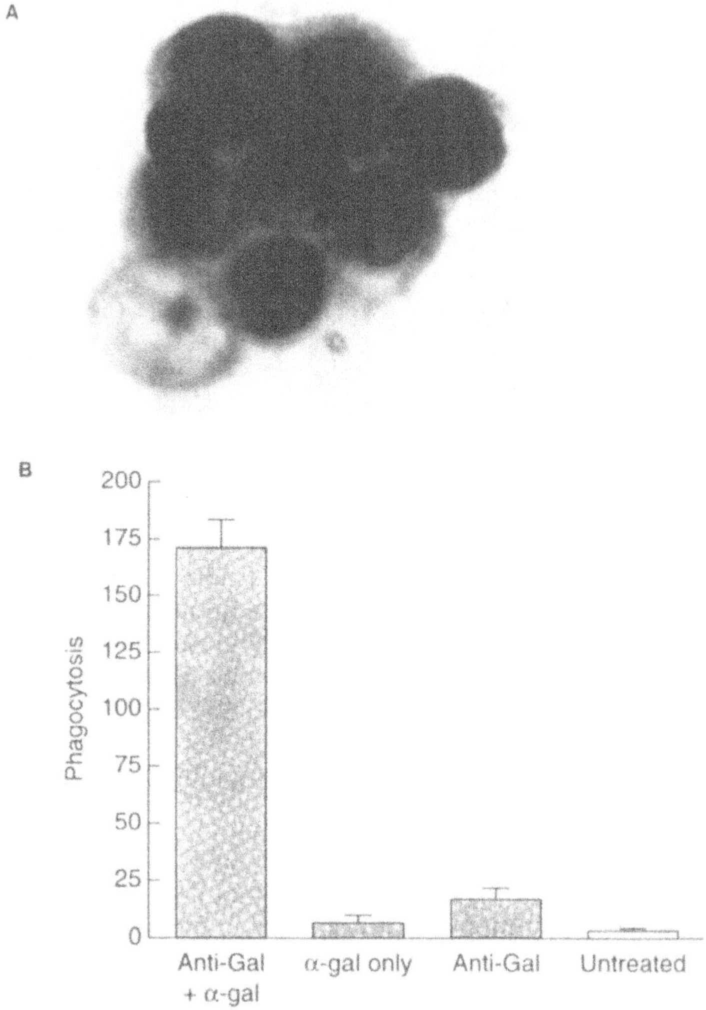


**Figure 4.** Anti-Gal-mediated phagocytosis of human pre-B ALL-1 leukemia cells by human macrophages. ALL-1 cells were subjected to α-gal epitope synthesis, then incubated with anti-Gal (4μg/ml) and adherent human macrophages for 3 hours at 37°C. The macrophages were then repeatedly washed to remove non-internalized cells and stained with Giemsa. (A) A macrophage containing seven phagocytosed tumor cells; the pale nucleus is that of the macrophage (magnification = x1000). (B) Phagocytosis as measured by the number of tumor cells phagocytosed per 100 macrophages (mean ± SE of three experiments; from LaTemple *et al.*, 1996 with permission).

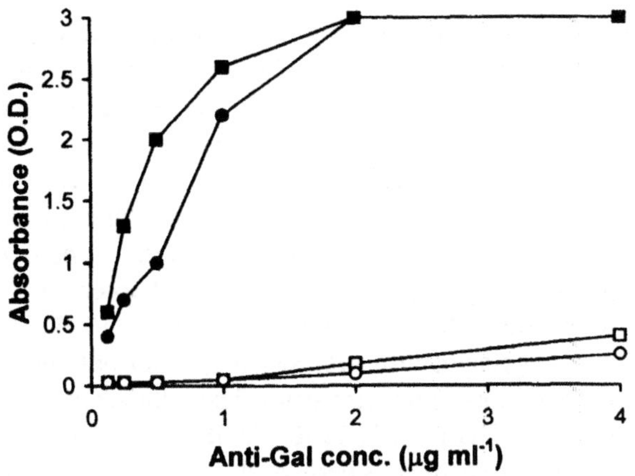


**Figure 5.** Binding of anti-Gal to α-gal epitopes synthesized on membranes of human mammary carcinoma (●) and colon carcinoma (■). Desialylated membranes from tumor homogenates were washed and attached to ELISA microtiter wells by drying of the membranes at 1 μg per well. Subsequently, membranes in wells were incubated with 5 μg/ml of recombinant α1,3 galactosyltransferase and 1 mM UDP-Gal. Open symbols represent control wells in which the attached membranes were incubated with the enzyme in the absence of UDP-Gal. Binding of anti-Gal was assessed by ELISA with anti-human IgG-HRP as secondary antibody, (from Galili and LaTemple, 1997 with permission).

epitope by the tumor cells. Many gene-transfer systems have been employed for efficient transduction of primary tumor tissue such as retroviral and adenoviral vectors. Successful gene-transfection of autologous human tumor vaccines relies on the establishment of primary tumor cell explants. It is important to use the most efficient system for gene-transfer since it may be difficult to obtain large amounts of live tumor cells from surgically removed tumor tissue for culture. Also, extended cultures of these primary tumor cell explants increases the chance that their antigenic composition may change (Klingemann and Dougherty, 1996; Pardoll, 1993; Hellstrom *et al.*, 1991). Therefore, the use of highly efficient systems may result in rapid and extensive expression of the transfected gene, thereby limiting the extent of tumor cell culture before administering the irradiated vaccinating autologous transduced tumor cells back into the patient.

### 3.3. Efficacy of α-Gal Expressing Tumor Vaccines in Vivo

To determine whether anti-Gal-mediated phagocytosis of α-gal epitope expressing autologous tumor vaccines, by APCs, translates into an anti-tumor immune response in vivo, an appropriate animal model is necessary. The reciprocal distribution pattern of anti-Gal production and α-gal epitope expression makes all non-primate mammals such as mice and rats unsuitable models. These animals exhibit α1,3 galactosyltransferase activity and thereby express α-gal epitopes on their tissues and lack anti-Gal. Similarly, a syngeneic tumor model that normally lacks α-gal epitopes is necessary in order to de novo express the α-gal epitopes. The modified tumor would serve as a tumor vaccine in an animal that produces anti-Gal. Monkeys produce anti-Gal similar to humans, yet there are no syngeneic tumor models available to study in these animals. Recently, the α1,3 galactosyltransferase knockout mouse was developed through the disruption of the catalytic domain of the α1,3 galactosyltransferase by homologous recombination with a specific targeting vector (see the chapter on "Generation of α1,3Galactosyltransferase Deficient Mice"; Thall *et al.*, 1995). As a result of the disruption of the α1,3 galactosyltransferase gene, knock out mice for this gene no longer express α-gal epitopes on their tissues. These knock out mice produce very low levels of anti-Gal (Thall *et al.*, 1996; Thall *et al.*, 1995). The relative low amounts of anti-Gal in these mice is perhaps due to the lack of antigenic stimulation since mice are normally housed in pathogen-free animal facilities. However, we were able to stimulate anti-Gal production to levels similar to those in humans by immunizing knockout mice with rabbit red blood cell (RRBC) membranes which contain abundant amounts of the α-gal epitope (LaTemple and Galili, 1998). Anti-Gal from these RRBC-immunized knockout mice was present as all isotypes in serum with the exception of IgA. Also, anti-Gal from knockout mice exhibited restricted specificity to α-gal epitopes as measured by anti-Gal binding to natural glycoproteins that express or lack α-gal epitopes (LaTemple and Galili, 1998). These knock out mice and the mouse B16-BL6 melanoma cells which lack α-gal epitopes (Gorelik et al., 1995), are currently studied as a model for evaluating the efficacy of α-gal expressing tumor vaccines in vivo.

## 4. CONCLUSIONS

The expression of α-gal epitopes on human tumor cell vaccines is a unique immunotherapeutic strategy which exploits the presence of the natural anti-Gal antibody in humans as an opsonizing antibody. In vivo formed complexes between anti-Gal and α-gal epitope-expressing tumor cells are targeted for phagocytosis by APCs through an Fcγ receptor-mediated mechanism. Targeting tumor vaccines to APCs in this manner will lead to increased processing and presentation of autologous TAA peptides and thus to effective activation of TAA-specific CD8+ and CD4+ T-cells. This strategy is based on the principle that antigens complexed with the corresponding specific IgG antibody are more effective at stimulating an immune response than when the antigen is administered alone. The advantage of this strategy is that it can potentially be used universally since anti-Gal is produced in abundant amounts by all humans. We have shown that α-gal epitopes can be synthesized on human tumor cells or cell membranes in vitro and that this expression mediates binding by anti-Gal and phagocytosis of the tumor cells by macrophages. α-Gal epitope expression on human tumor cell vaccines can be achieved by the in vitro synthesis of the epitope on tumor cell membranes using recombinant α1,3galactosyltransferase and UDP-Gal. Alternatively intact vaccinating tumor cells may be transduced with viral vectors containing the α1,3galactosyltransferase gene. It is likely that the efficacy of these vaccines will vary from one patient to another because of variations in potency of TAAs. Nevertheless, the anti-Gal-mediated targeting of autologous α-gal epitope-expressing tumor vaccines to APCs may be considered as an adjuvant therapy to improve vaccine immunogenicity. Ultimately, in some of the treated patients the elicited immune response may be potent enough to cause the eradication of all metastatic tumor cells, resulting in complete cure.

## REFERENCES

Asher A.L., Mule J.J., Kasid A., Restifo N.P., Salo J.C., Reichert C.M., Jaffe G., Fendly B, Kriegler M., and Rosenberg S.A., 1991, Murine tumor cells transduced with the gene for tumor necrosis factor alpha, *J. Immunol.* **146**: 3227-3234.

Baskar S., Ostrand-Rosenberg S., Nabavi N., Nadler L., Freeman G., and Glimcher L., 1993, Constitutive expression of B7 restores immunogenicity of tumor cells expressing truncated MHC class II molecules, *Proc. Natl. Acad. Sci. USA* **90**: 5687-5690.

Boehm U., Klamp T., Groot M., and Howard J.C., 1997, Cellular responses to interferon-γ, *Annu. Rev. Immunol.* **15**: 749-795.

Boon T., Gajewski T.F., and Coule P.G., 1995, From defined human tumor antigens to effective immunization?, *Immunol. Today* **16**: 334-336.

Browning M.J., and Bodmer W.F., 1992, MHC antigens and cancer: implications for T-cell surveillance, *Curr. Opinion Immunol.* **4**: 613-618.

Celis E., and Chang T.W., 1984, Antibodies to hepatitis B surface antigen potentiate the response of human T lymphocyte clones to the same antigen, *Science* **224**: 297-299.

Celis E., Abraham K.G., and Miller R.W., 1987, Modulation of the immunological response to hepatitis B virus by antibodies, *Hepatology* **7**: 563-568.

Chen L., Ashe S., Brady W.A., Hellstrom I., Hellstrom K.E., Ledbetter J.A., McGowan P., and Linsley P.S., 1992, Costimulation of antitumor immunity by the B7 counterreceptor for the T lymphocyte molecules CD28 and CTLA-4, *Cell* **71**: 1093-1102.

Chen L., Chen D., Block E., O'Donnell M., Kufe D.W., and Clinton S.K., 1997, Eradication of murine bladder carcinoma by intratumor injection of a bicistronic adenoviral vector carrying cDNAs for the IL-12 heterodimer and its inhibition by the IL-12 p40 subunit homodimer, *J. Immunol.* **159**: 351-359.

Collins B.H., Cotterell A.H., McCurry K.R., Alvarado C.G., Magee J.C., Parker W., and Platt J.L., 1994, Cardiac xenografts between primate species provide evidence of the α-galactosyl determinant in hyperacute rejection, *J. Immunol.* **154**: 5500-5510.

Colombo M.P., Modesti A., Parmiani G., and Forni G., 1992, Local cytokine availability elicits tumor rejection and systemic immunity through granulocyte-T-lymphocyte cross-talk, *Cancer Res.* **52**: 4853-4857.

Dranoff G., Jaffee E., Lazenby A., Golumbek P., Levitsky H., Brose K., Jackson V., Hamada H., Pardoll D., and Mulligan R.C., 1993, Vaccination with irradiated tumor cells engineered to secrete murine GM-CSF stimulates potent, specific, and long lasting anti-tumor immunity, *Proc. Natl. Acad. Sci. USA* **90**: 3539-3543.

Falo L.D., Kovacsovics-Bankowski M., Thompson K., and Rock K.L., 1995, Targeting antigen into the phagocytic pathway in vivo induces protective tumor immunity, *Nat. Med.* **1**: 649-653.

Fanger N.A., Wardwell K., Shen L., Tedder T.F., and Guyre P.M., 1996, Type I (CD64) and type II (CD32) Fc gamma receptor-mediated phagocytosis by human blood dendritic cells, *J. Immunol.* **157**: 541-548.

Fearon E.R., Pardoll D.M., Itaya T., Golumbek P., Levitsky H.I., Simons J.W., Karasuyama H., Vogelstein B., and Frost P., 1990, Interleukin-2 production by tumor cells bypasses T helper function in the generation of an antitumor response, *Cell* **60**: 397-403.

Foley E.J., 1953, Antigenic properties of methylcholanthrene-induced tumors in mice of the strain of origin, *Cancer Res.* **13**: 835-837.

Galili U., Rachmilewitz E.A., Peleg A., and Flechner I., 1984, A unique natural human IgG antibody with anti-α-galactosyl specificity, *J. Exp. Med.* **160**: 1519-1531.

Galili U., Macher B.A., Buehler J., and Shohet S.B., 1985, Human natural anti-α-galactosyl IgG. II. The specific recognition of α-(1.3)-linked galactose residues, *J. Exp. Med.* **162**: 573-582.

Galili U., Shohet S.B., Kobrin E., Stults C.L.M., and Macher B.A., 1988, Man, apes, and Old World monkeys differ from other mammals in the expression of α-galactosyl epitopes on nucleated cells, *J. Biol. Chem.* **263**: 17755-17762.

Galili U., 1993a, Evolution and pathophysiology of the human natural anti-α-galactosyl IgG (anti-Gal) antibody, *Springer Semin. Immunopathol.* **15**: 155-171.

Galili U., 1993b, Interaction of the natural anti-Gal antibody with α-galactosyl epitopes: a major obstacle for xenotransplantation in humans, *Immunol.Today* **14**: 480-482.

Galili U., and LaTemple DC. 1997, Natural anti-Gal antibody as a universal augmenter of autologous tumor vaccine immunogenicity, *Immunol. Today* **18**: 281-285.

Gansbacher B., Zier K., Daniels B., Cronin K., Bannerji R., and Gilboa E., 1990, Interleukin 2 gene transfer into tumor cells abrogates tumorigenicity and induces protective immunity, *J. Exp. Med.* **172**: 1217-1224.

Golumbek P.T., Lazenby A.J., Levitsky H.I., Jaffee E.M., Karasuyama H., Baker M., and Pardoll D.M., 1991. Treatment of established renal cancer by tumor cells engineered to secrete interleukin-4, *Science* **254**: 713-716.

Good A.H., Cooper D.K.C., Malcolm A.J., Ippolito R.M., Koren E., Neethling F.A., Ye Y., Zuhdi N., and Lamontagne L.R., 1992, Identification of carbohydrate structures which bind human anti-porcine antibodies: implications for discordant grafting in man, *Transplant Proc.* **24**: 559-562.

Gorelik, E., Duty, L., Anaraki F., and Galili, U. 1995. Alterations of cell surface carbohydrates and inhibition of metastatic property of murine melanomas by 1,3galactosyltransferase gene transfection. *Cancer Res.* **55**:4168-4173,

Gosselin E.J., Wardwell K., Gosselin D.R., Alter N., Fisher J.L., and Guyre P., 1992, Enhanced antigen presentation using human Fcγ receptor (monocyte/macrophge)-specific immunogens, *J. Immunol.* **149**: 3477-3481.

Grabbe S., Beisserts S., Schwartz T., and Granstein R.D., 1995, Dendritic cells as initiators of tumor immune response: a possible strategy for tumor immunotherapy?, *Immunol. Today* **16**: 117-121.

Greenberg P.D., 1991, Adoptive T cell therapy of tumors: mechanisms operative in the recognition and elimination of tumor cells, *Adv. Immunol.* **49**: 281-355.

Guinan E.C., Gribben J.G., Boussiotis V.A., Freeman G.J., and Nadler L.M., 1994, Pivotal role of the B7:CD28 pathway in transplantation tolerance and tumor immunity, *Blood* **84**: 3261-3282.

Gutterman J.U., Hersh E.M., Rodriguez V., McCredie K.B., Mavligit G., Reed R., Burgess M.A., Smith T., Gehan E., Bodey G.P., and Freireich E.J., 1974, Chemoimmunotherapy of adult acute leukemia: prolongation of remission in myeloblastic leukemia with BCG, *Lancet* **2**: 1405-1409.

Hellstrom K.E., 1991, In *Biologic Therapy of Cancer*. J.B. Lippincott.

Henion T.R., Macher B.A., Anaraki F., and Galili U., 1994, Defining the minmal size of catalytically active primate α1,3galactosyltransferase: structure function studies on the recombinant truncated enzyme, *Glycobiology* **4**: 193-201.

Herberman R.B., 1977, In *Mechanisms of Tumor Immunity* (Green I, Cohen S, McCluskey RT, eds.), pp. 175-191. J. Wiley and Sons.

Herberman R.B., 1992, Tumor Immunology, *JAMA* **268**: 2935-2939.

Herin M., Lemoine C., Weynants P., Vesiere F., Van Pel A., Knuth A., Devos R., and Boon T., 1987, Production of stable cytolytic T-cell clones directed against autologous human melanoma, *Intl J. Cancer* **39**:390-397.

Hollinshead A., Stewart T.H.M., Takita H., Dalbow M., and Concannon J., 1987, Adjuvant specific active lung cancer immunotherapy trials. Tumor-associated antigens, *Cancer* **60**: 1249-1262.

Hoover H.C., Surdyke M., Dangel R.B., Peters L.C., and Hanna M.G., 1984, Delayed cutaneous hypersensitivity to autologous tumor cells in colrectal cancer patients immunized with an autologous tumor cell Bacillus Calmette-Guerin vaccine, *Cancer Res.* **44**: 1671-1676.

Houston W.E., Kremer R.J., Crabbs C.L., and Spertzel R.O., 1977, Inactivated Venezuelan equine encephalomyelitis virus vaccine complexed with specific antibody: enhanced primary immune response and altered pattern of antibody class elicited, *J. Infect. Dis.* **135**: 600-610.

Huang A.Y., Golumbek P., Ahmadzadeh M., Jaffee E., Pardoll D., and Levitsky H., 1994, Role of bone marrow-derived cells in presenting MHC class I-restricted tumor antigens, *Science* **264**: 961-965.

Hui K., Grosveld F., and Festenstein H., 1984, Rejection of transplantable AKR leukaemic cells following MHC DNA-mediated cell transformation, *Nature* **311**: 750-752.

Kast W.M., Offringa R., Peters P.J., Voordouw A.C., Meloen R.H., Van der En A.J., and Melief C.M., 1989, Eradication of adenovirus E1-induced tumors by E1A specific cytotoxic T lymphocytes, *Cell* **59**: 603-614.

Klein G., and Klein E., 1977, Rejectability of virus-induced tumors and non-rejectability of spontaneous tumors: a lesson in contrasts, *Transplant Proc.* **9**: 1095-1104.

Klingemann H.G., and Dougherty G.J., 1996, Site-specific delivery of cytokines in cancer, *Molec. Med. Today* **2**: 154-159.

Knuth A., Danowski B., Oettgen H.F., and Old L., 1984, T-cell mediated cytotoxicity against autologous malignant melanoma: analysis with interleukin-2-dependent T-cell cultures, *Proc. Natl. Acad. Sci. USA* **81**: 3511-3515.

Knuth A., Wolfel T., Klehmann E., Boon T., and Meyer Zum Bueschenfelde K.H., 1989, Cytolytic T-cell clones against an autologous human melanoma: specificity study and definition of three antigens by immunoselection, *Proc. Natl. Acad. Sci. USA* **86**: 2804-2808.

Lanzavecchia A., 1985, Antigen specific interactions between T and B cells, *Nature* **314**: 537-539.

Lanzavecchia A., 1993, Identifying strategies for immune intervention, *Science* **260**: 937-944.

Lanzavecchia A., 1996, Mechanisms of antigen uptake for presentation, *Curr. Opin. Immunol.* **8**: 348-354.

LaTemple D.C., Henion T.R., Anaraki F., and Galili U., 1996, Synthesis of α-galactosyl epitopes by recombinant α1,3galacotysltransferase for opsonization of human tumor cell vaccines by anti-galactose, *Cancer Res.* **56**: 3069-3074.

LaTemple D.C., and Galili U., 1998, Adult and neonatal anti-Gal response in knock-out mice for α1,3 galactosyltransferase, *Xenotransplantation* **5**: 191-196.

Liu C., Gosselin E.J., and Guyre P.M., 1996, Fc gamma RII on human B cells can mediate enhanced antigen presentation, *Cell. Immunol.* **167**: 188-194.

Lord E.M., and Frelinger J.G., 1998, Tumor immunotherapy: cytokines and antigen presentation, *Cancer Immunol. Immunother.* **46**: 75-81.

Maass G., Schmidt W., Berger M., Schilcher F., Koszik F., Schneeberger A., Stingl G., Birnstiel M.L., and Schweighoffer T., 1995, Priming of tumor-specific T cells in the draining lymph nodes after immunization with interleukin 2-secreting tumor cells: three consecutive stages may be required for successful tumor vaccination, *Proc. Natl. Acad. Sci. USA* **92**: 5540-5544.

Mackay C.R., Marston W.L., and Dudler L., 1990, Naive and memory T cells show distinct pathways of lymphocyte recirculation, *J. Exp. Med.* **171**: 801-817.

Mackay C.R., Marston W., and Dudler L., 1992, Altered patterns of T cell migration through lymph nodes and skin following antigen challenge, *Eur. J. Immunol.* **22**: 2205-2210.

Manca F., Fenoglio D., LiPira G., Kunkl A., and Celada F., 1991, Effect of antigen/antibody ratio on macrophage uptake, processing and presentation to T cells of antigen complexed with polyclonal antibodies, *J. Exp. Med.* **173**: 37-48.

McCune C.S., O'Donnell R.W., Marquis D.M., and Sahasrabudhe D.M., 1990, Renal cell carcinoma treated by vaccines for active specific immunotherapy: correlation of survival with skin testing by autologous tumor cells, *Cancer Immunol. Immunother.* **32**: 62-66.

Mitchison N.A., 1955, Studies on the immunological response to foreign tumor transplants in the mouse: I. The role of lymph node cells in conferring immunity by adoptive transfer, *J. Exp. Med.* **102**: 157-177.

Moller P., and Hammerling G., 1992, The role of surface HLA-A, B, C molecules in tumor immunity, *Cancer Surv.* **13**: 101-128.

Nossal G.J.V., 1993, Tolerance and ways to break it, *Ann. NY Acad. Sci.* **690**: 34-41.

Old L.J., 1996, Immunotherapy for Cancer, *Scientific American* **275**: 136-143.

Pardoll D.M., 1993, Cancer vaccines, *Immunol.Today.* **14**: 310-316.

Pardoll D.M., 1994, A new look for the 1990s, *Nature* **369**: 357-358.

Powles R.L., Russel J., and Lister T.A., 1977, Immunotherapy of acute myelogenous leukemia: a controlled clinical study of 2.5 years after entry of the last patient, *Br. J. Cancer.* **35**: 265-272.

Prehn R.T., and Main J.M., 1957, Immunity to methylcholanthrene-induced sarcomas, *J. Natl. Cancer Inst.* **18**: 769-778.

Restifo N.P., Spiess P.J., Karp S.E., Mule J.J., and Rosenberg S.A., 1992, A nonimmunogenic sarcoma transduced with the cDNA for interferon gamma eleicits CD8+ T cells against the wild-type tumor: Correlation with antigen presentation capability, *J. Exp. Med.* **175**: 1423-1431.

Rosenberg S.A., Spiess P., and Lafreniere R., 1986, Anew approach to the adoptive immunotherapy of cancer with tumor-infiltrating lymphocytes, *Science* **233**: 1318-1321.

Roth C., Rochlitz C., and Kourilsky P., 1994, Immune response against tumors, *Adv. Immunol.* **57**: 281-351.

Sallusto F., and Lanzavecchia A., 1994, Efficient presentation of soluble antigen by cultured human dendritic cells is maintained by granulocyte/macrophage colonly-stimulating factor plus interleukin 4 and downregulated by tumor necrosis factor alpha, *J. Exp. Med.* **179**: 1109-1118.

Sandrin M., Vaughan H.A., Dabkowski P.L., and McKenzie I.F.C., 1993, Anti-pig IgM antibodies in human serum react predominantly with Gal($\alpha$1-3)Gal epitopes, *Proc. Natl. Acad. Sci. USA* **90**: 11391-11395.

Schmitt D.A., Hanau D., Bieber T., Dezutter-Dambuyant C., Schmitt D., Fabre M., Pauly G., and Cazenave J.P., 1990, Human epidermal Langerhans cells express only the 40-kilodalton Fc gamma receptor (FcRII), *J. Immunol.* **144**: 4284-4290.

Schrier P.I., and Peltenburg L.T.C., 1993, Relationship between myc oncogene activation and MHC class I expression, *Adv. Cancer Res.* **60**: 181-246.

Schwarz R.H., 1990, A cell culture model for T lymphocyte clonal anergy, *Science* **248**: 1349-1356.

Schwarz R.H., 1992, Costimulation of T lymphocytes: the role of CD28, CTLA-4, and B7/BB1 in interleukin-2 production and immunotherapy, *Cell* **71**: 1065-1068.

Schweighoffer T., Schmidt W., Buschle M., and Birnstiel M.L., 1996, Depletion of naive T cells of the peripheral lymph nodes abrogates systemic antitumor protection conferred by IL-2 secreting cancer vaccines, *Gene Therapy* **3**:819-824.

Stoner R.D., and Terres G., 1963, Enhanced antitoxin responses in irradiated mice elicited by complexes of tetanus toxoid and specific antibody, *J. Immunol.* **91**: 761-770.

Sulitzeanu D., 1993, Immunosuppressive factors in human cancer, *Adv. Cancer Res.* **60**: 247-267.

Sznol M., and Holmlund J., 1997, Antigen-specific agents in development, *Seminars in Oncology* **24**: 173-186.

Thall A.D., Maly P., and Lowe J.B., 1995, Oocyte Galα1-3Gal epitopes implicated in sperm adhesion to the zona pellucida glycoprotein ZP3 are not required for fertilization in the mouse, *J. Biol. Chem.* **270**: 21437-21440.

Thall A.D., Murphy H.S., and Lowe J.B., 1996, α1,3galactosyltransferase-deficient mice produce naturally occurring cytotoxic anti-Gal antibodies, *Transplant. Proc.* **28**: 556-557.

Townsend S.E., and Allison J.P., 1993, Tumor rejection after direct costimulation of CD8+ T cells by B7-transfected melanoma cells, *Science* **259**: 368-370.

Unkeless J.C., 1989, Function and heterogeneity of human Fc receptors for immunoglobulin G, *J. Clin. Invest.* **83**: 355-361.

Van den Eynde B., and Brichard V.G., 1995, New tumor antigens recognized by T cells, *Current Opin. Immunol.* **7**: 674-681.

Vogler W., Bartolucci A.A., and Omura G.A., 1978, A randomized clinical trial of remission induction, consolidation, and chemoimmunotherapy maintenance in adult myeloblastic leukemia, *Cancer Immunol. Immunother.* **3**: 163-170.

Wallich R., Bulbuc N., Hammerling G.J., Katzav S., Segal S., and Feldman M., 1985, Abrogation of metastatic properties of tumour cells by de novo expression of H-2K antigens following H-2 gene transfection, *Nature* **315**: 301-305.

Wang R.F., and Rosenberg S.A., 1996, Human tumor antigens recognized by T lymphocytes: implications for cancer therapy. *J. Leukocyte Biology* **60**: 296-309.

Zinkernagel R.M., Ehl S., Aichele P., Oechen S., Kundig T., and Hengartner H., 1997, Antigen localisation regulates immune responses in a dose- and time-dependent fashion: a geographical view of immune reactivity, *Immunological Reviews* **156**: 199-209.

Zuhrie S.R., Harris R., and Freeman C.B., 1980, Immunotherapy alone vs. no maintenance treatment in acute myelogenous leukemia, *Br. J. Cancer* **41**: 372-337.

# INDEX

GPSR Compliance

*The European Union's (EU) General Product Safety Regulation (GPSR) is a set of rules that requires consumer products to be safe and our obligations to ensure this.*

*If you have any concerns about our products, you can contact us on ProductSafety@springernature.com*

In case Publisher is established outside the EU, the EU authorized representative is:

Springer Nature Customer Service Center GmbH
Europaplatz 3
69115 Heidelberg, Germany

**Batch number: 10755284**

Printed by Printforce, the Netherlands